Chilled foods

Related titles from Woodhead's food science, technology and nutrition list:

Managing frozen foods (ISBN: 1 85573 412 5)

Edited by Christopher J Kennedy

Maintaining quality throughout the food chain is a key issue for the frozen food industry. This book provides a unique overview of the whole supply chain and key quality factors at each stage in the production, distribution and retail of frozen foods. It identifies the key quality parameters in production and distribution as well as describing the technology and working practices necessary to attain these standards. It is an invaluable resource for manufacturers, distributors and retailers.

Yoghurt – Science and technology Second Edition (ISBN: 1 85573 399 4)

A Y Tamime and R K Robinson

In its first edition this book quickly established itself as the standard reference in its field for both industry professionals and those involved in research. This completely revised and updated second edition is 40% longer than the first and includes developments such as the new 'bio-yoghurts' as well as all other recent changes and technological developments in the industry including: the production of strained yoghurt by ultrafication, the latest developments in mechanisation and automation and the implementation of HACCP.

Food processing – Principles and practice Second Edition (ISBN: 1 85573 533 4)

P J Fellows

The first edition of *Food processing and technology* was quickly adopted as the standard text by many food science and technology courses. The publication of a completely revised and updated new edition is set to confirm the position of this textbook as the best single-volume introduction to food manufacturing technologies available.

'. . . a well written and authoritative review of food processing technology – the essential reference for food technologists and students alike.' *Food Trade Review*

Details of these books and a complete list of Woodhead's food science, technology and nutrition titles can be obtained by:

- visiting our web site at www.woodhead-publishing.com
- contacting Customer Services (e-mail: sales@woodhead-publishing.com; fax: +44 (0)1223 893694; tel.: +44 (0)1223 891358 ext. 30; address: Woodhead Publishing Ltd, Abington Hall, Abington, Cambridge CB1 6AH, England)

If you would like to receive information on forthcoming titles in this area, please send your address details to: Francis Dodds (address, tel. and fax as above; e-mail: francisd@woodhead-publishing.com). Please confirm which subject areas you are interested in.

Chilled foods

A comprehensive guide

Second edition

Edited by
Mike Stringer and Colin Dennis

CRC Press
Boca Raton Boston New York Washington, DC

WOODHEAD PUBLISHING LIMITED
Cambridge England

Published by Woodhead Publishing Limited, Abington Hall, Abington
Cambridge CB1 6AH, England
www.woodhead-publishing.com

Published in North and South America by CRC Press LLC, 2000 Corporate Blvd, NW
Boca Raton FL 33431, USA

First edition 1992, Ellis Horwood Ltd
Second edition 2000, Woodhead Publishing Limited and CRC Press LLC

British Library Cataloguing in Publication Data
A catalogue record for this book is available from the British Library.

Library of Congress Cataloging in Publication Data
A catalog record for this book is available from the Library of Congress.

Woodhead Publishing Limited ISBN 1 85573 499 0
CRC Press ISBN 0-8493-0856-9
CRC Press order number: WP0856

Cover design by The ColourStudio
Project managed by Macfarlane Production Services, Markyate, Hertfordshire
Typeset by MHL Typesetting Limited, Coventry, Warwickshire
Printed by TJ International, Padstow, Cornwall, England

Contents

Preface

During the last 40 years, consumer choice has been transformed by developments in the production, distribution and retailing of food, which with improvements in the design and equipment of the domestic kitchen have facilitated a major change in our lifestyle.

Perhaps the most striking development is the marketing of a wide and expanding range of chilled perishable foods. Convenience, easy preparation and the 'fresh' and 'healthy' image and an extensive choice of different culinary tastes are attractive features. Many products are made by industrial processes using technology which has no parallel in the domestic kitchen. Others, notably cooked ready meals of all kinds, require skills, time and patience to prepare. These developments have occurred by the application of technology to the production, packaging, distribution and retailing of food.

The integrity and safety of chilled foods is multifactorial. Care is required at every stage in the food chain, from primary production of raw materials, through manufacture, distribution, retail and consumer use.

This book provides a comprehensive guide to the many important aspects necessary to provide the consumer with safe, high quality products, and includes recent developments in legislation. Since the introduction of the first coordinated chilled distribution chain in the late 1960s, substantial developments have occurred in the refrigeration equipment available and in the temperature monitoring and control systems.

Product, process and packaging developments which have occurred over the last 20 years have resulted in chilled foods representing a larger and increasing proportion of weekly purchases for home consumption as well as in institutional and service catering. Their safety and reliability have resulted from the application of scientific principles of food technology and depend on a series of safety factors in their preparation, processing, distribution and retail sale. The overriding requirement is for the reliable 'chill chain' to control the temperature at every stage from the final process of production to the moment of purchase and absence of abuse by the consumer. Hygienic preparation and production areas are a further essential requirement for chilled food manufacture. This has resulted from a greater understanding and awareness of hygienic design of equipment and buildings, together with appropriate cleaning and sanitation regimes. Developments in methods of detection of microorganisms have greatly assisted in improving approaches to hygiene practices and monitoring the microbiological status of raw materials and finished products.

The establishment of the shelf-life of chilled foods requires a full appreciation of the microbiological, chemical, physical and biochemical aspects which influence the sensory acceptability of products. These factors are discussed in relation to the safety and quality of products, together with methods of determining shelf life of such products. The importance of the application of HACCP as part of quality management systems in the production and distribution of chilled foods brings together the many aspects of chilled foods which are covered in this book.

Further developments in processing and packaging technology will undoubtedly contribute to the continued development and innovation in the chilled food sector. High pressure processing, electric-field sterilisation and active and intelligent packaging all offer potential in this respect.

As editors, we have between us over 35 years experience in chilled food science and technology. Our contacts with the contributing authors have been built up over these years. They all have wide research and industrial experience and are recognised experts in their fields. We consider ourselves fortunate to have secured their cooperation in providing a major and unique contribution to the scientific and technical understanding of the chilled food market.

We are grateful to all the authors for applying themselves so diligently to give the best of their knowledge and skills.

Mike Stringer
Colin Dennis

Contributors

Introduction

Professor Colin Dennis and Dr Mike Stringer
Campden & Chorleywood Food Research Association
Chipping Campden GL55 6LD

Tel: +44 (0)1386 842001 (CD)
+44 (0)1386 842003 (MS)
Fax: +44 (0)1386 842100
E-mail: c.dennis@campden.co.uk
m.stringer@campden.co.uk

Chapter 1

L. Bedford
Campden & Chorleywood Food Research Association
Chipping Campden GL55 6LD

Tel: +44 (0)1386 842013
Fax: +44 (0)1386 842100
E-mail: l.bedford@campden.co.uk

Chapter 2

Dr Ralph Early
Harper Adams University College
Newport TF10 8NB

Tel: +44 (0)1952 820280
Fax: +44 (0)1952 814783
E-mail: rearly@harper-adams.ac.uk

Chapter 3

Dr Steve James
Food Refrigeration & Process Engineering Research Centre
University of Bristol
Churchill Building
Langford
Bristol BS18 7DY

Tel: +44 (0)117 928 9239
Fax: +44 (0)117 928 9314
E-mail: steve.james@bristol.ac.uk

Chapter 4

Mr Robert D. Heap MBE
Cambridge Refrigeration Technology
140 Newmarket Road
Cambridge CB5 8HE

Tel: +44 (0)1223 365101
Fax: +44 (0)1223 461522
E-mail: crt@crtech.demon.co.uk

Chapter 5

Dr Mark Woolfe
Food Labelling, Standards and Consumer Protection Division
Food Standards Agency
PO Box 31037
Ergon House
London SW1P 3WG
Tel: +44 (0)20 7238 6168
Fax: +44 (0)20 7238 6763
E-mail:
mark.woolfe@foodstandards.gsi.gov.uk

Chapter 6

Dr Brian P F Day
Campden & Chorleywood Food Research Association
Chipping Campden GL55 6LD

Tel: +44 (0)1386 842082
Fax: +44 (0)1386 842100
E-mail: b.day@campden.co.uk

Chapter 7

Dr Steven Walker and Dr Gail Betts
Campden & Chorleywood Food Research Association
Chipping Campden GL55 6LD

Tel: +44 (0)1386 842011 (SW)
+44 (0)1386 842071 (GB)
Fax: +44 (0)1386 842100
E-mail: s.walker@campden.co.uk
g.betts@campden.co.uk

Chapter 8

Dr Roy Betts
Campden & Chorleywood Food Research Association
Chipping Campden GL55 6LD

Tel: +44 (0)1386 842075
Fax: +44 (0)1386 842100
E-mail: r.betts@campden.co.uk

Chapter 9

Dr Helen Brown and M. N. Hall
Campden & Chorleywood Food Research Association
Chipping Campden GL55 6LD

Tel: +44 (0)1386 842016 (HB)
+44 (0)1386 842014 (MNH)
Fax: +44 (0)1386 842100
E-mail: h.brown@campden.co.uk
m.hall@campden.co.uk

Chapter 10

Dr Linda Everis and Dr Gail Betts
Campden & Chorleywood Food Research Association
Chipping Campden GL55 6LD

Tel: +44 (0)1386 842063 (LE)
+44 (0)1386 842071 (GB)
Fax: +44 (0)1386 842100
E-mail: l.everis@campden.co.uk
g.betts@campden.co.uk

Chapter 11

Professor Martyn Brown
Microbiology Department
Unilever Research
Colworth Laboratory
Colworth House
Sharnbrook
Bedford MK44 1LQ

Tel: +44 (0)1234 222351
Fax: +44 (0)1234 222277
E-mail: martyn.brown@unilever.com

Chapter 12

Dr Stephane Durand
HP Foods Ltd

E-mail:
Stephane_DURAND@hpfoods.com

Chapter 13

Dr John Holah and R. H. Thorpe
Campden & Chorleywood Food Research Association
Chipping Campden GL55 6LD

Tel: +44 (0)1386 842041
Fax: +44 (0)1386 842100
E-mail: j.holah@campden.co.uk

Chapter 14

Dr John Holah
Campden & Chorleywood Food Research Association
Chipping Campden GL55 6LD

Tel: +44 (0)1386 842041
Fax: +44 (0)1386 842100
E-mail: j.holah@campden.co.uk

Chapter 15

Dr David Rose
Campden & Chorleywood Food Research Association
Chipping Campden GL55 6LD

Tel: +44 (0)1386 842088
Fax: +44 (0)1386 842100
E-mail: d.rose@campden.co.uk

Chapter 16

Ms Kaarin Goodburn
11 Yewfield Road
London NW10 9TD

Tel: +44 (0) 20 8451 0503
Fax: +44 (0) 20 8459 8061
E-mail: kgoodburn@bigfoot.com

Introduction

The chilled foods market

C. Dennis and M. Stringer, Campden and Chorleywood Food Research Association

Chilled food technology has had a very significant impact on the types of food eaten by consumers during the 1980s and 1990s. This method of food preservation has satisfied the desires of people for safe, reliable, 'fresh' products providing convenience despite the limited shelf-life. This introduction reviews the definition, range and market size of chilled food and indicates trends for the future.

I.1 Definition

Foods distributed under refrigeration and sold from refrigerator cabinets have been available for many years. Although there were many new chilled product introductions made during the 1970s, it was not until the 1980s that significant numerous and major technological developments for chilled foods occurred (Bond 1992). This trend continued throughout the 1990s with the major emphasis on value added, convenience and increasing consumer choice. This unprecedented activity during the 1980s and 1990s stimulated the production of good practice guides related to refrigerated or chilled foods. In 1990, the Institute of Food Science and Technology (IFST) defined chilled foods as 'perishable foods which, to extend the time during which they remain wholesome, are kept within specified ranges of temperature above −1°C' (IFST 1990). More recently the UK Chilled Foods Association (CFA) restricted the term to 'prepared foods' with the following definition 'prepared foods that, for reasons of safety and/or quality, are designed to be stored at refrigerated temperatures (at or below 8C but not frozen) throughout their entire life' (CFA 1997). Whereas the IFST definition emphasises 'perishability', the CFA

Table I.1 Number of chilled product introductions in the UK market-place

1972	35
1975	86
1980	249
1983	535
1985	605
1988	774
1990	945
1992	1578
1994	2385
1997	2920
1998	3616
1999	3365

definition excludes non-prepared materials such as raw meat, poultry and fish and commodity dairy products such as butter and cheese which are also not considered 'prepared'. The more general and broader definition of chilled foods has been used in compiling the content of this book, although some sections more appropriately only deal with prepared chilled foods as defined by the CFA.

Food Products Intelligence (FPI) at Campden and Chorleywood Food Research Association (CCFRA) has been monitoring new UK food and drink product introductions since 1969. Its records of new products for the 1980s and 1990s illustrates the dramatic increase in the number of chilled products identified as new (Table I.1) with almost fourfold increases in the 1980s and again in the 1990s. The definition of 'new' used by FPI is one that appears in a major food retail outlet and is previously unknown to FPI or has been recorded by FPI but has been packaged in a different size/format which creates a new eating occasion or new consumer purchase. Frequent updates on new chilled foods entering the UK market-place are available from FPI together with full details of each product on the NewFoods CD-ROM[1] or via the internet (www.newfoods.com).

Unlike other major technology sectors (e.g. frozen foods, ambient foods) the development of the chilled foods market in the UK has been dominated by own label brands for example Marks & Spencer, Tesco and Sainsbury. Approximately 80% of new chilled foods introduced during the 1980s and 1990s were own label. In particular, Marks & Spencer with the St Michael brand is recognised by both the trade and the consumer as having pioneered the early development of chilled foods and for initiating many innovative product concepts which have since been built on and expanded by other brands.

Chilled foods can be designed to be ready to eat, to be reheated (minimal heating before serving for organoleptic purposes) or to be cooked (thorough and

1 NewFoods: The UK new product database and visual guide on CD-ROM published by Blackwell Science, ISSN 1359–2971.

prolonged heating before serving for both safety and quality). Cooked chilled foods are sometimes wrongly referred to as 'cook-chill'. This specific category of foods has been defined as a catering system based on the full cooking of food followed by fast chilling and storage in controlled temperature conditions (0–3°C) and subsequent thorough reheating before consumption (Department of Health (UK) 1989). Cook-chill foods have a maximum recommended shelf-life of five days, inclusive of the day of cooking.

I.2 Drivers in the chilled food sector

The background to the market in the UK is provided by demographic trends shown in Table I.2. This shows the biggest areas of growth to be the 10–14, 35–44 and 55–64-year-old age ranges. Against this background, there are a number of forces driving the market:

- convenience
- snacking
- healthy eating
- variety and choice
- taste
- the origins of food
- competition.

It is reported that for 40% of eating occasions, convenience is the most important factor. The average home-cooked meal is estimated to take about 30 minutes to prepare. This has halved over the last decade (Anon. 1999a).

Table I.2 Demographic trends in the UK 1992–2002 (Source: Office for National Statistics)

	1992	1997	2002 (estimate)		
Age group	Number (millions)	Number (millions)	Number (millions)	(Total population) (%)	Percentage change 1997–2002
0–4	3.8	3.7	3.5	6	−4
5–9	3.7	3.9	3.7	6	−5.4
10–14	3.6	3.7	3.9	7	5.7
15–24	7.9	7.2	7.4	13	2.9
25–34	9.2	9.1	8.1	14	−12.7
35–44	7.8	8.3	9.2	16	11.7
45–54	6.9	7.7	7.7	13	0.4
55–64	5.8	5.8	6.5	11	13.1
65+	9.1	9.2	9.2	16	0.2

Table I.3 Different categories of chilled food product introductions in the UK market-place

	1997		1998		1999	
	Total	%	Total	%	Total	%
Microwaveable	586	20%	818	23%	763	23%
Healthy eating	296	10%	485	13%	526	16%
Vegetarian	790	27%	1163	32%	1143	34%
Organic	7	0%	78	2%	177	5%
Childrens	90	3%	139	38%	128	4%
Total	2920		3616		3365	

The drive towards more convenience foods, such as ready meals, has reflected the decline in traditional home-prepared meals and the associated skills. Convenience is a specific need of the cash-rich time-poor consumer. This development is in part the result of the increase in the number of working women, single-parent and single-person households with limited time available for home cooking. Government estimates suggest that there are over 12 million women in full or part-time work in the UK, representing 45% of the total workforce. Also in the UK, demographic trends show particular growth in numbers of older children and their parents. It is these families in particular, with children at school and with a greater degree of independence, where there is most scope for women to undertake part- or full-time work. It is also in these families in particular that members have developed more independent and flexible patterns of eating. Such patterns have also been stimulated by the increase in microwave ownership and the dramatic rise in the availability of microwaveable foods, especially chilled foods (see Table I.3). This can be seen in the increase in 'snacking', eating more frequent small meals at varying times in the day. Research in 1998 suggested that 31% of UK housewives snacked between meals (Anon. 1998a). The trend towards snacking is also reflected in the decline of the single family evening meal, and the shift towards differing members of the family eating at different times. This development has also prompted consumers to look for a range of light, easy to use ingredients such as cheese spreads or salads, for example, which can be used to prepare a variety of quick snacks or meals tailored to the requirements and preferences of individual family members. Recent research also shows that the total lunch box occasions (i.e. prepared and packed lunch) have grown steadily over the past seven years, up by 21%, with sandwiches featuring in 81% of these. Children are estimated to consume 675 million sandwiches in the UK or 7% of total consumption (Anon. 1999b).

A survey of 25,000 adults in the UK in 1997 found that 17% of all those surveyed and 22% of women in the survey were concerned about counting calories in their diet, whilst over half of all respondents claimed to have reduced fat intake in their diet. In the US, surveys also suggest that consumers are concerned about nutritional issues, particularly information on calorie and fat content of food products (Bender 1992, Rodolfo 1998). The trend in attitudes in the UK has been for a slight reduction in concern about healthy eating since the early 1990s, but a large number of consumers remain concerned about levels of fat intake. This is noticeably true of the 35–44-year-old age range, which has a particularly high proportion of dieters. The overall picture is of consumers feeling more in control of their diet, looking for low-fat and low-calorie products but with a greater tendency to allow the occasional treat in what they eat.

Consumers have also shown an increased interest in the origin and composition of food products and in their methods of production. After adverse publicity in the 1980s, consumers demonstrated an increased concern about the health implications of synthetic additives, related to a general fear of chemicals and their possible links to disease or allergy (Sloan 1986, Crowe 1992, Wandel 1997). This is reflected in increasing consumer pressure for fresh-tasting products with fewer preservatives and minimal preparation. During the 1990s there has also been a growing interest in more environmentally friendly and 'natural' methods of production, reflected in demand for organic foods (Jolly 1989). The FPI at Campden and Chorleywood Food Research Association recorded approximately 5% of the new chilled food products in 1999 as organic.

Exposure to a wider range of cuisine, stimulated in part by the growth in overseas holidays, has encouraged consumers to look for greater variety and novelty in the food they eat. There has been much greater interest in ethnic food, reflected for instance in the growth in popularity of Chinese, Indian and other ethnic chilled ready meals. Britain's younger generations are reported to be as familiar with ethnic food as with roast beef with nearly three-quarters of British households buying ethnic food (Anon. 1999c). Consumers are now more adventurous than ever, demanding variety and authenticity from the dishes selected. The flavours for the start of the new millennium are predicted to originate from South America (e.g. Cuba, Argentina, Brazil) to extend the hot and spicy trend (Sloan 1999). In tandem with this interest in variety, consumers continue to put a premium on taste and enjoyment of food. Recent research in the UK suggests that 21% of housewives ranked taste over other factors in what they ate, with 14% of respondents claiming to buy treats for themselves and their families at the weekend (Anon. 1998b). This interest reflects the relative decline in health concerns as a dominant factor in food purchases. This renewed emphasis on food as a treat or an indulgence has, for example, fuelled the expansion of the chilled desserts market.

Competition in the chilled food sector has intensified, with the leading retailers increasing the range of own-label products, and an increasing emphasis throughout the sector on quality and value-added products. Table I.4 indicates the new product introductions by the major UK retailers during recent years and not only particularly highlights the leading position of Marks & Spencer in this sector, but also the high level of activity by some of the other major players.

Table I.4 New chilled foods introduced by different UK retailers

Retailer	1997	1998	1999
Marks & Spencer	528	575	502
Tesco	402	526	544
J Sainsbury	367	504	498
Safeway	222	390	326
Asda	364	408	414
Waitrose	306	433	354
Wm Morrison	200	222	213
Somerfield	179	196	192
CWS/CRS Retail	146	155	159

Table I.5 summarises the key features of the UK chilled food market in the 1990s and highlights the importance of drivers such as convenience including microwaveability, snacking, healthy eating, ethnicity and vegetarianism. In addition, notable inclusions are products designed for home entertainment such as items for dinner parties as well as more informal type snacks and products specifically for children. These trends have developed from the extensive range of products available during the 1980s (Bond 1992).

I.3 Overall market size

In the UK, the chilled foods market for dairy products, meat products, ready meals, pizzas and prepared salads was valued at £5 billion in 1997, representing 6% of total grocery sales (Anon. 1998b). Forecasts are for continued growth at as much as 6% per annum, reaching a market value of over £6 billion by 2002 (Anon. 1998a; Anon. 1998b, Anon. 1997). The various chilled foods categories had the following shares of overall UK chilled foods sales in 1997 by value (Anon. 1998a, Anon. 1997: figures have been rounded to the nearest whole number):

- dairy products 70%
- meat products (excluding raw meat, poultry and fish) 15%
- ready meals, pizzas and prepared salads 15% (chilled ready meals 9%; pizzas 5%; prepared salads 1%).

These categories show differing levels of growth in value over the period 1993–7 ((Anon. 1998a; Anon. 1998b, Anon. 1997, IDF 1995):

- dairy products 17%
- meat products 30%
- ready meals 30–50% (differing estimates from Keynote and Mintel)
- pizzas 50%
- prepared salads 40%

Table I.5 Product features of UK chilled foods market in the 1990s

Product area	Key product development
Yellow fats	*Butter* • Increasing numbers of organic butters • Biggest innovation has been spreadable butter *Fat blends and spreads* • Spreads have been one of the success stories of the 1990s • The aim to produce a 'butter tasting' product with low/or no cholesterol • Olive oil products e.g. Olivio • The move to functional products e.g. Benecol
Cheese	• Cheddar still popular, the stronger flavours being most popular • Regional varieties and flavour additions continue • Healthy eating (specifically low fat) • Innovation with cheese snacks e.g. Lunchables, Dunkers and Cheestrings – ideal for lunch boxes. • Now large variety of imported cheeses • Cottage cheeses follow the ethnic trends for flavour additions and healthy eating trends • Organic varieties available
Milk	• Overall, decline in milk consumption • Health-conscious consumers have switched to lower-fat milks • Few with added vitamins and aimed at specific groups of consumers (age groups) • Flavoured milks are growing, aimed at younger consumers (in competition with soft drinks) • Organic milks also available
Poultry	• Gained more popularity during BSE crisis • Processed poultry products, and crumb coated are popular with children • Added value, marinaded chicken, particularly popular for the barbecue season • Free range and organic also available at a premium
Meat	• Leaner, smaller joints developed, more modern image • Quick cook joints and presentations, component meals and recipe dishes – convenience • Sausages declined during BSE crisis, but since then sausages have moved more upmarket and traditional – now more of a premium image • Marinaded/added-value cuts available, particularly during the barbecue season • RSPCA freedom food, animal husbandry, organic meats available
Fish and seafood	*Fish* • Perceived as healthy, benefited from BSE crisis. • Processed and crumb coated or battered aimed at encouraging children to eat more fish • In store fishmongers, with trained staff to 'educate' and help consumers • Salmon, once seen as a speciality, now farmed, are used in recipe dishes

Table I.5 Continued

Product area	Key product development
Fish and seafood *(continued)*	*Seafood* • Increased in popularity. Convenience and further processing make seafood more attractive
Dips, pâté and spreads	• Products are convenient, no preparation and reflect trends seen across other food groups e.g. vegetarian, healthy eating, ethnic and traditional • Lend themselves to the snacking culture • *Pâtés* – especially fish are ideal for starters and entertaining, with the emphasis on presentation • *Sandwich spreads* – quality and fresh ingredients to compete with the huge range of sandwiches now available • *Dips* – ideal complementary products for crisps and snacks
Pizzas	• A success story of the 1990s • Huge range of different bases and toppings, reflecting some of the trends seen across all food groups i.e. vegetarian, healthy eating, children's, ethnic • Perceived as a snack, but also popular as a main meal • Suits today's lifestyles of minimum cooking and no preparation time
Pastry products	• Pastry products still popular with the fillings reflecting vegetarian, healthy eating, traditional – poultry meat, fish and vegetables • Many different formats and use of different pastries e.g. filo pastry • Lattice topped became popular • Individual pies as well as family pies – ideal for single-person households and snacking
Recipe dishes and ready meals	• Indian, Italian and Oriental meals are outstripping traditional meals as popular alternative for consumers • Kit meals such as Fresh Creations and Just Cook (packed individually and arranged together in one cabinet, so combinations can be selected) (from Sainsbury's) enable customers to create restaurant-quality meals for two in 10 minutes • More indulgent and up-market foods e.g. Tesco's Finest and Marks & Spencer Café Specials for eating in rather than eating out • Snack meals suit today's lifestyles with family members eating different foods at different times. Vegetarian, healthy eating, children's products. Easy, convenient, microwave, reheat • Meat alternatives such as Quorn giving choice in basic raw material e.g. minced, cubed and in recipe dishes, and complete meals • Poultry recipe dishes and ready meal new products now outnumber red meat
Pasta	• Fresh pasta has grown in popularity, with the perception of it being 'quicker to cook' and 'because it's "fresh" it's better' • Pasta and sauces create quick, convenient meals with that Mediterranean healthy image • Filled pasta with an Italian-style bread is a quick-to-prepare meal
Rice	• Many different ethnic styles to accompany the many different dishes from India, China, Thailand. Ready prepared and convenient for quick reheating in either the microwave or the oven • More authentic styles as consumers become more knowledgeable

Table I.5 Continued

Product area	Key product development
Rice (continued)	• Arborio rice (Italian risotto rice) eaten as a ready meal. The traditional dish is quite time consuming – microwave reheating makes this a very quick meal
Salads, layered salads and deli dressings	• Ready prepared salad combinations now incorporate many different salad ingredients, e.g. raddichio, endive, rockette, lambs lettuce • Fresh oil-based salad dressings are now also a feature of the chilled cabinet • Pre-washed salads offer healthy, convenient product which can be tailored to today's eating requirements • All-year-round availability • Ideal for sandwiches and replicating restaurant-style foods and presentations • Organic possibilities • Dressed salads, layered salads and deli-style salads reflect choice of ingredients, different styles of eating, snacking, healthy, vegetarian and ethnic flavours and styles • *Salad dressings* – a huge variety of many flavours, reflecting ethnic influences and healthy eating
Sauces	*Savoury* • Pasta sauces no longer a niche market – to accompany pasta which is convenient, easy to prepare, healthy and extremely versatile – fitting today's lifestyles • Many new flavour sauces - microwaveable in the pot • Stir fry sauces, in many flavours and ideal for quick cook stir fries – again convenient, easy to prepare, healthy and extremely versatile *Sweet* • Custard sauces – indulgent with cream and healthy eating versions now commonplace. Some flavoured varieties available • Microwave reheat in the container for real convenience
Cream	• Additions to the standard presentations of double, single and whipping cream have been clotted, extra thick, spooning, alcohol flavoured and low-fat cream alternatives. All these presentations are now commonplace – many alcohol varieties, some using dual branding, available at Christmas time
Vegetables	• Again, convenience and quick preparation of fresh products at a price premium – suiting today's lifestyles • Ready prepared and ready to cook, some with added value sauces, dressings and flavoured butters – microwaveable in the pack • May be just bunches of carrots or upmarket variety mixes, or stir fry combinations using ethnic ingredients • No preparation, no waste, quick and easy and versatile
Vegetable accompaniments	• Convenient, no preparation quick reheat times and most often microwaveable • Mix and match with main ingredient (i.e. meat, poultry) to create variety e.g. flavoured mashed potato, roasted vegetables, cauliflower cheese

Table I.5 Continued

Product area	Key product development
Cooked poultry	• Varieties of joints e.g. breast, leg or wings or whole birds in many ethnic flavours – now commonplace • Can be reheated or eaten cold with salads or ideal for snacks and sandwiches • Also available sliced and flavoured and cut into strips or chunks for cooking • Not only prepacked but also available on deli counter • Home Meal Replacement market is growing – one of the most popular purchases is the rotisserie chicken
Cooked meats	• Availability of ethnic meats e.g. proscuito, salami • Many styles of cooked meats, straight and flavoured • Different formats for different uses – thick sliced and wafer thin • All suit sandwich making, snacking
Bakery	*Savoury* • Started with the garlic bread concept, of which there are many flavour extensions • The popularity of ethnic meals has led to development of ethnic-style breads e.g. flavoured topped foccacia *Sweet* • Cream-filled traditional and ethnic cakes, muffins, desserts and pastries. Sold individually packaged – ideal as a snack or whole for entertaining, for a dessert
Sandwiches	• Innovation in styles of bread, ethnic and traditional e.g. filled rolls, baguettes, croissants, wraps • Innovation in fillings – children's style, ethnic, traditional, healthy eating, vegetarian • Suits snacking and grazing
Soups	• Reflecting the ongoing need for convenience, quality (particularly in terms of taste) and choice, this market has seen massive growth • The flavours and combinations reflect traditional ethnic, vegetarian and healthy eating versions • They range from cold summer soups for entertaining to hearty winter recipes, as meal replacements • Many are microwaveable
Fruit juices and drinks	• Juices, drinks, nectars, smoothies – freshly squeezed, smooth, 'with bits', some with added vitamins • Many exotic flavour combinations • Seen as healthy and consumption increases in hot weather • Some organic varieties available
Yoghurts	• One of the oldest products in the supermarket but the range of products on offer is astounding • The biggest market is the children's sector • Low-fat and very low-fat yoghurts constitute a big market • Luxury yoghurts are also important, as are plain natural • Bio yoghurts also have a following and it is expected that in the future functional yoghurts will grow e.g. Maval, Benecol • Organic also have a niche sector • Sold as individuals, multipacks and split pots – ideal for snacking

Table I.5 Continued

Product area	Key product development
Yoghurts (continued)	• Some have spoons included in the pack to encourage eating on the go • Some are in pouches e.g. Yo-to-go from Yoplait, again ideal for lunchboxes and eating on the go
Desserts	• Pot desserts and fromage frais suit snacking, luxury, healthy eating and children's trends, similar to the yoghurts • Packaged in individual pots, some with spoons, multipacks, twinpots and tubes e.g. Frubes • Some of the indulgent varieties are dual branded with famous confectionery brands e.g. Cadbury's flake • Some up-market desserts are ideal for entertaining and are packaged in glass reusable containers • Amongst chilled desserts are such products as cheesecakes, fools, mousses, rice puddings and trifles
Ethnic accompaniments	• As people travel more and experience different cuisines they are demanding more authentic foods. The accompaniments for these are widely available as prepacked products and also sold on the deli e.g. onion bhajis, samosas, spring rolls. They are ideal for parties, snacking and as part of an Indian or Chinese meal
Meat alternatives	• There are many reasons for people becoming vegetarians or decreasing their intake of meat

Whilst the more mature dairy products market has shown steady growth, the most dynamic categories have been in ready meals, pizzas and salads, reflecting the increasing importance of convenience foods in consumer purchasing.

These figures may be compared to data on two dairy products for Europe shown in Table I.6. The biggest per capita consumption of yoghurt in 1992 was in the Netherlands, followed by France, Switzerland, Finland and Sweden. Average per capita consumption of yoghurt in Europe in 1992 was 9.76 kg. The biggest per capita consumption of chilled desserts in 1992 was Denmark, followed by France, Germany, Austria and Luxembourg, with an average per capita consumption across Europe in 1992 of 2.86 kg. The UK spends less on food as a percentage of total consumption, and around half of the European average for per capita consumption of yoghurt and chilled desserts. It is perhaps not surprising that, from such a modest base, it has shown some of the most dynamic growth. Whilst average annual growth in yoghurt and chilled desserts consumption by value for Europe as a whole between 1987 and 1992 was 6% and 14% respectively, comparable growth in the UK was 17% and 50%. This comparison suggests that these sectors of the UK chilled foods market are catching up rapidly with Europe and that growth will settle down in the coming decade to more modest levels reflecting the European average.

Table I.6 Consumption of yoghurt and chilled desserts in Europe (Sources: *Eurostat Yearbook 1997; European Food Databook 1994*)

Country	Consumption of food as a percentage of total consumption in 1995 (value)	Per capita consumption of yoghurt in 1992 (kg)	Per capita consumption of chilled desserts in 1992 (kg)
Europe	14.7	9.76	34.34
UK	10.4	4.91	1.38
Ireland	16.6	3.55	
France	14.2	18.39	6.96
Germany	10.9	9.47	2.86
Austria	13.0	8.15	2.57
Italy	16.3	3.54	0.45
Spain		9.54	2.10
Portugal		7.76	
Greece	28.8	7.78	
Netherlands	10.8	22.0	1.71
Luxembourg		8.11	2.58
Belgium	13.0	7.38	1.78
Denmark	14.3	9.15	10.12
Sweden	14.3	10.01	
Finland	14.1	11.74	0.95
Norway		7.66	

I.4 Individual categories within the chilled food sector (see also Table I.5)

I.4.1 Dairy products

Dairy products remain the biggest single category in the chilled foods sector. The various components of this category by value in 1997 in the UK were (Anon. 1998a; Anon. 1998b):

- cheese 40%
- butter, margarine and spreads 26%
- yoghurts/fromage frais 22%
- desserts 7%
- cream 5%.

Growth rates in value for these segments between 1993 and 1997 in the UK were:

- cheese 15%
- butter, margarine and spreads 4%
- yoghurts/fromage frais 19% (yoghurt 15%; fromage frais 44%)
- desserts 74–77% (Keynote and Mintel estimates)
- cream 20%.

Cheese remains both the biggest segment of dairy products and the biggest single sector in the chilled foods market in the UK (just under 30% of the chilled foods market as a whole). The sector has demonstrated steady growth in recent years. The strength of the market has been due, in part, to the emergence of low-fat cheeses, but it has also been boosted by consumers' desire for greater variety and improved taste, reflected in the popularity of mature cheddar and speciality cheeses. The soft cheese market has also benefited from the growth in snacking with successful brands such as Kraft Dairylea and Philadelphia.

The overall market for butter, margarine and spreads has remained generally static. Consumption of butter and margarine has declined overall, balanced by an increased use of low-fat spreads influenced by consumer concern for healthier eating. More recently, sales of butter have shown some improvement, suggesting consumers might be showing a renewed interest in taste as a factor in purchasing decisions. Sales have also been boosted by the introduction of spreadable butter aimed in part at the snacks market.

The yoghurt and fromage frais markets have benefited both from the move to snacking and the concern for healthier eating. Manufacturers such as Müller have met this demand through such developments as split pots and bio yoghurts which have helped it move out of the dessert category in consumers' eyes. Because they combine nutritional value with sweetness and are in conveniently small portions, yoghurt and fromage frais have become especially popular as a children's snack. The growth in sales has also been driven by increased interest in the range of Continental food, boosting sales of fromage frais and Greek yoghurt, for example. Natural fromage frais is widely seen by consumers as a low-fat alternative to cream, and this, together with its popularity as a children's snack, has kept sales expanding significantly. Indeed, some 39% of housewives in the UK claimed to have purchased fromage frais at some point during 1998 (Anon. 1998a). Similarly, the biggest areas of growth for yoghurt have been in bio yoghurts, low-fat and children's yoghurts, reflecting in part the growing importance of the 10–14 age group and their diet-conscious middle-aged parents. Per capita consumption of yoghurt in the UK is 4.8 kg per annum, under half of the total for Germany and just over one-fifth of French consumption, suggesting scope for further expansion of the UK market. However, with an annual growth rate of 17% between 1987 and 1992, compared to a European average of 6%, The UK market has been catching up rapidly. At a per capita consumption of 2.1 kg in 1993, the United States lags behind even the UK market, suggesting potential for growth (IDF 1995).

The most dynamic single sector of chilled dairy products has been desserts. This category is highly diverse, including rice pudding, mousse, trifles, cheese cake and gateaux. The biggest single categories are mousse and trifles. The success of this category has been partly the result of the development of low-fat desserts, but it has been fuelled by manufacturers' emphasis on luxury, premium products meeting consumer demands for taste and enjoyment. The biggest areas of expansion in this sector have been in sales of mousse, non-cream topped desserts such as tiramisu, rice pudding and fools. Sales of cream represent a small proportion of overall chilled dairy product sales, but sales have also benefited from a renewed consumer interest in taste and a sense of enjoyment and indulgence in eating.

I.4.2 Chilled meat products

Even when more narrowly defined to exclude raw meat, the chilled meats sector is the second largest sector of the dairy market, and has shown significant recent growth, despite concern over BSE. Growth has been stimulated by consumer demand for prepacked meats for snacks. Ham remains the most popular meat at the delicatessen, outselling other meats such as beef, pork and poultry. However, sales in this sector have also benefited from consumer interest in Continental, premium-priced products such as pâté and salami.

I.4.3 Chilled ready meals, pizzas and prepared salads

Although they still represent only 15% of the overall UK chilled foods market, these sectors have seen some of the most dramatic growth as consumers have looked for an alternative to home cooking. Improvements in quality have made ready meals a cheaper and more convenient option than eating out or ordering a takeaway. Almost half the sector is made up of ethnic dishes, primarily Chinese and Indian, reflecting the interest in more exotic tastes, with a further third made up of Continental recipes, particularly Italian. Indeed, Italian ready meals represent the biggest single variety of ready meal given the healthy image of pasta and its appeal to children. Over the period 1993–98 the value of the chilled ready-meals market has been estimated to have grown by as much as 50% in real terms, with recent growth rates of 7–10% a year. Within the ready-meals market, the biggest growth has come from vegetable-based ready meals, followed by fish-based meals and then those with a meat base, reflecting the healthier profile of vegetables and fish. The UK is the most developed market in Europe for chilled ready meals, with per capita consumption of 1 kg in 1994, compared to 0.3 kg for Germany, 0.2 kg for France and 0.1 kg for Spain (Anon. 1996). If UK trends are reflected in Europe, the European market for ready meals may well have significant future potential.

Pizza sales have shown the most dramatic growth of all, with an annual growth rate of over 10% in the period 1994–97. Manufacturers have sought to stimulate demand in part by introducing a wider variety of flavours to exploit consumer interest in more exotic tastes, and also by extending the range of vegetarian pizzas to respond to consumers' healthier life styles. At the premium end of the market manufacturers have sought to provide a more authentic taste by refocusing on traditional Italian recipes and ingredient quality. Consumer interest in convenience, healthier eating and more exotic tastes has fuelled strong growth in the prepared salads sector. Salads are now widely seen as a convenient replacement for vegetables in a main meal, as well as a popular ingredient in snacks. Manufacturers have responded by producing a wider range of mixed salads, using new varieties of leaves as well as other ingredients such as peppers.

Another important sector of the chilled food market not mentioned above is sandwiches, a major feature of the snacking habit. According to the British Sandwich Association sandwiches are now the UK's most popular fast food with a market of £3.25 billion and growing at 13% a year (Anon. 1999d). Sandwiches are estimated to represent 41% of the fast food market compared to burgers at 18% and fish and chips at 12%. The UK has the highest per capita consumption of sandwiches in the world, with every man, women and child munching their way through 37 bought sandwiches a year. Sandwiches are perceived as a healthy snack meal, they are portable, yet nutritious and offer infinite variety in terms of combinations of ingredients which can be used. This variety and choice in availability of sandwiches is highlighted in the data of new introductions in the major retail outlets recorded by FPI at CCFRA. The standard varieties, however, still account for 80% of the total sales and there has been little change in the most popular varieties according to the British Sandwich Association. The top selling sandwiches by fillings are tuna, chicken, egg mayonnaise/salad, ham, cheese, prawn and bacon, lettuce and tomato. The ethnic trend is also a feature of the sandwich sector with for example Spanish, Italian and Moroccan varieties.

I.5 Conclusion

The chilled foods market has been successful because it has met a number of customer needs. Chilled foods have been seen as fresh and healthy, and they have been ideally suited to meet the growing demand for ready meals and snacks. At the same time they have been rapidly adapted to cater for ever more cosmopolitan tastes and consumers' desire for variety, quality of sensory experience, and even indulgence in what they eat.

Product innovation will remain essential in an increasingly competitive sector of the industry. Market analysts suggest that continued growth may be restricted to the 'extremes' of the market, that is extra low-fat products at one end and premium indulgent products at the other (Anon. 1998b). Manufacturers will need to differentiate their products even more on the basis of quality, emphasising the fresh and 'authentic' taste consumers look for, and their concern for more environmentally friendly and 'natural' methods of production. Such trends suggest that the whole supply chain will need to pay attention to a range of factors, including:

- the importance of raw material selection in final product quality
- the quality of packaging and temperature control technologies across the chill chain in maintaining product quality
- the complexity and interdependence of technologies across the chill chain – the improved control of the microbiological and other factors affecting product safety.

1.6 References

ANON (1996) *The European ready meals market.* Leatherhead Food Research Association. Leatherhead.

ANON (1997) *The UK Food and Drinks Report.* Leatherhead Food Research Association. Leatherhead.

ANON (1998a) *Chilled ready meals/Yoghurt/Chilled pot desserts: market intelligence reports.* Mintel International Group. London.

ANON (1998b) *Chilled foods: 1998 market report.* Keynote: BMRB International. London.

ANON (1999a) 'Convenience is everything' *The Grocer*, May 29, p. 10.

ANON (1999b) *Supermarketing*, August 27, p. 20.

ANON (1999c) 'Ethnic Foods' *The Grocer*, August 21, pp. 53–66.

ANON (1999d) *The Grocer*, December 11, p. 12.

BOND S, (1992) 'Marketplace product knowledge – from the consumer viewpoint'. In *Chilled Foods: A comprehensive Guide*, 1st eds C. Dennis and M. F. Stringer, Ellis Horwood Ltd, Chichester, West Sussex. UK.

BENDER M M and DERBY B M, (1992) 'Prevalence of reading nutritional information and ingredient information on food labels among adult Americans: 1982–1988', *Journal of Nutrition Education*, 1992 **24** (6) 292–7.

CHILLED FOOD ASSOCIATION, (1997) *Guidelines for good hygienic practice in the manufacture of chilled foods.* ISBN 1901798003 Chilled Food Association, PO Box 14811, London, NW10 0ZR.

CROWE M, HARRIS S, MAGGIORE P and BINNS C 'Consumer understanding of food-additive labels', *Australian Journal of Nutrition and Dietetics*, 1992 **49** 19–22.

DEPARTMENT OF HEALTH, (1989) *Chilled and Frozen – Guidelines on cook-chill and cook-freeze catering systems.* HMSO, 49 High Holborn, London, WC1V 6HB.

IDF, (1995) In *Consumption Statistics for Milk and Milk Products 1993*, Doc. No. 301, International Dairy Federation, Brussels, Belgium, pp. 4–6.

INSTITUTE OF FOOD SCIENCE & TECHNOLOGY (UK), (1990) *Guidelines for the handling of chilled foods*, 2nd edn, IFST, London.

JOLLY D, SCHULTZ H, DIAZ-KNAUF D and JOHAL J, (1989) 'Organic foods: consumer attitudes and use', *Food Technology*, 1989 **November** 60–6.

RODOLFO M N, F R, LIPINSKI D and SAVUR N, (1998) 'Consumers' use of nutritional labels while food shopping and at home', *Journal of Consumer Affairs*, 1998 **32** (1) 106–120.

SLOAN A, POWERS M and HOM B, (1986) 'Consumer attitudes toward additives', *Cereal Foods World*, 1986 **31** (8) 523–32.

SLOAN A E, (1999) 'Top Ten Trends to watch and work on for the millennium'. *Food Technology* 53 (8) pp. 40–60.

WANDEL M, (1997) 'Food Labeling from a consumer perspective', *British Food Journal*, 1997 **99** (6) 212–9.

Part I

Raw materials

1

Raw material selection – fruits and vegetables

L. Bedford, Campden and Chorleywood Food Research Association

1.1 Introduction

Fresh fruits and vegetables are utilised in a wide range of chilled products. They may be sold whole, or peeled (for example peeled potatoes and onions) or further prepared (e.g. carrot batons). After washing or further preparation they form ingredients for mixes such as mixed fruit or salad packs or for further processing in a wide range of products. From this it can be seen that raw material requirements can be very varied but are specific to each end use. Sourcing of suitable raw material is essential for the production of final products of consistently high quality and for this both the producer and the user need to have a clear understanding of the requirements.

In this chapter some of the criteria for selection will be discussed. The examples are drawn from a range of fruit and vegetable crops. Most whole fruits and vegetables are best stored at chill temperatures and thus come naturally under the scope of this book. Some however, such as potatoes and tomatoes and some other fruits are low-temperature sensitive and should ideally be kept at higher temperatures. However even these items, once they are cut and prepared need to be kept chilled, to avoid enzyme-mediated changes and disease-related spoilage, both of which proceed more rapidly at higher temperatures.

Supply of suitable raw material requires collaboration between the grower and the purchaser of the produce. In effect, the grower forms the first link in the food chain. The decisions made, including variety selection and agronomic practices and the grower's skill in harvesting and where appropriate in storing the crop are crucial steps in the supply of high-quality raw material for chilled fruit and vegetable products.

1.2 Criteria for selection

First of all it is necessary to consider the factors which contribute to product variability and what makes raw material suitable for different purposes, particularly for use in chilled food products. Some of the factors are genetically controlled. Thus varieties may differ in size, shape and other characteristics. Many plant characteristics are also influenced by environmental factors, such as site and climate and seasonal weather patterns. These genetic and environmental aspects interact, contributing to the variability of the produce at harvest.

The parts of the plant which are consumed may be leaves, stems, roots, flowering heads, fruits or seeds, all requiring harvesting at the correct stage of maturity. Further factors apply after harvest, such as the handling and storage of the products before they are sold or prepared for further processing. Post-harvest factors can have a considerable impact on shelf-life and quality.

1.2.1 Variety

Plant species have recognisable inherited characteristics, which can be used to distinguish them from other species. Members of a species are generally able to interbreed easily, but much less easily, or not at all with other species. Within a species, natural variation gives rise to groups of individuals with small but definite differences, which are known as 'varieties'. When the variations are brought about by human intervention as in plant breeding then botanists use the term 'cultivars'. However in common parlance the term 'variety' is used for the man-made products of plant breeding as well.

When selecting raw material for particular purposes, one of the most important criteria under human control, is the choice of suitable varieties. There are many different ways in which varieties can differ. There may be obvious differences in colour, shape and size. There may be differences in field characteristics, such as yield, plant growth habit and disease resistance. In some cases flavour and other sensory characteristics may differ.

For most crops, a range of varieties can be used for any specific purpose, for example, a number of different varieties of Dutch white cabbage all possess the thick leaf texture and white colour required by the chilled salad producer. In other cases, the choice of a variety suitable for a specific purpose may be more limited; for instance an apple variety such as Cox may be specifically selected for a chilled fruit salad, because of its skin colour, which will enhance the overall appearance. A good variety has to meet the requirements of the primary producer, processor, retailer and ultimately the consumer.

Agronomic characteristics

Traditionally, farmers and growers have selected varieties for their field or 'agronomic' performance. They have been concerned to achieve high yields, this being a major factor affecting profitability. This may be in terms of total yield or, more importantly, of marketable yield. The latter refers to the saleable produce after all waste and defective material has been removed. Taking a root crop, such as carrots, all the roots will be lifted from the soil and grading will remove undersized and misshapen roots.

Disease resistance is another major issue. Field diseases can cause complete crop losses, or they may produce blemishes, which cause produce to be downgraded. Chemical pesticides can be applied to control some, but not all diseases. Genetic resistance is preferable, reducing the need for chemicals. Disease resistance may be controlled by a single gene or by several genes acting together (multigene resistance). Single gene resistance is easier to work with. However the disadvantage is that plant pathogens often rapidly produce new races able to overcome this type of resistance. An example is found in downy mildew of spinach (*Peronospora farinosa f sp spinaciae*). The disease is difficult to control chemically and varieties have been bred with specific genes for resistance, originally to races 1, 2 and 3 of the disease. Following the appearance in 1995 of the new race 4, whole crops were wiped out. Plant breeders have now produced a series of new varieties with resistance to race 4[1] but the resistance of these varieties may also have a limited life. Multigene resistance (field tolerance) gives greater stability. Some varieties of lettuce show this type of resistance to lettuce downy mildew (*Bremia lactucae*).

Plant habit is another varietal characteristic, which is particularly important when the flowering parts of the plant or the seeds are consumed. Plants should have sturdy stems to enable them to remain erect in windy conditions or when the foliage is wet. Varieties are said to have 'good standing ability' and not to 'lodge' or become flattened. In the green bean crop, plant breeders have given considerable thought to the plant architecture. Stems need to be sufficiently strong to support the combined weight of pods produced and, as yields increase, this becomes more of a problem. If all the pods were to be bunched together at the tops of the plants this might make for easier harvesting, but it could cause increased lodging. Plants on which the pods are distributed more evenly throughout the plant canopy are more likely to remain upright.

A major change in varieties over the last fifty years is associated with partial or complete mechanisation of harvesting. Older varieties of many crops, such as cauliflower and lettuce would have individual plants producing their heads over a period of time. In crops such as peas and beans individual pods were also produced over a long period. With the introduction of machine harvesting, the objective was once-over harvesting and the result, over years of breeding, has been varieties where individual heads mature very evenly. Peas and beans are said to be more 'determinate', that is to say that after producing a certain number of flowering nodes growth tends to stop. The pods then all develop and can be harvested at one time.

While all these field characteristics are necessary for the growth of crops, successful marketing of the crop requires attention to characteristics required by the processor, the retailer and ultimately by the consumer. These relate to quality of the final product.

Shape and size

Within some crops there is a range of shapes and sizes. For instance there is a range of types of carrots with different shapes and size of roots. Nantes varieties have cylindrical roots and so are preferred for pre-packing. Berlicum varieties are also cylindrical. They are generally larger and mature later in the season than Nantes varieties. The other common UK types are the conical Chantenays and Autumn King varieties which have very large roots tapering to a point. Larger roots of cylindrical or slightly conical varieties are suitable for slicing in chilled food products.

A recent development has been the breeding of specific varieties for 'mini vegetable' production. The concept is a response to consumer perception that small size equates to high quality and is seen as a way of adding value to vegetable products such as cauliflower. The varieties may also require special growing techniques. Varieties may also be selected according to their ability to produce the correct size of portion for prepared foods. Cauliflowers are often presented ready cut into florets and varieties differ in the ease with which they can be cut up and the size of florets produced.[2] For chopping or dicing, large size is required and the overall yield of prepared product is an important consideration. Large cabbages are specified for processing uses such as coleslaw production, the total yield of cabbage shreds being the major concern. This contrasts with retail sale where smaller varieties are required.

Colour and appearance

It is often said that the consumer buys by eye, so an attractive colour and appearance is essential. Varieties often vary in colour. The characteristic colours of different apple varieties make them easily identified. They differ not only in the ground colour (Coxes are yellow and red, Bramleys are green) but also in the markings known as 'russetting'. Chilled food manufacturers will consider the flesh colour, as well as the skins and the ability to resist grey discoloration when selecting a variety to provide colour to a mixed fruit salad.

Modern commercial carrot varieties are orange, but they were arrived at by careful selection from a varied ancestral gene pool in which yellow and purple colours were common. More recently a series of large-rooted varieties were developed for dicing. They had a deeper orange colour that was particularly evenly spread across the core and flesh of the roots. Colour may not be so important where the product is chopped or otherwise prepared. White cabbage for retail sale needs to have bright colour and fresh appearance. Slight greyness may be acceptable if the cabbage is used for coleslaw, as the colour will be masked by the mayonnaise.

Flavour and texture

To the consumer, of course, the ultimate requirement is good eating quality, and food products should have good natural flavour and texture. Varieties of some crops such as apples have very distinctive flavours and consumers commonly select their own favourites e.g. Cox, Golden Delicious, Russet. In other crops, many varieties have quite similar flavour. A consumer would be unlikely to be able to tell the difference between many varieties of, say, iceberg lettuce.

On the other hand, flavour has been a particular issue with Brussels sprouts. Varieties differ in flavour, particularly in levels of bitterness and excess bitterness has caused some varieties to be unacceptable. The flavour and aroma characteristics of varieties are governed by their chemical constituents. Breeders may be able to manipulate these, in order to improve varietal quality. The chemicals responsible for bitterness in Brussels sprouts have been identified as the glucosinolates, sinigrin and progoitrin.[3] In the 1970s and 80s deep green coloured varieties appeared on the market, which had been selected for resistance to insect pests. However, taste panel assessments showed these varieties to be bitter and chemical analysis confirmed that they were high in glucosinolates.

Some varieties, such as Topline, Rasalon and Lunet had consistently good quality over several years.[4] Van Doorn *et al.*[5] conducted consumer studies with a range of different Brussels sprout varieties and suggested that at a level of glucosinolates (sinigrin plus progoitrin) of above 2.2g per kg negative consumer reaction was registered. Breeders can now select for lower levels of these chemicals either by analysis or by tasting.

Flavour can vary even within the portion of the crop that is consumed. Both leeks and celery have white tissue at the base of their stems and greener tissue and leaves higher up. There are differences in flavour between white and green portions.[6] EU regulations[7] dictate the proportions of white and green in celery for sale and varieties vary in their ability to satisfy this requirement.

The texture of a product also contributes to eating quality. This is often related to maturity, with an over-mature product becoming tough. However, there may be variety differences. Potatoes have obvious texture differences being either waxy or floury. Waxy varieties are firm and will retain their shape after cooking. They are preferred for salad use or for products such as potato scallops. Floury varieties, which soften on cooking, are used for mashed potatoes.

Chilled food manufacturers use large volumes of sliced and diced onions to add flavour to their products. Onion varieties differ in pungency. In the UK the majority of main crop onions are of the Rijnsburger type which are relatively high in pungency.[8] Some of the varieties grown in Spain are less pungent and hence 'Spanish' onions are used if the onion is to be eaten raw. The American Vidalia type is noted for its sweetness.

Sweetness is an important flavour attribute of most crops but sugar levels can also be important for other reasons. Amounts of reducing sugars in potatoes influence colour after frying and there is a strong variety-related component to this effect.[9] Maris Piper is a preferred variety for the chipping trade because it is lower in reducing sugars. Flavour variations in tomatoes are related to differences in the amounts of sugars and acids in the fruit. If both are at low levels, the flavour will be bland. Commercial varieties with high levels of both acid and sugar are preferred.

Recently there has been interest in the health-promoting effects of some of the component substances found in fruit and vegetables. Varieties may differ in amounts of these so called bioactive substances. Schonhof *et al.*[10] working on calabrese determined concentrations of carotenoids, chlorophylls and glucosinolates in three different varieties. They reported that although there were seasonal influences on the concentrations of these, the genetic differences were constant.

1.2.2 Crop maturity

Another major consideration in raw material production and selection is harvesting of crops at the optimal stage of maturity. Visual indications often reflect the stage of development of the plant and aid the grower in timely harvesting. When bulb onions are mature the leaves begin to wilt and bend over. Growers refer to the 'fall-over' date. Assessment criteria are different for vegetables and fruits and assessment of the correct stage depends on the part of the plant that is consumed.

Leafy crops are harvested according to size and firmness. Most types of cabbage and lettuce are allowed to produce a heart and the heads should be firm and of the size required by the particular outlet. Cabbage greens are harvested before the hearts have had time to form. Lettuce and spinach for leafy salads are harvested only a few weeks after drilling, when leaf size is very small. In crops such as cauliflower and calabrese, where the flowering heads are consumed, assessment is also by eye and development is monitored until firm heads have formed. Once this stage is passed the heads will open out as flower development proceeds. In over-mature calabrese, the yellow flower petals may be seen.

Harvest time is less critical with root crops such as carrots, swedes, turnips, or parsnips. They may be harvested when they reach the size required by a particular outlet but are often stored in the field for several months. Carrots may be considered to have reached maturity when there is no white coloration at the tip of the root but will continue to grow after this. Bunching carrots may be harvested immature, with some white roots. Other types are harvested at a later stage when roots have achieved a sufficient size to meet specific market requirements. Similarly early potatoes are harvested immature, but other 'maincrop' or 'ware' potatoes are harvested after the leaves have died down, this being a sign that the crop has completed bulking up.

In some crops the immature pods or immature seeds are consumed (e.g. runner beans, dwarf French beans, peas, sweetcorn). As the seeds develop and swell they accumulate carbohydrates, first of all in the form of sugars and these are then converted to starch as the seed dries out. At the same time the pods are drying out and the texture becomes tough or mealy. Maturity assessment is particularly critical for these crops as they rapidly become tough if left in the field after optimum maturity. In beans the seed development is monitored and maturity is related to pod and bean size for specific varieties.

Fruits are divided into two ripening types. Climacteric fruits such as 'pome' fruits (apples and pears) apricots, peaches, plums and kiwifruit continue to ripen naturally after harvest. As the fruit matures, carbohydrates are accumulated as starch and as ripening proceeds starch is broken down to sugars. When they are allowed to mature on the tree the climacteric begins, respiration rate increases and there is an increase in the production of ethylene gas, which promotes ripening. If allowed to ripen fully on the tree the crop is only suitable for immediate sale. Fruit destined for long-term storage is harvested immature and storage conditions are designed to control ripening. Non-climacteric fruit, for example raspberries, strawberries and cherries, accumulate sugars as they ripen on the plant and do not continue ripening after they are harvested. They are harvested at maturity.

Maturity of both types of fruit can be assessed using a refractometer, which measures total soluble solids, and in the case of fruits this is mostly sugar. Where applicable, starch can be measured using the iodine test. This has been used for apples and pears but has been most successful as a maturity indicator for pears.

1.2.3 Growing and environmental influences on suitability

In addition to variety and maturity, in the production of high-quality raw material, the other major influences are environmental. Growing site and season and the growers' production techniques may in fact be as, or more, important to the final crop quality but the effects are more difficult to define or control.

Site

When selecting suitable sites, consideration must be given to climatic conditions and soil type. The former influences the ability to grow crops successfully. In England, some crops such as sweetcorn are best suited to growing in southern counties where overall temperatures are warmer and there are sufficient frost-free days. Low temperatures slow down growth of the plants and cob development. Frosts after sowing will kill young seedlings and at the end of the season will destroy the cobs. Other crops, more suited to growing in the UK are able to grow at lower temperatures, e.g. cabbages, potatoes, or to complete their development in a shorter time period, e.g. peas. Research on carrots, mainly in Scandinavia and the USA has also shown latitude and other site differences to be influential[11, 12] Martens *et al.*[13] found that season and site had a major influence on sensory quality and chemical composition, with variety being much less important.

Selection of suitable soil type is important, particularly for root crops. When carrots are grown on stony soils the result is fanged and twisted carrots. Hence in England either sand or peat soil types have historically been chosen. Root growth has been found to be different on these two soil types and roots from peat soils are longer than the same variety from sand sites. Site may also influence chemical composition and through this sensory quality. Heany and Fenwick[14] reported differences in glucosinolate concentrations and hence bitterness of Brussels sprout varieties from five UK sites. In Ireland, Gormley[15] found that soil type influenced chemical composition and sensory quality. Carrots from mineral soil had higher dry matter and higher levels of carotene and, after cooking, taste panels rated them higher for flavour and softer in texture. However Day[16], in a three-year project on sand and peat soils, found neither type to be consistently better for yield or quality.

Season

Crop production is greatly influenced by seasonal weather conditions. For optimum crop growth, the right balance of temperature and moisture are required. This not only has a great influence on crop growth and yield, but may also affect post-harvest quality. In wet conditions, leafy plant material often takes in a lot of water and results in soft tissue. If this occurs at harvest it can lead to tissue which bruises easily and leads to shorter shelf-life. Conversely, too little water leads to plants suffering from moisture stress and they may wilt or fail to develop properly as in cauliflowers which are said to 'button'. They produce only very small heads, too small to market.

Disease and pest development also require very specific temperature and moisture conditions and sophisticated prediction methods have been developed, based on weather data collected nationally, as in potato blight prediction, or locally. Some diseases prefer wet (e.g. powdery mildew) and some dry conditions (downy mildew) but this is to oversimplify the topic. Different regimes are required for different parts of the life cycle of the disease.

The yield of tomatoes is dependent on light levels during the growing season. Light levels also influence the growth of lettuce crops and the accumulation of nitrates in the plants.[17] Manufacturers of chilled baby foods will be particularly concerned that their raw materials are low in nitrates.

For chopped lettuce products a dense head will give the most efficient throughput for use in a chilled salad product. Recent research has shown that day and night temperatures have large effects on the shape and density of heads of iceberg lettuce.[18]

Growing techniques

Production techniques constitute the one area where the grower can have a major impact on the yield, quality and suitability for purpose of his crops. This includes time of drilling, plant spacing and plant protection throughout the growing season and the provision of all the nutrients needed to optimise production. The application of fertilisers to provide the plants with sources of nitrogen, phosphorus and potassium and other major and minor nutrients has been researched over many years. Nitrogen is essential for crop growth, but excess nitrates in the tissues are perceived as a possible health hazard. There are EU prescribed levels for lettuce and spinach. Field studies on a range of crops showed that some of the variation in nitrate levels may be attributed to rate of fertiliser application, but that there is likely to be an interaction with other factors affecting absorption, translocation and assimilation. Nitrate concentrations also varied in different parts of the plant, being higher for instance in the outer leaves of lettuce than at the heart.[19] New research on lettuce aims to investigate the extent to which nitrate accumulation of protected lettuce can be adjusted by different crop management and fertiliser treatments.[20]

The primary producer can also influence the suitability for purpose of particular crops by growing and harvesting suitably sized material. A major requirement of batoning carrots is that they are of suitable size and shape to maximise throughput of the batoning equipment. The optimum size is produced by a combination of suitable varieties and growing techniques and this has to be related to the requirements of different makes of equipment .

1.2.4 Post-harvest handling and storage

Once a crop is harvested its quality cannot be improved. At this stage, the objective must be to maintain the produce in good condition through any short- or long-term storage until it is delivered to the customer. Thus, the ultimate quality and shelf-life of a final product depends not only on growing conditions but also on harvesting and on post-harvest handling. Avoidance of handling damage at this stage is important. Rough handling leads to bruises which spoil the appearance of produce and can become a focus of infection by spoilage diseases. Bruising can be a major reason for losses of fruit in store as well as for vegetable crops such as Dutch white cabbage.

Maintenance of suitable post-harvest temperature is extremely important to maximise shelf-life, both for produce for immediate use and for that to be stored. Crops continue to respire after harvesting using up reserves and shortening shelf-life through wilting and yellowing. Respiration rate is temperature related and is roughly halved for every 10°C that the temperature is reduced. The general rule is to remove field heat as quickly as possible after harvest and then to maintain the produce at chill temperature. This is achieved by various methods such as vacuum cooling for lettuce, hydrocooling of carrots and tomatoes and storage in various types of refrigerated stores (see Ch. 4). Increasingly, use of the 'cool chain' aims to retain the produce at the required low temperature during packing and transport to the retailer.

Some delicate crops such as lettuce are not suitable for other than very short-term storage. Others such as root crops, cabbage and many fruit crops can be stored for many months to provide the continuity of supply required by the chilled food manufacturer and customer. Storage may be in field, for example for carrots, which are covered with straw, or in ambient stores for beetroot and other root crops. For longer storage, refrigerated stores are used. Many use moist air cooling to maintain the desired relative humidity and prevent dehydration and positive ventilation to draw air through the stored crop.

Specialised storage methods are required for some crops. For example there is a three-stage procedure for onion curing and storage, each stage requiring a different temperature and humidity.[21] In the first stage the aim is to dry the surface of the bulbs. At the next stage moisture is removed from their necks at high temperature. The temperature is then reduced to prepare the onions for long-term storage.

Controlled atmosphere stores have been developed for climacteric fruits. The aim is to retard the chemical changes which cause ripening. In store the balance of oxygen and carbon dioxide gases is controlled and ethylene scrubbers are used to prevent the build up of this gas. On removal from store the fruit all complete the final stages of ripening evenly.

The storage of crops can also affect the quality of raw material for further processing.[22] O'Bierne studying carrots for batons, found that there was a more rapid reduction in quality of harvested carrots in a crop left in the soil over winter than in an autumn harvested crop. He found that this was due to higher microbial levels in the overwintered crop.

1.2.5 Shelf-life

The most common conditions limiting post-harvest shelf-life are dehydration and wilting, yellowing of leafy material, browning of cut surfaces and disease development. Wilting occurs rapidly in produce after harvest. Storage at high humidity or packing whole produce in ice helps to reduce dehydration. Covering of retail packs also helps to restrict water loss.

Research has shown that ethylene is found, not only in climacteric fruits, but in other fruits and vegetables as well. It is involved together with respiration in the process of senescence and yellowing. Wills *et al.*[23] working with 23 different products demonstrated considerable extension of post-harvest life by reducing ethylene in the storage atmosphere. Produce should not be stored with fruits which are producing ethylene.

Diseases can also reduce shelf-life and post-harvest quality. Some of these spoilage diseases are specific to certain groups of crops. Others have a wide host range. *Botrytis cinerea* (grey mould) is one of the most common affecting a wide range of fruits and vegetables. Bacterial soft rots (*Erwinia* and *Pseudomonas* spp) are also ubiquitous.

1.3 Specifications

Having considered some of the factors influencing the characteristics of raw material, the primary producer needs to understand the requirements of the market sector and the individual companies to be supplied. To do this he needs to obtain a 'specification'. Major fruit and vegetable products appearing on the open market are covered by EU common standards of quality.[7,24,25] These contain minimum standards for legal sale of the product. Basically these ensure that the consignment is fresh and wholesome and at a suitable stage for consumption. The standards also provide a classification of produce, known as 'The Class System'. The highest quality produce is 'Extra Class' and below this are 'Class 1', 'Class 2' and, in some cases, also 'Class 3'. These standards are policed by the Marketing Inspectorate. Inspectors visit markets, ports and retail outlets throughout the UK. Produce sold for further processing is not subject to these regulations.

Many companies issue their own individual company specifications, which are normally more stringent than the EU standards. They also enable the company to be more specific in their individual requirements. Specifications should give definitions of different defects and tolerances for each. The following categories are usually included.

1. Foreign Matter (FM) – material of non-plant origin. This includes stones, soil, wood, glass, insects, etc. and any other material such as plastic which may have become included in the load. This category also includes any toxic material of plant origin such as potato or nightshade berries in vegetable crops. There is usually a nil tolerance for all these items, with the possible exception of insects. It may be impossible to achieve complete absence of insects, as even after chemical treatment they may remain in the dead state often hidden within the leaves of such items as lettuce and calabrese.
2. Extraneous Vegetable Matter (EVM) – parts of the crop plant other than that to be consumed, e.g. bits of stem in a consignment of Brussels sprouts or leaf in green beans.
3. Foreign EVM (FEVM) – parts of plants other than the crop species. This category does not include any toxic material (see FM above).
3. Blemishes – this section defines discoloured areas of product, most likely caused by disease or pest attack. However, there is usually no attempt to identify the cause.
4. Damage – bruising or mechanical damage.

The specification will also define any particular requirements, such as size of products and in some cases the variety or type to be used. As an example, carrots are produced for a range of different products. For retail sale of early or small rooted varieties as loose carrots, EU minimum standards are for roots with minimum diameter greater than 10 mm or weight greater than 8 g. For the larger maincrop varieties they are greater than 20 mm or 50 g. For prepacked carrots individual retailers will have their own specific size specifications. The carrots need to be of uniform diameter and length in order to fill a standard pack size. The type of carrots known as Nantes are most widely used. For diced carrots, the user may require a specific size of dice. The overall appearance of an end product which may comprise a number of other ingredients as well as the carrots, has to be considered. The raw material supplier will look for a large-rooted variety, in order to maximise yield and reduce wastage during dicing. Similarly, large-rooted varieties are also most suitable for batton production and different retailers may have slightly different size specifications. Straight-sided varieties produce less wastage but some of the waste can be used in the production of carrot shreds.

The search for uniform quality in prepared products can lead to very tight specifications. For sandwiches, for instance, the ideal tomato produces slices of even size, four of which will just cover a standard sized slice of bread.

1.4 New trends in raw material production

In the supply of raw material, the grower forms the first link in the chain of food production and in recent years more growers have become involved in the packing and marketing of their own produce. A large proportion of this produce is sold through major retailers, either whole or prepared or processed in various ways. The standards set are high in terms of product specification and one of the requirements is for uniform supply for twelve months of the year. This often involves grower packers in importing raw material to cover those times of year when UK product is not available. Some grower groups have formed partnerships in other countries in order to have greater control over the standard of the produce they are supplying outside the UK season

Consumers are now more concerned with the way crops are grown, e.g. the use of pesticides and effects on the environment. There is a requirement to provide evidence of traceability of specific batches of produce throughout the food chain, right back from the retailer to the grower. In case of complaints, retailers also need to be able to trace problem loads and to demonstrate due diligence. This has led to a philosophy of traceability from 'plough to plate' and to quality-management systems for growers and suppliers of produce, linking in with those of their customers. A number of quality-assurance schemes have been developed in the UK, including the 'Assured Produce Scheme'. The scheme, which is run by an independent organisation, sets standards for good agricultural practice and provides independent auditing of its members. Growers who sign up to this or similar schemes, have to be able to provide evidence of the methods they use in crop production, including use of pesticides and fertilisers and harvesting and post-harvest procedures. There are moves in Europe to implement similar standards, e.g. the EUREP standard. There is a demand for primary producers to provide evidence of its safe production. Techniques such as 'Hazard Analysis Critical Control Points' (HACCP) are used by companies to help with this. HACCP enables users to identify possible safety hazards and to ensure that systems are in place to control them.

Organic production is also increasing. Organic growers in the UK must comply with the standards of one of a number of approved organisations, the best known of which is probably the Soil Association. They must also register with the 'UK Register of Organic Food Standards' (UKROFS), which is the UK certifying authority and whose role is to regulate the production and marketing of foodstuff produced to organic standards. Similar systems exist in other EU countries.

Among non-organic growers there is increasing interest in integrated crop management (ICM) techniques. ICM has been defined as 'A combination of responsible farming practices which balance the economic production of crops with measures which conserve and enhance the environment'.[26] Research in this area is helping growers in decision-making as they seek to reduce the use of pesticides and other artificial inputs into their growing procedures and it forms a part of assured produce schemes.

1.5 New trends in plant breeding

Man has been engaged in altering plant characteristics for thousands of years. Originally this consisted in observing naturally occurring variability in plant populations and selecting those types that had the desired characteristics. In this way we have arrived at many commonly grown crops. From a species of *Solanum*, larger tubers and plants with lower levels of toxic glycosides have been developed which eventually produced modern cultivated types of potatoes. From the original wild cabbage, a plant with a rosette of leaves, a whole range of types have developed, including cabbage, cauliflower, Brussels sprouts, broccoli and kale. These are all genetically so similar as to be classified as the same species and yet morphologically are very different. In cabbage we eat the leaves, Brussels sprouts are swollen leaf buds, in cauliflower and calabrese it is the flowering heads that are swollen.

Classical plant-breeding techniques developed later and enabled plant breeders to manipulate the natural breeding process to obtain superior characteristics. The genetic material within the plant cells (the genotype) governs the characteristics of a plant, which we observe (known as the phenotype). During sexual reproduction genes from male and female plants are exchanged and new individuals are produced whose characteristics differ from those of their parents. Plant breeders manipulate this process. The first step is to identify individuals with the desired characteristics. Often the objective is to introduce a specific character such as disease resistance from one parent into a commercial variety, which would be used as the other parent. A cross is made between the two parents to produce the F1 generation. Individuals from this generation are allowed to interbreed to produce the F2 generation. From the F2, plants with the desired combination of characteristics are selected to produce a new variety. This method allows transfer of genes between members of the same species and in some cases between related species. For instance, onions (*Allium cepa*) and shallots (*Allium ascalonicum*) can easily be crossed.[27] Others such as various bean species have been more difficult to work with. These methods have produced the new varieties with higher yield, disease resistance and improved quality used today. Plant breeding is an ongoing process so that there continues to be a stream of new varieties from which the grower can choose.

Many new varieties are now F1 hybrids. These tend to be more vigorous than the parents as well as being more uniform in size and shape. The seed and pollen producing parents are induced to self-pollinate to produce inbred lines. These are crossed to produce commercial seed. Seed of F1 hybrid varieties is more expensive as, in order to maintain the variety, the cross has to be repeated each year. From this it can be seen that man has been able to manipulate plants to select the characteristics required, but that breeding has only been possible between members of the same, or closely related, species. Now genetic modification (GM) techniques enable geneticists to manipulate genes in a far more sophisticated way. These new techniques allow individual genes to be identified and copied by biochemical techniques and then to be transferred to a different species. Once the initial transfer of genes has been made, normal plant-breeding techniques are employed to produce the new varieties. It also allows the genetic material to be manipulated in such a way as to allow expression of particular genes to be controlled in very specific ways (turned up or turned down). While the technique allows genes to be introduced from completely unrelated species, animals as well as plants, it should be said that in many cases it has actually been used simply to alter genes within a species.

GM technology has been used to produce new varieties of some major agricultural crops, notably soya beans and maize. The first horticultural crop in which GM varieties became commercially available was the tomato. Initial research at Nottingham University led to the identification of many of the genes involved in ripening. The enzyme polygalacturonase (PG) is involved in the softening process. A 'slow softening' tomato was produced by altering the PG gene by reducing its expression. This was used in processing tomatoes to produce a firmer fruit and give a thicker paste. Tomato paste made from GM tomatoes was approved for sale in the UK in 1995, and was available from retailers before consumer concern caused them to revise their policy. Similar methods were used to produce a so called 'vine ripe' tomato variety for fresh market use. The fruit could be left on the vine to develop full colour without becoming too soft for commercial transportation and storage. The Flavr Savr™ tomato received UK approval in 1996 but has not to date been marketed in the UK. Meanwhile, vine ripe tomatoes developed by conventional breeding have become popular. Use of biotechnology is also increasing the understanding of the mode of functioning of genes.[28] It is possible to alter just one gene in a plant and then to observe the changes that occur.

The development of commercial varieties has now become a matter of major public concern on ethical grounds and on those of perceived safety. Future developments in this field must therefore be a subject of debate. To date, in the UK there are no commercial GM varieties of fruits or vegetables for the fresh market.

1.6 Conclusion

Growers of fresh fruit or vegetable raw material form the first link in the supply chain, which includes packers, processors and retailers. Their role is an important one. High-quality finished product cannot be produced unless the raw material is of good quality. Variety selection and timely harvesting must be combined with suitable growing conditions to give good produce at harvest. The intrinsic quality of the crop cannot be improved beyond this point and the objective in the later stages of storage, packing, transport and use is to maintain quality by suitable storage and transport conditions. Removal of field heat and maintenance of suitable temperature is particularly important for chilled products.

In order to maintain supplies of consistent quality for 52 weeks of the year, the grower/packer may now be responsible for sourcing supplies outside the UK season. This may involve partnerships with growers and suppliers outside the UK. Growers are increasingly likely to belong to assured produce schemes, which provide standards of best practice and are independently audited. The organic sector is growing and many non-organic growers are using integrated crop management systems.

Plant breeders seek to improve the quality of varieties available to commercial growers, drawing on scientific research to breed superior varieties. GM technology is already helping research workers to improve the understanding of how genes function in plants. The extent of future use in commercial breeding may well depend on public acceptance or otherwise.

1.7 Sources of further information and advice

Raw material supply starts with the selection of the seed. The grower can obtain information from seed companies specialising in horticultural crops or, in the case of fruits, from companies supplying young plants. Many seed companies have active plant-breeding programmes and new varieties appear in their catalogues each year. In order to weigh up the merits of particular varieties the grower can refer to the results of independent trials. In the UK, the National Institute of Agricultural Botany (NIAB) undertakes trials of all the major vegetable crops and publishes information on yields and field performance. Campden and Chorleywood Food Research Association also undertakes trials, with the emphasis on suitability of varieties for specific uses. Trials of top fruit and soft fruit are carried out at various sites, in trials funded by the Horticultural Development Council (HDC), who also fund part of the trial work on vegetables. This organisation administers funds collected by a levy on growers.

1.8 References

1. BEDFORD L V, *Spinach Varieties for Fresh Market and Freezing – Spring Sown Trials 1996*, CCFRA R & D Report no 30, 1996.
2. BEDFORD L V and BOND S, *Quality of Fresh Market Cauliflowers*, CCFRA Agrofood Report no. 3, 1992.
3. FENWICK G R, GRIFFITHS N M and HEANY R K, 'Bitterness in Brussels sprouts (*Brasssica oleracea* var. *gemmifera*): the role of glucosinolates and their breakdown products', *J Sci Food Agric*, 1983 **34** 73–80.
4. BEDFORD L V, *Sensory Quality of Brussels Sprouts: Survey of Varieties 1998*, CCFRA Review no. 11, 1988.
5. VAN DOORN J E, *et al.*, 'The glucosinolates sinigrin and progoitrin are important determinants for taste preference and bitterness of Brussels sprouts', *J Sci Food Agric*, 1998 **78** 30–8.
6. BEDFORD L V, 'Sensory quality of white and green portions of drilled and transplanted leeks', *Processing and Quality Assessment of Vegetables from*

Trials at Ministry Centres, d, Miscellaneous Autumn and Winter Crops', CCFRA Technical Memorandum no. 413, 1986.

7. *EC QUALITY STANDARDS FOR HORTICULTURAL PRODUCE: FRESH VEGETABLES*, MAFF Publications, London (PB05201) 1996.
8. BEDFORD L V, 'Dry matter and pungency tests on British grown onions', *J. natn. Inst. Agric. Bot.*, (1984) **16** 581–91.
9. STANLEY R and JEWELL S, 'The influence of source and rate of potassium fertiliser on the quality of potatoes for French fry production', *Potato Research*, 1989 **32** 439–46.
10. SCHONHOF I, KRUMBEIN A, SCHREINER M and GUTEZEIT B, 'Bioactive substances in cruciferous products, *Agri-food Quality I: Quality Management of Fruits and Vegetables*, Royal Society of Chemists, 1999.
11. ROSENFELD H J, BAARDSETH P and SKREDE G, 'Evaluation of carrot varieties for production of deep fried carrot chips – iv. The influence of growing environment on carrot raw material', *Food Research International*, 1997 **30** 611–18.
12. ROSENFELD H A, RISVIK E, SAMUELSEN R T and RODBOTTEN M, 'Sensory profiling of carrots from northern latitudes', *Food Research International*, 1997 **30** 593–601.
13. MARTENS M, ROSENFELD J and RUSSWURM Jr H, 'Predicting sensory quality of carrots from chemical, physical and agronomic variables: A multivariate study', *Acta Agric. Scand.*, 1985 **35** 407–20.
14. HEANEY R K and FENWICK G R, 'Glucosinolates in *Brassica* vegetables. Analysis of 22 varieties of Brussels sprout (*Brasssica oleracea* var. *gemmifera*).' *J Sci Food Agric*, 1980 **31** 785–93.
15. GORMLEY T R, ORIORDAIN F and PRENDIVILLE M D, 'Some aspects of the quality of carrots on different soil types, *J. Fd Technol*, 1971 **6**, 393–402.
16. DAY M J, 'The effects of soil type, delayed harvest and variety on canning carrot yield and quality,' *J. natn Inst. Agric. Bot.*, 1984 **16** 567–9.
17. NATIONAL ACADEMY OF SCIENCES (1981) *The Health Effects of Nitrate, Nitrite and N-Nitroso Compounds*, publ. National Academic Press, Washington DC.
18. 'A NEW GENERATION OF CROP FORECASTING SYSTEMS' *Agriculture Link Newsletter* Oct 1999, MAFF PB4659.
19. *NITRATE, NITRITE and N-NITROSO COMPOUNDS IN FOOD: SECOND REPORT*, MAFF Food Surveillance Paper no. 32 1992, HMSO.
20. 'NITRATE UPTAKE AND ACCUMULATION IN PROTECTED LETTUCE' *Agriculture Link Newsletter* Feb 2000, MAFF PB4826.
21. LANCASTER D, 'Bulb Onion Storage', *Onion Quality*, CCFRA Seminar Abstracts, 1996.
22. O'BEIRNE D, 'Modified atmosphere packaging of vegetables and fruits – an overview', *International Conference on Fresh Cut Produce Conference Proceedings*, CCFRA, 1999.
23. WILLS R B H, KU V V V, SHOHET D and KIM G H, 'Importance of low ethylene levels to delay senescence of non-climacteric fruit and vegetables',

Australian Journal of Experimental Agriculture, **39** 221–4.
24. *EC QUALITY STANDARDS FOR HORTICULTURAL PRODUCE: FRESH FRUIT*, MAFF Publications, London, (PB05191) 1996.
25. *EC QUALITY STANDARDS FOR HORTICULTURAL PRODUCE: FRESH SALADS*, MAFF Publications, London, (PB05211) 1996.
26. KNIGHT C, 'Introduction to integrated crop management', *Integrated Crop Management: A System for Fruit and Vegetable Crop Production*, CCFRA/HRI Conference Notes, 1999.
27. WATTS L, *Flower and Vegetable Plant Breeding*, Grower Books, 1980.
28. DOMONEY C, MULLINEAUX P and CASEY R, 'Nutrition and genetically engineered foods', *Nutritional Aspects of Food Processing and Ingredients*, Chapman and Hall, 1998.

2

Raw material selection: dairy ingredients

R. Early, Harper Adams University College

2.1 Introduction

As the first food of infant mammals, milk provides an important source of fat, protein, carbohydrate, vitamins and minerals, essential to the development of tissue and bone, and the growth of young. Milk is also a substance used beneficially by humans of all ages, both as a food in its own right and as a material for the production of milk products and milk-based food ingredients. The composition of milk varies significantly among species and bovine milk is most widely used world-wide for consumption as milk and for conversion into other products. Ovine and caprine milks are not without significance, particularly within the realms of cheesemaking.

The relevance of milk to chilled foods is found in the milk products which are chilled foods in their own right and in the range of milk-based ingredients used in the manufacture of chilled foods. Many milk products such as cheese and yogurt have a long heritage. In contrast, most milk-based ingredients are relatively recent innovations. Their existence is linked to the development of the modern food market place and the presence of convenience foods and ready meals, many of which are chilled foods.

2.2 Milk composition

Water is the main component of milk and most manufacturing techniques employed by the dairy industry concern methods of water control. With a water

Thanks go to Melanie Hooper (Dairy Crest Ingredients) and Steve Timms (Fayrfield Foodtec Ltd) for providing information on milk-based ingredients and their uses, and to David Jefferies (Oscar Meyer Limited) for giving advice on the use of dairy products in chilled ready meals.

Table 2.1 The major nutrients (A) and major components (B) contained in cows' whole milk, and major components on a dry basis (C)

Component	A per 100 ml	B %	C %
Fat (g)	4.01	3.9	30.8
Protein (g)	3.29	3.2	25.3
casein		2.6	20.6
whey proteins		0.6	4.7
Lactose (g)	4.95	4.8	37.9
Ash		0.75	5.9
Calcium (mg)	119		
Iron (mg)	0.05		
Sodium (mg)	56.7		
Vitamin A (retinol equivalent) (mg)	57.2		
Thiamin (mg)	0.03		
Riboflavin (mg)	0.17		
Niacin equivalent (mg)	0.83		
Vitamin B_{12} (μg)	0.41		
Vitamin C (mg)	1.06		
Vitamin D (μg)	0.03		
Energy (kJ)	283.6		
(kcal)	67.8		

Source: Adapted from Fox, P.F. and McSweeney, P.L.H. 1998. *Dairy chemistry and biochemistry.* Blackie Academic and Professional, London; and MAFF. 1995. *Manual of nutrition.* Stationery Office, London.

content of typically 87.5% cows' milk has a high water activity (a_w) of about a_w 0.993 (Fox and McSweeney 1998) and is prone to rapid microbial spoilage, unless adequately heat treated, packaged and stored. The manufacture of many milk products involves the removal of water, either partially or significantly, to help generate the characteristics of products and preserve the nutritional value of the milk solids that constitute them. The nutrients in whole milk are given in Table 2.1 along with the proportions of the major milk solids components: being milkfat, lactose (the milk sugar), the milk proteins (casein and the whey proteins), and the minerals or ash.

2.3 Functional approach

The different components of the milk solids exhibit what are termed 'functional properties', meaning that they fulfil specific roles within food systems, e.g. emulsification, gelation and water binding. Disagreement exists about the logic of the term 'functional properties', as all foods and food materials are functional (Anon. 1995a). With the development of so-called 'functional foods' or foods with health-giving/enhancing properties, the word functional when applied to food seems destined to create confusion. This said, the dairy industry and food

Table 2.2 Functional properties of the major milk components

Casein	Whey proteins	Milkfat	Lactose
Fat emulsification	Foaming	Air incorporation	Browning
Foaming	Gelation	Anti-staling	Free-flow agent
Precipitation by Ca^{2+}	Heat denaturation	Creaming	Humectant
Precipitation by chymosin	Solubility at any pH	Flavour carrier	Low sweetening power (27–39% of sucose)
Soluble at pH > 6		Gloss	Suppresses sucrose crystallisation
Water binding		Layering	
		Shortening	
		Unique flavour	

Source: Adapted from Early, R. 1998b. Milk concentrates and milk powders. In R. Early. (ed.). 1998b. Second edition. *The Technology of dairy products*. Blackie Academic and Professional, London.

manufacturers using milk-based ingredients recognise the functional properties of milk's components and dairy products and they are selected and modified accordingly for specific applications (Kirkpatrick and Fenwick 1987). The properties of dairy products which are foods in their own right, e.g. cream, butter, cheese, yogurt, are significantly a consequence of the functional properties of the milk solids of which they are comprised. The composition and proportion of milk solids varies according to the product concerned and gives rise to the characteristics that typify the product. On the other hand, the formulations of many milk-based food ingredients are regulated to maximise specific functional properties, or concentrate the functional value of certain milk components to benefit particular applications. The functional properties of the major components of milk are given in Table 2.2.

2.4 Sensory properties

The sensory properties of milk and milk products are a consequence of composition, which may be manifested in ways that relate to notions of quality. The components of milk products, a consequence of the chemistry of milk, give rise to the physical properties of products and both chemical and physical properties influence consumer sensory perceptions. The chemical and physical properties of milk products are influenced by raw milk quality, manufacturing processes, storage conditions and associated process controls. Manufacturers aim to assure the quality of products, and, hence, maximise consumer acceptability. However, the actions of microbes and chemical reactions such as oxidation may (and in time usually do) adversely affect the chemical and physical properties of products, leading to the loss of quality and a reduction in consumer acceptability. Consumers judge the sensory properties of milk products and products incorporating milk based ingredients by sight, smell, taste and feel (texture). Product attributes which stimulate a particular sense, or senses, are often regarded as the characterising attributes of a product. For example, blue Stilton cheese is judged by appearance, aroma, texture and flavour, whereas the flavour of butter is of critical importance to its acceptability and yogurt is judged principally by its clean, sharp acid flavour and smoothness on the palate.

The whiteness of liquid milk is caused by the light scattering of milkfat globules, colloidal calcium caseinate and colloidal calcium phosphate (Johnson 1974) though the presence of carotenes is important to the yellow colour of milkfat. The flavour of milk is a consequence of the major milk constituents as well as minor components. The milkfat globule, comprising lipids, phospholipids and caseins, is significant in creating the characteristic flavour of milk. The flavour of butter is a composite of the milkfat and serum (McDowall 1953), though its flavour is attributed to the relatively high proportions of short chain fatty acids that constitute butter triacylglycerols. Unfermented milk products are often described as having characteristic, clean, milky flavours, whereas the flavours of fermented products are mainly attributed to the conversion of lactose to lactic acid. The use of homofermentative bacteria gives rise to a clean, lactic taste, while heterofermentative bacteria produce aldehydes, ketones and alcohol in addition to lactic acid, causing a wide variety of flavour notes. The aromas of milk products are due mainly to short chain fatty acids with fewer than 12 carbon atoms, conventionally known as 'volatile fatty acids' (Berk 1986). Butyric acid, a C4 fatty acid with a melting point of −7.9°C, constitutes 5–6% of milkfat and is significant in creating the unique flavour and aroma of butter.

The flavours and aromas of milk products may be influenced intentionally or unintentionally (on the part of man) by microbial activity. The biochemical activity of bacteria and, in some instances, the action of moulds and yeasts gives rise to the wide variety of flavours and aromas of cheese. This is evidenced, for example, by ripe Camembert, the smell and taste of which arises partly from the hydrolysis of triacylglycerols and the liberation of short chain fatty acids, as well as the breakdown of proteins to ammonia and other products. The textures of milk products are influenced by moisture and fat contents, as well as factors such as pH where, as in yogurt, acidification to the isoelectric point of casein causes the formation of a gel. In the case of cheese the lower the moisture content the harder the product. Fat content and chemistry influence directly texture perceptions and 'mouth-feel', because the fatty acid profile of milkfat is subject to seasonal variation, with summer milkfat generally softer and yellower than winter milkfat. This is commonly experienced when butter is used as a spread, but the effect can also be important with other products though it may not be so obvious. There is not the space here to review fully the factors affecting the sensory perception of milk products, and reference to standard dairy chemistry texts is advised. A detailed consideration of the sensory judging of dairy products is made by Bodyfelt *et al.* (1988).

2.5 Microbiological criteria for milk products

Dairy products manufacturers provide microbiological criteria within their product specifications. Although manufacturers may have derived their own product standards, the microbiological criteria for some dairy products are generally accepted, as defined by IFST (1999). Table 2.3 lists the IFST recommendations for milk, cream and dairy products, while Table 2.4 addresses requirements for milk powders. The indicators and spoilage organisms for milk, cream, dairy products and milk powders are given in Table 2.5.

2.6 Chilled dairy products and milk-based ingredients used in chilled foods

The dairy industry makes many dairy products which exist as chilled foods in their own right and numerous milk-based ingredients which find application in chilled foods. It is not possible to consider all chilled dairy products and milk-based ingredients in detail here, though the principal products are briefly reviewed.

2.6.1 Pasteurised milk

Pasteurised milk is consumed widely as market milk. The fat contents of products are legally defined in the UK and descriptions are given in Table 2.6. It is also

Table 2.3 Microbiological criteria for milk, cream and dairy products

Organism	GMP	Maximum
Salmonella spp.	ND in 25 ml or g	ND in 25 ml or g
L. monocytogenes	ND in 25 ml or g	10^3 per g
S. aureus	< 20 per g	10^3 per g
E. coli O157*	ND in 25 ml or g	ND in 25 ml or g

Source: IFST. 1999. *Development and use of microbiological criteria for foods.* Institute of Food Science and Technology, London.

* Raw milk-based products.

Table 2.4 Microbiological criteria for powders

Organism	GMP	Maximum
Salmonella spp.	ND in 25 ml or g	ND in 25 ml or g
S. aureus	< 20 per g	10^3 per g
B. cereus	< 10^2 per g	10^4 per g
C. perfringens	< 10^2 per g	10^3 per g

Source: IFST. 1999. *Development and use of microbiological criteria for foods.* Institute of Food Science and Technology, London.

Table 2.5 Indicators and spoilage organisms for milk, cream, dairy products and milk powders

Product	Organism	GMP	Maximum
Soft cheese (raw milk)	*E. coli*	$< 10^2$	10^4
Processed cheese	Aerobic plate count	$< 10^2$	10^5
	Anaerobic plate count	< 10	10^5
Other cheeses	Coliforms	$< 10^2$	10^4
	Enterobacteriaceae	$< 10^2$	10^4
	E. coli	< 10	10^3
Pasteurised milk and cream	Coliforms	< 1	10^2
	Enterobacteriaceae	< 1	10^2
Other pasteurised milk products	Coliforms	< 10	10^4
	Enterobacteriaceae	< 10	10^4
	E. coli	< 10	10^3
	Yeasts (yogurt)	< 10	10^6
Milk powders	Aerobic plate count	$< 10^3$	Product dependent
	Enterobacteriaceae	$< 10^2$	10^4
	E. coli	< 10	10^3

Source: IFST. 1999. *Development and use of microbiological criteria for foods*. Institute of Food Science and Technology, London.

Table 2.6 Descriptions of pasteurised market milks in the UK

Milk type	**Description**
Natural whole milk	Milk with nothing added or removed
Homogenised whole milk	Homogenised milk with nothing added or removed
Standardised whole milk	Milk standardised to a minimum fat content of 3.5%
Standardised, homogenised whole milk	Milk standardised to a minimum fat content of 3.5% and homogenised
Semi-skimmed milk	Milk with a fat content of between 1.5 and 1.8%
Skimmed milk	Milk with a fat content of less than 0.1%

used in the manufacture of chilled products, particularly as a base for the production of sauces, such as bechamel, cheese and white sauces used in chilled ready-meals. In the production of pasteurised milk, raw milk is centrifugally clarified to remove insoluble particles and somatic cells. In accordance with UK dairy regulations (Anon. 1995b) it is then heat treated at not less than 71.1ºC for not less than 15 seconds. A negative phosphatase test confirms adequate heat treatment and a positive peroxidase test confirms the milk has not been overheated (taken above 80ºC). Semi-skimmed and skimmed milks are produced by centrifugally separating cream using a hermetic separator, as described by

Early (1998a) and Brennan *et al.* (1990). High-pressure homogenisation (Early 1998a, Brennan *et al.* 1990) is used to reduce the size of milkfat globules from as large as 20μm down to 1–2μm, thereby preventing the development of a cream layer, and the possible formation of a cream plug in glass bottles. Market milk is packaged in glass bottles, laminated paperboard cartons and plastic (high-density polyethylene) containers (Paine and Paine 1992).

For industrial use pasteurised milk may be delivered by stainless-steel road tanker or in 1-tonne palletised containers (pallecons). Pasteurisation does not destroy all the microbes present in raw milk and pasteurised milk must be stored at <8°C to retard microbial growth. The spoilage of short shelf-life dairy products is usually due to microbial activity and post-pasteurisation contamination with Gram negative psychrotrophic bacteria is often of significance (Muir 1996a). Frazier and Westhoff (1988) record the possible survival of heat-resistant lactic organisms (e.g., enterococci, *Streptococcus thermophilus* and lactobacilli) as well as spore-forming organisms of genuses *Bacillus* and *Clostridium*. Various quality defects are possible with pasteurised milk, including lactic souring, proteolysis (which is favoured by low-temperature storage) due, for example, to a protease produced by *Pseudomonas flourescens*, which survives pasteurisation even though the organism does not, and bitty cream caused by *Bacillus cereus*.

2.6.2 Cream

Market cream is produced for domestic use with a range of minimum fat contents, as given in Table 2.7. In the manufacture of chilled products, cream finds application in soups, sauces and toppings. The fat content of cream for manufacturing use will be determined by various factors, e.g., whippability, pumping, packaging/transport and storage limitations. Cream is an oil-in-water emulsion. The milkfat globules in unhomogenised cream range in diameter from 0.1μm to 20μm with an average of 3–4μm. They are stabilised by the milkfat globule membrane which is comprised of phospholipids, lipoproteins, cerebrocides, proteins and other minor materials. The membrane has surface active, or surfactant properties. Most of the lipid in milkfat is triacylglycerols though small amounts of diacylglycerols and monoacylglycerols may be present.

Table 2.7 Minimum fat contents of market creams in the UK

Half cream	12%
Cream or single cream	18%
Sterilised half cream	12%
Sterilised cream	23%
Whipped cream	35%
Whipping cream	35%
Double cream	48%
Clotted cream	55%

The fat soluble vitamins A, D, E and K are also present. Cream is separated from milk by centrifugal separation, nowadays using hermetic separators which are capable of producing product in excess of 70% fat. Cream is pasteurised at not less than 72ºC for not less than 15 seconds (or equivalent) and must be phosphatase negative and peroxidase positive. Half and single cream require high-pressure homogenisation to prevent phase separation and double cream may be homogenised at low pressure to increase viscosity. Whipping cream remains un-homogenised in order to assure its functionality. Clotted cream is a traditional product of the south-western counties of England. A number of methods of production exist, as described by Wilbey and Young (1989). In general they involve the heating of milk (from which clotted cream is skimmed) or 55% fat cream at moderately high temperatures (usually 75–95ºC) to cause the cream to form a solid material, or 'clot'.

Market cream is commonly packaged in injection moulded polystyrene flat-topped round containers. Cloake and Ashton (1982) note that in the packaging of cream it is important to exclude light, which may promote auto-oxidation of the milkfat, and prevent tainting and the absorption of water. For manufacturing use, pasteurized cream is delivered in bulk stainless-steel road tankers or one-tonne pallecons. Both market and industrial pasteurised creams require chilled storage at −8ºC. Rothwell *et al.* (1989) review a number of quality defects possible with cream. Poor microbiological quality can reduce shelf-life below 10–14 days and lipolysis, due to indigenous or microbial lipases, can result in rancidity. Physical defects concern poor viscosity, serum separation and poor whipping characteristics.

2.6.3 Sour cream

Sour cream is used both domestically and industrially, mainly in the preparation of sauces. It is produced by the lactic fermentation of single cream (not less than 18% fat) with organisms such as *Lactococcus lactis* subsp. *lactis*, *Lactococcus lactis* subsp. *cremoris* and *Leuconostoc mesenteroides* subsp. *cremoris*. Fermentation causes the precipitation of casein at its isoelectric point (pH 4.6) and the formation of a set product. Market sour cream can be fermented in the pot but for industrial use agitation is necessary to produce a pumpable product which can be transported in 20kg lined buckets or one-tonne pallecons. Pasteurisation of the cream prior to fermentation and the presence of lactic acid serve as preservation factors for the product, but chilled storage is also necessary and storage at −5ºC will enable a 20-day shelf-life. Crème fraîche is a variant of sour cream made by culturing homogenised cream with a fat content of 18–35% with LAB such as *Lactococcus lactis* subsp. *lactis*, *Lactococcus lactis* subsp. *cremoris* and *Lactococcus lactis* subsp. *lactis* var *diacetylactis*. Incubation at 30–32ºC for 5–6 hours gives a set product with a pH in the range 4.3–4.7. Stirred or set crème fraîche is supplied to the retail market to be eaten as a chilled dessert, though for industrial use in dips, sauces, desserts and ready meals, it is suppled in various forms including 20 kg pergals and one-tonne pallecons.

2.6.4 Butter

The domestic use of butter is well known, but as an industrial ingredient used in chilled foods manufacture it finds application in soups and sauces. It is a constituent of roux, along with flour, used in the preparation of sauces. Garlic butters and herb butters are used as garnishes in chilled ready meals and savoury dishes, fillings for garlic bread and as toppings for cooked meats, e.g., steak, chicken and fish. The production of sweet-cream butter involves inducing phase inversion of the oil-in-water emulsion of cream to create a water-in-oil emulsion, or butter. A number of methods exist, as reviewed by Lane (1998). A commonly used method is the Fritz and Senn method, which involves rapidly cooling 42% fat cream to 8ºC and holding for 2 hours, then raising the temperature to 20–21ºC for 2 hours and cooling to 16ºC, or the churning temperature. The tempering process reduces the level of mixed fat crystals in the milkfat globules, ensuring that the high melting point triacylglycerols crystallise as pure fat crystals. This improves the spreadability of the butter, particularly when the milkfat has a low iodine value and it is hard.

Tempered cream is processed in a continuous buttermaker with four sections: the churning cylinder beats the cream and causes the milkfat globule membrane to rupture, whereupon the fat crystals coalesce; the separation section drains buttermilk from the butter; the squeeze-drying section expels remaining buttermilk; the working section smooths the product. In the production of salted butter, salt is added in the working section. It dissolves in the aqueous, or discontinuous phase of butter, and at a rate of 1.6–2.0% results in a salt-in-water content of around 11% – sufficient to inhibit microbial activity. Microbial activity is not the sole cause of quality deterioration in butter. Evaporation causing surface colour faults and the development of oxidative rancidity caused by exposure to light are cited as possible problems (Richards 1982). Market butter is packed in tubs or foil which exclude light and possess high moisture barrier properties. Butter as an industrial ingredient is supplied in 25kg units, packed in corrugated fibreboard cases lined with polyethylene film. Lactic butter may be made by the fermentation of cream with lactic acid bacteria, though the flavour of cultured butter may be replicated by the addition of certain compounds to sweet cream used for butter making (Nursten 1997). Garlic and herb butters are made by blending butter with the relevant ingredient and extruding to produce the required portion shape and size.

2.6.5 Skimmed milk concentrate and skimmed milk powder

Skimmed milk concentrate and skimmed milk powder find application in custards, toppings, soups, sauces, dips and desserts. Skimmed milk is the by-product of cream separation and contains around 91% water. The skimmed milk solids are the milk proteins, lactose and minerals, with a trace of fat. Skimmed milk concentrates are made by vacuum evaporation and products of 35–40% total solids are common for bulk supply to food manufacturers. Higher solids levels give rise to problems of viscosity, age gelation and lactose crystallisation.

Pasteurised skimmed milk concentrate requires chilled storage. Skimmed milk powder is made by spray drying skimmed milk concentrate from around 60% total solids. Milk powder quality is influenced by the solids content of the dryer feed. A high solids level gives a high bulk density, and densities of >0.65 kg l^{-1} give the best handling properties with least tendency to create dust. Skimmed milk powder has a moisture content of less than 3.5% and a water activity of around 0.2. It can be stored for many months at ambient temperature without experiencing a deterioration in quality. Agglomerated skimmed milk powder with good water dispersion characteristics can be made using two-stage drying processes, often in spray dryers with integrated fluid beds, and the pre-heat treatment of the milk before evaporation and drying can be important to the heat stability of the milk proteins (Early 1998b).

2.6.6 Whey concentrate and whey powder

Whey concentrate is used in the production of margarine and non-dairy spreads, while whey powder may be used in the formulation of soups, sauces and desserts. Sweet whey is the by-product of enzyme-coagulated cheese production. The material has a pH of 5.8–6.6 and a titrateable acidity (TA) of 0.1–0.2%. Medium-acid whey and acid whey are respectively, the products of fresh acid cheese and acid casein manufacture. Sweet whey is most commonly used for the production of whey concentrate and whey powder. With a water content of over 94% and the presence of lactic acid bacteria whey is very perishable. It is pasteurised immediately after production to allow storage without deterioration. The solids content of sweet whey is 5.75% of which 75% is lactose. Whey concentrate is produced by vacuum evaporation and the low solubility of lactose sets the practical total solids limit to around 30%. The bulk product, delivered by road tanker, has a short shelf-life and is stable for 2–3 days at <8ºC. Non-hygroscopic whey powder is made by crystallising the lactose in whey at temperatures below 93.5ºC to ensure α-lactose monohydrate predominates. Lactose crystallisation allows the dryer feed to be concentrated to 58–62% total solids to achieve a dense powder. Demineralised whey powders can be made by ion-exchange and electrodyalysis (Houldsworth 1980) and nanofiltration techniques. Like skimmed milk powder, whey powder can be stored for many months at ambient temperature.

2.6.7 Lactose

Lactose, the milk sugar, can be used in the formulation of soups and sauces. Lactose yields the monosaccharides, D-glucose and D-galactose on hydrolysis with the enzyme β-galactosidase, and is designated as 4-0-β-D-galactopyranosyl-D-glucopyranose. It occurs in alpha and beta crystalline forms, though an amorphous form is also possible. The carbohydrate is less sweet than sucrose, and as a reducing sugar it is used in some foods to provide colour in the form of Maillard browning. For industrial food uses, anhydrous α-lactose monohydrate is preferred as it is a free-flowing, non-hygroscopic material and is easy to store without a deterioration in quality. To ensure the formation of α-lactose monohydrate crystals, whey is concentrated to about 65% total solids to form a supersaturated lactose solution at a temperature not exceeding 93.5°C. During step-wise cooling α-lactose monohydrate crystals are formed. Seed lactose is used to control crystal size below 25μm, or the threshold of detection by the palate. Lactose crystals are commonly separated from whey concentrate by decanter centrifuge, washed, dried in a fluid-bed dryer and packaged in multi-ply paper sacks lined with a heat sealed polyethylene bag.

2.6.8 Yogurt and Greek-style yogurt

Yogurt is a very popular chilled dairy product, and numerous brands and flavours can be found in supermarket chiller cabinets. A number of variants of yogurt also exist, including Greek-style yogurt. Industrial uses of yogurt include chilled dips, sauces, soups, desserts, toppings and ready meals such as curries and other ethnic dishes. Yogurt is made by the lactic fermentation of whole, standardised and skimmed milks. The pH of normal milk is 6.5–6.7, however the fermentation of lactose to lactic acid by lactic acid bacteria (LAB) reduces the pH to 2.6 causing the formation of an acid set gel. By the time fermentation is arrested the pH is often in the range 3.8–4.2. The organisms *Lactobacillus delbrueckii* subsp. *bulgaricus* and *Streptococcus salivarius* subsp. *thermophilus* have traditionally been used in yogurt production, though retail yogurts also incorporate other organisms such as *Lactobacillus acidophilus*, *Lactobacillus casei* subsp. *casei* and *Bifidobacterium* species to give mellow, fruity and less acid flavours. For industrial use yogurt may be based on milks of varying fat content, though retail products tend to be either skimmed milk or full cream milk varieties. In the production of yogurt skimmed milk powder is added to increase product viscosity and the non-fat milk solids level is raised from around 8.5% to 12–14%. However, gelatin and hydrocolloids such as modified starch, guar and pectin may also be added to influence viscosity and texture. Yogurt is made by heating milk which has been standardised for milk solids non-fat (MSNF)and fat levels, typically to 90–95°C for five minutes. The heat treatment kills vegetative contaminant bacteria, though higher heat treatments are required to kill spore-forming organisms. Bacteriophage which can retard or prevent LAB activity are destroyed and the whey proteins are denatured to the benefit of viscosity. The milk is cooled to around 42°C, starter is added at a rate of around 2% and fermentation occurs in a closed vessel. The starter may be a culture of LAB or a direct vat inoculation (DVI) freeze-dried or frozen concentrate (Stanley 1998). Following incubation the pH drop is arrested by striking (stirring) and cooling the coagulum, which may then be mechanically smoothed to form a base material for the production of plain yogurt, fruited yogurt or for use as an industrial ingredient. When destined for industrial use, yogurt is usually pasteurised to kill LAB and prevent a further, gradual change in acidity. Often supplied in 20 kg pergal containers, pasteurised yogurt has a shelf-life of some ten weeks stored at <16°C.

Market yogurts vary widely in style and include solid set yogurts fermented in the pot, to stirred and fruited products, and products which contain separate portions of yogurt and fruit or cereal in the now ubiquitous twin-pot format. The fat contents of the yogurt base of retail products also varies, from zero fat to high fat products. Greek-style yogurt is, conventionally, a strained yogurt. Whey is removed to increase the solids content to 22–26% to give a thick product, resembling cream cheese in texture. In recent years the market for biofermented yogurts has increased with increasing interest in health foods. Products fermented with organisms such as *Lactobacillus acidophilus* and *Bifidobacterium* spp. possess apparent probiotic properties and various starter cultures are used to effect specific textures and flavours, as considered by Marshall and Tamime (1997).

2.6.9 Cheese

As a product group, cheese presents many widely different types within the retail market-place. A number of products used domestically also find application in the manufacture of chilled foods, and particularly in the production of sauces and savoury ready meals. For example, Cheddar, Gruyere, Parmesan, Pecorino and Monterey Jack are used to garnish products such as lasagne dishes, baked potatoes, and grilled products requiring a cheese topping. Mascarpone can be used to thicken sauces, while Gorgonzola is used to flavour sauces.

Generally considered to be chilled foods, cheese falls into two categories (IFST 1990). Mould ripened soft cheese and cream cheese are classed as category 2 products, requiring storage at 0°C to +5°C, while hard cheese and processed cheese are category 3 products and must be stored at temperatures not exceeding 8°C. The UK Food Safety (Temperature Control Regulations) 1995 state the need for chilled foods to be kept at, or below 8°C. It should be recognised that many hard cheeses will not mature properly at chill temperatures and storage above 8°C may be needed. In such cases the scientific assessment of food safety is recommended (SCA 1997) based on an intelligent interpretation of appropriate microbiological safety factors, such as those suggested by CCFRA (1996). Cheese is made from milk or milk derived materials, by fermentation with LAB and, importantly, the use of a proteolytic enzyme, usually chymosin, to form a curd from which whey is drained to yield a solid product which may or may not be ripened. In some instances a secondary microbial flora, such as moulds, may be used to promote the ripening process. A large number of cheese varieties exist world-wide and Fox (1993) suggests the number exceeds 1000. Difficulties are encountered in classifying cheese varieties and Scott (1981) proposes systems of classification based on composition, ripening characteristics and moisture contents. Broadly, cheese can be defined as 'ripened cheese', 'mould ripened cheese' and 'unripened or fresh cheese' (Bylund 1995).

Ripened cheese

Cheese of this type sold as retail chilled foods includes varieties such as Parmesan, Cheddar, Cheshire, Lancashire, Double Gloucester, Red Leicester, Dunlop, Emmental, Gruyere, Edam, and Queso Manchego. Ripened cheese may be sold for domestic use as cut cheese or pre-packed in vacuum or modified atmosphere packaging. For industrial use ripened cheese is usually provided as block cheese, which is then cut, bowl chopped or milled according to requirements. However, cheese grated and packed in pouches is also supplied to the industrial and catering markets. Though ripened cheeses vary considerably in appearance, flavour, texture, etc., each type shares common basic steps in production, as follows:

1. Raw milk may or may not be pasteurised (Emmental is an unpasteurised ripened cheese).
2. Starter (LAB) is added to the milk which is then ripened at the temperature required to achieve the appropriate rate of starter activity.
3. Rennet (containing the proteolytic enzyme, chymosin) is added to form a coagulum.
4. The curd is cut to release the whey.
5. The whey is removed.
6. The curd is textured.
7. The curd is salted (this may be either by dry salting prior to, or after hooping, or by brining after removal from hoops).
8. The curd is contained in a hoop or mould to shape the cheese.
9. The cheese is stored and ripened.

An outline of Cheddar cheese production is given in Table 2.8. The quality of ripened cheese is dependent on composition and in the case of long-hold Cheddar cheese the factors of moisture in non-fat solids, fat in dry matter, pH and salt in moisture are all critical to achieving the best quality (Muir 1996b). During manufacture, cheese-makers will effect necessary process adjustments to control these factors and maximise quality. Frazier and Westhoff (1988) state that cheese may present mechanical or microbial defects. Microbial defects of ripened cheese may occur during manufacture, e.g., spore-forming species of *Clostridium* and *Bacillus* may produce gas holes and acid-proteolytic bacteria may cause off flavours. During ripening lactate-fermenting *Clostridium* spp. and heterofermentative lactics may give rise to 'late-gas' defects, and putrefaction may be caused by putrefactive anaerobes such as *Clostridium tyrobutyricum* and *Clostridium sporogenes*. Finished cheese can be spoiled by a variety of organisms, including, *Oospora caseovorans* causing white mould growth in the eyes of Swiss cheese, *Penicillium* spp. giving rise to a variety of mould discolorations and *Brevibacterium linens* causing red/orange spots. During the manufacture of ripened cheese care is taken to monitor the rate of acid development. In cases where acid production is slow, the risk exists that *Staphylococcus aureus* may develop to numbers sufficient to cause a toxin

Table 2.8 Outline of Cheddar cheese manufacture

Day 1	Raw milk	(a)	Standardise milk to give a fat to casein ratio of 1:0.7
		(b)	Pasteurise at 71.9°C for 15 seconds
		(c)	Cool to 29.5°C and fill cheese vats
		(d)	Add cheese starter at 1.5–3% of the milk volume – usually a mixed culture of LAB based on *Lactococcus lactis* subsp. *lactis* and *Lactococcus lactis* subsp. *cremoris*
		(e)	Ripen for 45 to 60 minutes – sufficient time for the titrateable acidity (TA) of the milk to rise from 0.15–0.17% to 0.20–0.22%
		(f)	Add rennet and form coagulum over 45 to 60 minutes
		(g)	Cut curd
		(h)	Scald by raising temperature to 39°C over 45 minutes
		(i)	Hold curd at 39°C for 45 to 60 minutes, agitate to encourage syneresis
		(j)	Drain whey from the curd when TA reaches 0.20–0.24%
		(k)	Cheddar curd for 90 minutes, until TA reaches 0.65–0.85%
		(l)	Mill curd
		(m)	Salt at 2.0–3.5% of curd weight
		(n)	Hoop
		(o)	Press for 16 hours (overnight)
Day 2		(a)	Remove cheese from hoop – pH between 4.95–5.15
		(b)	Bind/pack cheese to exclude air – vacuum packing in heat-sealed polyethylene bags is common
		(c)	Box vacuum packed cheese to retain shape
		(d)	Ripen at 10–12°C for 3 to 18 months
		(e)	Pre-pack or sell as block cheddar

Adapted from: Banks, J.M. 1998. 2nd edition. Cheese. In, R. Early (ed). *The technology of dairy products*. Blackie Academic and Professional, London.

hazard. Consequently, cheese-makers often operate a 'slow cheese' procedure, which ensures such cheese is identified and tested for toxin prior to release to the market.

Mozzarella is an important cheese in the industrial preparation of chilled foods such as pizza and pizza ingredients supplied to the retail market and food service outlets. Mozzarella is a 'pasta filata' cheese, which means that it is characterised by an elastic, stringy curd. Traditionally made from the milk of water buffalo, mozzarella is commonly made from cows' milk and both kinds of product can be found in supermarket chill cabinets. The initial stages of mozzarella production are similar to those of Cheddar manufacture, however at the point of milling a divergence occurs. Milled mozzarella curd is not salted, but passes to a cooker-stretcher where it is worked mechanically under water at a temperature of 65–80°C to obtain its laminar, 'chicken breast' qualities. It is extruded into moulds and cooled to around 40°C to retain its rectangular block shape, then placed in brine at 8–10°C and 15–20% salt concentration for sufficient time to achieve a level of 1.6% salt.

Mould ripened cheese
These cheeses utilise both bacterial cheese starters and moulds (fungi) to achieve the required product characteristics. Because of the use of moulds, which are biochemically very active compared with cheese starter bacteria, mould ripened cheeses mature more quickly than ripened cheeses and, generally, have a shorter shelf-life. Mould ripened cheeses are typified by Stilton and Roquefort which are 'blue cheeses' by virtue of the activity of *Penicillium roqueforti*. In contrast, Camembert is a white mould cheese, ripened by *Penicillium cambemberti* and *Geotrichum candidum*. The manufacture of mould ripened cheese is similar to the production of ripened cheese though, often, much less starter is used and the curd is either only lightly scalded or not at all. A key difference is, of course, the addition of moulds which may be added to the milk or mixed into the drained curd, the latter being the case in Stilton manufacture. The presence of the blue mould within the curd structure of maturing Stilton means that at the appropriate time, usually 8 to 12 weeks after the day of make, the cheese can be penetrated with cheese wires to allow air into the structure to stimulate mould growth and blueing. In the case of Roquefort, the blue mould is carried on bread crumbs which are mixed into the curd (Simon 1956). Alternatively, mould spores may be dispersed in water and sprayed onto the surface of cheese to produce a product such as Brie. Though not correctly mould ripened cheese, some products utilise bacteria and yeasts in surface ripening. Smear ripened and washed rind cheeses such as Munster obtain their orange/red coat from *Brevibacterium linens*, and St. Nectaire is surface ripened by both bacteria and yeasts.

Unripened cheese
This category includes cottage cheese, quarg (fromage frais), cream cheese and full fat soft cheese. All are found in the retail chilled foods market and are used in the industrial production of chilled foods. Traditionally, cottage cheese is made without the use of rennet. An acid coagulum is made from whole or skimmed milk by fermentation with LAB, e.g., *Streptococcus lactis*, *Streptococcus cremoris* and *Leuconostoc citrovorum* at a temperature of 20–22°C for up to 16 hours. A short incubation time of 6 hours is possible when rennet is used. The coagulum is cut and scalded at 49–55°C, the curd is washed with water at 49°C and then drained of diluted whey to yield cottage cheese with a granular, shotty texture. The production of quarg is similar to the quick set method of cottage cheese manufacture. Skimmed milk is used to produce a coagulum which is not scalded, as whey is removed by centrifugal separation in a quarg separator to yield a product of typically 17.5% total solids. The use of cheese starters gives a characteristic flavour which contrasts with that of yogurt. Cream may be blended with quarg to produce high fat products, though traditionally quarg is a low fat cheese. Fromage frais is a marketing appellation intended to make quarg more attractive to consumers. Cream cheese and full fat soft cheese are made by fermenting standardised milk with cheese starters and adding rennet to form a weak coagulum. Centrifugal separation is used to reduce the moisture content of the curd in each case, yielding a product of typically

51% total solids and 46% fat in the case of cream cheese, and 39% total solids and 30% fat in the case of full fat soft cheese.

Processed cheese
Often made from ripened cheese which has not achieved the required standard at grading, processed cheese is produced by cooking cheese in steam-jacketed vessels at 80–110ºC. Different cheeses are combined according to flavour requirements, though enzyme modified cheese is also used. Phosphates and citrates function as emulsifier and stabiliser systems and regulate the pH of products. Processed cheese with a pH of 5.2–5.6 is hard to firm and forms block cheese which may be sliced or grated for use as a topping or garnish in savoury meals and pizzas. A pH of 5.6–5.9 gives a spreading cheese of the type used in sandwiches.

2.7 Chilled desserts

Many chilled desserts utilise milk-based ingredients to provide flavour, texture and colour. Gelled desserts and mousses often use pasteurised milk, though milk concentrate is an option. Milk proteins contribute to viscosity, help to emulsify fat and contribute colour. Cream is a common ingredient and whipping cream assists in the development of an aerated structure as well as providing a rich flavour. In products containing chocolate, the addition of milk solids is important to the development of a milk chocolate colour. Crème brulée (originally Cambridge burnt cream) can be based on crème fraîche and skimmed milk, which provide a golden yellow colour in the form of Maillard browning, as well as texture, viscosity and a richness of flavour. Milk is a base ingredient of custard desserts. It provides the water to hydrate the modified maize starch used to thicken the product and provide viscosity, and contributes to flavour and mouthfeel, especially when full cream milk is used.

As with most products in the chilled desserts sector, variations on product types are possible, for instance milk may be substituted with low fat yogurt to make a 'reduced calorie' custard style product. Milk is also one of the two main ingredients in chilled rice puddings. It provides the moisture needed to swell rice grains, acts as a heat transfer medium during cooking and imparts flavour and colour to the product. Cheesecakes, as their name implies, incorporate cheese, though the type of cheese used varies according to the type of cheesecake made. Baked cheesecake can be based on cottage cheese or cream cheese, or a blend of the two. The cheese is used along with other ingredients such as sugar, starches and gums to thicken and stabilise (in traditional products egg yolk, cornflour and milk would be used) to produce a filling for a pastry case, which, when cooked, gels to form the smooth textured, cheesy-rich tasting centre of the product. Fresh cheesecakes are based on quarg, fromage frais and cream cheese, and blends of these cheeses, often filled into or onto a biscuit base. A thickening and stabilising system is required to control the viscosity of the cheese mix and various hydrocolloids such as starches and alginates are used. Many chilled desserts are topped with whipped cream and an obvious example is trifle, which is usually based on layered jelly and sponge cake beneath a whipped cream topping.

2.8 Ready meals

Many ready meals contain milk based ingredients. In pasta-based meals such as lasagne dishes milk and cream are used in the production of sauces where the whitening properties of milkfat globules and milk proteins can be of benefit. The casein also functions as an emulsifier, helping to incorporate added fat or butter. Various cheeses such as Parmesan, Cheddar, and Gruyere may be used as lasagne toppings, to give a golden brown finish. Dough-based products such as pizzas and tortilla-style products use cheese to provide an attractive finish and to enhance aroma. In such instances it is important that the cheese provides the correct melt characteristics. Mozzarella is the genuine pizza cheese, but blends of Mozzarella and Cheddar are sometimes used, and processed cheese may also find its way into some products.

Cheese is an important ingredient in the manufacture of some quiches, where it provides colour, texture and flavour. It is also used to enhance the consumer appeal of products such as potato gratin, by providing surface finish and, in this instance, Gruyere can often be found listed in the ingredients declaration alongside cream and butter. Many meat dishes contain dairy products, e.g., Lamb Provençal may contain Cheddar cheese as a garnish and cream and creme fraîche to thicken the sauce and provide flavour. Yogurt and butteroil are used in the preparation of sauces for ethnic dishes and pseudo-ethnic foods such as Chicken Tikka Masalla, and, for instance, the sauce of Chicken Pasanda may contain cream to provide richness, whitening and gloss. Savoury pouring sauces such as pasta sauces utilise cream and butter for flavour and gloss, as well as skimmed milk solids for whitening and emulsification purposes. The same is so of soups. Fresh soups such as cream of tomato contain cream and butter for flavour and gloss and sodium caseinate to increase viscosity, while asparagus soup may contain reconstituted skimmed milk powder and double cream for similar reasons.

2.9 Maximising quality in processing

2.9.1 Effect of heat on milk proteins

The heat stability of milk is influenced mainly by the heat stability of whey proteins. At temperatures above 65°C for more than a few seconds the denaturation of whey proteins proceeds quickly and at 90°C for five minutes it is complete. In contrast, casein is not, strictly, heat denaturable. The order of heat stability of whey proteins seems to depend on the method of assessment (Fox and McSweeney 1998), but generally, α-lactalbumin is considered to be more heat stable than β-lactoglobulin, followed by blood serum albumin and then the immunoglobulins. With the heat denaturation of whey proteins, β-lactoglobulin complexes irreversibly with κ-casein via disulphide bridges. This phenomenon interferes with the action of chymosin in cheese making, but it can increase the apparent heat stability of products such a skimmed milk powder. The effect of heat on whey proteins, causing them to gel, can be of benefit to food manufacturers as gelled whey proteins can be used to modify the textural properties of foods and bind water. Though various factors affect the gelling properties of whey proteins and particularly whey protein concentration, pH, salt concentration and the presence of fat (Mulvihill and Kinsella 1987). Concentrated whey protein products with protein levels exceeding 90% are made for use as food ingredients and can find application in chilled products such as soups and sauces.

2.9.2 Mechanical damage to milk and cream

The quality and function of liquid milk and cream may be impaired by mechanical action causing the destabilisation of the emulsion. The degree of destabilisation is dependent on a number of factors, including the shear rate, the fat content, the milkfat globule size and the ratio of solid to liquid fat. Mechanical destabilisation, e.g., as a consequence of over pumping or poor transport line design, can lead to an increase in free fat in raw milk, which is then susceptible to hydrolysis by lipoprotein lipases (Harding 1995).

2.9.3 Microbially induced proteolysis and lipolysis

Raw milk stored under refrigerated conditions is subject to microbial proteolysis and lipolysis due to *Pseudomonas* spp. and proteolysis due to psychrotrophic sporeformers of the genus Bacillus: mainly *B. cereus*, *B. circulans* and *B. mycoides*. While pasteurisation will destroy vegetative organisms, *Bacillus* spp. can survive and give rise to proteolysis in chilled liquid products. Additionally, though pasteurisation may denature indigenous lipases, bacterial lipases are considerably more heat resistant and can survive, giving rise to hydrolytic rancidity and the development of rancid flavours caused mainly by free butyric and caproic acids. The quality of dairy products is dependent on the quality of raw milk and processors must safeguard against the growth of bacteria that promote proteolysis and lipolysis. Post-pasteurisation contamination must also be prevented, as the reintroduction of these contaminant organisms (amongst others) can lead to the development of off-flavours and the loss of functionality in products.

2.9.4 Oxidative rancidity of milkfat

The development of oxidative rancidity in liquid milk and milk products containing milkfat is dependent on the presence of oxygen. The primary substrate for oxidation is the polyunsaturated fatty acids, linoleic and arachidonic, and those contained in the phospholipides and glycerides (Jenness and Patton 1959). Oxygen attacks the methylene groups adjacent to the carbon chain double bonds, giving rise to the formation of hydroperoxides. These compounds are unstable and oxidation proceeds by a free-radical mechanism. Sunlight and particularly light from fluorescent tubes can cause the auto-oxidation of milkfat, as can the presence of iron and copper salts. In milk products with low water activities the reaction rate of oxidation is highest at a_w 0.6, falling to a_w 0.4 and then rising again as a_w reduces (Fellows 1997), which is of significance to low-moisture, low-water-activity products.

2.9.5 Maillard reaction

The Maillard reaction is a non-enzymic browning reaction. It is the result of the interaction between a carbonyl group and an amino group to form a glycosamine and ultimately, melanoidins. The Maillard reaction occurs when milk is heated sufficiently to induce a reaction between lactose and the amino acid, lysine. It results in the development of a brown discoloration, often described as a cooked colour, and associated strong flavours.

2.10 Food safety issues

Though dairy products have a very good food safety record, the main public health concerns associated with milk and milk products relate to food-borne disease organisms and food-poisoning organisms. *Mycobacterium tuberculosis* and *Coxiella burnetti* are the most heat-resistant vegetative microbes found in milk and both are food-borne disease organisms. Jay (1996) gives the *D* 65.6°C as 0.20–0.30 minutes for the former and 0.50–0.60 minutes for the latter. This compares with a *D* value at the same temperature of 1.6–2.0 seconds for *Listeria monocytogenes*, a food-poisoning organism associated with soft cheese. The UK Dairy Products (Hygiene) Regulations 1995 state microbiological standards for raw cows' milk for processing, requiring a plate count at 30°C of $\leq$100,000 colony-forming units (cfu) per ml, and a somatic cell count of $\leq$400,000. The standards for some dairy products are also given and Jay (1996) reports milk product standards in the USA. The vegetative food-poisoning organisms of main concern in raw cows' milk are *Salmonella* spp., *Listeria monocytogenes*, *Staphylococcus aureus*, *Campylobacter* spp. and pathogenic *E. coli*. The emergent pathogen *E. coli* O157 is of growing concern due to its association with cattle (Buchanan and Doyle 1997). This is not, however, to say that raw milk is automatically contaminated with these organisms. A survey of raw milk carried out in England and Wales in 1995/96 found that of 1674 samples, 2% were positive for *Listeria monocytogenes*, 6.7% were positive for *Staphylococcus aureus* (one of the main causative organisms of mastitis) and 62% were positive for *Escherichia coli*, an indicator of faecal contamination (Anon. 1999).

Other possible hazards associated with raw milk include chemical contamination, such as antibiotics and other veterinary drugs residues, cleaning chemicals, environmental contaminants and mycotoxins arising from animal feeds, and physical contaminants such as wood, glass and metal.

The Hazard Analysis Critical Control Point (HACCP) system is now recognised as the best method of food safety control. Its use is encouraged by the World Health Organization and the International Commission on Microbiological Specifications for Foods. Within the EU all food businesses are required to implement the first five of the seven principles of HACCP (EEC 1993). In the UK the European directive is manifested in food safety regulations (HMSO 1995). It is not intended to describe the use of HACCP here. Early (1997) addresses its use in the dairy industry, while Leaper (1997) and Mortimore and Wallace (1998) give a detailed description of its general application. Miller *et al* (1997) consider HACCP in relation to quantitative risk assessment in the control of *Listeria monocytogenes*. In milk processing the control of hazards concerns (a) the prevention of contamination by hazards and (b) either the elimination of hazards in milk products or their reduction to acceptable levels. Effective herd management and good milking practice serve to reduce the contamination of raw milk by potential hazards to human health arising on the farm. Standards for the production, storage and transport of raw milk, such as those laid down in the UK's National Dairy Farm Assurance Scheme should be observed. Organisations charged with the responsibility of milk collection must guard against the contamination of raw milk by transit vehicles and good hygienic practice must be observed in the management and maintenance of milk tankers. Ideally the temperature of raw milk should not exceed 4ºC at the point of collection, though milk above this limit may be processed in the UK according to requirements stated in the regulations (Anon. 1995b). The chilling of milk immediately after milking prevents the growth of mesophilic pathogens and spoilage organisms. Psychrotrophic organisms will still grow at low temperature, however, and the storage life of raw milk must be limited prior to processing in order to avoid quality defects arising from proteolysis and lipolysis caused by biochemically active species such as *Pseudomonas* spp. At the factory raw milk is stored in thermally insulated silos to await processing. Commonly the milk passes through coarse filters on entry to silos, which serve to remove physical contaminants which may be potentially hazardous in nature.

In the manufacture of most dairy products pasteurisation is the key critical control point for the elimination or reduction of microbial pathogens. The foodborne disease organisms, *Mycobacterium tuberculosis* and *Coxiella burnetti*, and the vegetative food poisoning organisms, *Salmonella* spp., *Listeria monocytogenes*, *Staphylococcus aureus*, *Campylobacter* spp. and pathogenic *E. coli* spp. are all destroyed by pasteurisation at 71.7ºC for 15 seconds. Spore-forming bacteria such as *Clostridium* and *Bacillus* spp. will, of course, survive. Because no further heat treatment occurs after milk pasteurisation in the production of many chilled dairy products extreme care must be taken to prevent post-pasteurisation contamination by pathogenic microbes and spoilage organisms.

Within dairy factories high standards of hygiene and housekeeping are maintained to prevent the possibility of contamination from plant, equipment and the manufacturing environment. Barrier hygiene is commonly practised to preserve the integrity of hygiene standards and especially to prevent the transmission of microbial contaminants from raw milk to pasteurised milk areas. While some dairy products used as ingredients in chilled foods manufacture will receive heat treatments in chilled foods processing, the standards and systems used to make the ingredients are no less stringent than those used in the manufacture of chilled dairy products. Indeed, in the manufacture of products such as milk powders extreme care is taken to prevent the post-pasteurisation contamination of product, as considered by Mettler (1989; 1992; 1994).

2.11 Future trends

The traditional milk-based ingredients used in chilled products manufacture and traditional chilled dairy products will not change in the future. Developments are most likely to involve the innovation and diversification of non-traditional consumer products such as milk-based desserts, and formulated chilled foods (desserts, soups, sauces and ready meals) which utilise the functionality of milk based ingredients. Advances in processing technologies will lead to new kinds of milk based ingredients tailored for use in specific applications, as will improvements in the empirical understanding of the behaviour of milk components in food systems. Notions of competitive advantage and market share drive companies operating in the chilled dairy products and dairy ingredient sectors. The quest for sustainable economic growth, the 'Holy Grail' of all companies operating in Western economies, will stimulate the creation of 'technological partnerships' in food product innovation. By fusing the intellectual processes of customers and suppliers new possibilities and, hence, new opportunities will be created. By precisely matching customers' requirements and technical capabilities with the abilities of suppliers, milk-based ingredients will be finely tuned to function in specific chilled foods systems and to create attributes which set products apart from those of competitors. To a certain extent this is already happening, as ingredients suppliers already select raw materials from specific sources for processing in specific ways, for use in specific products. Whey powder derived from different sources, e.g., Emmental, Cheddar or acid casein whey, can offer different functional behaviour when blended with stabilisers and thickeners, for use in chilled dessert applications. The demineralisation of whey by different techniques, or varying mineral balances resulting from finely controlled demineralisation, can yield 'selected functional performance'. Skimmed milk from different countries offers different performance standards, as does butter which varies widely in flavour and melting characteristics according to breed of cow and country of origin.

To a certain extent innovation in chilled dairy products and milk-based ingredients for use in chilled foods will be restricted by the limits of raw material sources and processing technologies. Genetic engineering may, however, create new and so far unimagined possibilities. The concept of value-added has become a corner-stone of the consumer society, but free-market competition acts constantly to erode the value which food manufacturers and retailers are able to attach to their products. Consequently, new opportunities to add value to food products are continually sought and genetic engineering may hold the key to a new treasure chest in food product innovation. Kuzminski (1999) states that increased earnings will come through enhanced value which justifies higher prices and margins, and that biotechnology offers the potential to create enhanced value. This idea is echoed by Shelton (1999), who reports predictions that the food we will eat in the future will increasingly be based on raw materials and ingredients tailored (engineered) to our individual tastes, lifestyles and medical needs. Chilled foods manufacturers may be able to benefit from genetic engineering technology in a number of ways. For example, the insertion of genes from mammals not normally used in milk production, into bovine genomes might give rise to milks with unusual functional properties which can be exploited as novel ingredients in food manufacture. Also, bacteria used for the production of fermented chilled dairy products may be genetically engineered to yield new potentials, such as self-stabilising yogurts (no need to add stabilisers and thickeners as they are produced 'naturally', *in situ*), or bacterially vitaminized fermented milks and cheese, or products which contain vaccines produced by starter bacteria fermentation. There may even be the possibility of doing away with cows altogether. What is science fiction today frequently becomes science fact tomorrow. So, what about genetically engineering different strains of *E. coli* to produce the various components of milk which can then be combined in preferred proportions to yield products tailored for specific food applications? But if such a fiction becomes a fact, will the products still be milk products?

2.12 References

ANON 1995a, What does it mean? *Food Science and Technology Today*, **9** 93–100.

ANON 1995b, *Dairy Products (Hygiene) Regulations 1995*, S.I. No. 1086. HMSO, London.

ANON 1999, Raw cows' drinking milk. *Health and Hygiene*, 20, 115–6.

BERK Z, 1986. *Braverman's introduction to the biochemistry of foods*, Elsevier, Amsterdam, p. 170.

BODYFELT F W, TOBIAS J and TROUT G M, 1988. *Sensory evaluation of dairy products*, Van Nostrand/AVI, New York.

BRENNAN J G, BUTTERS J R, COWELL N D. and LILLEY A EV, 1990. 3rd edn. *Food engineering operations*, Elsevier Applied Science, London, pp 187–8, 119–21.

BUCHANAN R L and DOYLE M P, 1997 Foodborne disease significance of

Escherichia coli O157:H7 and other enterohemorrhagic *E. coli. Food Technology*, **51**, 69–76.

BYLUND G, 1995 *Dairy processing handbook*, Tetra Pak Processing Systems AB, Lund, p. 288.

CCFRA, 1996 *A code of practice for the manufacture of vacuum and modified atmosphere packaged chilled foods*, Guideline 11. Campden and Chorleywood Food Research Association, Chipping Campden, Gloucestershire, p. 18.

CLOAKE R R and ASHTON T R, 1982 Cream In: *Technical guide for the packaging of milk and milk products*, 2nd edn. Bulletin of the International Dairy Federation, Brussels, No. 143.

EARLY R, 1997 Putting HACCP into practice, *International Journal of Dairy Technology*, **50** 7–13.

EARLY R, 1998a 2nd edn. Liquid milk and cream. In: R. Early (ed) *The technology of dairy products*. Blackie Academic and Professional, London, pp. 1–49.

EARLY R, 1998b 2nd edn. Milk concentrates and milk powders. In: R. Early (ed) *The technology of dairy products*, Blackie Academic and Professional, London, pp. 228–300.

EEC, 1993 Council Directive 93/43/EEC (June 14, 1993) on the hygiene of foodstuffs, *Official Journal of the European Communities*, July 19, 1993, No. L 175/I.

FELLOWS P J, 1997 *Food processing technology*, Woodhead, Cambridge, p. 66.

FOX P F, 1993 2nd edn. Cheese: an overview. In: P. F. Fox. *Cheese: chemistry, physics and microbiology*, Volume 1, General aspects, Chapman and Hall, London, pp. 1–36.

FOX P F and MCSWEENEY P L H, 1998 *Dairy chemistry and biochemistry*, Blackie Academic and Professional, London, pp. 364.

FRAZIER W C, and WESTHOFF D C, 1988 4th edn. *Food microbiology*, McGraw-Hill, New York, pp. 287–93, 367–70.

HARDING F, 1995 Impact of raw milk quality on product quality In, F. Harding. (ed.) *Milk quality*, Blackie Academic and Professional, London, pp. 102–11.

HMSO, 1995 *The Food Safety (General Food Hygiene) Regulations 1995* HMSO, London.

HOULDSWORTH D W, 1980 Demineralization of whey by means of ion exchange and electrodyalysis, *Journal of the Society of Dairy Technology*, **33** 45–51.

IFST, 1990 2nd edn. *Guidelines for the handling of chilled foods*. Institute of Food Science and Technology, London, pp. 43–4.

IFST, 1999 *Development and use of microbiological criteria for foods*, Institute of Food Science and Technology, London, pp. 42, 49.

JAY J M, 1996 5th edn, *Modern food microbiology*, Chapman and Hall, New York, pp. 354, 422.

JENNESS R and PATTON S, 1959 *Principles of dairy chemistry* Robert E. Krieger, Huntington, pp. 58–60.

JOHNSON A H, 1972 The composition of milk. In: B H Webb, A H Johnson and J A Alford (eds) *Fundamentals of dairy chemistry*, AVI, Westport, pp. 1–57.

KIRKPATRICK K J and FENWICK R M, 1987 Manufacture and general properties of dairy ingredients, *Food Technology*, **41** 58–65.

KUZMINSKI L N, 1999 Food R&D in the 21st Century, *Food Technology*, **53** (2): 122.

LANE R, 1998 Butter and mixed fat spreads. In: R. Early *The technology of dairy products*, Blackie Academic and Professional, London, pp. 158–97.

LEAPER S, (ed.) 1997 2nd edn. *HACCP: a practical guide*. Technical Manual No. 38. Campden and Chorleywood Food Research Association, Chipping Campden, Gloucestershire.

MCDOWALL F H, 1953 *The buttermaker's manual*, Volume 1, New Zealand University, Wellington, p. 727.

MAFF, 1995 *Manual of Nutrition*, Stationery Office, London.

MARSHALL V M and TAMIME A Y, 1997 Starter cultures employed in the manufacture of biofermented milks, *International Journal of Dairy Technology*, **50** 35–41.

METTLER A E, 1989 Pathogens in milk powders – have we learned the lessons? *Journal of the Society of Dairy Technology*, **42** 48–55.

METTLER A E, 1992 Pathogen control in the manufacture of spray dried powders *Journal of the Society of Dairy Technology*, **41** 1–2.

METTLER A E, 1992 Present day requirements for effective pathogen control in spray dried milk powder production *Journal of the Society of Dairy Technology*, **47** 95–107.

MILLER A J, WHITING R C and SMITH J L, 1997 Use of risk assessment to reduce Listeriosis incidence *Food Technology*, **51**, 100–3.

MORTIMORE S and WALLACE C, 1998 2nd edition, *HACCP: A practical approach*, Chapman and Hall, London.

MUIR D D, 1996a The shelf-life of dairy products: 2. Raw milk and fresh products, *Journal of the Society of Dairy Technology*, **49** 44–8.

MUIR D D, 1996b The shelf-life of dairy products: 2. Intermediate and long life dairy products, *Journal of the Society of Dairy Technology*, **49** 119–24.

MULVIHILL D M and KINSELLA J E, 1987 Gelation characteristics of whey proteins and β-lactoglobulin, *Food Technology*, **41** 102–11.

NURSTEN H E, 1997 The flavour of milk and dairy products: I. Milk of different kinds, milk powder, butter and cream, *International Journal of Dairy Technology*, **50** 48–56.

PAINE F A and PAINE H Y, 1992 2nd edn. *A handbook of food packaging*, Blackie Academic and Professional, Glasgow, pp. 222–6.

RICHARDS E, 1982 Butter and allied products – quality control, *Journal of the Society of Dairy Technology*, **35** 149–53.

ROTHWELL J, JACKSON A C and BOLTON F, 1989 Trouble shooting. In, J. Rothwell, *Cream processing manual*, Society of Dairy Technology, Huntingdon, pp. 120–4.

SCA, 1997 *The specialist cheesemakers' code of best practice*, Specialist Cheesemakers' Association, Newcastle under Lyme.

SCOTT R, 1981 *Cheesemaking practice*, Applied Science, London, pp. 28–37.

SHELTON E, 1999 A Taste of Tomorrow, *The Grocer*, **222** 42–2.
SIMON A, 1956 *Cheeses of the world*, Faber and Faber London. pp. 74–5.
STANLEY G, 1998 Microbiology of fermented milk products. In: R. Early (ed). *The technology of dairy products*, Blackie Academic and Professional, London, pp. 50–80.
WILBEY R A, and YOUNG, P. 1989 2nd edn. Scalded and clotted cream. In, J. Rothwell, *Cream processing manual*, Society of Dairy Technology, Huntingdon, pp. 74–82.

3

Raw material selection: meat and poultry

S. J. James, Food Refrigeration and Process Engineering Research Centre

3.1 Introduction

Much poultry and red meat is sold in a chilled unprocessed state. However, an increasing proportion is used as a basic raw material for chilled meat products and ready meals. A growing trend is the development of added-value convenience meals, especially ethnic products, many of which are pre- or part-cooked and necessitate chilled storage. Many of these products contain meat as a key ingredient. This meat for further processing can be supplied chilled, as boneless blocks of frozen material or increasingly as minced or diced material. The dice or mince can be chilled or increasingly it will be supplied as bags of individually quick frozen (IQF) product. This chapter discusses a number of issues which influence the quality of meat as a raw material in high added-value chilled foods.

The quality of meat is judged by its bacterial condition and appearance. Bacterial condition is subjectively assessed by the presence or absence of odour or slime. Quantitative tests can also carried out to determine the total viable counts and the presence of specific pathogens or indicator organisms. Appearance criteria are primarily; colour, percentage of fat and lean and amount of drip exuding from the meat. Any unacceptable change in the microbial or appearance criteria will limit the shelf-life of the meat. After cooking its eating quality is partially judged by its appearance but mainly by its tenderness, flavour and juiciness.

Red meat and poultry are very perishable raw materials. If stored under ambient conditions, 16–30°C, the shelf-life of both can be measured in tens of hours to a few days. Under the best conditions of chilled storage, close to the initial freezing point of the material, the storage life can be extended to

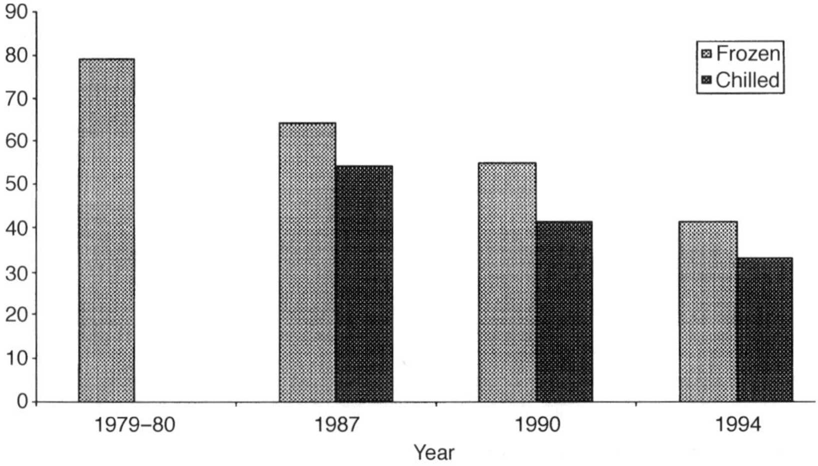

Fig. 3.1 Percentage of chilled and frozen poultry carcasses contaminated with salmonella found in UK surveys carried out in 1979–80 to 1994.

approaching six weeks for some red meat. Even under the best commercial practice (strictly hygienic slaughtering, rapid cooling, vacuum packing and storage at super chill (-1 ± 0.5°C)) the maximum life that can be achieved in red meat is approximately 20 weeks, however, freezing will extend the storage life of meat to a number of years.

In a perfect world, red meat and poultry would be completely free of pathogenic (food poisoning) micro-organisms when produced. However, under normal methods of production pathogen-free meat cannot be guaranteed. For example, salmonella contamination of chilled and frozen poultry carcasses has been significantly reduced in the UK (Fig. 3.1). However, over one-third was still contaminated in 1994.[1] While the internal musculature of a healthy mammal or bird is essentially sterile after slaughter, all meat animals carry large numbers of different micro-organisms on their skin/feathers and in their alimentary tract. Only a few types of bacteria directly affect the safety and quality of the finished carcass. Of particular concern are food-borne pathogens such as *Campylobacter* spp., *Clostridium perfringens*, *Salmonella* spp., and pathogenic serotypes of *Escherichia coli*. The minimum and optimum growth temperatures for some of the pathogens associated with red and poultry meat are shown in Table 3.1.

Inevitably, small numbers of pathogens will be present on meat and cooking regimes are designed to eliminate their presence. Most red meat and poultry food poisoning is associated with inadequate cooking or subsequent contamination after cooking and poor cooking and storage.

Normally it is the growth of spoilage organisms that has the most important effect in limiting the shelf-life of meat. The spoilage bacteria of meats stored in air under chill conditions include species of *Pseudomonas*, *Brochothrix* and *Acinetobacter/Moraxella*. Varnam and Sutherland state that in general, there is

Table 3.1 Chilled storage life of meat and meat products at different storage temperatures

	Storage time (days) in temperature range:							
	−4.1 to −1.1°C		−1 to 2°C		2.1 to 5.1°C		5.2 to 8.2°C	
Food	Mean	sd	Mean	sd	Mean	sd	Mean	sd
Bacon			45	6	15	3	42	20
Beef	40	26	34	32	10	8	9	9
Lamb	55	20	41	46	28	34		
Pork	50	58	22	30	16	16	15	18
Poultry	32	18	17	10	12	11	7	3
Veal	21		10	6	49		49	
Rabbit			9	7	13	6		
Offal	7		7	6	14	7		
Bacon			45	6	15	3	42	20
Sausage	80	43	21	16	36	28	24	10

little difference in the microbial spoilage of beef, lamb, pork and other meat derived from mammals.

The presence of exudate or 'drip', which accumulates in the container of pre-packaged meat, or in trays or dishes of unwrapped meat, substantially reduces its sales appeal.[3] Drip can be referred to by a number of different names including 'purge loss', 'press loss' and 'thaw loss' depending on the method of measurement and when it is measured. Drip loss occurs throughout the cold chain and represents a considerable economic loss to the red meat industry. Poultry meat is far less prone to drip. The potential for drip loss is inherent in fresh meat and is influenced by many factors. Some of these, including breed, diet and physiological history, are inherent in the live animal. Others, such as the rate of chilling, storage temperatures, freezing and thawing, occur during processing. Meat colour can be adversely affected by a variety of factors, including post-mortem handling, chilling, storage and packaging.[4]

In Australia, CSIRO[5] stated that 'Toughness is caused by three major factors – advancing age of the animal, 'cold shortening' (the muscle fibre contraction that can occur during chilling) and unfavourable meat acidity (pH).' There is general agreement on the importance of these factors, with many experts adding cooking as a fourth equally important influence.

3.2 The influence of the live animal

Some of the factors that influence the toughness or meat are inherent in the live animal. Church and Wood[4] state that it is now well established that it is the properties of the connective tissue proteins, and not the total amount of collagen in meat, that largely determines whether meat is tough or tender. As the animal grows older the number of immature reducible cross-links decreases. The mature cross-links result in a toughening of the collagen and this in turn can produce tough meat. Increasing connective tissue toughness is probably not commercially significant until a beast is about four years old.[6]

The pigment concentration in meat which governs its colour is affected by many factors affecting the live animal. These include species – beef, for example, contains substantially more myoglobin than pork; breed and age – pigment concentration increases with age; sex – meat from male animals usually contains more pigment than that from female animals; muscle – muscles that do more work contain more myoglobin.

There are also two specific meat defects; dark, firm, dry (DFD) and pale, soft, exudative (PSE) associated with the live animal that result in poor meat colour. DFD meat has a high ultimate pH and oxygen penetration is low. Consequently, the oxymyoglobin layer is thin, the purple myoglobin layer shows through, and the meat appears dark. In PSE meat the pH falls while the muscle is still warm and partial denaturation of the proteins occurs. An increased amount of light is scattered and part of the pigment oxidised so that the meat appears pale.

3.2.1 Between species and breeds

In all species the range of storage lives found in the literature is very large (Table 3.1) and indicate that factors other than species have a pronounced effect on storage life. Overall, species has little effect on the practical storage life of meat. In general, beef tends to lose proportionately more drip than pork and lamb. Unfrozen poultry meat looses little if any drip. Since most of the exudate comes from the cut ends of muscle fibres, small pieces of meat drip more than large intact carcasses. In pigs, especially, there are large differences in drip loss from meat from different breeds and between different muscles. Taylor[7] showed that there was a substantial difference, up to 2.5 fold, in drip loss between four different breeds of pig (Table 3.2). He also showed that there was a 1.7 to 2.8 fold difference in drip between muscle types (Table 3.3).

Although there is a common belief that breed has a major effect on meat quality CSIRO[8] state 'although there are small differences in tenderness due to

Table 3.2 Drip loss after two days storage at 0°C, from leg joints from different breeds of pig cooled at different rates

Breed	Drip loss (% by wt.)	
	Slow	Quick
Landrace	0.47	0.24
Large White	0.73	0.42
Wessex X Large White	0.97	0.61
Pietrain	1.14	0.62

Table 3.3 Drip loss after two days storage at 0°C from four muscles from two breeds of pig cooled at different rates

	Drip (as % muscle weight)					
	Cooling rate	Semi-tendinosus	Semi-membranous	Adductor	Biceps femoris	Combined (4 muscles)
Pietrain	Quick	2.82	4.40	5.52	2.69	3.86
	Slow	3.99	6.47	6.61	4.11	5.30
Large White	Quick	1.69	2.01	2.92	1.04	1.92
	Slow	1.95	3.50	5.07	2.32	3.21

breed, they are slight and currently of no commercial significance to Australian consumers.' That said, there are substantial differences in the proportion of acceptable tender meat and toughness between *Bos indicus** and *Bos taurus** cattle. The proportion of acceptable tender meat has been found to decrease from 100% in Hereford Angus crosses, to 96% in Tarentaise, 93% in Pinzgauer, 86% in Brahman and only 80% in Tsahiwal.[9] Toughness of meat increases as the proportion of *Bos indicus* increases.[10]

3.2.2 Animal to animal variation

There is little data on any relationship between animal to animal variation and chilled storage life. However, it is believed to cause wide variations in frozen storage life; differences can be as great as 50% in the freezing of lamb.[11, 12] Differences would appear to be caused by genetic, seasonal or nutritional variation between animals, but there is little reported work to confirm this view. Variations were found between the fatty acids and ratio of saturated/unsaturated fatty acids in lambs from New Zealand, America and England.[13] Differences related to sex, geographical area and cut were mainly a reflection of fatness, with ewes having a greater percentage of body fat than rams. However, differences between areas were found to produce larger variations between animals than sex differences. A number of other trials have detailed differences between animals.

There can also be significant differences in texture within a breed. Longissimus dorsi shear force values for double muscled Belgium Blue bulls were significantly higher than those of the same breed with normal conformation.[14] Calpain I levels at 1 h and 24 h *post mortem* were also much lower. It was suggested that the lower background toughness in the double muscle was compensated for by reduced *post mortem* proteolytic tenderisation.

Sex of the animal appears to have little or no influence on tenderness. Huff and Parrish[15] compared the tenderness of meat from 14-month-old bulls and

* *Bos indicus* are tropical and semitropical breeds of cattle primarily Brahman and *Bos taurus* are temperate breeds such as Hereford or Aberdeen Angus.

steers, and cows (55 to 108 months old). No differences were found between the tenderness of bulls and steers. Tenderness decreased with the age of the animal. Hawrysh *et al.*[16] reported that beef from bulls may be less tender than that from steers. Sex can have a substantial influence on flavour. For example, cooking the meat from entire male pigs can produce an obnoxious odour known as 'boar taint'. Problems can also occur with meat from intact males of other species. However, they can still be attractive to industry because of their higher rate of growth and lower fat content.

3.2.3 Feeding

The way in which an animal is fed can influence its quality and storage life. It has been reported that chops from pigs fed on household refuse have half the frozen storage life of those fed on a milk/barley ration.[17, 18] Again, pork from pigs that had been fed materials containing offal had half the practical storage life (PSL) and higher iodine numbers in the fat than that of pigs which had not been fed this type of diet.[19] Conversely, Bailey *et al.*[20] did not find any differences between meal- and swill-fed pigs after 4 and 9 months at −20°C. Rations with large amounts of highly unsaturated fatty acids tend to produce more unstable meat and fat.

The type of fatty acid composition of 'depot fat' in poultry and its stability have been shown to be directly related to the fatty acid composition of ingested fats.[21, 22] The feeding of fish oils or highly unsaturated vegetable oils (such as linseed oil) to poultry is known to produce fishy flavours in the meat.[21, 23]

The use of vitamin E supplements is recommended for both beef and turkey. This will 'result in delayed onset of discoloration in fresh, ground and frozen beef and in suppression of lipid rancidity, especially in fresh, ground and frozen beef and less so in cooked beef'.[24] With turkey, vitamin E supplements have been shown to improve oxidative stability of cooked and uncooked turkey burgers during six months frozen storage at −20°C.[25]

3.2.4 Variations within an animal

Reports of variations in the storage life of different cuts of meat are scarce and primarily deal with dark and light meat. Both Ristic[26] and Keshinel *et al.*[27] have found that poultry breast meat stores better than thigh meat. Ristic states that frozen breast meat will store for 16 months while thigh meat can be stored for only 12 months due to its higher fat content. Judge and Aberle[28] also found that light pork meat stored for a longer time than dark meat. This was thought to be due to either higher quantities of haem pigments in the dark muscle (which may act as major catalysts of lipid oxidation), or to higher quantities of phospholipids (which are major contributors to oxidised flavour in cooked meat).

3.3 Pre- and post-slaughter handling

3.3.1 Red meat

The way animals are handled and transported before slaughter affects meat quality and its storage life. Increased stress or exhaustion can produce PSE (pale soft and exudative) or DFD (dark firm and dry) meat, which is not recommended for storage, mainly due to its unattractive nature and appearance. Jeremiah and Wilson[29] found that the use of PSE muscle produced low yields after curing and it was concluded that PSE meat was unsuitable for further processing.

Experiments designed to determine the effect of treatments immediately before or at the point of slaughter appear to show that they have little effect on meat texture. Exercising pigs before slaughter has been shown to have no effect on texture parameters, i.e. muscle shortening and shear force.[30] The use of different stunning methods (both electrical and carbon dioxide) does not seem to have a significant effect on the quality of pork.[31]

Consumers' surroundings influence their appreciation of tenderness.[32] Consumers were more critical of the tenderness of beef steaks cooked in the home than those cooked in restaurants. The Warner-Bratzler force transition level for acceptable steak tenderness was between 4.6 and 5.0 kg in the home and between 4.3 and 5.2 kg in the restaurants.

Overall, there appears to be little correlation between chilling rates or chilling systems and bacterial numbers after chilling. The microorganisms that usually spoil meat are psychrotrophs; i.e. bacteria capable of growth close to 0°C. Only a small proportion of the initial microflora on meat will be psychrotrophs; the majority of microorganisms present are incapable of growth at low temperatures. As storage temperature rises the number of species capable of growth will increase. The growth rate of microorganisms also accelerates with increasing temperature. In the accepted temperature range for chilled meat, −1.5 to 5°C, there can be as much as an eightfold increase in growth rate between the lower and upper temperature. For any particular treatment the maximum chilled storage life will be obtained by holding the meat at −1.5°C. Chilled storage life is halved for each 2–3°C rise in temperature.

Odour and slime cause by the growth of microorganisms will be apparent after approximately 14.5 and 20 days respectively with beef sides stored at 0°C (Fig. 3.2). At 5°C the respective times are significantly reduced to 8 and 13 days, respectively.

Rapid chilling reduces drip loss (Table 3.2 & 3.3). However, chilling has serious effects on the texture of meat if it is carried out rapidly when the meat is still in the pre-rigor condition, that is, before the meat pH has fallen below about 6.2.[33] In this state the muscles contain sufficient amounts of the contractile fuel, adenosine triphosphate (ATP), for forcible shortening to set in as the temperature falls below 11°C, the most severe effect occurring at about 3°C. This is the so-called 'cold-shortening' phenomenon, first observed by Locker and Hagyard[34] and its mechanism described by Jeacocke.[35] The meat 'sets' in the shortened state as rigor comes on, and this causes it to become extremely

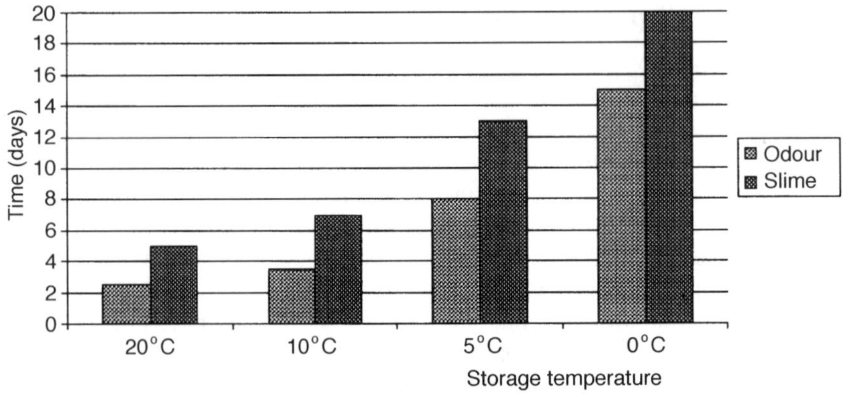

Fig. 3.2 Time for odour and slime to develop on beef carcasses at different storage temperatures.

tough when it is subsequently cooked.[36] If no cooling is applied and the temperature of the meat is above 25°C at completion of rigor then another form of shortening 'rigor' – or 'heat-shortening' will occur.[37]

Electrical stimulation (ES) of the carcass after slaughter can allow rapid chilling without much of the toughening effect of cold shortening. However, Buts *et al.*[38] reported that in veal ES followed by moderate cooling affected tenderness in an unpredictable way and could result in tougher meat. Electrical stimulation will hasten rigor and cause tenderisation to start earlier at the prevailing higher temperature. In meat from carcasses given high or low voltage stimulation and slow cooling, adequate ageing in beef can be obtained in about half the time of non-stimulated beef[38] This will therefore reduce the requirement and cost of storage.

When meat is stored at above freezing temperatures it becomes progressively more tender. This process, known as ageing, conditioning or maturation is traditionally carried out by hanging the carcass for periods of 14 days or longer. The rate of ageing differs significantly between species and necessitates different times for tenderisation. Beef, veal and rabbit age at about the same rate and take about ten days at 1°C to achieve 80% of ageing (Table 3.4). Lamb ages slightly faster than beef but more slowly than pork. The ultimate tenderness will depend on the initial 'background' tenderness of the meat and the tenderisation that has occurred during chilling. In veal acceptable tenderness can be obtained after five days at 1°C compared with 10 days for beef.

Red colour is more stable at lower temperatures because the rate of oxidation of the pigment decreases. At low temperatures, the solubility of oxygen is greater and oxygen-consuming reactions are slowed down. There is a greater penetration of oxygen into the meat and the meat is redder than at high temperatures.

The major improvement in tenderness has been shown to occur in less than 14 days. In a study by Martin *et al.*,[39] in which more than 500 animals were examined, it was concluded that for beef carcasses, a period of six days is

Table 3.4 Time taken to achieve 50 and 80% ageing at 1°C for different species[8]

Species	Time (d) taken to achieve 50%	80%
Beef	4.3	10.0
Veal	4.1	9.5
Rabbit	4.1	9.5
Lamb	3.3	7.7
Pork	1.8	4.2
Chicken	0.1	0.3

sufficient for a consumer product of satisfactory tenderness. Buchter[40] also showed that no significant increase in tenderness occurs after 4–5 days for calves and 8–10 days for young bulls at 4°C. The ageing process can be accelerated by raising the temperature, and the topic was well studied in the 1940s, 50s and 60s. Ewell[41] found that the rate of tenderising more than doubled for each 10°C rise. Meat from a three-year-old steer requiring ten days at 0°C to reach the same tenderness as two days at 23°C. Sleeth *et al.*[42] showed that the tenderness, flavour, aroma and juiciness of beef quarters and ribs aged for 2–3 days at 20°C were comparable to those aged 12–14 days at 2°C. Busch *et al.* demonstrated that steaks from excised muscles held at 16°C for two days were more tender than those stored at 2°C for 13 days.

The microbiological hazards of high-temperature ageing were well recognised and several investigators used antibiotics and/or irradiation to control bacterial growth.[42, 44, 45] Although high-temperature ageing in conjunction with ultraviolet (UV) radiation has been used in the US its use has not expanded owing to its high cost.[46] With the use of irradiation gaining more acceptance in the US its use to accelerate ageing in conjunction with modified atmosphere packaging and high-temperature storage has been investigated by Mooha Lee *et al.*[47] Irradiated steaks stored for two days at 30°C were more tender than unirradiated controls stored at 2°C for 14 days (Table 3.5).

In red meat, there is little evidence of any relationship between chilling rates and frozen storage life. However, there is evidence for a relationship between

Table 3.5 Shear values for steaks after post mortem ageing

	Treatment		
Storage time (d)	Unirradiated 2°C	Irradiated 15°C	Irradiated 30°C
1	4.47±1.76	4.89±1.40	4.24±1.65
2	–	4.81±0.73	3.44±1.34
3	–	4.07±1.03	3.33±1.21
7	3.62±1.00	–	–
14	3.52±1.51	–	–

storage life and the length of time that elapses before freezing occurs. Chilled storage of lamb for one day at 0ºC prior to freezing can reduce the subsequent storage life by as much as 25% when compared to lamb which has undergone accelerated conditioning and 2 hours storage at 0ºC.[12] It has been shown that pork which had been held for seven days deteriorated at a faster rate during storage than carcasses chilled for one and three days.[48] Ageing for periods greater than seven days was found by Zeigler[49] to produce meat with high peroxide and free fatty acid values when stored at −18ºC or −29ºC. Although shorter ageing times appear to have a beneficial effect on storage life there is obviously a necessity for it to be coupled with accelerated conditioning to prevent any toughening effects. Whilst there has been significant research in such areas as these, little appears to be known about the relationship between the frozen storage life of meat as a raw material and the chilled life of the product in which it is used.

3.3.2 Poultry

After bleeding and death, poultry carcasses are scalded by immersing them in hot water for approximately three minutes. Scalding loosens the feathers so that they can be easily removed. Carcasses can either be soft scalded at 52–53ºC or hard scalded at 58ºC. Hard scalding removes the cuticles on chicken skin, which gives an unattractive appearance after air chilling. Generally if the poultry is marketed in a chilled state it will be soft scalded and air or spray chilled.

Spray washing is used at numerous points during processing to remove visual contamination. It also has some small role in reducing bacterial contamination. The EU Poultry Meat Directive[50] requires poultry to be washed inside and out immediately prior to water chilling. The amount of water to be used (i.e. 1.5 litres for a carcass up to 2.5 kg) is defined in the directive. Water chillers are designed to operate in a counter-current manner to minimise cross-contamination. The carcasses exit from the chillers at the point where the clean, chilled water enters the system. Again, the Directive defines a minimum water flow through the chiller, i.e. at least one litre per carcass for carcasses up to 2.5 kg in weight.

Chicken breast muscle ages ten times faster than beef (Table 3.4). Hence, ageing in poultry carcasses occurs during processing and is usually accomplished before they reach the chiller/freezer. Pool *et al.*[51] have shown that there were no detectable flavour differences over an 18-month period between turkey that had been frozen immediately and turkey that had been held at +2ºC for 30 hours.

3.4 Conclusions

Even with the best practice, the maximum shelf-life of chilled meat can be measured in weeks. Freezing will extend the storage life of meat to a number of years. If the frozen storage life is not exceeded, freezing and frozen storage of meat has little (if any) effect on the main quality parameters. Drip is the only exception, as it is substantially increased by freezing.

Numerous experiments and blind testings have shown that consumers cannot tell the difference between frozen and chilled meat. However, they are still willing to pay up to twice as much for the equivalent chilled product. Despite a large price differential in favour of the frozen product it is the 'fresh' chilled market that is growing fastest. For example, there is a growing trend for retailers to sell bulk packs of chilled chicken breasts specifically for home freezing. The breasts are individually packed and the consumer is happy to pay a premium price and slowly freeze them in a domestic freezer. There are no obvious technical developments that are likely to change the trend towards chilled products with poultry carcasses and portions. Chilled product is more convenient for the consumer even if there are no quality advantages over frozen poultry. If food safety becomes much more of an issue then freezing directly after decontamination may become more common. The low temperature will stop any growth of pathogenic or spoilage organisms that survive the decontamination process.

In the red meat industry there is an increasing demand for meat of a consistent, guaranteed high eating quality. Specifications already take into account, breed, age, feed and handling of the live animal. Slaughter, chilling and ageing conditions are also carefully controlled. However, at the end of this carefully controlled process we have a quality-chilled product with a very limited shelf-life. Small changes in levels of initial contamination or ageing temperatures or times can result in unacceptable levels of pathogenic and spoilage micro-organisms. As with poultry one way of overcoming this problem would be to combine decontamination with subsequent freezing.

3.5 References

1. ANON *Safer eating – Microbiological food poisoning and its prevention*, London, Parliamentary office of science and technology, 1997, ISBN 1 897941 56 0.
2. VARNAM A H and SUTHERLAND J P, *Meat and Meat Products*, Chapman & Hall, London, 1995, ISBN 0 412 49560 0.
3. MALTON R and JAMES S J, Drip loss from wrapped meat on retail display. Meat Industry, 1983 **56** (5) 39–41.
4. CHURCH, P N and WOOD, J M, *The manual of manufacturing meat quality*, London & New York, Elsevier Applied Science, 1992, ISBN 1 85166 628 1.
5. CSIRO, *Tender beef*, Meat Research News Letter, 1988 88/4.
6. HUSBAND P M and JOHNSON B Y, Beef tenderness: the influence of animal age and postmortem treatment, *CSIRO Food Res, Q*. 1985 **45** 1–4.
7. TAYLOR A A, Influence of carcass chilling rate on drip in meat. In, *Meat Chilling: Why and How? Meat Research Institute Symp. No. 2*, (ed. C. L. Cutting). 5.1–5.8 1972.

8. CSIRO, Does breed influence the tenderness of beef? *CSIRO Newsletter* 92/4, 1992.
9. KOCH R M, DIKEMAN M E and CROUSE J D, *J.Anim.Sci.*, 1982 **54**, 35.
10. CROUSE J D, CUNDIFF L V, KOCH R M, KOOHMARAIE M and SEIDEMAN S C, *J.Anim.Sci.*, 1989 **67**, 2661.
11. WINGER R J, Storage life of frozen foods. New approaches to an old problem. *Food Technology in New Zealand* 1984 **19** 75, 77, 81, 84.
12. WINGER R J, Storage life and eating related quality of New Zealand frozen lamb: A compendium of irrepressible longevity. In *Thermal processing and quality of foods*. Elsevier Applied Science Publishers. 541–52, 1984.
13. CRYSTALL B B and WINGER R J, Composition of New Zealand Lamb as influenced by geographical area, sex of animal and cut. Meat Industry Research Institute of New Zealand (MIRINZ). Publication no. 842. 1986.
14. UYTTERHAEGEN L, CLAEYS E, DEMEYER D, LIPPENS M, FIEMS L O, BOUCQUÉ C Y, VAN DE VOORDE G and BASTIAENS A, Effects of double-muscling on carcass quality, beef tenderness and myofibrillar protein degradation in Belgian Blue White bulls. *Meat Sci.*, 1994 **38** 255–67.
15. HUFF E J and PARRISH Jr F C, Bovine longissimus muscle tenderness as affected by postmortem ageing time, animal age and sex. *J.Fd.Sci.* 1993, **58** 4, 713–16.
16. HAWRYSH Z J, PRICE M A and BERG R T, The influence of cooking temperature on the eating quality of beef from bulls and steers fed three levels of dietary roughage, *J.Inst.Can.Sci.Technol Aliment*, 1979, **12** 2 72–7.
17. WISMER-PEDERSON J and SIVESGAARD A, Some observations on the keeping quality of frozen pork, *Kulde*, 1957 **5** 54.
18. PALMER A Z, BRADY D E, NAUMAN H D and TUCKER L N, Deterioration in freezing pork as related to fat composition and storage treatments, *Food Technology*, 1953 **7** 90–5.
19. BOGH-SORENSEN L and HOJMARK JENSEN J, Factors affecting the storage life of frozen meat products. *International Journal of Refrigeration*, 1981 **4.3** 139–42.
20. BAILEY C, CUTTING C L, ENSER M B and RHODES D N, The influence of slaughter weight on the stability of pork sides in frozen storage, *Journal of Science Food and Agriculture*, 1973 **24** 1299–304.
21. KLOSE A A, HANSON H L, MECCHI E P, ANDERSON J H, STREETER I V and LINEWEAVER H, Quality and stability of turkeys as a function of dietary fat, *Poultry Science*, 1953 **32** 83–8.
22. KUMMEROW F A, HITE J and KLOXIN S Fat rancidity in eviscerated poultry, *Poultry Science*, 1948 **6** 689–94.
23. WEBB J E, BRUNSEN C C and YATES J D, Effects of dietary fish meal level on flavour of pre-cooked frozen turkey meat, *Poultry Science*, 1974 **53** 1399–404.
24. LIU, Q, LANARI, M C and SCHAEFER, D M A review of dietary vitamin E supplementation for improvement of beef quality, *J. Anim. Sci*, 1995 **73** 3131–40.

25. WEN, J, MORRISSEY, P A, BUCKLEY, D J and SHEEHY, P J A Oxidative stability and α-tocopherol retention in turkey burgers during refrigerated and frozen storage as influenced by dietary α-tocopheryl acetate, *British Poultry Science*, 1996 **37** 787–95.
26. RISTIC, M Influence of the water cooling of fresh broilers on the shelf life poultry parts at −15ºC and −21ºC, *Lebensm*, 1982 **15** 113–6.
27. KESKINEL, A, AYRES, J C and SNYDER, H E Determination of oxidative changes in raw materials by the 2-thiobarbituric acid method, *Food Technology*, 1964 101–4.
28. JUDGE, M D and ABERELE, E D Effect of pre-rigor processing on the oxidative rancidity of ground light and dark porcine muscles, *Journal of Food Science*, 1980 **45** 1736–9.
29. JEREMIAH, L E and WILSON, R The effects of PSE/DFD conditions and frozen storage upon the processing yields of pork cuts, *Canadian Institute of Food Science and Technology Journal*, 1987 **20** 25–30.
30. IVENSEN, P, HENCKEL, P, LARSEN, L M, MONLLAO, S and MØLLER, A J Tenderisation of pork as affected by degree of cold-induced shortening, *Meat Science*, 1995 Vol. 40, pp. 171–81.
31. GARRIDO, M D, PEDAUYE, J, BANON, S, MARQUES, F and LAENCINA, J Pork quality affected by different slaughter conditions and post mortem treatment of the carcasses, *Food Science and Technology – Lebensmittel-wissenschaft & Technologie*, 1994 **27** (2) 173–6.
32. MILLER, E, HOOVER, L C, GUERRA, A L, HUFFMAN, K L, TINNEY, K S, RAMSEY, C B, BRITTIN, H C and HUFFMAN, L M. Consumer acceptability of beef steak tenderness in the home and restaurant, *J.Fd.Sci.*, 1995 Vol. 60, 5, pp. 963–5.
33. BENDALL, J R The influence of rate of chilling on the development of rigor and 'cold shortening', *Meat Chilling – Why and How?*, Meat Research Institute, Langford, UK, 1972 3.1–3.6.
34. LOCKER, R H and HAGYARD, C J Cold shortening in beef muscles, *J. Sci. Fd. Agric.*, 1963 **14** 787.
35. JEACOCKE, R E The mechanism of cold shortening, *IIR Meat Chilling 86*, Bristol, UK, 1986 13–16.
36. MARSH, B B and LEET, N G Meat tenderness 3, *J Fd Sci*, 1966 **31** 450–60.
37. DRANSFIELD, E Optimisation of tenderisation, ageing and tenderness, *Meat Sci.*, 1994 **36** 105–21.
38. BUTS, B, CASTEELS, M, CLAEYS, E and DEMEYER, D Effects of electrical stimulation, followed by moderate cooling, on meat quality characteristics of veal *Longissimus dorsi*. *Meat Sci.* 1986 **18** 271–79.
39. MARTIN, A H, FREDEEN, H T and WEISS, G M Tenderness of beef longissimus dorsi muscle from steers, heifers and bulls as influenced by source, post-mortem ageing and carcass characteristics, *J. Fd Sci*, 1971 **36** 619.
40. BUCHTER, L Development of a standardised procedure for the slaughter of experimental beef animals from the Danish Progeny Station 'Egtved', *Proc. 16th Meeting European Meat Res. Workers*, Bulgaria, 1970 45.
41. EWELL, A W The tenderising of beef, *Refrig. Engng*, 1940 **39** 237–40.

42. SLEETH, R B, KELLEY, G G and BRADY, D E Shrinkage and organoleptic characteristics of beef aged in controlled environments, *Fd Technol*, 1958 **12** 86.
43. BUSCH, W A, PARRISH, F C Jr and GOLL, D E Molecular properties of post-mortem muscle. 4. Effect of temperature on adenosine triphosphate degradation, isometric tension parameters, and shear resistance of bovine muscle, *J. Fd Sci.*, 1967 **32** 390.
44. DEATHERAGE, F E and REIMAN, W (1946) Measurement of beef tenderness and tenderisation of beef by Tenderay process, *Fd Res*, 1946 **11** 525.
45. WILSON, G D, BRAWN, P D, CHESBRO, W R, GINGER, B and WEIR, C E The use of antibiotics and gamma irradiation in the ageing of steaks at high temperatures, *Fd Technol*, 1960 **14** 143.
46. MARAIS, G J K Aspects of the meat trade in the USA, *Meat Ind. Pretoria*, 1968 October–December 33.
47. MOOHA LEE, SEBRANEK, J and PARRISH Jr F C Accelerated post-mortem ageing of beef utilising electron-beam irradiation and modified atmosphere packaging, *J.Fd.Sci.*, 1995 **1** (5) 133–6.
48. HARRISON, D L, HALL, J L, MACKINTOSH, D L and VAIL, G E Effect of post mortem chilling on the keeping quality of frozen pork, *Food Technology*, 1956 **10** 104–108.
49. ZEIGLER, P T, MILLER, R C and CHRISTIAN, J A Preservation of meat and meat products in frozen storage. The Pennsylvania State College, School of Agriculture, State College, Bulletin no. 530 1950.
50. EU POULTRY MEAT DIRECTIVE. Council Directive 71/118/EEC.
51. POOL, M F, HANSON, H L and KLOSE, A A Effect of pre-freezing hold time and anti-oxidant spray on storage stability of frozen eviscerated turkeys. *Poultry Science*, 1950 **29** 347–50.

Part II

Technologies and processes

4

The refrigeration of chilled foods

R. D. Heap, Cambridge Refrigeration Technology

4.1 Introduction

Chilled foods are foods which are cooled to a temperature above their freezing point and which need to be maintained at that temperature to preserve quality. Generally such foods will lose value if frozen, and in many cases freezing will destroy them. From the refrigeration viewpoint, the range of foods regarded as chilled is very wide. In this chapter they are taken to include fresh fruits and vegetables, both temperate and tropical in origin, the whole range of meat, fish and dairy products, and prepared complete meals. Frequently a narrower definition covering only prepared foods is used (Anon. 1997).

It is immediately obvious that refrigeration is essential for the production, storage and distribution of chilled foods. However, the range and variety of refrigeration equipment required is less readily apparent. Consider, for example, the operation of a cook-chill catering facility. Raw materials from around the world are cooled in distant pack houses and transported across the oceans in highly developed refrigerated transport systems. They then pass through refrigerated port stores and via refrigerated road transport to distribution depots from which, either directly or indirectly, they are despatched to the catering facility. This is shown diagrammatically in Fig. 4.1.

Here, refrigerated stores maintain quality prior to use. Some raw materials may be frozen rather than chilled, and will require thawing equipment. Following the cooling operation, the food is chilled using blast chillers or, in some cases, immersion chillers, and will then be stored under refrigeration before distribution in insulated or refrigerated vehicles. It may then be held in refrigerated storage or display cabinets before re-heating. In addition, the waste produced during the food preparation may be stored under refrigeration. The

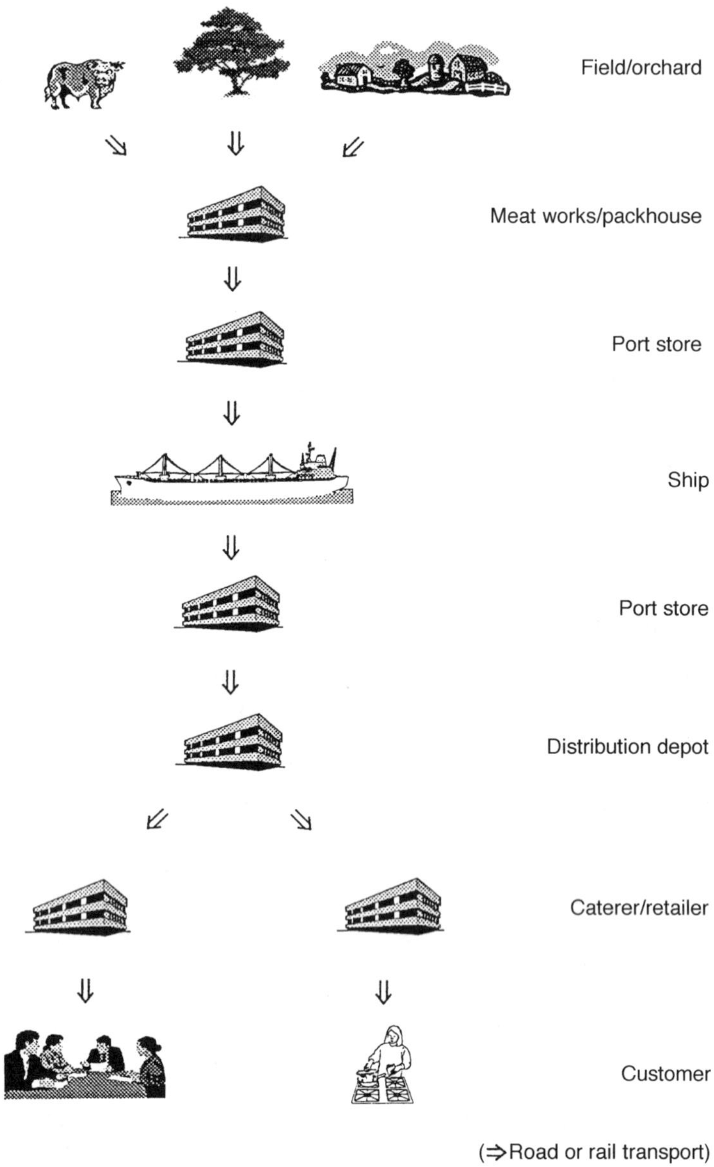

Fig. 4.1 The chill chain.

general lack of problems in refrigeration machinery, which is so essential, is a tribute to the reliability of the technology. Refrigeration is almost a forgotten part of the chilled food preparation process in the mind of the consumer; it is taken for granted.

Food refrigeration is not new. Natural ice and evaporative cooling have been used for millennia, and the relatively recent use of mechanical refrigeration to store foodstuffs at chill temperatures in fact dates back to the US apple stores of the 1870s (Thévenot 1979). Refrigerated transport of chilled (as distinct from frozen) meat started between the US and the UK around 1875, and longer-distance chilled transport from Australasia to Europe dates from 1895 (Critchell and Raymond, 1969). By 1901, the UK was importing over 160,000 tonnes of chilled beef annually.

4.2 Principles of refrigeration

The basic principles of vapour compression refrigeration were established in the 19th century, and this form of refrigeration is almost universally adopted nowadays. At its simplest, such a refrigeration system has four interlinked components (Fig. 4.2). A refrigerant fluid in the vapour state is compressed to a higher pressure, and consequently a higher temperature. The high temperature gas is cooled and liquefied in a condenser. The cool liquid then passes through a restrictor to a lower pressure area, cooling further in the process. The cold liquid can then be used to extract heat from a storage space or cooling area, this heat vaporising the cold, low pressure liquid in an evaporator. The cold vapour is then fed back to the compressor to complete the cycle.

The compressor, condenser, expansion restrictor, and evaporator form the basic components. Whilst heat is extracted from a process at the evaporator, the extracted heat plus the heat equivalent of the compression energy must be rejected at the condenser. This means that any refrigeration device must reject a quantity of heat, which is greater than the heat energy removed from the product or space being cooled. The energy used by a vapour compression refrigeration machine depends on its design, but generally the larger the temperature difference between evaporator and condenser, the greater the energy used in the compressor for a given amount of cooling duty. Also, the greater this temperature difference is, the smaller will be the refrigerating capacity of the system.

Theoretical analysis of refrigeration cycles and full details of components may be found in numerous refrigeration textbooks (Gosney 1982, ASHRAE hand-books, Alders 1987) and would be out of place in the present publication. Nevertheless, a basic appreciation of the principles outlined above is useful for all users of refrigeration equipment.

4.3 Safety and quality issues

Food safety is concerned with freedom from pathogens and toxins – food should not make people ill, nor should it poison them. Food quality is the nutritional value and the perception of taste, texture and appearance of a foodstuff that is safe. Ideally, food safety is subject to legislative controls, whereas food quality is an issue best left to market forces.

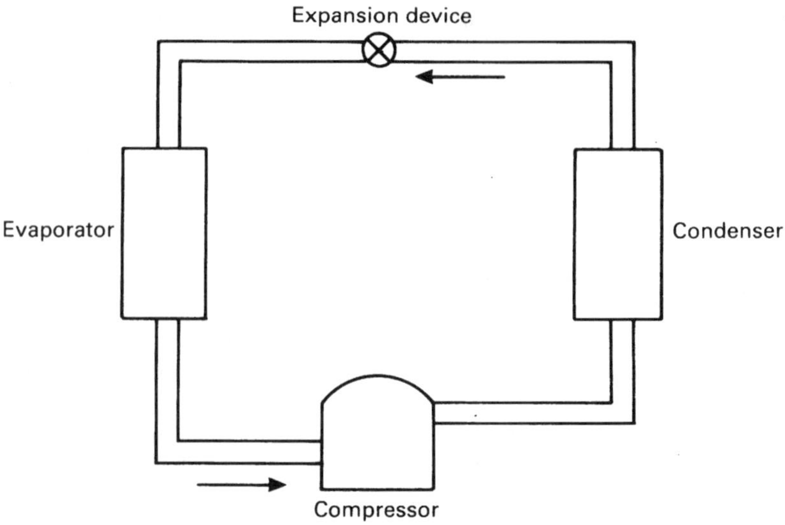

Fig. 4.2 The basic vapour compression refrigeration circuit.

For chilled foods, safety and quality may or may not be interlinked. For fresh fruits and vegetables, spoilage will make the food unpalatable but there is unlikely to be a health risk. For many prepared foods including cooked meats, growth of food poisoning pathogens will take place to an extent dependent on temperature and time, leading to injurious food which may look and taste satisfactory. For some dairy products, pathogen growth and off-flavours may develop together. In every case, the maintenance of safety and quality depends on maintaining as low a temperature as is possible without damaging the food. This has to be achieved throughout all stages of the chill chain shown in Fig. 4.1.

4.4 Refrigerant fluids and the environment

Until the early 1990s, the choice of refrigerant fluids for use within the closed vapour compression refrigeration cycle was a matter of little concern to equipment users. Unfortunately, it is now realised that those fluids developed over the years for efficiency and for safety have unexpected environmental side-effects when they are released into the atmosphere.

Ozone depletion and global warming are two quite separate environmental problems. The ozone layer, which protects the surface of the earth from excessive ultraviolet radiation, is damaged by the emission of stable chemicals containing chlorine or bromine. These chemicals include CFC (chlorofluorocarbon) and HCFC (hydrochlorofluorocarbon) refrigerants, which contribute to ozone depletion in the stratosphere, and also to atmospheric global warming. Global warming is a natural phenomenon, in which heat from the sun is trapped in the atmosphere by, particularly, carbon dioxide and water vapour. Fears of excessive global climate change are associated with high emissions of carbon dioxide (mainly from burning fossil fuels in power stations and elsewhere) and of other more powerful but much less abundant 'greenhouse' gases including HFCs (hydrofluorocarbons).

Under the auspices of the Montreal Protocol (Anon. 1987), the developed world ceased the production of ozone-depleting CFCs during the 1990s. This was made possible by the substitution of less environmentally harmful HCFCs. The HCFCs themselves are expected to be phased out by 2010–20, if not earlier, and in some applications there are no known effective substitutes at present available. In Europe, the use of CFCs in existing equipment will be banned, and the supply of new equipment using HCFCs will be prohibited. At the time of writing, dates for these limitations are still uncertain. These matters have two direct impacts on users of chilling equipment. Firstly, every change of technology costs money, and may in some cases result in an increase in running costs as well as re-equipment costs. Secondly, it may be necessary in the future to move from locally safe but globally harmful CFCs and HCFCs to globally safe but potentially locally hazardous substances such as ammonia and propane. These latter substances can be used safely, but there are added costs and added needs for proper training of equipment users. HFCs (hydrofluorocarbons) have been developed as alternatives. These do not deplete ozone and are widely available, but are being targeted by some environmentalists, as they are greenhouse gases within the Kyoto Protocol.

The purchaser of equipment needs to be aware of these matters, as he may otherwise obtain machinery that will have to be modified or even replaced long before its expected economic life is over. There could also be financial implications at the time of machinery disposal. The reduction of CFC and HCFC use in insulating foams in storage cabinets and stores has been well publicised, but at the time of writing the implications for refrigeration machinery are insufficiently widely appreciated.

A further related issue is the relation between global warming, energy use, and energy efficiency. New refrigerant fluids may be less efficient, but future environmental concerns may penalise excessive energy use. The equipment specifier is likely to face some difficult choices over the next few years. The user will face new responsibilities for minimising refrigerant leakage, ensuring efficient operation, and using only properly qualified maintenance staff. For a fuller discussion see (Heap 1998).

4.5 Chilled foods and refrigeration

The benefit of chilled storage is the extension of life of the foodstuff in good condition, by slowing down the rate of deterioration. Chilling, it must be emphasised, cannot improve the quality of a poor product; neither can it stop the processes of spoilage – it can only slow them down (see Chapters 7, 9, 10).

For the international land transport of chilled (and frozen) foods, the UNECE Treaty Agreement on the International Carriage of Perishable Foodstuffs and on the Special Equipment to be used for such Carriage (ATP) lays down various provisions. Foods are classified and maximum temperatures are stated in the ATP agreement as follows (UNECE 1998):

Red offal	+3°C
Butter	+6°C
Game	+4°C
Milk for immediate consumption	+4°C
Industrial milk	+6°C
Yoghurt, kefir, cream, fresh cheese	+4°C
Fish, molluscs, crustaceans	in melting ice (0°C)
Unstabilized meat products	+6°C
Meat (not offal)	+7°C
Poultry, rabbits	+4°C

This list excludes prepared vegetable foods with or without dressings and fresh fruit and vegetables.

There are two quite distinct applications of refrigeration to chilled foods. These are the chilling operation itself, in which the foodstuff is cooled from either an ambient temperature of maybe 30°C or a cooking temperature of over 70°C, and the chilled storage, at a closely controlled temperature of between −1.5°C and +15.0°C depending on the foodstuff. Chilling equipment and chilled storage equipment are quite different in their requirements and their design, and although some chilling equipment may be used for chilled storage, storage equipment is not designed to cool products, only to maintain temperature. Transport refrigeration for chilled food distribution is a special case of storage, and transport equipment should not be expected to provide rapid cooling.

4.6 Chilling

The rate at which heat can be extracted during chilling is dependent on many factors. The size and shape of the pack or container will affect the rate of heat transfer to the cooling air (or, in some cases, water). The temperature and speed of the air will also affect this. Within the pack, the weight, density, water content, specific heat capacity, thermal conductivity, latent heat content, and initial food temperature will each play a part.

In the case of unpackaged foods, the factors leading to rapid cooling also lead to rapid loss of moisture, so it may seem that slow cooling is better. Generally, this is not the case, as the extended cooling time is also an extended drying-out time. More rapid chilling is possible with thinner packs, with higher airspeeds, and with lower air temperatures. All these lead to higher operating costs, so equipment design has to be a compromise to give the best overall operating system. This means that a range of equipment is available for different applications, and an appropriate choice must be made, dependent on the planned operation.

4.7 Chilling equipment

4.7.1 Cooling systems

For most prepared foods, air blast cooling chambers or tunnels are used. Water immersion (hydrocooling) is used for some vegetables; and for fresh, leafy produce, vacuum coolers may be appropriate. For some fresh produce which has a relatively long storage life, cooling may be achieved using storage chambers, but frequently cooling rates will be enhanced by the use of special air circulation arrangements. Each of these systems will be considered in turn below.

4.7.2 Blast chillers

Blast chillers operate by passing cold air over foodstuffs at high speed. For cook-chill catering and similar operations, there are various guidelines, such as those issued by the DHSS in the UK. These recommend that equipment should be capable of chilling foods of up to 50 mm thickness from 70°C down to a core temperature of 3°C or below within 90 minutes. This requires an air speed of at least 4 metres per second and an air temperature of around −4°C.

Small, 'reach-in' chillers taking batches of up to 30 kg are available, for 'buffer' supplies in catering and for teaching and research. Larger models with capacities of up to a quarter of a tonne of foodstuffs are designed to accommodate wheel-in trolleys of trays. A typical single trolley unit might have a nominal capacity of 45 kg, typically accommodated on a trolley taking 20 trays of food. The evaporator and fans are located to the side of the interior chamber, and the compressor and condenser may be located either in the top of the unit or remotely, depending on whether the heat and noise emitted can be accommodated locally or not. Controls will permit the unit to be used as a chilled store at 0–3°C, or will operate the chilling cycle using any combination of air temperature, product probe temperature, or simple timer. At the end of the chilling cycle, a defrosting cycle to remove ice and frost from the evaporator is operated. Total power draw for a 45 kg unit is about 7 kW.

With a two-hour load/chill/defrost cycle, it is convenient to operate a four-batch shift, with the final batch being left in the cabinet as overnight storage. Optionally, temperature recorders may be fitted to monitor operation. For larger units, there may be doors at each side, so that chilled trolleys may be rolled through into a chilled food holding store at 0–3°C. It is also possible to obtain combination units with a frozen food storage cabinet alongside a chiller/chill storage unit. This allows a caterer to remove frozen food, cook and portion it, then chill and finally store the completed portions.

Other forms of blast chillers have been developed for the chilling of fresh poultry, which use a carbon dioxide tunnel in which CO_2 snow is used to provide cooling. Although this can achieve good results, there is considerable risk of surface freezing which would be unacceptable for many products. Liquid nitrogen is another 'total-loss' refrigerant that may be used to cool cabinets. As the temperature of liquid nitrogen at atmospheric pressure is −196ºC, careful control is necessary. An alternative may be synthetic liquid air (SLA) (Waldron and Pearce 1998), which overcomes the danger of asphyxiation that exists with other cryogens.

All 'total-loss' systems depend on the availability of compressed, liquefied gases, and it should be noted that the total energy use of such systems (including that needed for liquefaction) is much greater than that of equivalent mechanical refrigeration systems, so running costs may be high. In some applications, either reduced capital costs or increased chilling speed may make such systems attractive.

4.7.3 Hydrocoolers

The use of chilled water, either sprayed down through a chamber or in an immersion tank, provides very rapid cooling with no risk of freezing. It is normally only applicable to fresh fruits and vegetables that can withstand water immersion, and so is a little outside the scope of the general range of chilled foods, though it may be applied to vacuum packs of prepared foodstuffs. Water is normally recirculated in such systems, so great care is necessary to ensure continued cleanliness by regular flushing out, addition of fungicides, or whatever may be necessary for the particular product. It is of course possible to combine a degree of hydrocooling with normal cleaning operations for items such as root vegetables.

4.7.4 Vacuum coolers

Vacuum coolers are highly specialised and expensive pieces of equipment, well suited to the rapid cooling of pre-packaged leafy vegetables. They operate at low pressure with wet produce in a sealed chamber, under which conditions the cooling is mostly achieved by low temperature evaporation of moisture. The process is in batches, with cooling times of about 15–30 minutes, and typical equipment can accommodate several tonnes of produce, normally on pallets or trolleys.

4.7.5 Store cooling

For large volumes of live produce, particularly fresh fruits and vegetables, cooling may be achieved by placing cartoned or binned produce in a cool store and allowing the circulation of air in the store to provide all the cooling that is necessary. This is a slow process, taking several days and dependent on the store air circulation and the stacking of the produce. In many fruit stores, a combination of store extract fans, curtains, and planned stacking as shown in

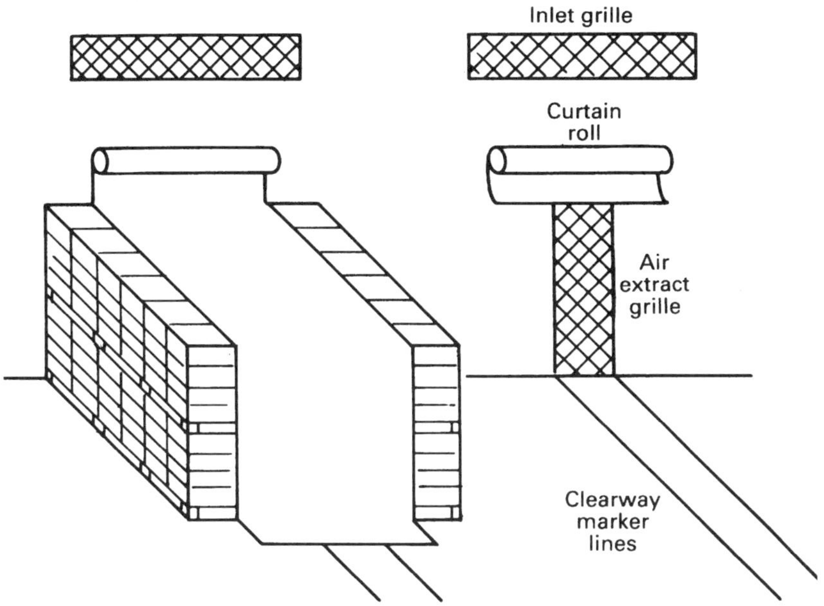

Fig. 4.3 Cooling tunnel arrangement within a store.

Fig. 4.3 is used to provide a simple form of something approaching a blast chiller. Air is extracted through a uniform thickness of cartons, with air entry other than through the cartons blocked by the curtain. If necessary, pallet bases are closed to air movement by the insertion of plastic foam or other convenient material.

4.7.6 Summary of equipment

These types of equipment may be summarised as follows:

Blast chillers:	preferred equipment for most applications, design must be matched to production requirements.
Hydrocoolers:	an excellent alternative in those cases where they may be used, especially fresh fruits and vegetables.
Vacuum coolers:	specialist equipment for limited application.
Store cooling:	with suitable stacking and curtains, a good method with wide application.

4.8 Chilled storage

Chilled storage equipment may be seen around the world in a wide range of sizes, each suited to the particular operation for which it is designed. At its smallest, it may be an absorption cycle refrigerator in a caravan or boat. There are larger domestic and commercial refrigerated storage cabinets, then small walk-in stores, and finally stores large enough to be served by forklift trucks handling pallets or bins, some of which can accommodate thousands of tonnes of produce. Some refrigerated fruit stores have the addition of atmosphere control in which low levels of oxygen and high levels of carbon dioxide can be maintained, which further enhances storage time of respiring fresh produce. This technology is not new, but is gaining wider acceptance in the search for better quality maintenance (Bishop 1996). Storage cabinets designed primarily for display are discussed in a separate section below, and the design of domestic refrigerators and slightly larger commercial storage cabinets is outside the scope of this book.

For most chilled food preparation and short-term storage areas, walk-in stores are appropriate. These can be constructed and designed as part of a total building, but more often are likely to be modular units sited within the overall structure. If pre-cooked chilled foods are to be stored, they should not be mixed with any other products requiring chilled storage. UK Electricity Association recommendations (Anon. 1989), suggested the following points should be taken into consideration in specifying a store for cook-chill products:

- first decide on the container type and handling method to be used, e.g. disposable containers in baskets to be stored on roll pallets, or whatever;
- consider frequency and size of consignments to be taken from store to calculate storage time required and size of store;
- thermal insulation should be adequate to maintain temperature at satisfactory running costs (with allowance for insulation deterioration with time and use);
- surface finishes (interior and exterior) must be durable and easily changed;
- twin compressor systems should be considered to give security in the case of breakdown;
- air curtains or secondary 'flap' doors should be specified for larger stores with frequent movements in and out;
- refrigeration evaporator coils should have capacity sufficient to allow for reduced efficiencies due to frosting or fouling;
- defrosting should be efficient, with adequate facilities for taking away defrost water;
- an alarm or safety system should be in place to prevent staff being accidentally locked in the store;
- continuous temperature monitoring and/or recording equipment should be installed, with some form of 'out-of-hours' monitoring in case of equipment failure.

Modular stores are available from small, self-contained units of about 2 m^3 volume up to 30 m^3 or more. The refrigeration condenser units may be mounted above or alongside the store, or may be remotely sited if there is insufficient ventilation to take away the heat rejected. If required, banks of multi-compartment stores can accommodate chilled, frozen or fresh produce. One particular arrangement for a modular store constructed within a building with a

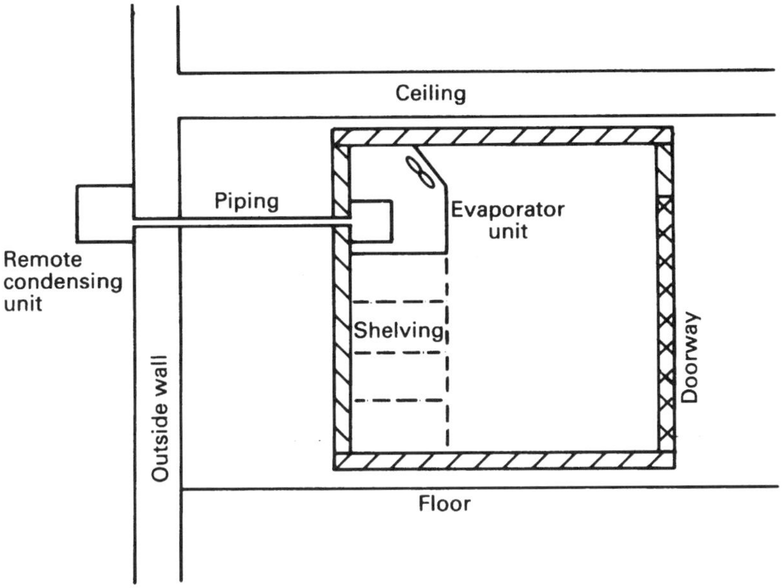

Fig. 4.4 Modular cold room for low ceiling height accommodation.

low ceiling height is illustrated in Fig. 4.4. Many other arrangements are possible, dependent on local requirements.

For larger stores, a basic requirement is the consideration of product movements and handling methods. This will dictate store height, whether or not a fixed racking system for pallet or carton storage is required, and how large the store should be. It is very common for large stores to be designed for specific applications, and for them to be used for something quite different after a few years, so flexibility of possible use should be designed in. Large stores should have provision for subdivision without major structural alteration.

Refrigeration evaporators, from which the cooled air is circulated, may be mounted at one end of the store, or at high level on one wall, or at high level within a central ridge. Whatever the arrangement, it is essential to consider the pattern of air movement for every likely store loading pattern, and ensure there is sufficient fan power to provide even temperatures. Most of the general considerations listed above for cook-chill stores also apply to larger stores.

Many stores have been constructed in which a steel structure is used to support an internal insulation barrier, but increasingly this design is being superseded by pre-formed panel constructions in which the interlocking panels are self-supporting. In suitable applications, the latter system offers appreciable cost savings.

In designing any large store, it is essential to consider protection of vehicles and products during loading and unloading operations. Protection against sun and rain is essential, and in more critical operations, temperature-controlled loading bays may be necessary.

4.9 Refrigerated transport

4.9.1 General requirements

Refrigerated transport of chilled foods must be seen as a total operation involving the movement of chilled goods from one fixed storage area to another. The operation involves a 'chain' of events, of which the actual movement of goods in a road vehicle, intermodal freight container, rail wagon, ship or aircraft is only a part. Temperature maintenance throughout the chain is essential for success, and the finest transport equipment cannot compensate for poor handling at loading, wrong packaging and stowage, or inadequate product cooling (Frith 1991). The term 'refrigerated transport' may itself be misleading, in that frequently it should be 'temperature-controlled transport'. In cold winter conditions, it may be necessary to heat chilled foods in order to prevent freezing damage, and for many fresh tropical fruits quite moderate temperatures can produce irreversible chilling damage. For example, bananas should not be allowed to cool below about 13°C. In areas of the world having severe winter conditions, heating requirements can be considerable. The distinction between 'refrigeration' and 'temperature control' is important for equipment users, who may not appreciate that a wrong temperature-setting on transport equipment may lead to foodstuffs being heated, whereas in many static stores it would only lead to lack of refrigeration.

In general, transport equipment is designed to maintain temperature, and not to provide cooling. Whilst foodstuffs can be cooled to some extent during transport, this is a slow and non-uniform method of attempting to cool, and it should not be depended upon. Pre-cooled foodstuffs should be loaded under temperature-controlled conditions wherever possible. In some cases, packaging designed for horizontal airflow coolers may not allow further cooling in transport, where vertical airflow is usual.

The range of transport refrigeration equipment is wide, as are the needs for transport. At its simplest, it could be an insulated box containing water ice. At its most complex, it might be an intermodal freight container with integral refrigeration machinery. This equipment is capable of maintaining either frozen or chilled goods at any selected temperature between −25°C and +30°C in ambient temperatures from −20°C to +50°C. Most frequently it will be a road vehicle designed either for local deliveries or for long distance or bulk distribution (Fig. 4.5).

The temperature control requirements for chilled foods are more difficult to achieve than those for frozen foods. Typically, it may be necessary to maintain cook-chill products between 0°C and 5°C, and for many products closer tolerances are required, whereas with frozen foods there will be an upper limit temperature, perhaps −18°C, but no lower limit. To ensure temperature uniformity in a load of chilled foodstuffs, relatively high rates of continuous air circulation and high levels of temperature control are necessary, and careful stowage within the vehicle may be needed to achieve this.

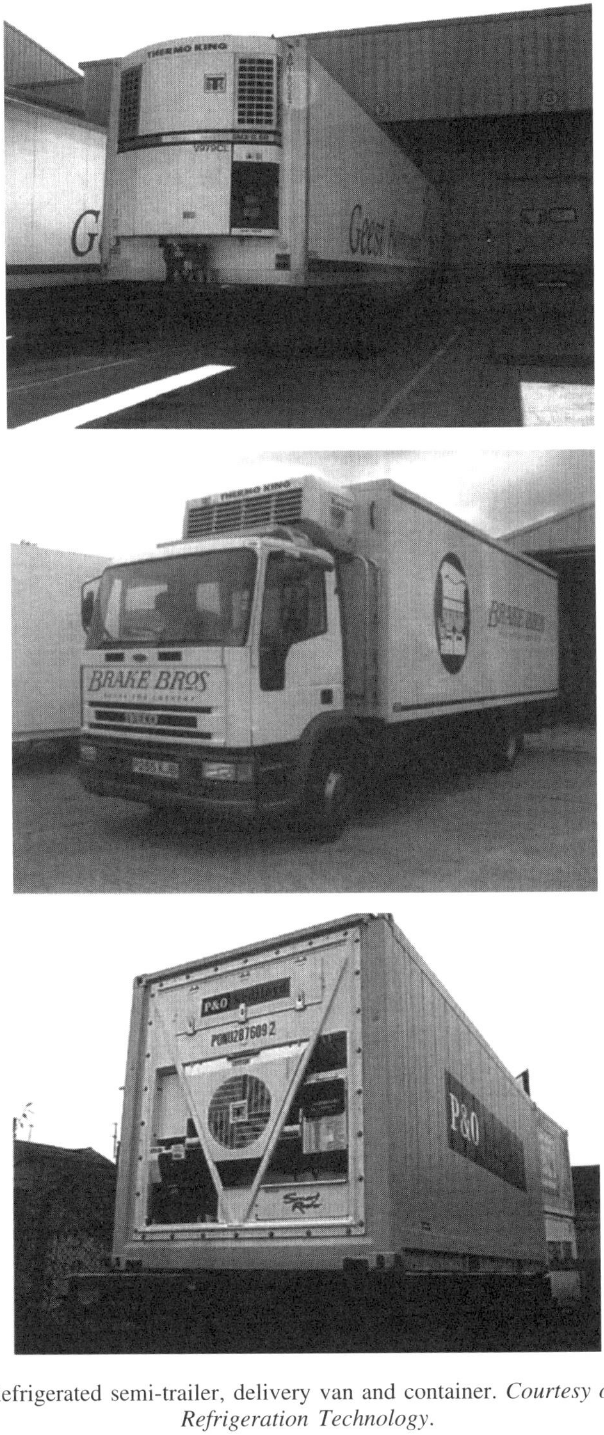

Fig. 4.5 Refrigerated semi-trailer, delivery van and container. *Courtesy of Cambridge Refrigeration Technology.*

4.9.2 Road vehicles

Refrigerated road vehicles fall into two basic categories. Firstly, there are large semi-trailers, with refrigeration units that can be run independently of the tractor unit. Secondly, there are rigid-bodied vehicles of various sizes, which may have independent refrigeration units, or may have units driven from the vehicle engine or axles, or may depend on eutectic storage media. Semi-trailers are used for long-distance or bulk movements, generally with only one or a few destinations. Journey times may vary from two hours for supermarket distribution to several days for fresh produce transportation. A typical arrangement is shown in Fig. 4.6. Whilst most such vehicles use diesel engine drives with optional electric alternatives, some use total-loss refrigerant tanks (liquid nitrogen or carbon dioxide) to reduce both capital cost and noise levels in sensitive areas.

In most developed countries, semi-trailers are designed for use in ambient temperatures of 30°C or above, with thermal insulation with overall value of 0.7 W/m^2K or better. If frozen goods may also be carried, insulation of less than 0.4 W/m^2K would be used. Increasingly, multi-purpose multi-compartment vehicles are being produced, capable of carrying frozen, chilled and fresh produce simultaneously in different compartments.

Rigid-bodied vehicles vary from large vehicles very similar in use to the semi-trailers to small delivery vehicles for multiple deliveries of chilled foods to corner shops. Refrigeration units may be driven by diesel or electric motors, or by hydraulic drive from the vehicle chassis, or may be based on either total-loss or eutectic systems. The latter two are more often used for frozen food transport, being relatively difficult to control at chilled temperatures. Delivery vehicles may require walk-in access for order selection from fixed shelving, and may have to operate with large numbers of daily door openings. Legislation such as the UK food safety regulations (Anon. 1995a) has provided the impetus for much development of vehicle design to meet increasingly stringent temperature requirements.

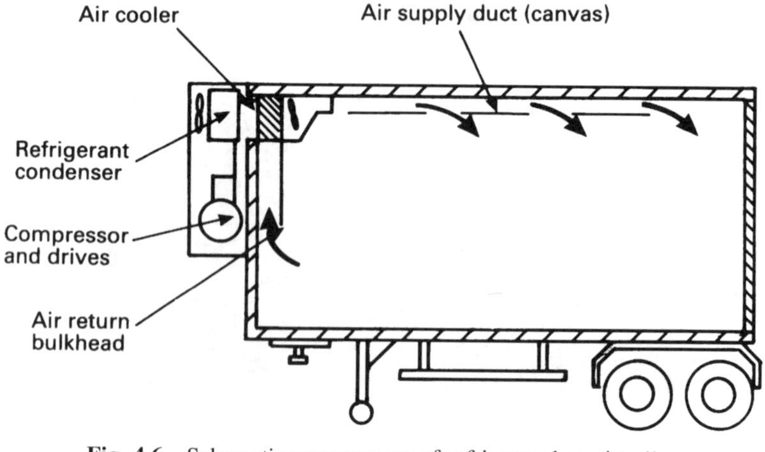

Fig. 4.6 Schematic arrangement of refrigerated semi-trailer.

Commercial requirements have led to the development of multi-compartment vehicles with independent temperature control in each compartment. These vehicles are suited to distribution from stores to retailers, as they can move frozen, chilled and ambient temperature goods simultaneously. Such vehicles may have separate cooling coils in each compartment, or may depend on fans to transfer a limited amount of cold air from the coldest to a warmer compartment. Generally, refrigerated vehicles control the temperature of the air supplied to the cargo space, and monitor the temperature of the air returning to the refrigeration units with an external gauge or display or, increasingly, a display within the vehicle cab. Some older vehicles primarily designed for frozen foods may only control the temperature of air returning to the refrigeration unit, with the risk of freezing chilled foods that are loaded too warm.

4.9.3 Intermodal freight containers

Intermodal freight containers ('ISO' containers) with integral refrigeration machinery are widely used for the long-distance transport of fresh fruit and vegetables and chilled meat. Because of journey times of up to six weeks, they have highly developed refrigeration and control systems, and they are capable of operating over a wide range of conditions. They are normally used only for point-to-point international transport involving a substantial sea journey (Frith 1991, Heap 1989, Anon. 1995b), though there are occasions on which the lease of such a container can be a very convenient way of providing a temporary chilled storage facility. Standard sizes are either 20 foot or 40 foot length with capacities of about 28 or 60 m^3, and refrigeration units are electrically driven from three-phase supplies from either mains or a diesel generator. They differ from most road vehicles in that air is supplied to the load space from a 'T'-section floor grating (Fig. 4.7).

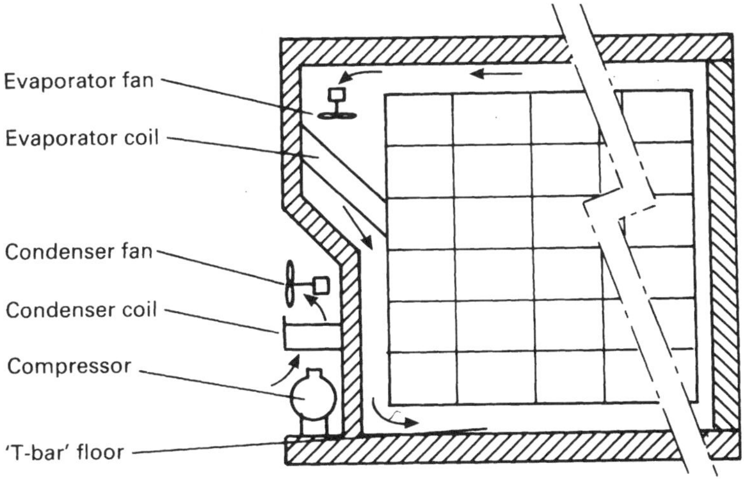

Fig. 4.7 Schematic arrangement of refrigerated container.

4.9.4 Summary of equipment

Note that the requirement is for temperature control, which may include heating.

Road transport:	large semi-trailers for international and national distribution, large rigid bodied vehicles mainly for distribution to retailers, smaller vehicles for local distribution.
Containers:	ISO integral refrigerated containers for long-distance movements primarily by sea.

4.10 Refrigerated display cabinets

The refrigerated display cabinets used in retail premises fall into two distinct groups. Most are vertical multi-deck cabinets for the display and self-service retailing of packaged chilled foods, fresh meat and poultry. Examples of these are shown in Fig. 4.8. There are also the delicatessen or 'serve over' cabinets for foods which are normally not packaged but cut and served. Multi-deck cabinets have a refrigeration evaporator in the base, and this may be supplied either from a self-contained condensing unit or, in larger installations, be piped to a central store cabinet refrigeration system. The evaporator coil is mounted in the lower part of the cabinet behind or under the display area, and fans blow cooled air both from behind the shelves in a forward direction and also downwards in an air 'curtain' from the top front of the cabinet. Warmed air is returned through a grille at the base of the cabinet.

Modern multi-deck cabinets may be designed to maintain food temperatures at 5°C or below, but some older cabinets will frequently have difficulty achieving temperatures below 10°C. Food temperatures are not just a function of cabinet design: they also depend on method of use. Very tight or untidy cabinet loading can prevent proper air circulation, as can indiscriminate placing of large price or advertising tickets. High store temperatures or excessive radiant heating from lights can lead to warm foodstuffs. Good housekeeping allied to the use of some type of night covers when the store is closed will give the best results. Cabinets are designed to maintain temperatures, and should not be loaded with foodstuffs which are warm. In some countries, cabinets with doors have largely superseded the use of open-fronted multi-deck cabinets, to provide more positive refrigeration at all times. Such cabinets have severe disadvantages to the retailer, both in loading time and in customer resistance. Some open-fronted cabinets also incorporate shelves for display of non-chilled goods related to the chilled products on display.

Serve-over display cabinets have food displayed on a base over which cold air flows, and normally have a glass front from behind which the food is served. Air from a rear evaporator may be gravity-fed or fan-assisted, but, as much of the food in these cabinets is not wrapped, excessive air speeds must be avoided to prevent dehydration and weight loss. For the same reason, these cabinets are usually used for display only whilst sales are in progress, and other storage

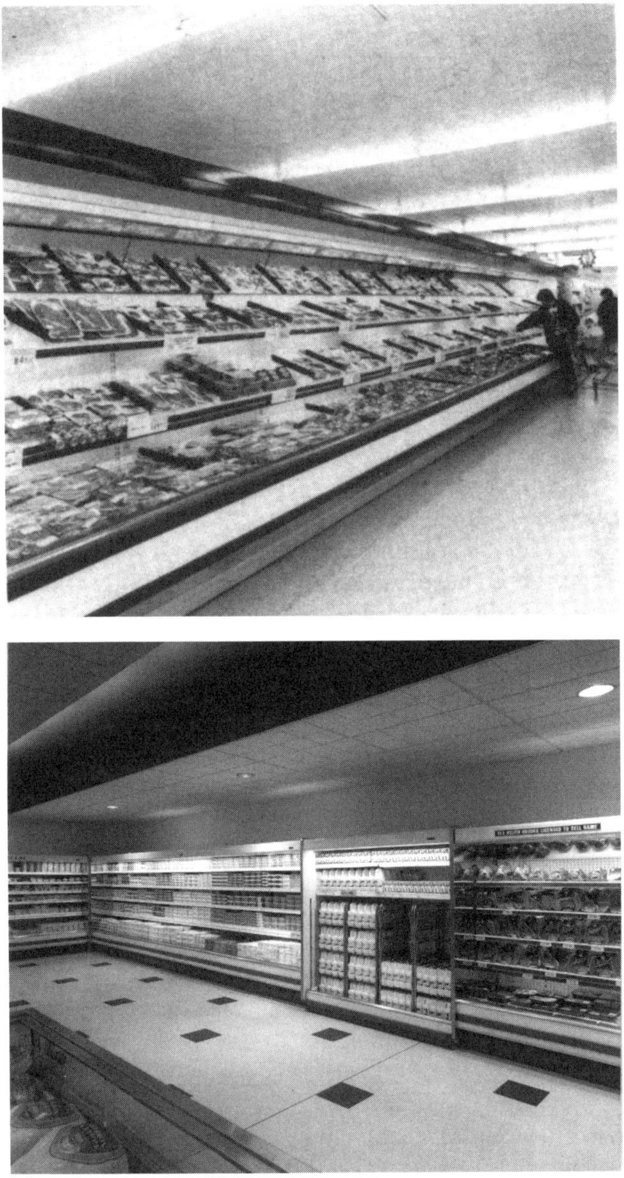

Fig. 4.8 Multi-deck display cabinets. (a) Freestanding module. (b) Extended display area. *Courtesy of George Barker and Company (Leeds) Ltd.*

cabinets are used to store food overnight. A diagrammatic representation of a serve-over cabinet is shown in Fig. 4.9. A variation on the serve-over cabinet is the chilled ingredient display and store used in some catering establishments. This type of cabinet stores refrigerated ingredients below the counter section, and cold air blown across the underside of the display pans keeps these cool,

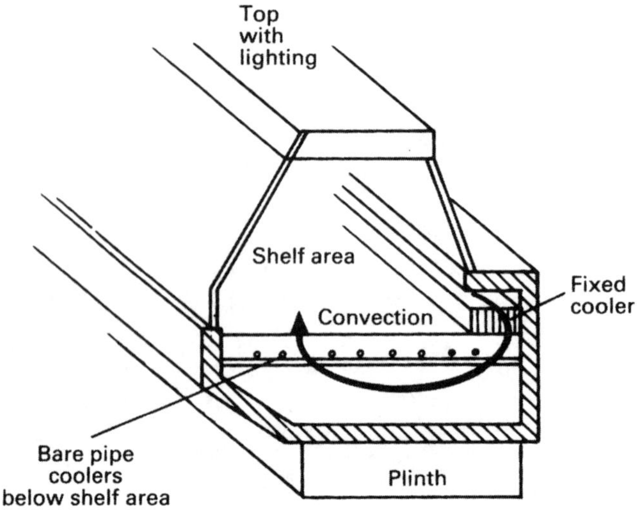

Fig. 4.9 Serve-over cabinet.

aided by a curtain of cool air blown across the surface of the pans. This is just one example of the way in which display equipment can be designed to meet specific requirements.

In selecting any cabinet, the method of use, standard of temperature control and cost will be major factors. Ease of maintenance and cost of running are also important. Choice of refrigerant must be considered in the light of consumers' views of the environmental issues discussed above. Cabinets using HFCs are favoured in many countries at the time of writing, with CFCs and HCFCs regarded as obsolete for new equipment. Some other countries favour indirect systems in which cabinets are cooled by cold liquids such as brine or glycol solutions, cooled by central plant which in large stores could operate on ammonia. Purchasers need to be aware of both current and pending legislation in their country of operation, and future trends are not always clear to see.

Summary.
Manufacturers produce a wide range of cabinets to meet all retail requirements, including combinations of goods with different temperature needs.

4.11 Regulations and legislation

There are various regulations and laws relating to the refrigeration of chilled foods, and this topic is covered in more detail in Chapter 16.

In some countries, there are general food safety requirements, and sometimes there are, as in the UK, specific requirements relating to temperatures of certain foodstuffs (Anon. 1995a). It must be appreciated that such requirements are dependent both on having suitable refrigeration equipment and on using it appropriately, and this has considerable implications for staff training for all those involved in preparation, storage, distribution and retailing of chilled foods. For international transport, the 'ATP' regulations (UNECE 1998) apply for journeys between countries which are party to that particular UN agreement. These regulations specify insulation and refrigeration requirements for vehicles, together with previously listed temperature requirements for various particular classes of foods. Some countries, e.g. France, Spain, Portugal and Italy, have national regulations related to the ATP requirements, but these internal regulations are not part of ATP itself. The ways in which the EC's single market development will lead to either harmonisation or mutual recognition of the various requirements have yet to be resolved.

In some cases, refrigeration of chilled and of frozen foods are closely linked, and it is necessary to distinguish between the quite distinct legislative requirements for the two groups. Requirements for common handling equipment may impose additional regulations on chilled foodstuffs refrigeration systems if they are to be suitable for use with frozen goods, for example in the case of multi-temperature cold rooms or distribution vehicles.

4.12 Sources of further information

The Institute of Refrigeration, Kelvin House, 76 Mill Lane, Carshalton, Surrey SM5 2JR, UK. Web site: www.ior.org.uk

British Refrigeration Association, FETA, Henley Road, Medmenham, Marlow, Bucks SL7 2ER, UK. Web site: www.feta.co.uk

International Institute of Refrigeration, 177 bd Malesherbes, F 75017, Paris, France. Web site: www.iifiir.org

Cambridge Refrigeration Technology, 140 Newmarket Road, Cambridge CB5 8HE, UK. Web site: www.crtech.co.uk

4.13 References

ALDERS A W C, (1987) *Marine refrigeration manual*, RMCA, The Hague.

ANON, (1987) UNEP, *Montreal protocol on substances that deplete the ozone layer*.

ANON, (1989) *Planning for cook-chill*, The Electricity Council, London, 5th edn.

ANON, (1995a) *Food safety (general hygiene) regulations*, HMSO S.I. 1763.

ANON, (1995b) *Guide to refrigerated transport*, International Institute of Refrigeration, Paris.

ANON, (1997) *Guidelines for good hygienic practice in the manufacture of chilled foods*, Chilled Food Association.

ASHRAE HANDBOOKS, ASHRAE, Atlanta, updated annually on a 4-year cycle for *Fundamentals, Refrigeration, Systems, Equipment*.

BISHOP D, (1996) *Controlled atmosphere storage. A practical guide.* David Bishop Design Consultants, Heathfield.

CRITCHELL J T and RAYMOND J, (1969) *A history of the frozen meat trade*, Dawsons, London, (original edition 1912).

FRITH J, (1991) *The transport of perishable foodstuffs*, SRCRA Cambridge.

GOSNEY W B, (1982) *Principles of refrigeration*, Cambridge University Press.

HEAP R D, (1989) Design and performance of insulated and refrigerated ISO intermodal containers, *Int. J. Refrig*., Vol. 12 (May), 137–45.

HEAP R D, (1998) *Global warming – considerations for the air-conditioning and refrigeration industry – Dreosti Memorial Lecture.* SAIRAC, Cape Town.

THÉVENOT R, (trans. Fidler, J. C.) (1979) *A history of refrigeration throughout the world*, International Institute of Refrigeration, Paris.

UNECE, (1998) *Consolidated text of the agreement on the international carriage of perishable foodstuffs and on the special equipment to be used for such carriage (ATP)*, UNECE Inland Transport Committee, Geneva.

WALDRON S N and PEARCE I A, (1998) *The application of synthetic liquid cryogens in the distribution, storage and production of food.* IIR Conference, *Refrigerated transport, storage and retail display*, Cambridge, 1998. IIR, Paris.

5

Temperature monitoring and measurement

M. L. Woolfe, Food Standards Agency, London

5.1 Introduction

The practice of measuring and keeping records of temperatures is not new to the food industry, and has been undertaken by certain sectors, e.g. canning, for many years. However, its widespread application in the refrigerated food sector, other than fitting temperature measurement equipment to chill stores, is relatively recent. The prime factor focusing attention on temperature monitoring was the concern about food poisoning and the introduction of new legislation covering temperature control of chilled foods, where temperature abuse and likely growth of pathogens could be a problem. However, national changes were overtaken by developments in the European Community where a harmonised hygiene directive was developed and agreed. This coupled with vertical hygiene directives on animal based foods laid more emphasis on risk management. Thus the practice and use of temperature monitoring has matured quite rapidly over the last ten years and become more integrated into quality and safety management systems.

5.1.1 Changes in legislation

Both the Food Hygiene (Amendment) Regulations 1990[1] and the Food Safety Act 1990[2] produced a significant change across the chill chain. The Food Hygiene (Amendment) Regulations 1990[1] introduced temperature controls for certain types of chilled foods which applied at all stages of the 'chill chain'. Further minor amendments were made in 1991.[3] Up to this date very few end users of refrigerated systems practised regular temperature monitoring, however,

The views expressed in this Chapter are those of the author and should not be regarded as a statement of official Government policy.

when they began to apply it they realised the concurrent benefits of process and quality control it brought.

The Food Safety Act 1990 gave Ministers additional powers to legislate in many new areas. One significant change in the 1990 Act is found in Section 21. This describes the conditions under which a defence to charges brought under the Act can be offered. The 'warranty' defence under the 1984 Act has been substituted by a 'due diligence defence'. In order to show 'due diligence', companies have to demonstrate that they took all 'reasonable precautions' and exercised 'all due diligence' in their operations. Many companies have moved to better systems of control and inspection on the basis of parallel case law of 'due diligence' in other legislation.

A harmonised horizontal Hygiene Directive 93/43/EEC[4] was agreed and implemented in the UK in 1995 (The Food Safety (General Food Hygiene) Regulations).[5] These laid emphasis on a risk or HACCP (Regulation 4(3)) approach to hygiene rather than giving prescriptive or detailed rules about hygienic requirements and practices. There is a general requirement that temporary premises and equipment for transport must be capable at maintaining the food at appropriate temperatures and where necessary their design must allow those temperatures to be monitored.

The requirements of the Directive relating to temperature control are enacted in the Food Safety (Temperature Control) Regulations 1995.[6] In addition, the Government was able to take advantage of the computer model developed over five years (MAFF Micromodel) to look at predictions of microbiological growth under the different temperature conditions to ensure safe food, and as a result the earlier temperature control regulations were simplified.

Controls already existed in European Community legislation in the trade of animal-based foods, e.g. meat, meat products, poultrymeat, etc. In order to implement the Single Market after January 1993 some ten vertical directives were agreed dealing with the hygienic production of animal products from fresh meat to bivalve molluscs. Some of these were new and others re-negotiated from intra-community trade directives. All of these Directives have some temperature control requirements. Work is in progress to consolidate all the vertical hygiene directives and the horizontal Hygiene Directive into one simplified directive. The only mandatory requirement for temperature monitoring and keeping records is based on a European Community measure which requires monitoring equipment to be fitted to cold stores and vehicles which store or transport quick-frozen foods (Commission Directive 92/1/EC).[7] This same requirement has also been adopted in the UN/ECE agreement which facilitates cross-frontier traffic in perishable foods (ATP[8]) in order to harmonise EC requirements for third-country vehicles.

5.1.2 Risk and quality management systems

Once companies began to investigate and implement temperature monitoring systems, it soon became evident that there are derived benefits to offset the incurred capital costs and effort. Better control of temperature underpins both the safety and quality of the food product, and can bring economic benefits of energy efficiency.

The changes in legislation also necessitated the setting up of appropriate systems to ensure the safety of foods. All food businesses have the responsibility of identifying critical steps in their own processes. The approach adopted nationally and internationally is HACCP (Hazard Analysis and Critical Control Point). This identifies the risks and the critical control points in the process to control the risks. The important issue is that a HACCP plan is unique to a particular product and process, and it should be continually assessed. Help in implementing HACCP has been given by provision in the Hygiene Directive which encourages sectorial guidelines, and to date six guides[9] have been published. Temperature monitoring may or may not be part of the HACCP plan depending on the product and process involved. There is no specific requirement to keep records of temperature checks, but records may be helpful to show that legal requirements have been met. More importantly temperature monitoring is put into context with all the other control points, and integrated into the overall HACCP system.

It is clear that HACCP is rarely implemented in isolation but combined with quality systems to ensure that a production unit manufactures food that is safe and is of consistent quality for the consumer. There are many systems of quality assurance, and the most widely used ones are based either on ISO 9000 or on TQM (total quality management). ISO 9000[10] series has two main standards (ISO 9001 and 9002) and various guidelines, and companies achieve accreditation when they have implemented the standards. TQM is more a cultural approach involving all members of an organisation in achieving consistent quality and consumer satisfaction and also has the concept of continual improvement.

5.1.3 Improvement in technology

The ability to produce microelectronics relatively cheaply has enabled the manufacture of relatively small devices that store large amounts of data. These are now routinely integrated with computerised management systems. The last few years have seen an enormous advance in computer and communications technology. Now satellite tracking systems can follow a vehicle's position and give total information about the refrigeration and engine systems to its depot. Retail display cases can also have integrated temperature and humidity control to ensure full shelf-life of non-pre-packed foods. Thus where temperature measurement is part of the integrated safety and quality, new technology lends itself to the storage and processing of the data.

5.2 Importance of temperature monitoring

Temperature control requirements for England and Wales apply to food which is likely to support the growth of micro-organisms or the formation of toxins.

These foods need to be kept at 8°C or below. However, this requirement has to be implemented in combination with the others laid down in the general hygiene regulations (The Food Safety (General Food Hygiene) Regulations 1995[5]).

Obviously, if pathogenic organisms can be prevented from entering the food then temperature control is necessary only for extending the shelf-life of the product. However, this is rarely the case and the approach taken by HACCP is to identify at each stage of the preparation of a food where the hazards occur and how they can be controlled. Reducing the temperature does not kill micro-organisms, but it retards their growth. Hence keeping raw materials, intermediate and finished products at chill temperatures will play its part in ensuring the food is safe. The other important areas are proper hygiene training of operatives, prevention of physical contaminants, suitable fittings and equipment, and good cleaning regimes and pest control.

Refrigeration equipment is built to function for long periods without attention, however there are many events apart from breakdown which can affect temperature control. The defrost cycles need attention to ensure they are at the correct frequency, and loading of food into refrigerated systems is often crucial to its operation and proper air flow. Air temperature monitoring can indicate whether refrigerated equipment is functioning correctly and is being operated correctly, even though it may be more difficult to extrapolate food temperatures. In some circumstances air temperature monitoring is not possible and product temperature or product simulant temperature is required.

5.3 Principles of temperature monitoring

5.3.1 Choice of system

There are an enormous number of different temperature monitoring systems available commercially, from a simple thermometer to a fully computerised system linked to a local refrigeration system or even central control system. The choice of system will depend on exactly the amount of detail the operator requires and the cost at which this information is provided. If the monitoring system is to provide detailed information on the operation of a system linked with other reactive management systems, then obviously a more elaborate and complex system is required. This may include a large number of sensors to enable a very complete picture of the temperature distribution within a refrigerated system to be obtained. It may also include other information such as defrost cycles, compressor and expansion valve pressures, door openings, and energy consumption, and may be linked to an alarm system (and even telephone) stock keeping and batch codes of product. On the other hand, if monitoring is being carried out only to ensure that food is being kept within certain temperatures as a critical control point, then the amount of information which is collected may be reduced.

Very little formal advice has been given in the previous literature on temperature monitoring (IIR,[11] SRCRA,[12] BRA,[13] RFIC[14]). Guidelines published by the IFST[15] give details about air temperature monitoring, and these were further amplified in the Department of Health's Guidelines.[16] These Guidelines have been superseded by Industry Codes of Practice.[9] Practical advice on temperature monitoring has been included as an advisory Annex in some of the Codes (Retail and Catering Codes in particular), but these do not form part of the Codes.

5.3.2 Which temperature to monitor?

When designing a monitoring system, there are certain considerations in the choice of temperatures to be measured in the refrigerated system. These are:

- The choice of whether to monitor air temperatures, product temperatures or simulated product temperatures will depend on the individual system and the way it operates.
- The sensors should preferably be fixed in a position where they will not be damaged during commercial activity. If manual readings are used, these should be taken from accessible positions.
- The temperatures chosen should be representative of the refrigerated system and give a picture of its functioning, and therefore be linked indirectly with the product temperature.

5.3.3 Air temperature monitoring

In terms of regulatory compliance and as part of HACCP, the temperature of the food should be monitored. However, the storage or holding times of chilled food are relatively short, making product temperature monitoring difficult without disruption to normal commercial activity and requiring the intervention of trained operators. It is easier to fix sensors separate from food loads, which are connected to read-out systems, where temperatures can be recorded automatically or manually.

Most refrigerated systems function by circulating cold air over the system's evaporator, and then passing this cold air over the food load to remove heat from the food. Movement of air is by mechanical fans or in some cases gravity, which relies on density being greater for cold air than for warm air. In the case of mechanical circulation, the air returns to the evaporator after passing over the food, making the returning air the same temperature or warmer than the food it is cooling. Localised heating effects from lighting or other effects may give rise to 'hot spots' or uneven temperature distribution, and make a small part of the food load warmer than the return air. In general, the relationship between air temperature and product temperature is best established by examining the difference between the cold air leaving the evaporator and warmer air returning to the evaporator. This gives a measure of the performance of the refrigerated system and its effectiveness in keeping the food cold (BRA[13]). This differential is also used as the basis of air temperature monitoring. However, in order to relate the air temperatures to product temperature, it is necessary to carry out a load test. The load test involves examining the differential of air temperatures and comparing it to product temperature over a sufficient period of time to ensure the system is working under normal conditions.

With closed systems such as chill stores and vehicles, where the only perturbation derives from defrost cycles, door opening and changing loads, determination of the relationship between air and product temperature is simpler. The warmest locations in the system have to be determined, and product temperatures followed over a period of time in order to relate them to air temperatures.

With open systems such as display cabinets, their operation is more sensitive to environmental conditions and location. Room temperature and humidity variations, perturbation of the air curtain by draughts or customer movement can change the temperature distribution. Under these circumstances, load testing can be more difficult.

Cabinet manufacturers perform a load test to check cabinet's performance (BS EN 441-5: 1996[17]), using a set loading of standardised blocks of a gel (Tylose) (BS EN 441-4: 1995[18]) under controlled environmental conditions of temperature with a constant air flow across the front of the cabinet. Whether the manufacturer's load test will deviate from that *in situ* will depend on how close the conditions and load follow the actual working operation of the cabinet. The siting and environmental effects of draughts and lighting should be checked with a range of foods on display.

5.3.4 Alternatives to air temperature monitoring

There are some circumstances where air temperature monitoring is not appropriate or needs modification. In closed cabinet systems, such as chill storage cabinets using gravity cooling from an ice-box or backplate, the air temperature inside requires significant time to recover after door openings.[19] Thus, periodic readings of air temperatures would have little meaning and bear no relationship to the temperatures of the food being stored. In this case it would be better to monitor either a food sample or a 'simulated' food sample. The thermal mass of the sample would make it less sensitive to rapid air temperature changes. Also it is possible to match the 'food simulant' to have a similar cooling factor or similar thermal diffusivity to the food being monitored.[20] The use of such monitoring would be essential for example where cooling is by conduction such as cold plate (dole plate) serving units in catering, or where air flows are low velocity (gravity-fed serve-over cabinets).

Even where the system is cooled by forced air but the variations in air temperatures are large, e.g. small delivery vehicles and cabinet refrigerators, air temperature monitoring is still difficult to interpret. By increasing the response time or 'damping' the sensor or measuring system, the trends in air temperature can be followed, whilst removing the short-term variations. 'Damping' can be achieved physically by increasing the thermal mass of the sensor or electronically by alteration of the read-out circuitry.

5.4 Temperature monitoring in practice

5.4.1 Chill storage

Walk-in chill stores

Walk-in stores consist of an insulated store chamber cooled by one or more fan-assisted air cooling units, depending on their size. The position of cooling units around the chamber varies, but is usually at ceiling level (Fig. 5.1). Air circulation should be designed to give proper distribution throughout the chamber, and to eliminate any 'hot spots' or stratification of air layers. In nearly all cases, air temperature recovery after door openings or defrost is rapid, permitting air temperature to be used as the most convenient means of monitoring. Retention of cold air can be further improved with the use of strip plastic curtains, or an air curtain above the door, minimising the ingress of warm air on door openings.

The number of sensors to be used to monitor air temperatures in a chill store will depend on its size and the number of cooling units. Table 5.1 gives an indication of the minimum number of sensors related to volume of the store, with stores less than 500 m^3 being able to monitor air temperatures using one sensor. The positioning of the sensor is such that it gives an indication of the warmest air temperature and hence the warmest food in the store. This warmest location depends on the design of the store, especially the position of the air cooling unit in the store.

Figure 5.2 gives an example of air temperatures during the 24-hour operation of a large chill store. The graph shows temperature variations during peak activities of movement of chilled foods in the afternoon and evening compared

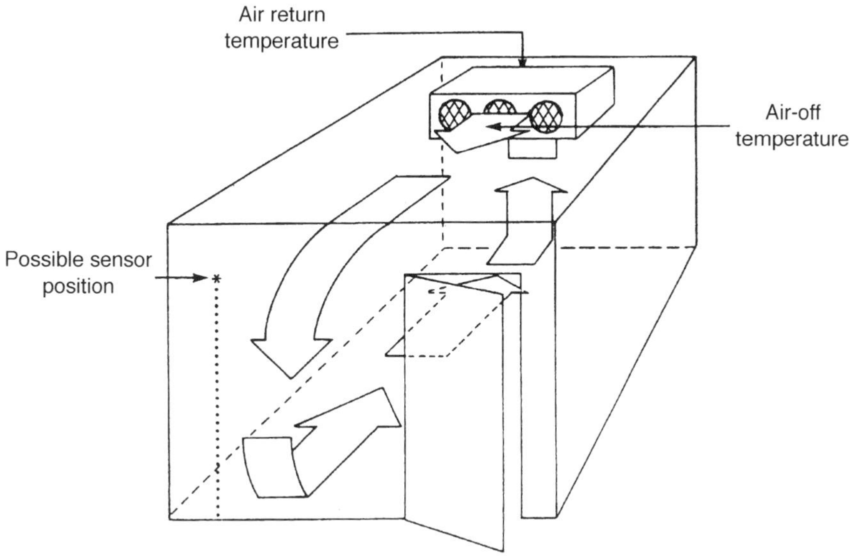

Fig. 5.1 Air circulation in a chill store.

Table 5.1 Number of sensors recommended in chill stores

Chambers volume above (m^3)	Number of sensors
500	2
5 000	3
20 000	4
50 000	5
85 000	6

to quieter loading activity in the morning. Differences between wall sensors and air return temperatures are very small in this case, and can be affected by their positioning in the store. For chill stores less than 500m^3, the single sensor could be placed in the air return of the cooling unit. In a closed system such as a store with adequate air distribution, the temperature reading of the air return approximates to the mean temperature of the food load. If there is not good air distribution, then it may be better to put the one sensor in a position more representative of the warmest air temperature. This may be located at the following positions:

- the maximum height of the food load, furthest away from the cooling unit
- at approximately two-thirds the height of the chamber, away from the door and the direct path of the cooling unit
- two metres above floor level, directly opposite the cooler unit.

If the cooling unit is placed above the door, the negative pressure produced by the fan can increase the amount of air drawn into the chamber during door openings. Thus, air return temperature monitoring is not often appropriate in this case. For larger stores, different sensors can be used to indicate the temperatures in different parts of the store. In addition, placing extra sensors in the air outlet and air intakes of one or more of the cooling units gives further information on the performance of the refrigeration system.

Cabinet refrigerators

Cabinet refrigerators are free-standing, small-sized units with single or double doors. They can be cooled by fan-assisted cold air or by gravity-circulated air from an integral icebox or backplate (Figs. 5.3, (a), (b) and (c)). As indicated earlier, air temperature monitoring is not as appropriate to these types of refrigerated systems as it is to walk-in chill stores.

Fan-assisted refrigerators will recover relatively quickly after door openings, but a large number of door openings, especially at most active periods of use, will make any temperature readings difficult to interpret. Air temperature monitoring can be more meaningful if a 'damped' sensor, with a response time of around 15 minutes, in the air return position is used (Fig. 5.3(a)). Damping can be achieved by using a metal or plastic sheath over the sensor or suspending the sensor in water, oil or glycerol. Figure 5.4 shows the effect of 'damping'

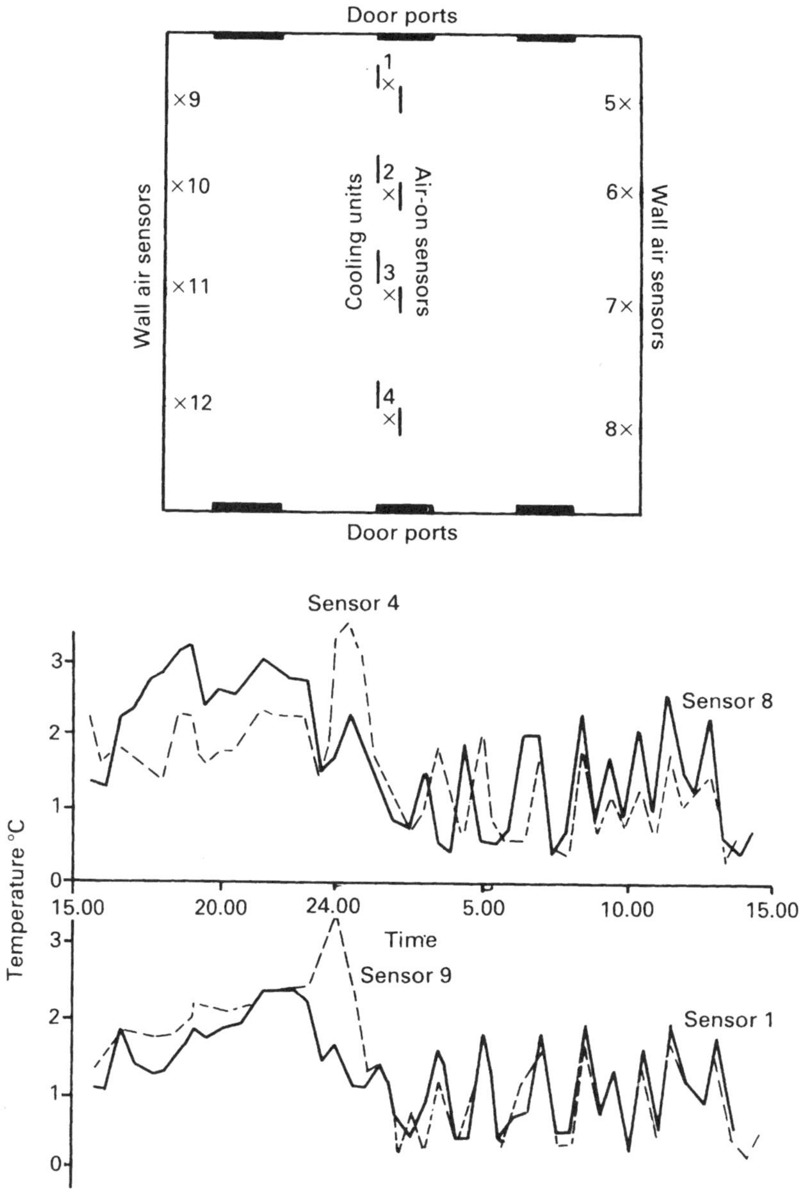

Fig. 5.2 Air temperature monitoring record of large chill store (40 000 m^3).

when the sensor is set at the centre of a plastic tub of water, and readings are compared to air temperatures after door openings.

Since cabinets cooled by a backplate or icebox have weak air circulation and long recovery times after door openings, it is more appropriate to monitor their temperatures using food temperatures or, even better, a simulated food

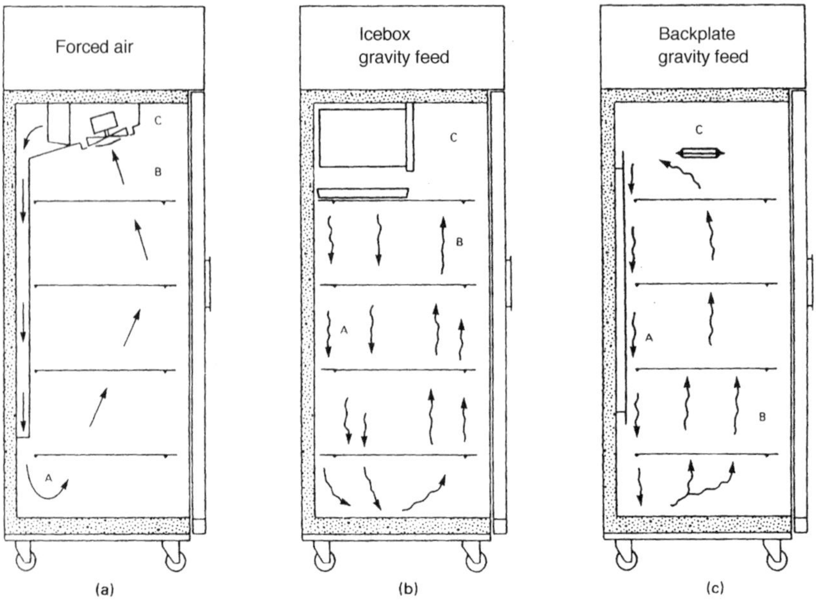

A: Air off. B: Air return (air-on). C: Load limit or warmest point.

Fig. 5.3 Cabinet refrigerators. (a) Forced air refrigerator. (b) Icebox refrigerator. (c) Backplate refrigerator.

temperature. As foods are microbiologically unstable, food temperature monitoring would require using different foods each day, and might lead to wastage. Permanent positioning of a sensor requires a stable food simulant. It is important when choosing a food simulant that it behaves similarly to the food being monitored, and it is robust to different working conditions. It is recommended to determine the cooling factor of the specific package or piece of food and match this with a particular food simulant, or match the thermal diffusivity of the food to that of the simulant.[20] Values for cooling factors of different foods, package sizes are also published,[20] as well as thermal diffusivities for a range of plastic materials. Regular checks should be made with a food simulant system to ensure that the sensor embedded in it is accurate and functioning properly, and that the simulant is performing as it should.

5.4.2 Chilled transport

Distribution of chilled foods is carried out in many different types of vehicle ranging from large 40-foot heavy goods vehicles with independent cooling units, to light goods vehicles relying on insulated containers to maintain temperature of pre-chilled foods. Pre-chilling to the correct temperature is essential given that most refrigeration units are designed to maintain temperature not cool the load down.

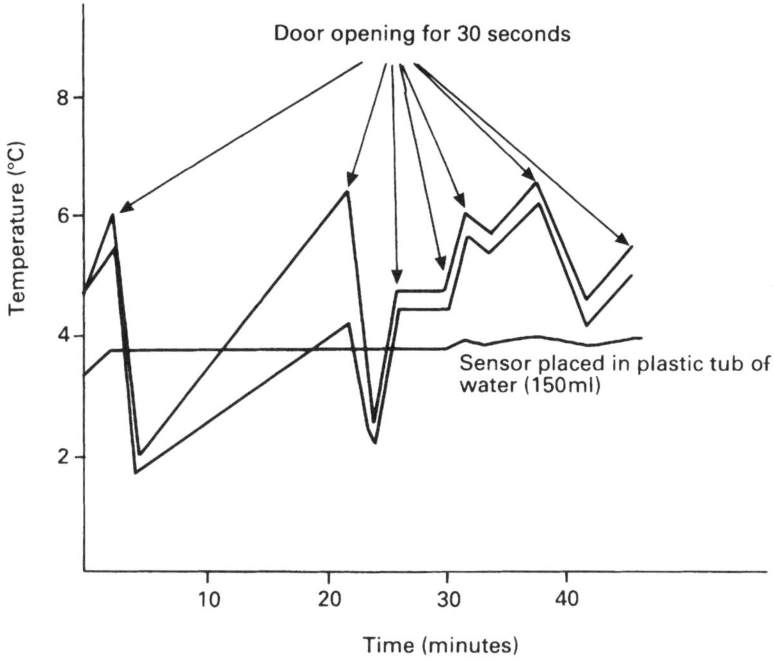

Fig. 5.4 Effect of 'Damping' air sensor.

Temperature controlled vehicles

An independent refrigeration unit usually powered by diesel, often with an auxiliary electric motor, is used to circulate cold air around the vehicle compartment from an evaporator unit at the front of the vehicle. A trend in many multiple distribution depots is to use vehicles with movable bulkheads so that one vehicle can carry frozen and chilled foods at different temperatures in the same load. Each compartment will have its own evaporator, which can control temperatures independently.

The cold air is distributed in different ways within the different vehicles, but the majority have cold air leaving from the top of the cooling unit near the roof, and returning via the base to the front of the vehicle and the return air intake (Fig. 5.5). Correct loading and spacing of the load within the vehicle is crucial to ensure adequate cold air distribution within the compartment. If the load is not spaced correctly, circulation can be restricted and 'hot spots' can occur. The maximum length and width of vehicles is set by regulation, and hence the free space available to loads within an insulated chamber, place further restrictions on achieving correct loading. Some vehicles are cooled by direct evaporation of liquid nitrogen from a reservoir tank on the vehicle. These vehicles have the advantage of being much quieter than mechanically refrigerated vehicles, and temperature control can be better. However, an adequate supply of liquid nitrogen is required for the journey, which can limit their range and number of stops.

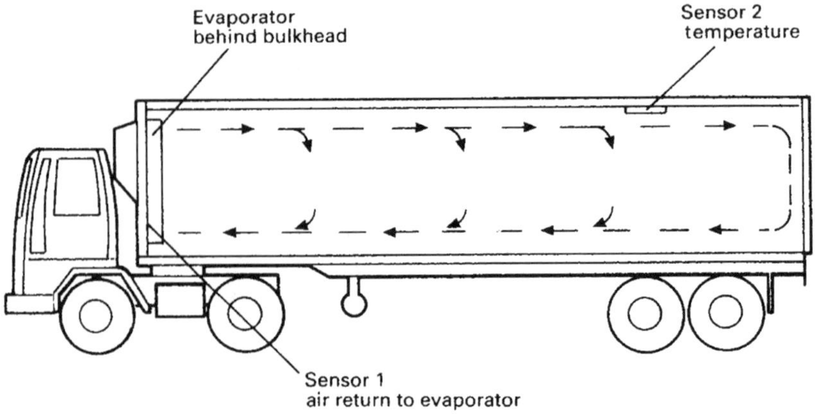

Fig. 5.5 Air temperature monitoring of temperature controlled vehicle.

Temperature read-out and single-channel chart records, which have been used for many years on refrigerated vehicles, placed the sensor so as to measure the air return temperature. This returning air should indicate the mean temperature in the load, provided that there is good distribution to all parts of the load. Short circuiting of air may result in colder return air temperatures.

Long vehicles, especially those without air duct distribution of cold air in the ceiling of the compartment, are advised to fit a second sensor placed nearer the rear of the vehicle (Fig. 5.5). The addition of a second sensor is not sufficient to give a full and accurate picture of temperature distribution within the chamber, but it will measure the cold air leaving the evaporator, and may give a better picture of cold air circulation inside the compartment. This second sensor will serve as a check on the functioning of the measuring system, and makes tampering more difficult. It should demonstrate that the evaporator and fan unit are functioning properly and that cold air is reaching the back of the vehicle. It will give a temperature baseline with which to measure the return air, and indicate more easily when the cooling unit has been switched off, or a load added which has been insufficiently cooled. Prevention of freezing of part of the load can also be more easily avoided. Comparison of the normal differential temperatures between the rear sensor and the return air sensor may also indicate poor air distribution within the compartment.

The frequency of recording for electronic loggers will depend on the length of the journey. A maximum interval of 15 minutes is recommended for journeys of up to 8 hours. Longer journeys may use longer intervals between recordings. Other information such as defrost cycles, door openings and load identification may be required. It is important that a driver be aware of any problem occurring with the temperature of the load. The temperature read-out is often visible to the driver in the vehicle's wing mirror, and in some cases the read-out is presented as a mirror image. Obviously, the driver should have complete concentration on the road, and it is better that an alarm system be fitted which warns the driver when something is wrong.

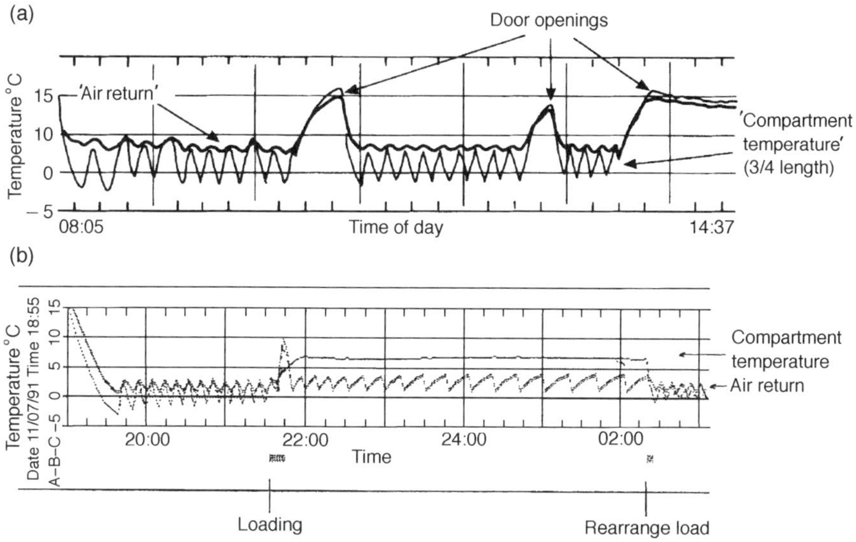

Fig. 5.6 (a). Normal air temperature record and (b). Poorly loaded vehicle air temperature record of chilled foods vehicle (By permission of Cold Chain Instruments.)

Figure 5.6 (a) shows an example of temperature monitoring in a vehicle fitted with two sensors, including the effect of door openings. Figure 5.6 (b) illustrates the care by which air temperature records must be interpreted. The system is operating normally until the chamber is loaded. From this point, the air return sensor gives the acceptable reading apart from slightly longer cycles. Whereas the compartment sensor at the rear of the vehicle shows a temperature rise indicating that the flow of cold air has been restricted by the load. This causes short-circuiting of cold air from the evaporator and hence the longer cycling period activated by the thermostat. As soon as the driver rearranges the load to restart the air flow to the rear of the vehicle, the temperature drops. This problem would not have been obvious had there been only one sensor on the air return.

The monitoring of vehicles with movable bulkheads would require more sensors to enable the temperature to be recorded in each separate compartment. This may be achieved in several ways. The easiest would be to monitor the air intake of each cooling unit. Alternatively, more sensors could be fixed to the roof of the chamber to enable compartment temperatures to be monitored, whatever the position of the bulkhead, and in addition to the air return measurements. The use of small temperature loggers, whose position can be changed to suit the bulkhead arrangement, may provide another solution.

For vehicles cooled by liquid nitrogen, the sensors have to be positioned in order to account for any temperature gradients occurring in the chamber. Forced circulation should eliminate gradients. If fans are not used then sensors should be placed above and below the load.

Small delivery vehicles
Many light goods vehicles delivering chilled foods are fitted with refrigerated units driven from the vehicle engine or transmission. This means that cooling is not possible whilst the vehicle is stationary. For vehicles less than 3 m^3, advances in refrigeration have allowed the retro-fitting of efficient refrigeration units powered by the vehicle's battery.

The main problem in maintaining good temperature control arises from the number of door openings and the amount of time doors may stay open whilst orders are prepared and delivered. Typical 'High Street' delivery patterns can result in doors being open for 40% of the working day. This can make temperature control very difficult and also the use of air temperature monitoring inappropriate. Employment of plastic strip curtains above the door can help to reduce warm air entry whilst doors are open. However, information can be obtained if the air temperature sensors are 'damped' by suspending them in a small bottle of liquid such as oil or glycerol. The large fluctuations are removed from the temperature records and the trends in overall chamber temperature followed. An example of this technique of monitoring is shown in Fig. 5.7.

Vehicles using eutectic plates or insulated boxes to carry foods normally use a food simulant or actual food to monitor temperature during a journey. The

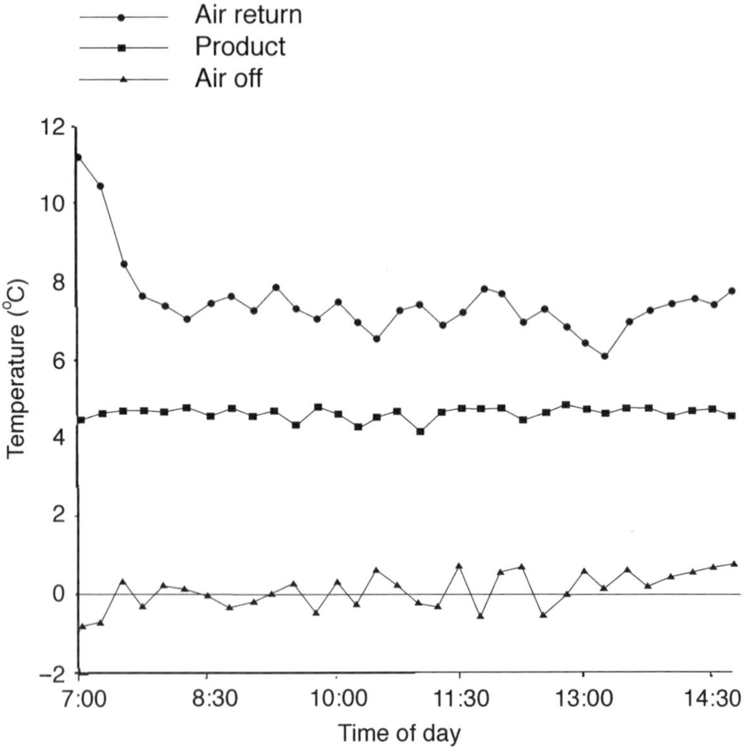

Fig. 5.7 Air temperature record of a small delivery vehicle.

positioning would have to be as representative of the load as possible. The temperatures could be read manually, but could also be connected to a chart recorder or logging system.

5.4.3 Display cabinets

The majority of chilled foods are displayed in open cabinets. Some sectors use display cabinets with closed doors; for the purposes of monitoring, these can be considered as storage cabinets (see section above on 'Cabinet refrigerators'). The open cabinets can be divided into two main groups, multi-deck open cabinets and serve-over cabinets.

Multi-deck cabinets

A fan draws air from the front grille of the cabinet, where it is cooled by passing through an evaporator. The cool air emerges from the back of the shelves to cool the food, and from the top grille to form an air curtain in front of the shelves (Fig. 5.8(a)). Advances have been achieved in the design of cabinets, which include the reduction of heat gain from internal lighting, and the stabilising of the air curtain by improved design or addition of a second curtain. Thus the ease of monitoring the temperature in multi-deck cabinets will be determined by their design and operation. In principle, the differential between the air returning from the shelves and the air emerging on the shelves is an indicator of the cabinet's performance. Positioning of sensors or manual reading of temperature is taken from the top air-curtain grille ('air-off') and the lower air return grille ('air return') (Fig. 5.8(a)).

If the normal pattern of variation in air temperatures can be linked to variations in product temperature on the shelves, then air temperature monitoring can be used on a routine basis. If other factors interfere with the air temperature relationship, such as excessive radiant heat absorption, or if a relationship cannot be established between product and air temperatures, then a product or food simulant may have to be used.

Figure 5.9 shows two different air temperature patterns. The first (Fig. 5.9(a)) shows regular cyclical changes in air temperature, whilst Fig. 5.9(b) indicates much steadier conditions except during a defrost cycle. In both cases, establishment of the range of air temperature with the warmest product temperatures would allow effective air temperature monitoring.

5.4.4 Serve-over display cabinets

There is a wide range of cabinets in this group for displaying meat, fish, delicatessen products, pâtisserie, cheeses and ready-to-eat products in catering establishments. In many cases the food is cooled by cold air from a refrigerated unit, but in some, especially in catering, the food is cooled by contact with a cold plate or well (dole plate) or crushed ice. The effect of radiant heat from lighting or sunlight can be more pronounced with serve-over cabinets and affect food temperature significantly.

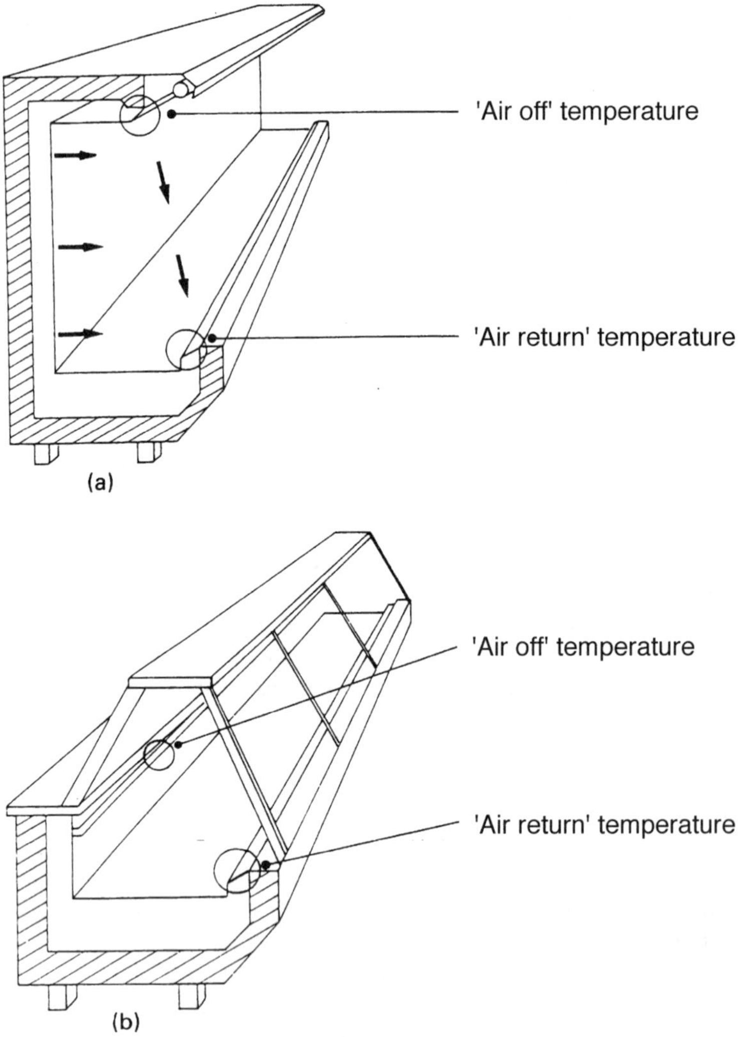

Fig. 5.8 Air temperature monitoring in retail display cabinets. (a) Multi-deck cabinet. (b) Serve-over cabinet.

Figure 5.8(b) shows a typical serve-over cabinet for retail sale of delicatessen products using fan-assisted cold air. Air emerging from the back grille cascades over the food and returns via a front grille. In the case of gravity-fed cabinets, where air enters a back grille and emerges at the base shelf, there is no air return grille. Air velocities are low in serve-over cabinets to reduce dehydration of the displayed products; this also makes air temperature measurement more difficult. Positions of sensors or manual measurements for air temperatures are also shown in Fig. 5.8(b). The relationship of air to product temperatures is necessary before air temperatures can be used routinely.

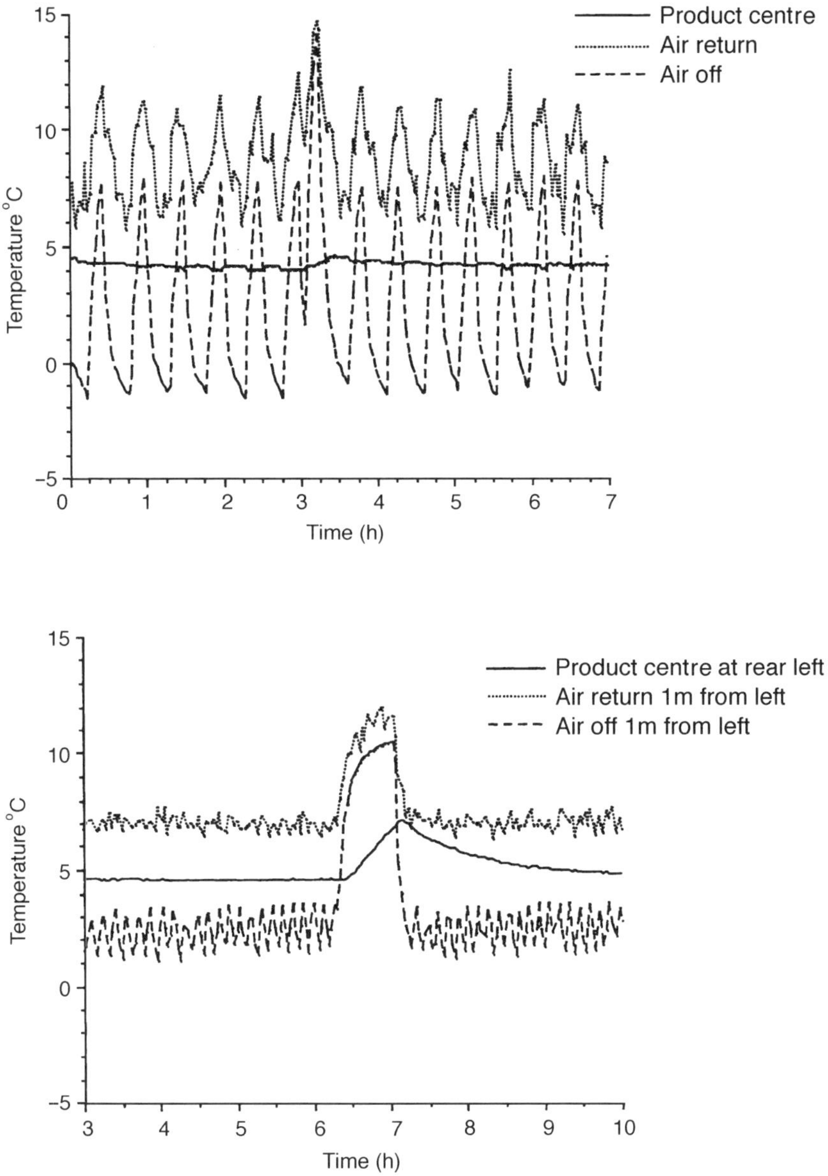

Fig. 5.9 Temperature monitoring records of two different display cabinets. (By permission of the University of Bristol.)

In many cases it will be easier to monitor the temperature of the cabinet using food temperatures or a food simulant. The temperature near the front of the cabinet will usually be indicative of the warmest locations and hence warmest foods within the cabinet. Air temperature monitoring is not suitable for cabinets cooled by conduction (dole plate or crushed ice). In this case, direct measurement of food temperatures is appropriate, but should be carried out, as with all determinations of this type, with a clean, well-disinfected probe.

5.5 Equipment for temperature monitoring

5.5.1 European Standard for temperature recorders and thermometers

In view of the multiplicity of national specifications and testing procedures for temperature recorders and thermometers for the transport of quick-frozen foods in Member States of the European Union under Directive 92/1/EC,[7] and the growing importance of temperature control in chilled food and frozen food sector, the European Centre for Standards (CEN) has agreed a standard for temperature recorders. One Standard, BS EN 12830: 1999,[21] and one draft Standard prEN 13485,[22] cover temperature recorders for the transport, storage and distribution of chilled, frozen, quick/deep-frozen food and ice cream and thermometers in the same field of application respectively. There is a third draft Standard, prEN13486[23] which lays down the procedure for periodic verification of the temperature recorders and thermometers in the preceding standards.

Temperature recorders

The CEN Standard[21] lays down requirements for electrical safety, resistance to mechanical vibrations, and operational performance in climatic conditions. It also sets a minimum specification for accuracy, response time, recording interval and the maximum relative timing error. Table 5.2 gives the climatic environments under which a temperature recorder and a thermometer measuring air temperature must operate even when stored or operated for short periods under those conditions. Obviously these vary as to whether the device is operating inside the refrigerated chamber or at the vagary of outside weather conditions compared to a device operating inside a building or vehicle. The CEN Standard[21] also lays down the test conditions under which the specification for temperature recorders are determined.

Thermometers

The draft Standard prEN13485[22] defines a specification for thermometers for measuring air temperature in transport, storage and distribution and product temperature of chilled and frozen food. Table 5.2 gives the environmental conditions under which thermometers measuring air temperature have to operate under different uses and Table 5.3 their response times. Table 5.4 gives the environmental conditions under which portable thermometers for air temperature and product measurement must operate. For product thermometers there is also a limit of 0.3°C change in accuracy when operating across the full range of ambient temperatures (−20°C to +30°C). The accuracy classes for thermometers measuring air and product temperature are given in Table 5.5. The draft

Table 5.2 The climatic environments under which a temperature recorder and a thermometer measuring air temperature must operate

	Recorder or thermometer for storage and distribution, located outside refrigeration case in heated or air conditioned premises and with external sensor	**Recorder or thermometer for transport** located in or outside vehicle with external sensor	**Recorder or thermometer for storage and distribution**, located inside refrigeration case and with external or internal sensor	**Recorder or thermometer for transport** located inside refrigeration case with internal or external sensor
Thermometer, recorder and displaying device rated **operating conditions***	+5°C + 40°C	−30°C + 65°C	−30°C + 30°C	−30°C + 30°C
Thermometer, recorder and displaying device **limiting conditions**†	0°C + 50C	−30°C + 70°C	−40°C + 50°C	−40°C + 70°C
Thermometer, recorder and displaying device and sensor **storage or transport conditions**‡	−20°C + 60°C	−40°C + 85°C	40°C + 60°C	−40°C + 85°C

* Conditions which device operates to specification.
† Conditions which device can withstand whilst in operation so that it will subsequently operate according to specification when under its rated conditions.
‡ Conditions which device can withstand whilst not operational so that it will subsequently operate according to specification when under its rated conditions.

Table 5.3 Response times* for the sensors in temperature recorders

Type of device	Transport	Storage	All uses
External sensor	10 min. max.	20 min. max.	
Internal sensor	–	–	60 min. max.
Fixed thermometers	10 min. max.	20 min. max.	
Portable thermometers	–	–	3 min. max.
Product temperature thermometer	–	–	3 min. max.

* Response time is the time needed for the measured or recorded value to reach 90% of the actual change of applied temperature under test conditions

Table 5.4 Climatic environment under which portable thermometers and thermometers for product temperature must operate

	Thermometers for product temperature
Rated operating conditions	−20°C + 30°C*
Limiting operating conditions	−30°C + 50°C
Storage operating conditions	−30°C + 70°C

* For measurements made in this ambient temperature range, the measuring accuracy shall not change more than 0.3°C.

Table 5.5 Accuracy classes for thermometers measuring air or product temperatures

	Air temperature		Product temperature	
Class	1	2	0.5	1
Maximum possible errors	±1°C	±2°C	±0.5°C	±1°C
Resolution	≤0.5°C	≤1°C	≤0.1°C	≤0.5°C

Standard[22] also lays down the test procedures for determining temperature measuring error and response time.

5.5.2 Sensors

Accuracy

Whatever the system for collecting or recording temperatures, the sensor or heat-sensitive part is the common factor between them. The three principal types of sensors in commercial use are thermocouples, platinum resistance and semiconductor (thermistor). The choice of which type of sensor will depend on the requirements for accuracy, speed of response, range of temperatures, robustness and cost.

Until recently, the majority of general-purpose thermometers and measuring systems used a thermocouple in the heat-sensitive part of the system. This is a pair of dissimilar metals joined together at one end, usually by a soldered joint. The circuit is completed by a second junction held at a known temperature (often referred to as a 'cold' junction). In food applications, where temperatures are relatively close to ambient, two types of thermocouple predominate: Type K thermocouples, which use wires of Chromel (a nickel–chromium alloy) and Alumel (a nickel–aluminum alloy); and Type T thermocouples, which use wires of copper and Constantan (a copper–nickel alloy). The advantages of the thermocouples are their low cost, facility to be hand-prepared from reels of wire, and a very wide range of temperature measurement (−184°C to 1600°C).

Table 5.6 shows the permissible sensor accuracies for the three types of sensors, which in the cases of thermocouples and the platinum resistance sensor conform to a standard requirement. The difference in instrumental error arises

Table 5.6 Sensor and system accuracies

	Type K	Type T	Pt resistance	Thermistor
Sensor accuracy (°C)	±1.5*	±0.5*	±0.2†	±0.1
Instrument accuracy‡ (°C)	±0.3	±0.3	±0.2	±0.2
System accuracy (°C)	±1.8	±0.8	±0.4	±0.3

* BS 4937: Class A[24]
† BS 1904: Class A[25]
‡ Includes cold junction compensation accuracy.[26]

from the fact that the electronic circuitry has to be compensated for changes in the reference or 'cold junction' (normally ambient temperature). This is measured by a built-in semi-conductor sensor, which automatically compensates for changes in ambient temperature.

Errors using thermocouples increase when the ambient temperature varies widely, i.e. moving from a cold to a hot environment. Other errors with thermocouples can be produced by induced voltages from motors or transmitters, moisture, and thermal gradients in other junctions. A move to greater accuracy for measuring and monitoring restricts the use of thermocouple sensors to Type T only, which will normally meet the basic specification for air temperature monitoring (CEN Standard[21]).

Thermistor sensors change resistance with temperature, but normally can be used to measure only a narrower range of temperatures than the thermocouple (−40°C to 140°C). Their use for measuring food temperatures has increased since the introduction of a basic specification for measuring systems for food temperatures to give an accuracy of ±1°C, which is reinforced by the draft CEN Standard for thermometers.[22] They are rugged, provide good accuracy and repeatability and are not unduly affected by changes in ambient temperatures.

Platinum resistance thermometers also have a system accuracy which meets the draft CEN Standard.[22] They may be used over a wider range of temperatures (−270°C to 850°C). Normally their response time (Table 5.7) is slower, unless they are specially constructed for a fast response. Corrections have to be made for resistance of leads, and the self-heating effect. Their higher cost has restricted their use to applications where high accuracy is required in fixed process control applications.

Calibration and periodic verification

During manufacture, each sensor and instrument is checked to ensure that it meets specification and achieves an accuracy within tolerances set by each manufacturer and in accordance with BS EN 12830:1999[21] and prEN 13485:1999.[22] In many applications of monitoring, different sensors are plugged into an instrumentation system, and they are normally regarded as interchangeable. However, in the case where more precise readings are required, an individual calibration is undertaken of the sensor and instrument together

Table 5.7 Typical response times (seconds) in air and water[26]

	Still air	Forced air	Water*
Exposed thermocouple	20	5	–
Shrouded thermocouple	150	40	6
Exposed thermistor	45	20	–
Shrouded thermistor	260	50	12
Shrouded platinum	365	65	15

* Mounted in 'chisel' probe in water, time for 20°C change to 99% level.

(system). This measures the system's reading against a range of applied temperatures. The applied temperatures have to be traceable to a national standard (e.g. National Physical Laboratory). The resulting table or graph of the calibration certificate enables the system reading to be corrected to a true reading (within the tolerances of the calibration).

Once a temperature monitoring system is installed, it is essential that periodic checks are carried out to ensure that the equipment is functioning correctly and meets the same specification as when it was purchased, and as described in prEN13486:1999.[23] The frequency of checks depends on the use of the equipment and will consist of routine checks on the functioning of the equipment and those carried out by the manufacturer (or a suitably qualified laboratory). The maximum period recommended is one year for a manufacturer's check or after a long period of non-use or operating incident. The equipment is normally checked against another thermometer, which has been calibrated against a standard. It is normal to check the accuracy, and functioning of the clock or verify the recording duration.

Sensor housing and probes

In monitoring applications, the sensor element has to be protected from damage or breakage. This can range from coating with an epoxy resin to embedding in a stainless-steel sheath. If fast response is required, the thermal mass has to be as low as possible. It is also important that sensors, which are mounted inside chambers or vehicles to measure air temperatures, are also protected from damage during commercial activity of loading or unloading food, but not in such a way as to restrict air flow.

Monitoring and measuring food temperatures often requires sensors mounted in hand-held probes. The design of the probe depends on its application. The most common probe is for insertion into foods, and therefore has a sharpened tip (see Fig. 5.10(a)). If a non-destructive temperature measurement is required then a probe is required which can be inserted between food packs or cases. Good contact between the packaging and the probe, together with an adequate period to allow readings to settle, are essential to minimise errors in this type of measurement. Examples of probes used to measure between packs and cases are shown in Figs 5.10((b) and (c)).

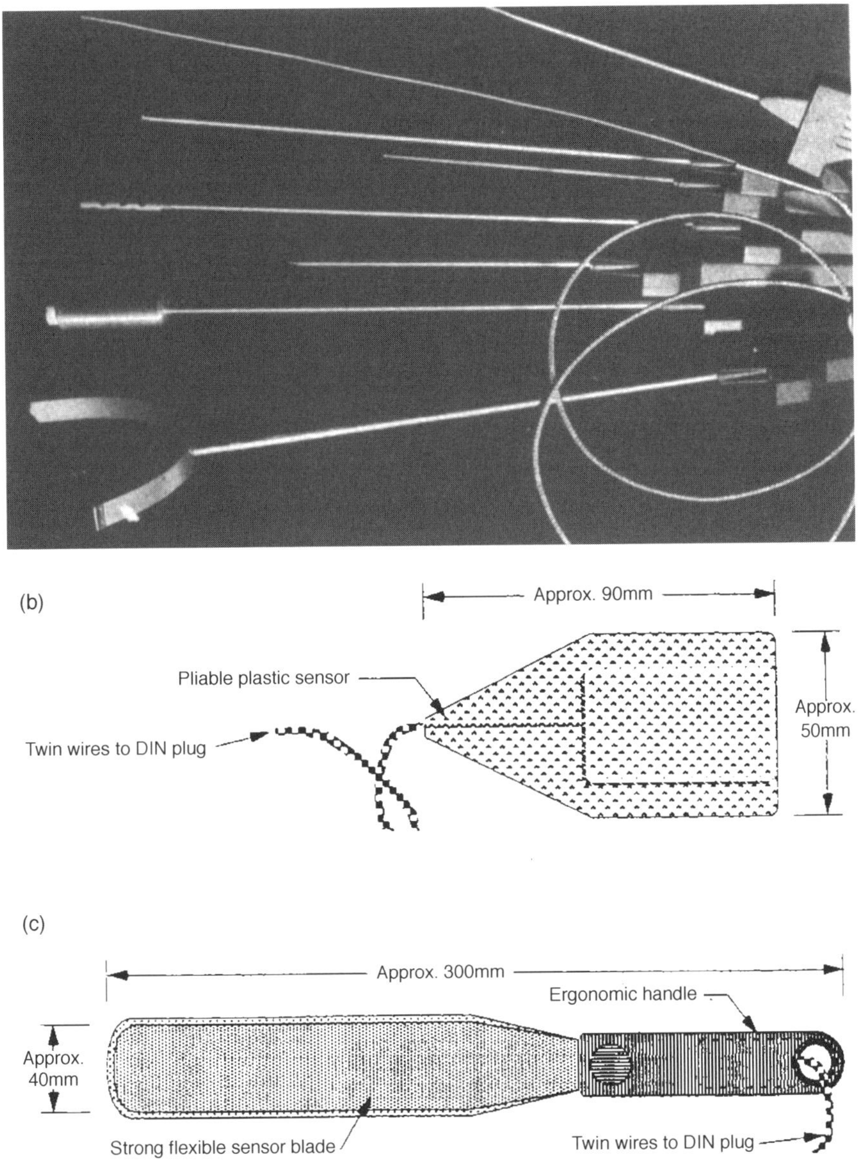

Fig. 5.10 Hand-held temperature probes. (a) Various air and product temperature probes. (b) A probe for between-pack temperatures. (c) A probe for between-case temperatures.

5.5.3 Read-out and recording systems

Single readout systems

Instrumentation has progressed from the original single read-out thermometer, the mercury- or alcohol-in-glass thermometer. The development of dial and stick thermometers with analogue or digital display removes the danger of breakage, but their use can be limited by their low accuracies especially those based on bi-metallic strips. Dial thermometers which were used to indicate air temperatures in display cabinets have largely been replaced by digital thermometers.

Thermochromic liquid crystals change in orientation and transparency depending on their composition and temperature. When set into strips they will display the appropriate temperatures printed under them. Their accuracy is limited, but can achieve ±1°C.

More common is the electronic digital readout instrument, which is powered by batteries. The resolution and interval of display temperature will vary with model and type of sensor. Temperatures can be stored and even printed out, and an alarm given if the temperature goes outside a preset limit.

Chart recorders

Historically, a trace on a moving chart was the only method available to produce a temperature history and record. The use of chart recorders is less common and they have been overtaken by electronic instruments but some may still be found in fixed system applications such as cold or chill stores and vehicles. The charts can be circular or mounted on a roll to give a rectangular chart, and a trace obtained with ink or pressure or heat-sensitive paper. Circular chart recorders have the advantage that the temperature history is visible and abrupt changes apparent, and the chart can be easily stored for future reference. The timescale of the chart is usually over 24 hours, 7 or 31 days, but some long-distance marine recorders may operate for 6–8 weeks. The chart clock and electronics can either be battery driven for mobility or driven from the mains for fixed applications. The accuracy of the recording duration according to BS EN 12830: 1999[21] has to be 0.2% of the recording time when less than 31 days and 0.1% of the recording time when over 31 days. Accuracy of the system varies with sensor, but more modern chart recorders are better than 0.5°C in the range 0–25°C. Their limitation is often the resolution on the chart divisions and the thickness of the trace. Fixed system charts can be sophisticated instruments with the possibility of recording 30 or more different channels in different colours and print modes.

Chart recorders which are fitted to vehicles (or more often to the trailers) have to be more robust in their construction. They need to withstand the rigours of the road for all types of terrain and weather conditions. Recorders giving two or more traces are available, and event markers (e.g. noting door openings) can be added.

Fixed processing system for chill stores

The difficulty in interpretation of a large number of different traces and the rapid developments in microelectronics and computer technology have encouraged the replacement of chart recorders by data logging systems. This allows not only the storage of large amounts of data, but also its manipulation and analysis and integration into management systems.

In chill store operations where large numbers of temperature measurements are being taken every day throughout the year, it has become increasingly the practice to install computerised systems to handle the data. There may be a digital temperature display on a control unit situated near the refrigerated system, but more often information can be retrieved on a visual display unit situated in a control room. Alarm systems can be integrated with the system when any of the parameters being monitored are outside preset limits. The alarm can be transmitted to maintenance staff inside the system's locations or to outside premises through telecommunication networks.

Vehicle temperature logging systems

Several companies have developed dedicated temperature logging systems for vehicle monitoring. These have been designed to withstand the vibration and harsh conditions encountered in transport as stipulated in BS EN 12830: 1999.[21] Data is collected over the whole journey from loading to unloading, and alarm signals given if temperatures are outside preset limits. The equipment can either fit inside the vehicle cabin and is often the same size as a vehicle radio, or will be fitted outside the vehicle often next to the refrigeration control unit.

In addition, distribution customers are increasingly requiring a record of the temperature history of the food they receive. Systems have been developed which give an instant print-out of temperatures up to the point of delivery, to be attached with the delivery documentation. Other features found are ability for variable logging periods and up to 12 months memory, event recorders for defrost and door openings, and multiple channels for multi-compartment monitoring. The retrieval of the information is becoming more sophisticated with downloading facilities to office PCs via manual collection units or by radio, infrared, or satellite communications.

Portable data logging systems

The miniaturisation of circuitry has produced some very compact and powerful data logging systems, some of which are small enough to travel with food cases or pallets and record temperatures during passage through the chill chain. The devices can also be used as fixed systems in stores and vehicles. This is useful where the position of fixed sensors would have to be changed from time to time, e.g. temporary chill stores or movable partitions in multi-compartment vehicles. The choice of system will then depend on its particular application, convenience of its use, and price. An evaluation of two such devices has been reported by Kleer *et al.*,[27] where they were used in large-scale catering systems, and found useful in recording the critical control points of the process.

Another type of data logger ('electronic chicken') is useful for monitoring display cabinets. The logger is placed on a shelf and then records temperatures from a food simulant contained in the logger, which has the same thermal properties as the food displayed on the shelf. An alarm light is fitted to the logger so that problems can be easily noticed and remedied. The data is downloaded for display and analysis via an infrared remote reader.

The specification of commercial systems is changing as micro-electronics improve, and it is certain that miniaturisation of the loggers will continue. Most of the present systems are still too big to fit into cases without removing one of the food packs; much smaller and slimmer devices would be able to fit between food packs.

Remote sensing devices – non-contact thermometers

All objects emit energy at temperatures above absolute zero. This energy increases in intensity, but decreases in wavelength, with temperature. In the temperature range of interest to chilled foods, infrared radiation can be measured to determine the temperature. As temperature increases, the intensity increases and the peak energy moves to shorter wavelengths. Hence, most low-temperature, commercial, infrared thermometers filter a band (8–14 microns) out of the infrared spectrum and measure its intensity. Using such a band reduces atmospheric (water vapour, carbon dioxide) absorption and 'distance sensitivity' of the instrument. Very narrow bands (2.2, 5.2 and 7.9 microns) can be used to give greater accuracy at very high temperatures, but signals are very low requiring expensive high-gain amplifiers.

Not all materials emit the same energy at the same temperatures. The ratio of the energy radiated by a material compared with a perfect radiator or blackbody is known as the 'emissivity'. Emissivities vary from 0 to 1.0, with most organic substances having a value around 0.95. Different substances vary in the amount of energy they absorb, reflect and transmit. Infrared thermometers have emissivity compensators which have to be set for different values (0.1–1.0) to allow for these differences. The target size is also important. The instrument averages all the temperatures it sees in its field of vision. Unless the object fills all the field of vision, the temperature reading will be an average of the object and its surroundings. Focal distances vary with machine, from close-up to 50 metres. The further away, the more difficult it is to pin-point targets, and laser sighting is common on many models.

There are two main types of remote sensing equipment. In one type, pistol-shaped instruments are pointed at a target, and the temperature is read from a digital readout at the back of the instrument. Laser targeting can be built into the pistol to give through-the-lens sighting to locate the target, and long-distance devices are often assisted by optical telescopic sighting. Accuracies claimed for this type of instrument are around ±1°C.

Research carried out by the University of Bristol[28] on nine commercially available infrared thermometers revealed that care must be taken in using and interpreting results from these devices. Surface temperatures may be quite different from the internal food temperature. This is a more acute problem with frozen foods where the differential between the surface temperature and interior temperature can be large especially when the food is being transferred in a higher ambient temperature than −18°C. The infrared sensor not only measures the radiation emitted from the surface as a result of its temperature, but also radiation which is reflected from the surroundings of the food, e.g. lighting.

Table 5.8 Mean error in °C with standard deviation (in parentheses) of different packaging materials

Instrument	Clear MAP	Glossy Cardboard	Plastic bag	Printed laminate foil	Printed MAP	Printed vacuum pack	Moduli average
a	0.6 (0.1)	1.7 (0.1)	1.1 (0.6)	6.6 (0.6)	1.9 (0.5)	1.3 (0.1)	2.2
b	−0.3 (0.0)	0.7 (0.0)	0.8 (0.6)	5.3 (0.6)	0.6 (0.1)	1.4 (0.1)	1.5
c	0.7 (0.1)	0.6 (0.0)	0.5 (0.0)	6.0 (0.1)	0.4 (0.6)	0.4 (0.0)	1.4
d	−3.3 (0.3)	−4.5 (0.5)	−5.1 (0.4)	7.0 (0.2)	−9.1 (1.0)	−7.2 (0.2)	6.0
e	−1.9 (0.6)	−2.3 (0.1)	−2.5 (0.0)	4.1 (0.0)	1.8 (0.1)	−0.6 (0.1)	2.2
f	0.8 (0.1)	0.9 (0.4)	1.0 (0.5)	4.2 (3.0)	2.9 (0.3)	2.3 (0.1)	2.2
g	−0.1 (0.1)	−0.5 (0.6)	0.4 (0.6)	6.2 (0.5)	0.2 (0.2)	0.6 (0.1)	1.3
h	0.5 (0.4)	3.8 (0.3)	6.4 (0.7)	10.4 (0.9)	6.1 (0.8)	4.0 (1.4)	5.5
i	−2.2 (0.0)	−1.2 (0.0)	−0.8 (0.6)	3.2 (0.0)	−0.9 (0.6)	−1.0 (0.6)	1.5
Mod. Av.	1.4	1.3	2.1	5.9	2.7	2.1	

Depending on the type of packaging, the reflected radiation can be quite considerable and hence will give an incorrect surface temperature.

There was a large difference between the performance of the nine devices when used in a commercial retail cabinet in a retail outlet. Table 5.8 shows the performance of these using six different packaging materials. Infrared readings were compared with a calibrated thermocouple inserted under the surface of the pack. Out of the five instruments, two (b & g) had errors less than 1°C, and five less than 2.5°C, with two further instruments giving unsatisfactory errors. The highest errors of all of the instruments were found with a printed foil pack which gave the most reflected radiation. It is recommended that angled and brightly lit packs be avoided with an infrared thermometer and only horizontal or vertical positioned packs in a cabinet chosen, with the device perpendicular to the top surface. In order to improve the accuracy, lighting must be reduced as much as possible, and distances to take readings must be as short as possible with as consistent a product as the situation allows.

If the thermometer is moved from one ambient temperature to another, e.g. room temperature to a chill store, then for best repeatability of measurements it is advisable to allow at least 30 minutes for the instrument to adjust to the new ambient temperature. The thermometer should also be checked regularly against surfaces of known temperature. It is possible to make a relatively cheap black body calibration chamber with black PVC plastic tubing and a copper block. Alternatively commercial systems are available.

The other type is based on infrared video camera-type instruments. Thermal images are displayed on a video display unit, either in colour or monochrome. A temperature scale on the display gives the temperature which corresponds to the individual colour or shade. Cameras range from low-resolution types, often used for locating victims trapped in collapsed buildings, to very high-resolution, sophisticated systems which allow computerised manipulation of the data. Infrared systems have been found to be very useful for industrial control and energy efficiency, with the ability to pick out overheating components and heat losses. They are also being used increasingly in on-line applications in the food industry, either with the hand-held type or with specific models available for this. For example, sealing rollers on microwavable plastic trays can be monitored to ensure evenness of heating, and products emerging from cooking, heating or cooling tunnels can also be screened for evenness of heating or cooling. These are applications where the target is consistent and relative temperatures are more important than accurate temperatures. The readings are instantaneous and the information can be linked directly to control systems.

Infrared temperature measurement will never replace electrical temperature measurement for accurate determinations for enforcement of temperature control legislation. However, there are exciting possibilities for its use in routine monitoring and temperature auditing where relative temperatures are very important, and care needs to be exercised in the interpretation of results. Hand-held devices can be used to monitor the surface temperature of cases unloaded from a vehicle on an acceptance or rejection basis or scan a display cabinet for 'hot-spots'.

5.6 Temperature and time–temperature indicators

5.6.1 Performance of TTIs

Temperature monitoring has been discussed in terms of displaying temperature readings of the surrounding air or of the food or simulated food itself. However, it is possible to use a physico-chemical mechanism and a resulting colour change to display (a) a current temperature, (b) the crossing of a threshold temperature, or (c) an integration of the temperature and the time of exposure to temperature after activation. Such devices are called temperature indicators (TIs) in the first two cases or time–temperature indicators (TTIs) in the last case.

The indicators are normally integrated onto a packaging material which can be attached to the food packaging or the outside of the surrounding or bulk packaging, and can follow the food throughout the chill chain. The type of information than can be provided is one or more of the following:

- reject or accept on the basis of a colour change
- temperature abuse above a threshold temperature
- partial time–temperature history above a threshold temperature
- full time–temperature history linked to shelf life.

In order that the devices can be used in commercial situations for monitoring, they should have the following features and be supplied with the following information by the manufacturer.[29]

- ease of application to food packs;
- instructions on the activation of the TTI just before application, including the temperature at which the device must be stored if it has to be kept at low temperatures (frozen) once manufactured until applied;

- for TIs, the threshold temperature and its tolerance limits (3 × standard deviation) in °C, and a response time (inertia) in minutes;
- for TTIs the maximum and minimum temperature limits in °C over which the device will function and the time to end point with the tolerance at sufficient numbers of temperatures throughout the range stated by the manufacturer (above the critical reference temperature in the case of partial history TTIs). The number of temperatures and time to end point combinations shall not be less than five;
- the performance tolerance for the time to end point as in BS 7908: 1999[29]:

Category A (up to ±2.5%), Category B (up to ±5%), Category C (up to ±10%), Category D (up to ±20%);

- for partial history TTIs the critical reference temperature, i.e. the temperature at which the physico-chemical change is activated to produce an irreversible change in °C and its tolerance;
- the storage conditions of these devices should be given so that their performance is unaffected. Also any conditions which could affect the performance apart from temperature, e.g. light should be stated;
- the devices should be tamper-proof.

It is important to appreciate that TTIs are based on physical changes, or chemical and biochemical reactions. Their performance does not in general mimic microbiological changes in food but the biochemical and chemical reactions which cause deterioration in the sensory quality of foods. Normally, biochemical reactions change at a faster rate than chemical ones. However, each food will have a different combination of reactions and hence a different rate. In designing an indicator, it may be important that the activation energy of the device is similar to that of the food deteriorations as well as rate of deterioration.[30, 31]

Over 100 patents have been filed on processes which could be used as a basis for indicators. These include changes with temperature based on melting-point temperature, enzyme reaction, polymerisation, electrochemical corrosion, and liquid crystals. The result of the change is usually a colour difference, which can be represented as a static change or moving band. Available temperature and time–temperature devices have been reviewed.[32, 33, 34] Many temperature and time–temperature indicators have been launched commercially over the past 10–15 years, but very few have survived. Therefore only a few of the more successful devices will be described.

3M MonitorMark™

These TTIs consist of a paper blotter and track separated by a polyester film layer. The blotter is impregnated with chemicals of specific melting point and a blue dye. If a pre-determined threshold temperature is reached the chemicals melt, and hence the devices are partial history TTIs. The indicator is activated by removing the film, the chemicals and dye diffuse along the blotter and track. There are five windows which, as they turn blue, allow the exposure to temperature to be estimated. The diffusion rate increases with temperature above the melt point. Varying tags are produced to correspond to different lengths of time at different melt temperatures (these range from −15°C to ±31°C).

Lifelines

Lifelines has developed several indicators all of which show a full time–temperature history. The indicator part consists of polymeric compounds that change colour as a result of accumulated temperature exposure. The colour change is based on polymerisation of acetylenic monomers which proceeds faster at higher temperatures leading to a more rapid darkening of the indicator.

In the first type of label, there are two parts; a standard bar code and an indicator bar. A portable computer with an optical wand is used to read the indicator, giving a reflectance reading. Initially reflectance is high (95–100%); this falls during use, as the reaction proceeds and the colour darkens (50% reflectance). The computer correlates the change in reflectance to the time–temperature characteristics. This is linked to the information about the product in the bar code, and predictions on the quality life can be made.

Developments in Lifelines technology have resulted in the manufacture of the 'Fresh-Check' indicator, designed for consumer use. This device consists of two circles;® a small inner circle which contains the polymer, and a printed dark or black outer ring. The inner ring darkens when exposed to time and temperature combinations at a rate predetermined by the durability of the food. The consumer is advised not to consume the product when the inner ring has become darker than the printed outer one (see Fig. 5.11).

In order to try to link the indicator to microbiological safety as well as quality deterioration, a new development has been the incorporation of a second polymer system in the centre ring. If the indicator stays below a pre-set maximum then the polymer will change colour as above, and the change is linked to the durability. If the temperature rises above the maximum, then the second system starts to polymerise, and the centre will darken abruptly at the end of a predetermined length of time. Lifelines labels are not physically activated, and once manufactured respond to any temperatures to which they are exposed. Therefore before use, indicators must, at all times, be stored at −18°C or colder.

Vitsab®

A range of TTIs are manufactured by Visual Indicator Tab Systems in Sweden and are based on an enzymatic release of protons which changes the colour of a pH indicator from green to yellow. The rate of release is temperature related and the rate can be varied to match the shelf-life and temperature of chilled and frozen foods. The indicator can be stored at room temperature and is activated

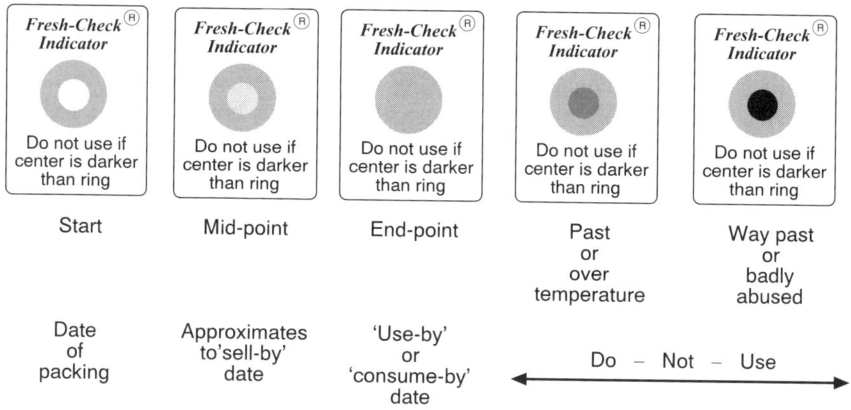

Fig. 5.11 The Lifelines 'Fresh-Check' indicator – staged examples.

by pressure which breaks an internal pouch allowing the components to mix. The circular indicator can be printed on flexible or semi-rigid packaging and can be incorporated into or positioned on the seal of the packaging. Activation can be post-sealing or during sealing. The TTI is also available mounted on a card which can either be placed between packs on a pallet or inside a bulk pack.

5.6.2 Practical use of time–temperature indicators (TTIs)

There are some technical difficulties in the use of TTIs compared to other methods of monitoring. The fact that most are applied to the outside of a food pack means that surface temperatures are being used to change the indicators. As long as food packs are in cases, this is probably still a good guide to the food temperature with a tolerance. However, for foods on display, this may give a false indication of shelf-life, owing to radiant heat absorption, unless the effect is eliminated or compensated for. Therefore, as a means of following the integrity of the chill chain from manufacture up to the point of display, TTIs may have a practical advantage in use over certain other types of monitoring in that they give a simple and individual indication of temperature abuse.

A survey of 511 consumers, carried out by the National Consumer Council,[35] indicated that almost all respondents (95%) thought that TTIs were a good idea, but only grasped their concept after some explanation, indicating that substantial publicity or an education campaign would be required. Use of TTIs would have to be in conjunction with the durability date, with clear instructions about what to do when the indicator changed colour.

The relationship and possible conflict between the indication of the TTI and the durability date on the food was considered a problem. In the retail situation, nearly half those questioned would trust the TTI response if it had not changed but the product was beyond its durability date. If the TTI changed before the end of the durability date when stored at home, the majority of respondents (57%)

would use their own judgement in deciding whether a food was safe to eat, with at least 25% putting some of the blame on the food suppliers. However, the value of TTIs was recognised for raising confidence in retail handling, and improving hygiene practices when food is taken home and stored in refrigerators. Concerns over their technical performance (accuracy and reproducibility), were expressed, and the question of whether they could be tampered with or interfered with was also raised. These concerns are shared by the food industry, and have been addressed by the publication of a technical specification for time–temperature indicators.[29, 36]

Reluctance by retailers to use an indicator on retail packs for consumer use is understandable because of the difficulties that are raised. To date, no permanent commercial use of TTIs on retail packs has been adopted in the United Kingdom. However, both French and Spanish supermarket chains have used Lifelines Fresh-Check indicators on selected items of chilled food for quite extended times, but have decided not to use them on a long-term basis. TTIs have found more extensive use in the medical field to ensure that vaccines and medicines are transported and stored correctly. In addition their use to ensure the integrity of the chill chain up to retail sale by having indicators on outer carton or pallets as a further check is being examined by the chill and frozen food industry and retailers. TTIs advantages over other types of monitoring equipment of giving easy and clear answers to whether temperature abuse has taken place makes them an attractive addition to assuring safety and quality to the consumer.

5.7 Temperature modelling and control

The use of computer modelling as an aid to predict what is happening in complex systems is well established and has been applied in the refrigerated food sector. It has enabled food temperatures to be predicted if the conditions of use of the refrigerated system are known.

5.7.1 Short-distance delivery vehicles

The difficulties in monitoring and maintaining food temperatures in small delivery vehicles with many stops at retail outlets have been examined by the University of Bristol,[37] and a commercial computer programme (CoolVan[38]) has been developed to aid the design and operation of these vehicles. The programme examines the changes to the temperature of the air inside the vehicle. The ingress of heat through the insulation from the outside air and solar radiation are taken into account as is the infiltration of air through the back door when the vehicle is travelling and stationary with the door open. The thermal properties of the new insulation of the vehicle and the age of the vehicle enables the reduced heat transfer coefficient of the walls to be predicted, and each side of the vehicle can be treated separately. The infiltration of air during door openings is the one of major factors in heat gain.

Transparent plastic strip curtains have been recommended as a way of reducing ingress of air, and measurements showed that blowing air from the cooling system directly at the curtains helped to counteract warm air entering the gaps at the top of the curtain.

At each stage of the programme's development it was tested against measured data. The programme was able to predict the mean temperature of the food inside the vehicle with an accuracy better than 1°C at any time throughout the journey. However, food temperature within the vehicle actually varied by more than 5°C at one time due to the uneven temperature within the vehicle.

5.7.2 Retail display

Computer modelling has been developed to examine the way retail cabinets behave and so improve their design. Computational fluid dynamics (CFD) is a tool which enables changes to be made to the computer model and see which effect produces the best results before checking this against actual measurements. It has been applied to study the effects of retail cabinets on supermarket environments, and in particular the cold air spillage from frozen food cabinets to the aisles.[39] The model predicted temperatures at floor level of between 5°C and 15°C, whereas measured values ranged from 13°C to 22°C. The model is better at showing trends than actually using it to predict or follow actual temperatures. There is obviously an exciting future ahead where computer models will be used to improve the design of all refrigerated equipment in the food chain and improve energy efficiency whilst maintaining food temperatures.

5.8 Further reading

JABLONSKI J R, TQM implementation. In: *Implementing Total Quality Management: an Overview*, San Diego, Ca, Pfeiffer and Co., 1991.

WEBB N B and MARSDEN J L, Relationship of the HACCP system to Total Quality Management. In: *HACCP in Meat, Poultry and Fish Processing*, Edited by Pearson A.M. and Dutson T.R. Advances in Food Research Vol. 10, 1995 Blackie Academic and professional, Chapman and Hall.

JAMES S.J., Controlling food temperature during production, distribution and retail, *New Food* 1999 **1** (3) pp. 35–45.

FRPERC Publication No. 584. MAFF Awareness Initiative: Computational Fluid Dynamics for the Food Industry, University of Bristol, 1994.

5.9 References

1. Food Hygiene (Amendment) Regulations 1990, SI 1990 No. 1431, London HMSO.
2. Food Safety Act 1990, Chapter 16, London, HMSO.

3. Food Hygiene (Amendment) Regulations 1991, SI 1991 No. 1343, London HMSO.
4. Council Directive 93/43/EEC on the hygiene of foodstuffs. OJ No L 175, 19.7.93 pp. 1–11.
5. The Food Safety (General Food Hygiene) Regulations 1995, SI 1995 No. 1763, London HMSO.
6. The Food Safety (Temperature Control) Regulations 1995, SI 1995 No. 2200, London HMSO.
7. Commission Directive 92/1/EEC on the monitoring of temperatures in the means of transport, warehousing and storage of quick-frozen foods for human consumption. OJ No.L 34, 11.2.92, pp. 28–9.
8. Agreement on the International Carriage of Perishable Foodstuffs and on the Special Equipment to be used for such Carriage (ATP) 1970 United Nations, New York. E/ECE 810 Rev. 1, E/ECE/TRANS/563 Rev. 1.
9. Industry Guides to Good Hygienic Practice: Baking Guide, Catering Guide, Markets and Fairs Guide, Retail Guide, Wholesale Distributors Guide, Fresh Produce Guide. Chadwick House Group Ltd 1997–99.
10. ISO (International Standards Organisation) 9000 Series of Standards 1994. ISO 9000: Quality Management and Quality Assurance Standards, Part 1: Guidelines for Selection and Use. ISO 9001: Quality Systems – Model for Quality Assurance in Design/Development, Production, Installation and Servicing. ISO 9002: Quality Systems – Model for Quality Assurance in production and Installation.
11. International Institute of Refrigeration, *Recommendations for chilled storage of perishable produce*. Paris 1979.
12. Shipowners Refrigerated Cargo Research Association, *The transport of perishable foodstuffs*. 2nd edn, Cambridge 1991.
13. British Refrigeration Association, *Testing of food temperatures in retail establishments*, 1986.
14. RFIC (Refrigerated Food Industry Confederation), *Guide to the Storage & Handling of Frozen Foods*, Methyr Tydfil, Stephens & George Ltd, 1994.
15. Institute of Food Science & Technology, *Guidelines for the Handling of Chilled Foods* 2nd edn, 1990
16. Department of Health, *Guidelines on the Food Hygiene (Amendment) Regulations 1990*, HMSO, 1991.
17. BS EN 441-5: 1996 *Refrigerated display cabinets: Part 5. Temperature test*. London, British Standards Institution.
18. BS EN 441-4: 1995 *Refrigerated display cabinets: Part 4. General test conditions*. London, British Standards Institution.
19. JAMES S J, EVANS J A and STANTON J, Performance of Domestic Refrigerators, *Proceedings of 11th International Conference on Home Economics*, 13–15 Sept., Middlesex Polytechnic, UK 1989.
20. TUCKER G, Guideline No. 1: Guidelines for the use of thermal simulation systems in the chilled food industry. Campden and Chorleywood Food Research Association, 1995.

21. British Standard BS EN 12830:1999, Temperature recorders for the transport, storage and distribution of chilled, frozen and deep-frozen/quick frozen food and ice cream – tests, performance and suitability. London, BSI. 1999.
22. Draft CEN Standard prEN 13485, Thermometers for measuring air and product temperature for the transport, storage and distribution of chilled, frozen and deep-frozen/quick frozen food and ice cream – tests, performance and suitability. Brussels CEN March 1999.
23. Draft CEN Standard prEN 13486, Temperature recorders and thermometers for the transport, storage and distribution of chilled, frozen and deep-frozen/quick frozen food and ice cream – periodic verification. Brussels CEN March 1999.
24. BS 4937: Part 20. Specifications for thermocouple tolerances. London, British Standards Institution, 1983.
25. BS 1904: 1984. Specification for industrial platinum resistance thermometer sensors. London British Standards Institution, 1984.
26. FAIRHURST D, Temperature monitoring in the cold and chill chain, (A one-day Seminar sponsored by MAFF, 30.01.1990.) *Food Science Division Report*, MAFF, London. 1990.
27. KLEER J, PASTARI A, WIEGNER J and SINELL H, Recording temperature patterns with modern recording systems (original German), *Fleischwirtschaft*, 1991 **71** (6) 698–704.
28. JAMES S J and EVANS J A, 'The accuracy of non contact temperature measurement of chilled and frozen food', *IChemE Food Engineering Symposium*, University of Bath, 19–21 Sept. 1994, Publication No. 106, FPERC, University of Bristol 1994.
29. British Standard BS 7908: 1999 Packaging – temperature and time–temperature indicator– performance specification and reference testing. London, British Standards Institution, 1999.
30. TAOUKIS P S and LABUZA T P, 'Application of time–temperature indicators as shelf-life monitors of food products', *Journal of Food Science*, 1989 **54** (4) 783–8.
31. TAOUKIS P S and LABUZA T P, 'Reliability of time–temperature indicators as food quality monitors under non-isothermal conditions', *Journal of Food Science*, 1989 **54** (4) 789–92.
32. BALLANTYNE A, An evaluation of time–temperature indicators, *Technical Memorandum No. 473*, Campden and Chorleywood Food Research Association, 1988.
33. SELMAN, J. D. and BALLANTYNE, A., 'Time–temperature indicators: Do they work?', *Food Manufacture*, 1988 **63** (12) 36–8, 49.
34. SELMAN, J. D., 'Time–temperature indicators: how they work' *Food Manufacture* 1990 **65** (8) 30–1 and 33–4.
35. Ministry of Agriculture, Fisheries and Food Publication, Time–temperature indicators: Research into consumer attitudes and behaviour, National Consumer Council, 1991.

36. GEORGE, R.M. and SHAW, R., A food industry specification for defining the technical standards and procedures for the evaluation of temperature and time–temperature indicators, *Technical Manual No. 35*, Campden and Chorleywood Food Research Association, 1992.
37. GIGIEL A.J., JAMES S.J. and EVANS J.A., Controlling Temperature During Distribution and Retail, Proceeding of the 3rd Karlsruhe Nutrition Symposium, *European Research towards Safer and Better Food*, 18–20 October 1998, edited by Gaukel V and Speiss W.E.L. pp. 284–92, 1998.
38. FRPERC Newsletter 'Predicting food temperatures in refrigerated transport' number 17, May 1997 pp. 4–5, University of Bristol.
39. FOSTER A.M. and QUARINI G.L., 'Using advanced modelling techniques to reduce the cold spillage from retail display cabinets into supermarket stores' ICR/IIR Conference, *Refrigerated Transport, Storage and Retail Display*, Cambridge, 29 March–1 April 1998, FRPERC Publication No. 586, University of Bristol.

6

Chilled food packaging

B. P. F. Day, Campden and Chorleywood Food Research Association

6.1 Introduction

During recent years there has been a greatly increased consumer demand for perishable chilled foods which are perceived as being fresh, healthy and convenient. The major food retailers have satisfied this consumer demand by providing an ever increasing range of value-added chilled food products. The wide diversity of chilled foods available is accompanied by a huge range of packaging materials and formats which are used to present attractively packaged foods in retail chill cabinets. This chapter overviews the requirements and types of packaging materials and formats which are commonly utilised for a broad variety of chilled food products. In addition, established and emerging packaging technologies for extending chilled food shelf-life, such as modified atmosphere packaging, vacuum packaging, vacuum skin packaging and active packaging, are described and new developments are highlighted. Process and packaging techniques that rely on heat treatments to achieve extended shelf-lives for chilled food products, such as hot-fill, sous-vide and in-pack pasteurisation, are outside the scope of this chapter, but are described in Chapter 11.

6.2 Requirements of chilled food packaging materials

Table 6.1 lists the main requirements of a chilled food package (Turtle 1988). Depending on the type of food packaged, not all of these requirements will need to be satisfied. The packaging material must contain the food without leaking, be non-toxic and have sufficient mechanical strength to protect the food and itself

Table 6.1 Main requirements of a chilled food package

• Contain the product	• Seal integrity
• Be compatible with food	• Prevent microbial contamination
• Non-toxic	• Protect from odours and taints
• Run smoothly on filling lines	• Prevent dirt contamination
• Withstand packaging processes	• Resist insect or rodent infestation
• Handle distribution stresses	• Be cost effective
• Prevent physical damage	• Have sales appeal
• Possess appropriate gas permeability	• Communicate product information
• Control moisture loss or gain	• Show evidence of tampering
• Protect against light	• Easily openable
• Possess antifog properties	• Be tolerant to operational temperatures

from the stresses of manufacture, storage, distribution and display. Certain packs require a degree of porosity to allow moisture or gaseous exchange to take place, and packaging materials used in these situations should possess appropriate permeability properties. Alternatively, most modified atmosphere packs require moisture and gases to be retained within the pack and hence the packaging materials used should possess appropriate barrier properties. The specific requirements for modified atmosphere packs are described later. Depending on the type of chilled food product, the packaging material may need to be tolerant of high temperatures experienced during hot filling, in-pack pasteurisation or re-heating prior to consumption. The packaging material, particularly with high-speed continuous factory operations, may need to be compatible with form–fill–seal machines. The pack closure must have seal integrity but at the same time should be easy to open. There may be a need for reclosure during storage after opening in the home. Also, with the increased incidence of malicious contamination, tamperproof or tamper-evident packaging is desirable. The package is the primary means of displaying the contained chilled food and providing product information and point-of-sale advertising. Clarity and printability are two pertinent features that require consideration in the choice of materials. Finally, the packaging must be cost-effective relative to the contained food. For example, a prepared ready meal retailing at a high price can support a considerably higher packaging cost than a yoghurt dessert selling at a fraction of that price.

6.3 Chilled food packaging materials

Once the requirements of a container for a particular chilled food product have been established, the next step is to ascertain which type of packaging material will provide the necessary properties. The answer is almost certain to be more than one type. Packaging materials consisting of paper, glass, metal or plastic

Table 6.2 Comparison of chilled food packaging materials

Material	Main technical advantages
Aluminium	Impermeability Lightweight Container axial strength Withstands internal pressure
Paper	Great variety of paper grades Ease of decoration Adjunct to all other packaging materials Lightweight
Semi-rigid plastics	Properties variable with type of plastic Choice of container shape In-house manufacture Lightweight
Flexible plastics	Properties variable by combination Very lightweight containers Tailor-made sizing
Glass	Chemical inertness Impermeability Product visibility Container axial strength Withstands internal vacuum pressure Reuse facility

have their individual advantages and these should be exploited when making a choice. The main technical advantages of current chilled food packaging materials are compared in Table 6.2, while the principal types of materials (and their abbreviations) are listed in Table 6.3 (Turtle 1988). For any particular product, a number of materials can generally be used, either as separate components or in the manufacture of a composite.

6.3.1 Paper-based materials

Paper and board are widely used in chilled food packaging. They are easy to decorate attractively and are complementary to all other packaging materials in the form of labels, cartons, trays or outer packaging. They are available with coatings such as wax, silicone and polyvinylidene chloride (PVDC) or as laminates with aluminium foil or flexible plastics. Such coating or lamination imparts heat-sealability or improves oxygen, moisture or grease barrier properties. For example, butter is traditionally packed in waxed paper or aluminium laminated paper.

Dual ovenable trays can be made of paperboard that is extrusion-coated with polyethylene terephthalate (PET). They can resist temperatures up to 220ºC and

Table 6.3 Chilled food packaging materials

Aluminium foil	EVOH (ethylene-vinyl alcohol)
Cardboard	HDPE (high density polyethylene)
Cellulose	HIPS (high impact polystyrene)
Cellulose fibre	LDPE (low density polyethylene)
Glass	LLDPE (linear low density polyethylene)
Natural casings	MXDE (modified nylon)
Paper	OPP (orientated polypropylene)
Metallised board	OPS (orientated polystyrene)
Metallised film	PA (polyamide-nylon)
Steel	PC (polycarbonate)
Plastics	PE (polyethylene)
ABS (acrylonitrile-butadiene-styrene)	PET (polyethylene terephtlalate)
APET (amorphous PET)	PETG (modified PET)
CA (cellulose acetate)	PP (polypropylene)
CPET (crystallised PET)	PS (polystyrene)
CPP (cast polypropylene)	PVC (polyvinyl chloride)
EPS (expanded polystyrene)	PVDC (polyvinylidene chloride)
EVA (ethylene-vinyl acetate)	UPVC (unplasticised polyvinyl chloride)

hence are suitable for microwave and conventional oven heating of chilled ready meals. Another application of paperboard in chilled food packaging is in the area of microwave susceptors which enable the browning and crisping of meat and dough products, e.g., pizza and pies, during microwave heating. A typical microwave susceptor is constructed of metallised PET film laminated to paperboard.

6.3.2 Glass

Glass jars and bottles are the oldest form of high-barrier packaging and have the advantages of good axial strength, product visibility, recyclability and chemical inertness. Returnable glass bottles are still used extensively for pasteurised milk in the UK. Aluminium caps and closures make opening simple, while tamper-evident features such as pop-up buttons provide an important consumer safety factor. Impact breakage of glass containers is a major disadvantage, but new glass technology and plastic sleeving with polyvinyl chloride (PVC) or expanded polystyrene (EPS) have helped to reduce glass breakage.

6.3.3 Metal-based materials

Pressed aluminium foil trays have a long history of use for prepared frozen meals and hot take-away food. They are also used for many chilled ready meals. Their temperature stability makes them ideal for conventional oven heating, but precautions should be taken to prevent arcing if used in microwave ovens. Guidelines have been developed for the successful use of foil containers in microwave ovens (Foil Container Bureau 1991). In some circumstances, aluminium foil enables more uniform heating than microwave-transparent trays (Bows and Richardson 1990). Aluminium foil or aluminium laminated paper are also used for many dairy products, such as butter, margarine and cheese. Aluminium foil is also used in cartonboard composite containers for chilled fruit juices and dairy beverages. In addition, aluminium or steel aerosol containers are used for chilled creams and processed cheeses.

6.3.4 Plastics

Plastics are the materials of choice for the majority of chilled foods. Chilled desserts, ready meals, dairy products, meats, seafood, pasta, poultry, fruit and vegetables are all commonly packed in plastics or plastic-based materials. Semi-rigid plastic containers for chilled foods are predominantly made from polyethylene (PE), polyproplyene (PP), polystyrene (PS), PVC, PET and acrylonitrile-butadiene-styrene (ABS). Other plastics such as polycarbonate (PC) are used in small quantities. Containers are available in a wide range of bottle, pot, tray and other shapes and thermoforming, injection moulding and blow moulding techniques give food processors the option of in-house manufacture. Flexible plastics offer the cheapest form of barrier packaging and may be used to pack perishable chilled food under vacuum or modified atmosphere. Multilayer materials are typically made by coextrusion or coating processes, using sandwich layers of PVDC or ethylene-vinyl alcohol (EVOH) to provide an oxygen barrier. Alternatively, plastics such as PE or PP may be metallised or laminated with foil to provide very high-barrier materials. The required technical properties and pack size and shape may be matched to a desired specification, thereby ensuring cost effectiveness. The oxygen and water vapour transmission rates of aluminium foil and selected monolayer plastic films are compared in Table 6.4.

Most polymers used for chilled food packaging are thermoplastics, i.e., they are reversibly softened by the application of heat, provided that no chemical breakdown occurs. PE is derived from the polymerisation of ethylene, whereas other thermoplastics such as PP, PVDC, PS, PVC, ethylene-vinyl acetate (EVA) and ABS are similarly polymerised from ethylenic monomers. In contrast, plastics such as polyamide (PA), PC and PET are manufactured by condensation reactions. For example, PET film is produced from PET resin, the polycondensation product of ethylene glycol and terephthalic acid, by a stretching process known as biaxial orientation.

6.4 Packaging techniques for chilled food

6.4.1 Modified atmosphere packaging (MAP)

MAP is an increasingly popular food preservation technique in which the gaseous atmosphere surrounding the food is different from air (Day 1989). The consumer demand for fresh, additive-free foods has led to the growth of MAP as

Table 6.4 Oxygen and water vapour transmission rates of selected packaging materials

Packaging film (25 μm)	Oxygen Transmission rate ($cm^3\,m^{-2}\,day^{-1}\,atm^{-1}$) 23°C: 0% RH	Water vapour transmission rate ($g\,m^{-2}\,day^{-1}$) 38°C: 90% RH
Al foil	neg.[a]	neg.[a]
EVOH	0.2–1.6[b]	24–120
PVDC	0.8–9.2	0.3–3.2
MXDE	2.4[b]	25
PET	50–100	20–30
PA6	80[b]	200
PETG	100	60
MOPP	100–200	1.5–3.0
UPVC	120–160	22–35
PA11	350[b]	60
PVC	2000–10 000[c]	200
OPP	2000–2500	7
HDPE	2100	68
PS	2500–5000	110160
OPS	2500–5000	170
PP	3000–3700	10–12
PC	4300	180
LDPE	7100	16–24
EVA	12 000	110–160
Microperforated	20 000–2 000 000[d]	–

[a] Dependent on pinholes.
[b] Dependent on moisture.
[c] Dependent on moisture and level of plasticiser.
[d] Dependent on degree of microperforation.

a technique to improve product image, reduce wastage and extend the quality shelf-life of a wide range of foods (Day 1992). Established chilled products now available in MAP include red meats, fish, seafood, poultry, crustaceans, offal, cooked and cured meats and fish, pasta, pizza, kebabs, cheese, cooked and dressed vegetable products, dairy and bakery goods, ready meals, and whole and prepared fresh fruit and vegetables (Day and Wiktorowicz 1999, Air Products Plc 1995).

Gases

The gas mixture used in MAP (see Table 6.5) must be chosen to meet the needs of the specific food product, but for nearly all products this will be some combination of carbon dioxide (CO_2), oxygen (O_2) and nitrogen (N_2) (Day and Wiktorowicz 1999). Carbon dioxide has bacteriostatic and fungistatic properties and will retard the growth of mould and aerobic bacteria. The combined negative effects on various enzymic and biochemical pathways result in an increase in the lag phase and generation time of susceptible spoilage microorganisms. However, CO_2 does not retard the growth of all types of

Table 6.5 Gas-mix guide for MAP of retail chilled food products

Chilled food item	% CO_2	% O_2	% N_2
Meat (red)	15–40	60–85	0–10
Meat (cured)	20–35	–	65–80
Meat (cooked)	25–30	–	70–75
Offal (raw)	15–25	75–85	–
Poultry (white)	20–50	–	50–80
Poultry (reddish)	25–35	65–75	–
Fish (white)	35–45	25–35	25–35
Fish (oily)	35–45	–	55–65
Crustaceans and molluscs	35–45	25–35	25–35
Fish (cooked)	25–35	–	65–75
Pasta (fresh)	25–35	–	65–75
Ready meals	25–35	–	65–75
Pizza	25–35	–	65–75
Quiche	25–35	–	65–75
Meat pies	25–35	–	65–75
Cheese (hard)	25–35	–	65–75
Cheese (mould-ripened)	–	–	100
Cream	–	–	100
Fresh fruit/vegetables	3–10	210	80–95
Vegetables (cooked)	25–35	–	65–75

microorganisms. For example, the growth of lactic acid bacteria is improved in the presence of CO_2 and a low O_2 content. CO_2 has little effect on the growth of yeast cells. The inhibitory effect of CO_2 is increased at low temperatures because of its enhanced solubility in water to form a mild carbonic acid. The practical significance of this is that MAP does not eliminate the need for refrigeration. The absorption of CO_2 is highly dependent on the water and fat content of the product. Excess CO_2 absorption can reduce the water-holding capacity of meats, resulting in unsightly drip. In addition, some dairy products can be tainted, and fruit and vegetables can suffer physiological damage owing to high CO_2 levels. If products absorb excess CO2, the total volume inside the package will be reduced, giving a vacuum package look known as pack collapse. In MAP, O_2 levels are normally set as low as possible to inhibit the growth of aerobic spoilage microorganisms and to reduce the rate of oxidative deterioration of foods. However, there are exceptions; for example, O_2 is needed for fruit and vegetable respiration, colour retention in red meats or to avoid anaerobic conditions in white fish MA packs.

Nitrogen is effectively an inert gas and has a low solubility in both water and fat. In MAP, N_2 is used primarily to displace O_2 in order to retard aerobic spoilage and oxidative deterioration. Another role of N_2 is to act as a filler gas so as to prevent pack collapse. Other gases such as carbon monoxide, ozone, ethylene oxide, nitrous oxide, helium, neon, argon, propylene oxide, ethanol vapour, hydrogen, sulphur dioxide and chlorine have been used experimentally or on a restricted commercial basis to extend the shelf-life of a number of food products. For example, carbon monoxide has been shown to be very effective at maintaining the colour of red meats, maintaining the red stripe of salmon and inhibiting plant tissue decay. However, the commercial use of most of these other gases is severely limited owing to safety concerns, regulatory constraints, negative effects on sensory quality or economic factors (Day 1992).

Argon (Ar) and nitrous oxide (N_2O) are classified as miscellaneous additives and are permitted gases for food use in the European Union. Air Liquide S.A. (Paris, France) has stimulated recent commercial interest in the potential MAP applications of using Ar and, to a lesser extent, N_2O. Air Liquide's broad range of patents claim that in comparison with N_2, Ar can more effectively inhibit enzymic activities, microbial growth and degradative chemical reactions in selected perishable foods. Although Ar is chemically inert, Air Liquide's research has indicated that it does have biochemical effects, probably due to its similar atomic size to molecular O_2 and its higher density and solubility in water compared with N_2 and O_2 (Brody and Thaler 1996). Hence, Ar is probably more effective at displacing O_2 from cellular sites and enzymic O_2 receptors with the consequence that oxidative deterioration reactions are likely to be inhibited. In addition, Ar and N_2O are thought to sensitise microorganisms to antimicrobial agents. This possible sensitisation is not yet well understood, but may involve alteration of the membrane fluidity of microbial cell walls, with a subsequent influence on cell function and performance (Day 1998). Clearly, more independent research is needed to better understand the potential beneficial effects of Ar and N_2O.

Packaging materials

Specifically with regard to MAP, the main characteristics to consider when selecting packaging materials are as follows:

Gas permeability. In most MAP applications, excluding fresh fruit and vegetables, it is desirable to maintain the atmosphere initially incorporated into the package for as long a period as possible. The correct atmosphere at the start will not serve for long if the packaging material allows it to change too rapidly. Consequently, packaging materials used with all forms of MA-packed foods (with the exception of fresh fruit and vegetables) should have barrier properties. Typically, the lidding film consists of 15μ PVDC-coated PET/60μ PE and the tray consists of 350μ PVC/PE (see Fig. 6.1). Alternatively, PA/PVDC/LDPE, PA/PVDDC/HDPE/EVA, OPP/PVDC/LDPE or PC/EVOH/EVA may be used for the lidding film, and HDPE, PET/EVOH/LDPE, PVC/EVOH/LDPE or PS/EVOH/LDPE used for the tray (Air Products Plc 1995).

The permeability of a particular packaging material depends on several factors such as the nature of the gas, the structure and thickness of the material, the temperature and the relative humidity (RH). Although CO_2, O_2 and N_2 permeate at quite different rates, the order $CO_2 > O_2 > N_2$ is always maintained and the permeability ratios CO_2/O_2 and O_2/N_2 are usually in the range 3 to 5. Hence, it is possible to estimate the permeability of a material to CO_2 or N_2

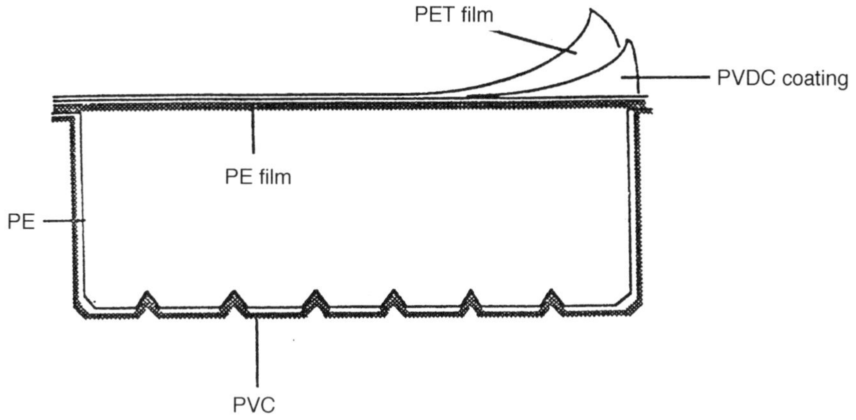

Fig. 6.1 Construction of a typical tray and lidding film MA pack.

when only the O_2 permeability is known. As a general rule, packaging materials with O_2 transmission rates less than $100\,cm^3\,m^{-2}\,day^{-1}\,atm^{-1}$ are used in MAP. Packaging materials are usually laminated or coextruded in order to have the necessary barrier properties required (Roberts 1990).

In contrast to other perishable foods that are packed in MA systems, fresh fruit and vegetables continue to respire after harvest and any packaging must take this into account. The depletion of O_2 and the accumulation of CO_2 are natural consequences of the progress of respiration when fresh fruit or vegetables are stored in a sealed package. Such modification of the atmospheric composition results in a decrease in the respiration rate, with a consequent extension in the shelf-life of fresh produce. However, packaging film of the correct permeability must be chosen to realise the full benefits of MAP of fresh produce (Day 1998). Typically, the key to successful MAP of fresh produce is to maintain an equilibrium MA (EMA) containing 2–10% O_2/CO_2 within the package. For highly respiring produce such as mushrooms, beansprouts, leeks, peas and broccoli, traditional films like LDPE, PVC, EVA, OPP and cellulose acetate (CA) are not sufficiently permeable. Such highly respiring produce is most suitably packed in highly permeable microperforated films. However, microperforated films are relatively expensive, permit moisture and odour loss, and may allow for the ingress of microorganisms into sealed packs during wet handling situations (Day 1998). A very interesting development for the packing of fresh prepared produce involves the use of high O_2 (70–100%) MAP which has been recently shown to overcome the many disadvantages of current air packing and low O_2 MAP. High O_2 MAP has been demonstrated to inhibit enzymic discolorations, prevent anaerobic fermentation reactions and inhibit microbial growth with the result of extending prepared produce shelf-life (Day, 1998; 1999a).

Water vapour transmission rate. Water vapour transmission rates are quoted in $g\,m^{-2}\,day^{-1}$ at a given temperature and RH. Similar to gas permeabilities, there is a wide variation between different packaging materials (see Table 6.4). However, there is no correlation between what is a good barrier to gas and what is a good barrier to water. A further complication is that some materials (e.g. nylons and EVOH) are moisture-sensitive and their gas permeabilities are dependent on RH (Day 1992).

Mechanical properties. Packaging materials used for MAP must have sufficient strength to resist puncture, withstand repeated flexing and endure the mechanical stresses encountered during handling and distribution. Additionally, if trays are to be thermoformed, the web must draw evenly and not thin excessively on the corners. Poor mechanical properties can lead to pack damage and leakage (Day 1992).

Sealing reliability. It is essential that an integral seal is formed in order to maintain the correct atmosphere within a MA pack. Therefore, it is important to select the correct heat-sealable packaging materials and to control the sealing operation. For example, in high-speed form–fill–seal operations, it is important to consider the hot tack of the material. Additionally, there is often a requirement for a peelable seal so that the consumer can gain easy access to the contents. However, the balance between peelability and integrity of the seal must be determined (Day 1992).

Transparency. For most MA packed foods, a transparent package is desirable so that the product is clearly visible to the consumer. However, high-moisture foods stored at chilled temperatures have the tendency to create a fog on the inside of the package, thereby obscuring the product. Consequently, many MAP films are treated with coatings or additives to impart antifog properties so as to improve visibility. These treatments only affect the wetability of the film and have no effect on the permeability properties of the film (Roberts 1990).

For some MA packed foods (e.g. green pasta and cured meats), it may be desirable to exclude light in order to reduce undesirable light-induced oxidation reactions. In these cases, light barriers such as colour-printed or metallised films may be used. Another influence of light is the possibility of a 'greenhouse effect' causing a temperature rise within chilled food packs (Malton 1976). However, Gill (1987) concluded that this effect was not an important factor in increasing temperatures of products displayed in chilled cabinets.

Type of package. The type of package used will depend on whether the product is destined for the retail or the catering trade. Popular options include flexible 'pillow packs', 'bag-in-box' and semi-rigid tray and lidding film systems (see Fig. 6.1).

Microwavability. The ability of MAP materials to withstand microwave heating is important, particularly in the case of ready-to-eat food products. For example, the low softening point of PVC makes the popular PVC/LDPE thermoformed trays unsuitable for microwave oven heating. Hence, materials with greater heat resistance, such as CPP, CPET and polystyrene-high temperature (PSHT) are used for MA-packed food products intended to be heated in a microwave oven.

6.4.2 Vacuum packaging (VP)

Vacuum packaging is an established technique for packaging chilled foods such as primal red meats, cured meats and cheese. Similar to MAP, VP extends shelf-life of food by removing O_2 and thus inhibiting the growth of aerobic spoilage microorganisms and reducing the rate of oxidative deterioration (Yamaguchi 1990).

Packaging materials

In order to maintain a vacuum around the food, high oxygen barrier materials are required. Although the requisite O_2 barrier for VP depends on the type of food packaged, O_2 transmission rates of less than $15\,cm^3\ m^{-2}\ day^{-1}\ atm^{-1}$ are generally required. Also, packaging materials with low water vapour transmission rates must be used. Typical VP materials consist of coextruded or laminated films such as OPP/EVOH/PE, PA/PE, PET/PE, OPP/PVDC/PE, OPP/ PVDC/ OPP and PVC/EVOH/PVC (Yamaguchi 1990).

6.4.3 Vacuum skin packaging (VSP)

Vacuum skin packaging is a technique which was developed to overcome some of the disadvantages of the traditional vacuum pack and MAP (White 1990). The VSP concept relies upon a highly ductile plastic barrier laminate which is gently draped over a food product, thereby moulding itself to the actual contours of the product to form a second skin. The product's natural shape, colour and texture are highlighted and, since no mechanical pressure is applied whilst drawing the vacuum, soft or delicate products are not crushed or deformed. Successes of VSP in the UK market include sliced cooked and cured meats, pâté and fish products (e.g. peppered mackerel). Unlike VP, VSP and MAP allow pre-sliced meats to be easily separated after pack opening. In VSP, the appearance of the product is enhanced and the wrinkle-free skin prevents product movement, thereby enabling vertical retail display. Also, since the bottom and top web films are sealed from the edge of the pack to the edge of the product, pack integrity is maximised and juice exudation is limited. Finally, VSP saves space in domestic refrigerators compared to MA packs and is ideally suited for freezing since the second skin prevents formation of ice crystals on the product surface, thereby eliminating freezer burn and dehydration (White 1990).

6.4.4 Active packaging

Active packaging refers to the incorporation of certain additives into packaging film or within packaging containers with the aim of extending product shelf-life

Table 6.6 Selected examples of active packaging systems

Active packaging system	Mechanisms	Actual and potential food applications
O_2 scavengers	1. Iron based 2. Metal/acid 3. Metal (e.g. platinum) catalyst 4. Ascorbate/metallic salts 5. Enzyme based	Bread, cakes, cooked rice, biscuits, pizza, pasta, cheese, cured meats and fish, coffee, snack foods, dried foods and beverages
CO_2 scavengers/ emitters	1. Iron oxide/calcium hydroxide 2. Ferrous carbonate/metal halide 3. Calcium oxide/activated charcoal 4. Ascorbate/sodium bicarbonate	Coffee, fresh meats and fish, nuts and other snack food products and sponge cakes
Ethylene scavengers	1. Potassium permanganate 2. Activated carbon 3. Activated clays/zeolites	Fruit, vegetables and other horticultural products
Preservative releasers	1. Organic acids 2. Silver zeolite 3. Spice and herb extracts 4. BHA/BHT antioxidants 5. Vitamin E antioxidant	Cereals, meats, fish, bread, cheese, snack foods, fruit and vegetables
Ethanol emitters	1. Encapsulated ethanol	Pizza crusts, cakes, bread, biscuits, fish and bakery products
Moisture absorbers	1. PVA blanket 2. Activated clays and minerals 3. Silica gel	Fish, meats, poultry, snack foods, cereals, dried foods, sandwiches, fruit and vegetables
Flavour/odour adsorbers	1. Cellulose triacetate 2. Acetylated paper 3. Citric acid 4. Ferrous salt/ascorbate 5. Activated carbon/clays/zeolites	Fruit juices, fried snack foods, fish, cereals, poultry, dairy products and fruit

(Day 1999b, Rooney 1995, Labuza and Breene 1989). Such additives or 'freshness enhancers' are capable of scavenging O_2, CO_2 and/or ethylene; releasing preservatives; emitting ethanol or CO_2; and absorbing moisture and/or flavours and odours (see Table 6.6). Many of the freshness enhancers claimed to extend shelf-life of food products (many of which are chilled foods) have yet to be approved for use with foods (Day 1999b, Rooney 1995).

Active packaging is an emerging and exciting area of food technology that is developing owing to advances in packaging technology, material science, biotechnology and new consumer demands. This technology can confer many preservation benefits on a wide range of ambient-stable and chilled food products. The intention is to extend the shelf-life of foods, whilst at the same

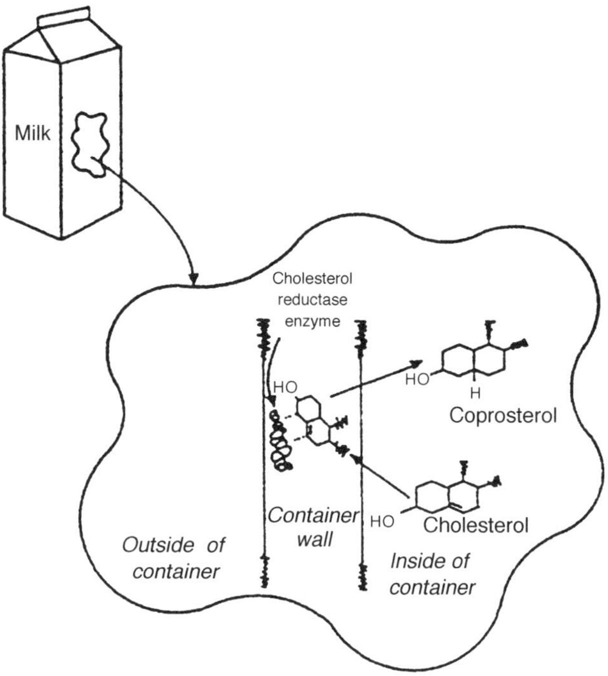

Fig. 6.2 In-package immobilised enzyme cholesterol removal from milk.

time maintaining nutritional quality and assuring microbial safety (Labuza and Breene 1989). The use of active packaging is becoming increasingly popular and many new opportunities will open up for utilising this technology in the future (Day 1999b).

Apart from the active packaging systems listed in Table 6.6, many other systems have been and are being developed by utilising a combination of various ceramics, enzymes, chemicals and materials to control in-pack atmospheres. These include photosensitive dyes which indirectly scavenge O_2 and antimicrobial packaging films and materials. Perhaps one of the most interesting developments involves the bringing together of packaging engineering and enzymology so that the package can actually change the chemical composition of packaged liquid foods (Brody 1990). For example, PharmaCal Ltd of California developed immobilised enzyme packages that are capable of removing lactose and cholesterol from milk (see Fig. 6.2).

6.5 Future trends

The following trends are likely to influence the chilled food packaging industry during future years:

6.5.1 Environmental factors

The packaging industry is going to face an increasing burden which has been placed on it by Packaging Waste legislation throughout Europe. The need to meet recycling and recovery targets is resulting in short term packaging solutions which may not be in the best interest of long-term sustainability. The legislation requires more packaging to be designed for reuse or recycling, yet under certain circumstances these may not be the most environmentally friendly solutions. Further more, the additional requirement for packaging minimisation may work against design for recycling since the lightest-weight materials may not be the easiest to recycle. Finally, future Packaging Waste legislation is likely to increase recycling and recovery targets, with the possibility of reuse targets being introduced (Stirling-Roberts 1999).

6.5.2 Consumer-driven packaging innovation

The increased focus on consumer needs provides the packaging industry with opportunities to innovate. Consumers respond well to added-value packaging innovations that improve the functionality and design of packaging, e.g. easy to open and resealing devices, easy to pour bottles, tamper-evident features, time/temperature indicating labels and microwave 'doneness' indicators. Packinging is increasingly being viewed as a strategic marketing tool. The retail supply chain is becoming more responsive and consumer-driven, and the effect on the packaging industry is that demand is for increasingly smaller quantities of consistent quality packagings delivered against tight schedules. In response to consumer demand, a wider range of packaging formats is now on offer and this range is likely to expand even wider in the future. For example, fresh chilled soups can be bought in glass bottles, plastic tubs, laminated paperboard cartons and flexible pouches, each of which has different technical and marketing advantages. Consumers have also responded well to convenient food packages and this trend will undoubtedly continue in the future, e.g. prepared fruits, vegetables and salads which are ready to eat or ready to heat in microwaveable packaging (Stirling-Roberts 1999).

6.5.3 New materials and technology

Driven by both environmental concerns and economics, new lighter-weight packaging materials are being developed throughout Europe and around the world. Examples include the introduction of superior performance plastics using metallocene catalysts and micro-flute corrugated paper and board materials. Research and development in the fields of edible and biodegradable packaging continues to expand as well as methods of reducing the cost of packaging recycling. Also, more strategic developments in the areas of barcode tagging, active and intelligent packaging and digital printing will continue to expand in the future (Stirling-Roberts 1999, Pugh 1998, Anon. 1998).

Regarding new materials, the invention of advanced catalyst technologies, such as metallocene technology, has enabled the design of new plastic resins, many of which allow for thinner, high-performance packaging materials which can be tailored for specific application requirements. Technical advantages claimed for metallocene plastic resins include improved rigidity, clarity and gloss; excellent heat seal and hot tack strength and puncture resistance; and light-weight or downgauging opportunities that were not previously available with traditional plastic resins (Pugh 1998, Anon. 1998).

6.6 Sources of further information

ROBERTSON G L, (1993). Food Packaging – principles and practice. Marcel Dekker, Inc., New York, USA.

KADOYA T, (ed.) (1990). Food Packaging. Academic Press, London, UK.

BRODY A L and MARSH K S, (eds) (1997). Wiley encyclopaedia of packaging technology, 2nd edn, J. Wiley and Sons, Inc., New York, USA.

6.7 References

AIR PRODUCTS PLC, (1995) The Freshline® guide to modified atmosphere packaging (MAP). Air Products Plc, Basingstoke, Hampshire, UK, pp. 1–66.

ANON, (1998) Plastics of the future. *Packaging News*, March edition, 63.

BOWS J R and RICHARDSON P S, (1990) Effect of component configuration and packaging material on microwave reheating of a frozen three component meal. Int. J. Fd. Sci. Technol. **25**, 538–50.

BRODY A L, (1990). Active packaging, *Fd Eng.*, **62** (4) 87–92.

BRODY A L and THALER M C (1996). Argon and other noble gases to enhance modified atmosphere food processing and packaging, *Proceedings of IoPP conference on 'Advanced technology of packaging'*, Chicago, Illinois, USA, 17th November.

DAY B P F, (1989) Extension of shelf-life of chilled foods, *Euro. Fd. Drink Rev.*, **4** 47–56.

DAY B P F, (1992) Guidelines for the good manufacturing and handling of modified atmosphere packed food products, *Technical Manual No. 34*, Campden and Chorleywood Food Research Association, Chipping Campden, Glos., UK.

DAY B P F, (1998) Novel MAP – A brand new approach, *Fd. Manuf.* **73** (11) 24–6.

DAY B P F, (1999a) High oxygen MAP for fresh prepared produce and combination products, *Proceedings of the International Conference on 'Fresh-cut produce'*, Campden and Chorleywood Food Research Association, Chipping Campden, Glos., UK.

DAY B P F, (1999b) Recent developments in active packaging, *South African Food & Beverage Manufacturing Review* **26** (8) 21–7.

DAY B P F and WIKTOROWICZ R, (1999) MAP goes online. *Fd. Manuf.* **74** (6) 40–1.

FOIL CONTAINER BUREAU, (1991) Foil in microwave ovens, *Packaging*

Magazine, **62** (684) 24.

GILL J, (1987) An investigation into the chemical and physical changes taking place in chilled foods during storage and distribution, Part A: Greenhouse effect, *Technical Memorandum No. 465*, Campden and Chorleywood Food Research Association, Chipping Campden, Glos., UK.

LABUZA T P and BREENE W M, (1989) Applications of active packaging for improvement of shelf-life and nutritional quality of fresh and extended shelf-life foods, *J. Fd Process. Pres.*, 13, 1–69.

MALTON R, (1976). Refrigerated retail display for fresh meat. *Institute of Meat Bulletin* **91** 17–19.

PUGH M, (1998) A catalyst for change, *Packaging Magazine*, March 12 edn, 30–31.

ROBERTS R, (1990) An overview of packaging materials for MAP. In: Day, B. P. F. (ed.), *International Conference on Modified Atmosphere Packaging*, Campden and Chorleywood Food Research Association, Chipping Campden, Glos., UK.

ROONEY M L, (ed.) (1995) Active food packaging. Chapman & Hall, London, UK, pp. 1–260.

STIRLING-ROBERTS A, (1999) Where to next? *Packaging News*, Dec. edition, 8–9.

TURTLE B I, (1988) Cost effective food packaging, *World Packaging Directory*, Cornhill Publications Ltd, London, pp. 67–72.

WHITE R, (1990) Vacuum skin packaging – a total packaging system. In: Monbiot, R. (ed.), *Flexible Packaging for Food Products Conference*, IBC Technical Services Ltd, Byfleet, Surrey, UK.

YAMAGUCHI N, (1990) Vacuum packaging. In: Kadoya, T. (ed.) *Food Packaging*, Academic Press, London, UK, pp. 279–92.

Part III

Microbiological and non-microbiological hazards

7

Chilled foods microbiology

S. J. Walker and G. Betts, Campden and Chorleywood Food Research Association

7.1 Introduction

Chilled foods represent a large and rapidly developing market with an extremely wide range of food types. Traditionally these were simple meat, poultry, fish and dairy products but recent trends have moved towards a greater variety and more complex products (Stringer and Dennis 2000). As more innovative products are produced, the variety of ingredients have also increased. Many of these ingredients are sourced around the world and relatively little may be known about their microbiological status. The numbers and types of microorganisms that may be isolated from the full range of chilled foods are very diverse. During the storage of chill products, the microbial flora of the product is not static but affected by many factors, principally the time and temperatures of storage. The spoilage and safety of chilled foods is a complex phenomenon involving physico-chemical, biochemical and biological changes. Often these interact and changes in one affect the rate of change in the others. This review will be concerned only with microbiological issues in relation to chilled foods.

With developments in the manufacture and transport of chilled foods, these items may now be rapidly disseminated over a wide geographical area, i.e. different countries and sometimes continents. Therefore should a microbiological issue arise it may be similarly widely spread. Consequently, the microbiological status of chilled foods has become more significant. Greater surveillance both within and between countries will allow such microbiological issues to be more rapidly identified, traced and resolved.

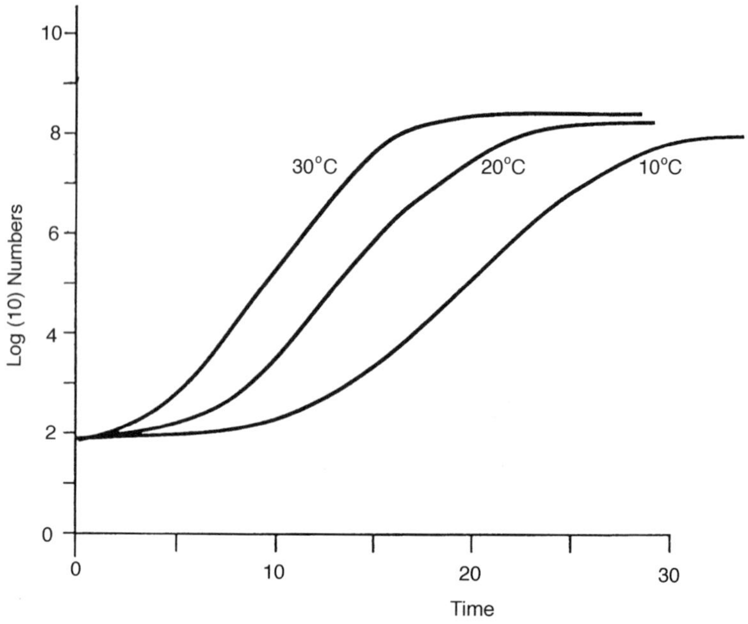

Fig. 7.1 Effect of temperature on the growth of microorganisms.

7.2 Why chill?

The effect of reducing temperature is to reduce the rate of food deterioration. This applies not only to the chemical and biochemical changes in foods but also to the activities of microorganisms. The effect of temperature on microbial growth is shown in Fig. 7.1. As the storage temperature decreases, the lag phase before growth (time before an increase in numbers is apparent) extends and the rate of growth decreases. In addition, as the minimum temperature for growth is approached, the maximum population size attainable often decreases. On a cellular basis, the effect of temperature on growth is a complex issue involving the cell membrane structure, substrate uptake, respiration and other enzyme activities. These have been discussed by Herbert (1989).

The range of temperatures over which microorganisms can grow is extremely wide. Michener and Elliott (1964) reported that a number of microorganisms, mainly yeasts, were able to grow below 0°C and a pink yeast isolated from oysters was reported to grow at −34°C. Therefore, chilling alone cannot be relied upon to prevent all microbial growth. The use of chill temperatures will, however, reduce the rate and extent of microbial growth.

7.3 Classification of growth

Microbiologists have attempted to characterise microorganisms based on their abilities to grow at various temperatures. Most commonly, the cardinal temperatures for growth (minimum, optimum and maximum growth temperatures) are used. With chilled foods, the factor of most concern is the *minimum growth temperature* (MGT), which represents the lowest temperature at which growth of a particular microorganism can occur. If the MGT of a microorganism is greater than 10°C, then this microorganism will not grow during chill storage. Whilst MGT values for microorganisms have been published, care is needed. If the time period for the investigation reporting this value was too short, or sampling intervals too widely spaced, the resultant value will be erroneous. For example, although an MGT of −0.4°C has been reported for *Listeria monocytogenes*, the lag phase before growth was in excess of 15 days (Walker *et al.*, 1990a). Had the study terminated before this time, the reported MGT would have been higher. The MGT is affected by other factors including the pH, salt, preservatives and previous heat treatments. A true estimate of the MGT can be determined only when other factors are optimal for growth.

If a microorganism is stored below its MGT, gradual death may occur, but often the microorganism will survive and growth will resume should the temperature subsequently be raised. It was noted by Alcock (1984) that the survival of salmonellae was worse at temperatures just below the MGT compared with lower temperatures. Storage at temperatures below the minimum for growth should not be considered to be a lethal process for microorganisms as in many cases, growth will resume if the temperature is subsequently raised.

The *optimum growth temperature* represents the temperature at which the biochemical processes governing growth of a particular microorganism are overall operating most efficiently. At this temperature, the lag phase before growth is minimised and the growth rate maximised. As the temperature rises above the optimum, the rate of growth decreases until the maximum growth temperature is reached. In general, the maximum growth temperature is only a few degrees (Celsius) higher than the optimum. With some specialised microorganisms, isolated from hot springs, the *maximum growth temperature* may exceed 90°C (Jay, 1978). At temperatures just above the maximum for growth, cell injury starts to occur. If the temperature is subsequently reduced, then growth may resume, although a period of time may be required to permit cell repair. At higher temperatures, the inactivation of one or more critical enzymes in the microorganism becomes irreversible and cell damage occurs, leading to cell death. Such microorganisms will not be able to repair and resume growth if temperatures are reduced. The concepts of cell injury and death have been discussed by Gould (1989b).

Based on the relative positions of the cardinal temperatures, microorganisms can be divided into four main groups, *viz.*, psychrophile, psychrotroph, mesophile and thermophile (Table 7.1). With chilled foods, the groups of most concern are the psychrophiles and psychrotrophs. In the past, these terms have been used synonymously, which has led to much confusion. It is now accepted that the term 'psychrophile' should only be used for microorganisms which have a low (i.e. ≤ 20°C) maximum growth temperature (Eddy, 1960). True psychrophiles are rare in food microbiology and generally limited to some

Table 7.1 Classification of microbial growth (Jay 1978, Walker and Stringer 1990, Morita 1973)

Temperature (°C)	Psychrophile	Psychrotroph	Mesophile	Thermophile
Minimum	< 0–5	< 0–5	(5 to)[a] 10	(30 to)[a] 40
Optimum	12–18	20–30 (35)[a]	30–40	55–65
Maximum	20	35 (40–42)[a]	45	(70 to)[a] > 80

[a] Figures in parentheses are occasionally recorded for microorganisms assigned to a particular classification.

microorganisms from deep-sea fish. The major spoilage microorganisms of chilled foods are psychrotrophic in nature.

7.4 The impact of microbial growth

Under suitable conditions, most microorganisms will grow or multiply. Bacteria multiply by the process of binary fission, i.e. each cell divides to form two daughter cells. Consequently, the bacterial population undergoes an exponential increase in numbers. Under ideal conditions some bacteria may grow and divide every 20 minutes and so one bacterial cell may increase to 16 million cells in 8 hours. Under adverse conditions, e.g. chilled storage, the generation time (doubling time) will be increased. For example with an increased time of two hours, the population obtained after 8 hours would be only 16 cells. Even under ideal conditions, growth does not continue unchecked and is limited by a range of factors including the depletion of nutrients, build-up of toxic by-products, changes to the environmental conditions or a lack of space.

7.4.1 Food spoilage

During growth in foods, bacteria will consume nutrients from the food and produce metabolic by-products such as gases or acids. In addition, they may produce a number of enzymes which results in the breakdown of the cell structure or of components (e.g. lipases and proteases). When only a few spoilage microorganisms are present, the consequences of growth may not be apparent. If however, the microorganisms have multiplied then the production of gases, acid, off-odours, off-flavours or deterioration in structure in the food may become unacceptable. In addition, the number of microorganisms may be apparent as a visible colony, production of slime or an increase in the turbidity of liquids. Some of the enzymes produced by spoilage bacteria may remain active, even when a thermal process has destroyed the causative microorganisms in the food.

The relationship between microbial numbers and food spoilage is complex and depends on the number, type and activity of the microorganisms present, the type of food and the intrinsic and extrinsic conditions. In some cases this is well understood, e.g. vacuum packed cod (Gram and Huss 1996). In general, a greater understanding is needed of the relationship between specific spoilage microorganisms in particular foods and the deterioration in sensory quality.

7.4.2 Food-borne pathogens

With many human pathogens, the greater the number of cells consumed, the greater the chance of microbial invasion, as the larger number of cells may be able to evade/swamp the body's defence mechanism. Higher numbers may also result in a shorter incubation period before the onset of disease. Consequently, control, and preferably inhibition, of growth in foods is essential. However, with some invasive pathogens (e.g. viruses, *Campylobacter*), the infectious dose is low and growth in the food may not be necessary. Other pathogenic microorganisms may produce a toxin in the food which results in disease. Preformed toxins are usually produced at high cells densities and so usually growth has occurred. If the toxin is heat stable, it may remain although all microorganisms have been eliminated from the food. Consequently it is important to control growth at all stages of the chill chain.

7.5 Factors affecting the microflora of chilled foods

7.5.1 Initial microflora

With healthy animal and plant tissues, microbial contamination is absent or at a low level except for the exterior surfaces. For example, fresh muscle from healthy animals is usually microbiologically sterile, and aseptically drawn milk from healthy cows contains only a few microorganisms (mainly streptococci and micrococci) derived from the teat canal. Similarly, the interior of healthy undamaged vegetables does not contain microorganisms although the exterior may be contaminated with a wide range of microorganisms of soil origin. During slaughter or harvesting, subsequent processing and packaging, these raw materials become contaminated from a wide range of sites. Typically, these sites include water, air, dust, soil, hides/fleece/feathers, animals, people, equipment and other food materials. Consequently, a large range of microorganisms can be isolated from foods. Those which are able to grow may potentially give rise to microbial spoilage or public health issues. The hygienic practices of all food operations, from slaughter/harvesting through retail sale to consumer use, will affect the level of microbial contamination of products. In general, the lower the initial level of contamination, the greater the time until microbial spoilage is evident.

7.5.2 Food type

The intrinsic properties (e.g. pH, water activity, acidity, natural antimicrobials) of different foods vary greatly. Such factors affect the ability of microorganisms to grow and the rate of growth and will be discussed in more detail in subsequent sections of this chapter. With different food types, the nutritional status varies although foods are generally not nutritionally limiting for microorganisms. Foods rich in nutrients (e.g. meat, milk, fish) permit faster growth than those with a lower nutritional status (e.g. vegetables) and so are more prone to spoilage. Slaughter and harvesting practices may affect the intrinsic properties of a food. For example, poor practices in the husbandry and slaughter of pigs may lead to pork being classified as DFD (dark, firm, dry) or PSE (pale, soft, exudative), both of which are more prone to spoilage than 'normal' pork. With DFD meat, the pH is higher, so permitting faster growth, whilst nutrient leakage and protein denaturation from PSE meat also allow more rapid microbial proliferation.

Even within a single food ingredient or product, variations in the pH, a_w and redox potential may occur and so affect the nature and rate of microbial multiplication. The situation may be further complicated in multi-component foods where migration of nutrients and gradients of pH, a_w and preservatives may occur. In addition, microorganisms unable to grow on one ingredient may come into contact with a more favourable environment and so permit growth.

7.5.3 Processing

Chill storage

The time of storage will affect microbial numbers. Generally, microbial numbers increase with time in chilled foods at neutral pH values, low salt concentrations and the absence of preservatives. However, low pH values or high salt concentrations in foods may cause microbial stasis, injury or even death. At chill temperatures however, the rate of death is often reduced and so the microorganism may survive for longer periods compared with higher (e.g. ambient) temperatures. In many cases a combination of processing and preservation factors may be used to achieve a safe, high quality product with an acceptable shelf-life. Such combination treatments have been reviewed by Gould (1996).

The ability of individual microorganisms to grow and their rates of growth are affected by temperature. As discussed previously, some microorganisms (mainly psychrotrophs) are better adapted to growth at chill temperatures. Therefore during chill storage not only will the total number of microorganisms change, but also the composition of the microflora will alter. For example, with freshly drawn milk, the microflora is dominated by Gram-positive cocci and rods, which may spoil the product by souring if stored at warm temperatures. At chill temperatures, these microorganisms are largely unable to grow and the microflora rapidly becomes dominated by psychrotrophic Gram-negative rod-shaped bacteria (most commonly *Pseudomonas* spp.) (Neill, 1974). A similar change in the microflora composition has also been reported for other chill-stored foods (Huis in't Veld, 1996).

Heating

As part of their manufacture, many chilled foods undergo a heating process. This will reduce microbial numbers, generally resulting in a pasteurised rather than a sterilised product, otherwise chill storage would be unnecessary. Food pasteurisation treatments have been reviewed by Gaze (1992). The degree of heat applied will affect the types of microorganisms able to survive. In general, the Gram-negative rod-shaped bacteria, which proliferate in chilled foods, are sensitive to heat and are readily eliminated. Although these bacteria may be isolated from, and even spoil, heated foods, their presence is usually attributable to post-heating contamination.

Some Gram-positive bacteria are tolerant to mild heat and classified as thermoduric (e.g. some *Lactobacillus*, *Streptococcus* and *Micrococcus* species, Jay 1978). However pasteurisation processes are designed to destroy all vegetative cells. Other bacteria however, produce heat-resistant bodies, called spores, which may survive. The genera of concern are *Bacillus* and *Clostridium* species, which include both pathogenic and spoilage strains. Whilst these bacteria are generally out-competed in chilled foods by the Gram-negative rod-shaped bacteria, *Bacillus* and *Clostridium* species may grow relatively unhindered in heated foods subsequently stored chilled.

Acidification

Several types of chilled foods are naturally acidic (e.g. fruit juices) or acidified using either a fermentation process (e.g. yoghurt) or by the direct addition of acids (e.g. coleslaw). As with temperature, microorganisms have pH limits for growth. The pH optima for most pathogenic bacteria is usually in the range 6.8–7.4 (Jay 1978), which is similar to the human body pH in which they are adapted to grow. Typical minimum pH values for growth are shown in Table 7.2. The minimum pH for the major spoilage bacteria of meat, poultry and dairy products is approximately 5.0 whereas other microbial types, in particular yeasts and moulds, may grow at pH values of 3.0 or less. Consequently, mildly acidified products may be spoiled by acid-tolerant bacteria (lactic acid bacteria and some Enterobacteriaceae) whilst more acid products are spoiled by yeasts and moulds. Both pH and temperature interact, and the minimum pH for growth at optimal temperatures may be significantly less than that at chill temperatures (George *et al.* 1988). At pH values below the minimum for growth, some microorganisms will die rapidly in the food, whilst others may persist for the life of the product. Of particular concern in acid foods is the pathogen *E. coli* O157: H7, which is more acid tolerant than other pathogens. It may grow at pH values of 4.0 or below and survive for considerable periods at lower pH values (Conner and Kotrola 1995, Deng *et al.* 1999).

In addition to the pH, the acid type used affects the microbial stability of the foods. The organic acids (lactic, acetic, citric and malic) are more antimicrobial than the inorganic acids (hydrochloric, sulphuric). Care is needed with published literature, as the minimum pH values reported often have used inorganic acids. Therefore the minimum pH for growth in foods is often higher than that quoted,

Table 7.2 Typical minimum pH and a_w values for growth of microorganisms (Anon. 1991b, Gould 1989, Mitscherlich and Marth 1984, ACMSF 1995)

Microorganism	Minimum pH	Minimum a_w
Bacillus cereus	4.9	0.91
Campylobacter jejuni	5.3	0.985
Clostridium botulinum (non-proteolytic)	5.0	0.96
Clostridium botulinum (proteolytic)	4.6	0.93
Clostridium perfringens	5.0	0.93
Escherichia coli	4.4	0.95
Escherichia coli O157:H7	3.8–4.2	0.97
Lactobacillus species	3–3.5	0.95
Pseudomonas species	5.0	0.95
Salmonella species	4.0	0.95
Staphylococcus aureus	4.0 (4.6)[a]	0.86
Many yeasts and moulds	< 2.0	0.8–0.6
Yersinia enterocolitica	4.6	0.95

[a] Minimum pH with toxin production.

as organic acids are present. Within the organic acids, the order of decreasing antimicrobial efficiency is usually acetic, lactic, citric then malic acid. With the organic acids, the undissociated form of the acid is effective against microorganisms and the degree of dissociation is dependent on the pH of the food. Organic acids and their use in food systems have been discussed by Kabara and Eklund (1991).

The pH and acid composition does not remain constant during the life of some foods. Changes in pH will affect the types of microorganisms able to grow and their growth rates. With some foods, fermentation results in a pH decreases during storage whilst in others an increase can be noted. For example, during maturation of mould-ripened cheeses, the pH value of cheese near the surfaces increases owing to proteolytic activity of the mould, and this has been related to the ability of *Listeria monocytogenes* to grow in these products, but not in the unripened cheeses (Terplan *et al.* 1987).

Reduced a_w

The a_w is a measure of the amount of water available in a food which may be used for microbial growth. As the a_w of a food is reduced, the number of microorganisms able to grow and their rate of growth is also reduced (Sperber 1983) (Table 7.2). The a_w of a food may be reduced either by the removal of water (i.e. drying) or by the addition of solutes (e.g. salt or sugar). In response to diet and health issues, many jam and sauce products have reduced their sugar content. Thus the intrinsic preservation system (i.e. low a_w) of the product has been compromised and some microorganisms, mainly yeasts, may now grow. These products generally recommend refrigeration after opening to prevent microbial growth. The a_w of a product may interact with other preservation factors, including temperature, to maintain the safety of chilled foods (Glass and

Doyle 1991). In general, yeasts and moulds are more tolerant than bacteria of low a_w values in foods (Jay 1978). As bacterial growth is largely inhibited, yeasts and moulds may then grow and cause the spoilage defects in such products.

Preservatives

In order to maintain their microbial stability, many chilled products contain natural or added preservatives, e.g. salt, nitrite, benzoate, sorbate. The presence of these compounds affects the type and rate of product spoilage that may occur. Their applications and mechanisms of action have been reviewed in Russell and Gould (1991). As discussed previously, *Pseudomonas* species tend to predominate on chilled fresh meat. The addition of curing salts (i.e. sodium chloride and potassium nitrite) to pork meat to form bacon, largely inhibits the growth of these microorganisms and spoilage is caused by other microbial groups (e.g. micrococci, staphylococci, lactic acid bacteria) (Gardner 1983, Borch *et al.* 1996). Similarly, the British sausage is largely a fresh meat product, but is preserved by the addition of sulphite. This prevents the growth of the *Pseudomonas* species, and microbial spoilage will be caused by sulphite-resistant *Brochothrix thermosphacta* or yeasts (Gardner 1983).

The number and type of microorganisms able to grow in preservative-containing chilled foods depend on the food type, preservative type, pH of the food, preservative concentration, time of storage and other preservation mechanisms in the food. Overall, yeasts and moulds tend to be more resistant to preservatives compared with bacteria and so may dominate the final spoilage microflora. Recent trends in food processing have tended to reduce or eliminate the use of preservatives. Care is needed with such an approach, as even small changes may compromise the product safety and microbiological stability.

Storage atmosphere

The use of modified atmospheres, including vacuum packaging, for the storage of chilled foods is increasing. Often these are chosen to maintain sensory characteristics of a product, but many will also inhibit or retard the development of the 'normal' spoilage microflora. *Pseudomonas* species, the major spoilage group in chilled proteinaceous foods, require the presence of oxygen to grow. Therefore, the use of vacuum packaging or modified atmospheres excluding oxygen will prevent the growth of this microbial group. Whilst other microorganisms can grow in the absence of oxygen, they generally grow more slowly and so the time to microbial spoilage is increased. The spoilage microflora of vacuum-packed meats is usually dominated by lactic acid bacteria or *Brochothrix thermosphacta* (Borch *et al.* 1996). In some cases, Enterobacteriaceae or coliforms may cause the spoilage of vacuum-packed and modified-atmosphere-packed (MAP) foods (Gill and Molin 1991).

Most commercial MAP gas mixtures for chilled food usually contain a combination of carbon dioxide, nitrogen and oxygen. The inhibition of bacteria becomes more pronounced as the amount of carbon dioxide increases. The effects of carbon dioxide on microbial growth have been discussed by Gill and Molin (1991). More recently, the use of other gases (including the noble gases) and high levels of oxygen have been used to extend the shelf-life of chilled foods (Day 2000).

Good temperature control is essential to obtain the maximum potential benefits of modified-atmosphere and vacuum packing. Should temperature abuse occur, the rate of spoilage will be similar to that without the atmosphere. It has been suggested that modified atmospheres will inhibit the 'normal' spoilage microflora of a food, but the growth of some anaerobic or facultatively anaerobic pathogens (e.g. *Clostridium* species, *Listeria monocytogenes*, *Yersinia enterocolitica*, *Salmonella* species and *Aeromonas hydrophila*) will be largely unaffected. Consequently the food may appear satisfactory but contain food-poisoning microorganisms. Of particular concern is the potential for growth of psychrotrophic *Clostridium botulinum*, which has been addressed by Betts (1996). In general, moulds require oxygen for growth and so are unlikely to create problems in vacuum-packaged or MAP (excluding oxygen) foods. Conversely, many yeasts can grow in the presence or absence of oxygen, although aerobic growth tends to be more efficient, thereby permitting more rapid growth.

Combinations

Many chill products do not depend on a single preservation system for their microbial stability, but a combination of the factors described above. These can be effective in controlling microbial growth (Gould 1996). With such foods, care is needed during their manufacture, distribution and sale because inadequate control of one factor may permit rapid growth. Furthermore, the use of two or more systems in combination may select for a particular microbial type (Gould and Jones 1989). For example, 'sous-vide' processing involves the vacuum packaging of foods, followed by a relatively mild heat treatment (pasteurisation). The heat treatment will eliminate vegetative microorganisms but not spore-forming bacteria. During subsequent chill storage (up to 30 days) in vacuum packaging, anaerobic spore-forming bacteria, including *Cl. botulinum*, may grow in the absence of other microorganisms. In order to prevent this happening, the product should be stored below the minimum temperature for growth of *Cl. botulinum*, the formulation of the product adjusted to prevent growth, or the heat treatment applied increased (Betts, 1992).

7.6 Spoilage microorganisms

Microbiological spoilage of chilled foods may take diverse forms, but all are generally as a consequence of growth which manifests itself in a change in the sensory characteristics. In the simplest form, this may be due to growth *per se* and often the production of visible growth, and this is common in moulds which produce large often pigmented colonies. Bacteria and yeasts may also produce visible (sometimes pigmented) colonies on foods. Other forms of spoilage including the production of gases, slime (extracellular polysaccharide material), diffusible pigments and enzymes which may produce softening, rotting, off-odours and off-flavours from the breakdown of food components. The taints produced by microbial spoilage have been reviewed by Dainty (1996) and Whitfield (1998).

Spoilage is usually most rapid in proteinaceous chilled foods such as red meats, poultry, fish, shellfish, milk and some dairy products. These products allow good microbial growth as they are highly nutritious, have a high moisture content and relatively neutral pH value. In an attempt to reduce the spoilage rates of these foods, they are often modified as discussed previously. For chilled products, these modifications may not entirely prevent microbial growth and spoilage, but do limit the rate and nature of spoilage. In general the microorganisms responsible for spoilage of a food are those which are best able to grow in the presence of the preservation mechanisms that are operating within that food. Care is needed to distinguish between those microorganisms present in spoiled food and those responsible for the spoilage defect (often called specific spoilage organisms or SSO) which may be only a fraction of the microflora (Gram and Huss 1996). Consequently, the relationship between sensory spoilage and microbial numbers is often only poorly correlated.

Traditional microbiology is often of limited value for control of spoilage microorganisms as the time taken to get results represents a significant proportion of the shelf-life. Recently more rapid and molecular techniques have become available for the detection of general or specific spoilage organisms (Venkitanarayanen *et al.* 1997, Gutiérrez *et al.* 1997). For discussion in this chapter, spoilage microorganisms have been arbitrarily divided into six categories: Gram-negative (oxidase positive) rod-shape bacteria; coliform enterics; Gram-positive spore-forming bacteria; lactic acid bacteria; other bacteria; yeasts and moulds.

1. Gram-negative (oxidase positive) rod-shaped bacteria

Overall, this group comprises the most common spoilage microorganisms of fresh chilled products. The minimum growth temperatures are often 0–3°C and they grow relatively rapidly at 5–10°C. Although they may represent only a small proportion of the initial microflora, they rapidly dominate the microflora of fresh proteinaceous chilled stored foods (Huis in't Veld 1996, Cousin 1982, Gill 1983). Within this general group, the genus *Pseudomonas* is most common although other genera include *Acinetobacter*, *Aeromonas*, *Alcaligenes*, *Alteromonas*, *Flavobacterium*, *Moraxella*, *Shewenella* and *Vibrio* species (Walker and Stringer 1990). These microorganisms are common in the environment, particularly in water, and so many easily contaminate foods. Often they may proliferate on inadequately cleaned surfaces of food processing plant or equipment and so contaminate foods.

The Gram-negative (oxidase positive) rods may spoil products by the production of diffusible pigments, slime material on surfaces and enzymes which result in food rots, off-flavours and off-odours (Jay 1978, Cousin 1982, Gill 1983). Some of the enzymes produced by *Pseudomonas* species are extremely heat-resistant and may produce long-term defects (e.g. rancidity or age-gelation) in thermally processed products with extended shelf-lives.

Although well adapted to grow at chill temperatures, this group tends to be sensitive to other factors such as the presence of salt or preservatives, lack of oxygen, low (< 5.5) pH and a low (< 0.98) a_w. Should these preservation mechanisms be present in a food, the Gram-negative (oxidase positive) rod-shaped bacteria compete less well and other microbial groups may cause spoilage. *Vibrio* species are unusual as they tolerate relatively high salt levels and so may cause spoilage on chilled stored bacon and other cured products. *Photobacterium phosphorum*, a very large marine vibrio is the dominant spoilage microorganism in vacuum packaged cod (Gram and Huss 1996). Overall, this group is not heat-resistant and so is readily removed by mild thermal treatments. Their presence in heat processed foods is usually as a consequence of post-process contamination.

2. Coliform/enteric bacteria

This bacterial group also consists of Gram-negative rods, but these may be distinguished from the above group by a negative oxidase reaction. Traditionally, microbiologists have tended to examine for these groups separately, as their sources, significance and factors affecting growth may differ. This group is frequently used as an indicator of inadequate processing or post-process contamination. Compared with the Gram-negative (oxidase positive) rod-shaped bacteria, the coliform-enteric group is generally less well adapted to growth at temperatures of less than 5–10°C although many may grow at temperatures as low as 0°C (Ridell and Korkeala 1997). However, they often dominate the flora at temperatures of 8–15°C (Huis in't Veld 1996, Cousin 1982). The coliform/enteric group is less sensitive to changes in pH compared with the Gram-negative (oxidase positive) rod-shaped bacteria and so of more significance in mild acid products. They are however, generally sensitive to low a_w, preservatives, salt and thermal treatments (Jay 1978).

The coliform/enteric group do not necessarily require the presence of oxygen for growth. In addition, they have a fermentative metabolism and so may break down carbohydrates to give acids, which may result in souring of milk (Cousin 1982). In contrast, the metabolism of the Gram-negative (oxidase positive) bacteria is oxidative and fermentation does not occur. Other types of spoilage include the production of pigmented growth, gases, slime, off-odours and off-flavours. Off odours have been described as 'grassy', medicinal, unclean and faecal (Walker and Stringer 1990).

Typical spoilage species include *Citrobacter*, *Escherichia*, *Enterobacter*, *Hafnia*, *Klebsiella*, *Proteus* and *Serratia* (Jay 1978, Walker 1988). These microorganisms are widely disseminated in the environment, including in animals. Poor slaughter and dressing practices may contribute to their presence in foods.

3. Gram-positive spore-forming bacteria

This group, of particular significance, can produce heat-resistant bodies (spores) which can survive many thermal processes. Such heating may destroy all vegetative cells, leaving the relatively slow growing spore-formers to dominate the microflora. The minimum growth temperatures are often 0–5°C, although growth is often slow below 8°C (Huis in't Veld 1996, Cousin 1982, Coghill and Juffs 1979).

The genera of concern in this group are *Bacillus* and *Clostridium* species. Again, these are common in the environment and spores may survive for considerable periods. The most common form of spoilage is the production of large quantities of gas which may result in pack or product blowing (Cousin 1982, Walker 1988). The heat resistance of psychrotrophic strains is considered to be lower than that of mesophilic strains (Reinheimer and Bargagna 1989), but the former group is of concern in chilled pasteurised foods.

4. Lactic acid bacteria

At chill temperatures, lactic acid-producing bacteria grow slowly if at all. Consequently, if they are to cause spoilage, growth of most other bacterial species must be inhibited. This group is more tolerant of low pH than other spoilage bacteria and may multiply at pH values as low as 3.6 (Jay 1978). The lactic acid bacteria are also more resistant than the previously discussed spoilage bacteria to slight reductions in the a_w and some *Pediococcus* species are salt-tolerant. Lactic acid bacteria usually predominate on vacuum-packed products and in some modified-atmosphere-stored foods, and may even grow in atmospheres containing 100% carbon dioxide (Gill and Molin 1991). This bacterial group comprises both rod- and coccus-shaped Gram-positive bacteria and typical genera include *Carnobacterium, Lactobacillus, Leuconostoc, Pediococcus* and *Streptococcus* species (Borch *et al.* 1996). Spoilage is generally by the production of acid which results in souring with or without concomitant gas production (Walker and Stringer 1990).

Lactic acid-producing bacteria are deliberately added during the manufacture of some chilled foods (e.g. cheese, yoghurts, some salamis) and are essential for the development of the desired product characteristics. In addition, there is much interest in the potential use of lactic acid bacteria as a novel preservation system, as many produce antimicrobial compounds in addition to acids (Lücke and Earnshaw 1991).

5. Other bacteria

Depending on the food type and preservation system operating, other microorganisms may also cause problems in chilled foods. For example, *Brochothrix thermosphacta* is a Gram-positive rod-shaped bacterium which is occasionally present on raw meats but does not normally create a spoilage problem. Products preserved with sulphite (e.g. fresh British sausage) may encourage the development of this bacterium (Gardner 1981). Furthermore, it can grow in atmospheres with a low oxygen level and/or high carbon dioxide concentration and so may cause problems in vacuum-packed or modified-atmosphere-packed meat products. In vacuum-packed sliced meats, this microorganism produces an objectionable pungent 'cheesy' odour.

Micrococcus species are Gram-positive cocci which can grow in the presence of high salt concentrations. They tend not to grow well at chill temperatures but can cause souring and slime production on cured meats and in curing brines should temperature abuse occur (Gardner 1983). Other microorganisms that may cause spoilage problems in cured meats and/or vacuum-packed meat products are *Corynebacterium*, *Kurthia* and *Arthobacter* species (Gardner 1983, Gould and Russell 1991).

6. Yeasts and moulds

Compared with bacteria, both yeasts and moulds grow more slowly in foods permitting good growth and so are generally out-competed. Therefore this group is seldom responsible for the spoilage of fresh proteinaceous foods. If, however, the conditions in the food are altered to limit bacterial growth, the role of yeasts and moulds may become more significant. Many yeasts can grow at temperatures less than 0ºC (Michener and Elliott 1964). Furthermore, yeasts and moulds are generally more resistant than bacteria to low pH, reduced a_w values and the presence of preservatives (Jay 1978). Moulds tend to require oxygen for growth whereas many yeasts can grow in the presence or absence of oxygen. Most yeasts and moulds are not heat-resistant and are readily destroyed by a thermal process. The mould genus *Byssochlamys* however, may produce relatively heat-resistant ascospores (Bayne and Michener 1979).

Freshly collected meat, poultry, fish and dairy products rarely contain yeasts or moulds but they rapidly become contaminated from the environment. In particular, air movements may be an important vector of transmission, especially with mould ascospores. Typical spoilage yeasts include *Candida*, *Debaryomyces*, *Hansenula*, *Kluveromyces*, *Rhodotorula*, *Saccharomyces*, *Torula* and *Zygosaccharomyces* species (Walker and Stringer 1990, Pitt and Hocking 1985). Moulds that may be isolated from spoiled chilled foods include *Aspergillus*, *Cladosporium*, *Geotrichum*, Mucor, *Penicillium*, *Rhizopus* and *Thamnidium* species (Pitt and Hocking 1985, Filtenborg *et al.* 1996). Fungal spoilage may be characterised by the production of highly visible, often pigmented, growth, slime, fermentation of sugars to form acid, gas or alcohol, and the development of off-odours and off-flavours. Odours and flavours have been described as yeasty, fruity, musty, rancid and ammoniacal.

As with the lactic acid bacteria, yeasts and moulds are sometimes deliberately added to food products. For example, the development of *Penicillium camembertii* on the surfaces of Brie and Camembert cheeses is essential for the desired flavour, odour and texture characteristics. This mould growing on other types of cheeses would be described as a spoilage defect.

7.7 Pathogenic microorganisms

Foods may be considered to be microbiologically unsafe owing to the presence of microorganisms which may invade the body (e.g. *Salmonella*, *Listeria monocytogenes, E. coli* O157:H7 and *Campylobacter*) or those which produce a toxin ingested with a food (e.g. *Clostridium botulinum*, *Staphylococcus aureus* and *Bacillus cereus*). The growth of pathogenic microorganisms in foods may not necessarily result in spoilage, and so the absence of deleterious sensory changes cannot be relied upon as an indicator of microbial safety. Furthermore, some toxins are resistant to heating and so may remain in a food after viable microorganisms have been removed. It is therefore essential that an effective programme is used to ensure the safety of foods from production, through processing, storage and distribution to consumption. Within the UK, the recent trends in food poisoning and the issues contributing to this have been extensively reviewed by Border and Norton (1997).

As discussed previously, storage at chill temperatures cannot prevent all microbial growth, but can prevent the growth of some types and retard the rate of growth in others. As far back as 1936, Prescott and Geer recommended that foods permitting growth of microorganisms should be stored at less than 10°C (50°F) and preferably ca. 4°C (39°F) to prevent the growth of pathogens or toxin production. That was sound advice in terms of the food-borne pathogens recognised at that time. The risk of growth by food-borne pathogens is a combination of the minimum growth temperatures, the growth rate at chill temperatures and the time and temperature(s) of storage. The minimum growth temperatures of pathogenic bacteria have been discussed by Walker and Stringer (1990).

Whilst the majority of food-borne disease is caused by relatively few bacterial types – mainly *Salmonella* and *Campylobacter* (Border and Norton 1997), the number of bacteria recognised as food-borne pathogens, however, has steadily increased. Whilst this may, in part, reflect a true underlying increase in the incidence, it may also be due to a greater awareness of these microorganisms and improvements in methodologies. For discussion in this chapter, the pathogenic bacteria of concern for chilled foods can be arbitrarily divided as follows.

Microorganisms capable of growth at temperatures below 5°C

This group is potentially of greatest concern as they continue to multiply even with 'good' refrigeration temperatures. Although growth may continue, temperature control is critical and the growth rate becomes increasingly slow as the temperature is reduced (see Fig. 7.1). In addition, temperature control can interact effectively with other factors to prevent or greatly limit growth.

Listeria monocytogenes

The bacterium now identified as *L. monocytogenes* was first recognised as a human pathogen in 1926 (Murray *et al.*), but its role in food-borne disease was not apparent until the late 1970s. Reported cases in the UK increased dramatically during the 1980s, and decreased during subsequent years. The symptoms of disease are protean and range from a mild flu-like illness to meningitis, septicaemia, stillbirths and abortions (Ralovich 1987). In general, the major symptoms of disease are restricted to the pregnant mother, foetus, elderly and immunocompromised. With the latter three groups, the mortality level can be high (McLauchlin 1987). The epidemiology of *L. monocytogenes* has been discussed by Schuchat *et al.* (1991).

A very wide range of foods including meat, poultry, dairy products, seafoods and vegetables have been reported to be contaminated with *L. monocytogenes* and have been reviewed by Bell and Kyriakides (1998b). Whilst the total absence of *L. monocytogenes* from raw meats, poultry and vegetables is difficult to ensure, the bacterium has been isolated from products which have undergone a listericidal thermal process (Lund 1990). Such isolations are of concern as many of these chilled foods may be consumed without further heating. The presence of *L. monocytogenes* on cooked foods suggests that post-process contamination may have occurred. Several studies have shown this bacterium has been isolated from a wide range of sites in several types of factory (Cox *et al.* 1989) and may be spread by some cleaning procedures (Holah *et al.* 1993). Sites of particular concern include those where water is present. Environmental control of *Listeria*, particularly in key areas of production (e.g. after cooking) is crucial to the prevention of product contamination. Whilst the number of cases of reported listeriosis in England and Wales peaked dramatically between 1986 and 1988 which was associated with contaminated imported pâté. Following public warnings about this, the number of cases declined to the annual rate prior to this (100–150 cases per year) (Border and Norton 1997).

The major concern with *L. monocytogenes* is its ability to grow at low temperatures, and a minimum growth temperature of –0.4°C has been reported (Walker *et al.* 1990). Temperature control will however, retard the rate of growth (Fig. 7.2). Conversely, temperature abuse during storage of a food can exacerbate problems. *Listeria monocytogenes* is more resistant than many other vegetative bacteria to some, but not all, of the preservation mechanisms used in food manufacture (e.g. chilling, reduced water activity) and these have been reviewed by Walker (1990). Whilst resistance may be noted to these preservation systems when examined individually, foods are complex and interactions may occur which effectively prevent growth. The use of predictive models (see Section 7.9) for microbiology is an efficient method to identify such interactions.

L. monocytogenes is not considered to be a classically heat-resistant bacterium. It is generally accepted that conventional HTST milk pasteurisation (71.7°C/15 seconds) will eliminate this microorganism when freely suspended in milk (Bradshaw *et al.* 1991). In other foods, decimal reduction times of 8–16 seconds have been reported at 70°C (Gaze *et al.* 1989). It has been recommended (Anon. 1989) that foods subject to a cook-chill process be heated to a minimum of 70°C for 2 minutes (or the thermal equivalent) to ensure the effective

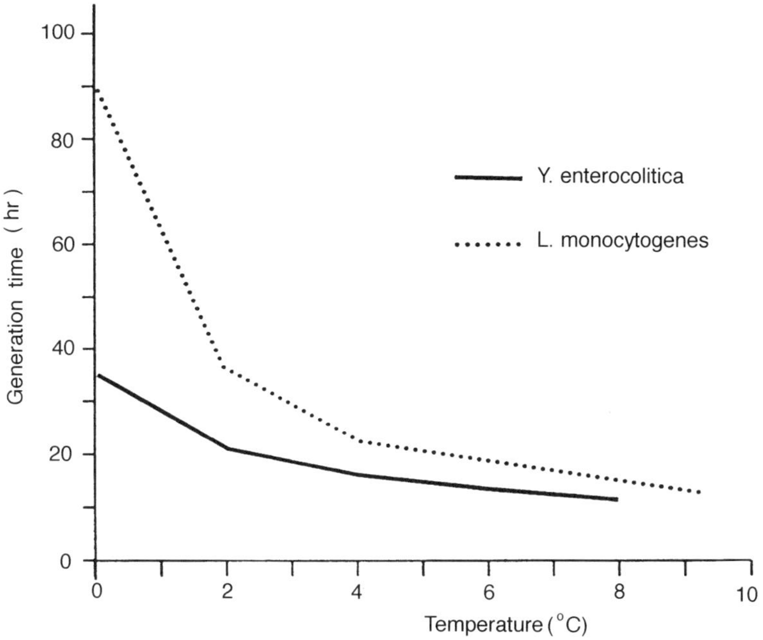

Fig. 7.2 Effect of temperature on the generation time of *L. monocytogenes* and *Y. enterocolitca*

elimination of this bacterium. Overall, the control of *L. monocytogenes* in foods and food environments (Holah 1999) is of concern to food processors.

Yersinia enterocolitica

Like *L. monocytogenes*, *Y. enterocolitica* was first described over 50 years ago (Schliefstein and Coleman 1939) but largely ignored as an agent of food-borne disease until the 1970s. Outbreaks of disease have implicated chilled foods such as pasteurised milk (Tacket *et al.* 1984), tofu (Tacket *et al.* 1985) and chocolate milk (Black *et al.* 1978). Whilst the reported incidence of *Y. enterocolitica* in gastrointestinal samples is generally low, it has increased, but as before, this may be due not only to a true underlying increase but also to a greater awareness of this bacterium, recognition of symptoms and improved methodologies. In some countries (e.g. Belgium and the Netherlands) disease by *Y. enterocolitica* has surpassed that of *Shigella* and even rivals that of *Salmonella* (Doyle 1990).

The symptoms of human yersiniosis are protean (Schiemann 1989). Overall, acute gastroenteritis is the most common symptom, particularly with children, and is characterised by diarrhoea, abdominal pain, fever and less commonly, vomiting. With adolescents, abdominal pain may be localised in the right iliac fossa area of the body and misdiagnosed as appendicitis. In the outbreak involving chocolate milk, 17/257 (6.6%) of the cases had their appendix removed (Black *et al.* 1978). The mortality rate from human yersiniosis is low and, other than cases involving appendectomies, the symptoms are generally self-limiting and rarely require treatment (Schiemann 1989). With adults, secondary symptoms may occur several weeks after the typical gastrointestinal symptoms disappear. Most commonly these are post-infectious polyarthritis and erythema nodosum (Schiemann, 1989).

A wide variety of foods have been reported to be contaminated with *Y. enterocolitica* including many chilled products, i.e. raw and cooked meats, poultry, seafoods, milk, dairy products and vegetables (Greenwood and Hooper 1989). Care is needed as the isolates responsible for disease generally belong to a few specific bio-serotypes, whilst those from foods and the environment belong to a wide range of bio-serotypes (Gilmour and Walker 1988, Logue *et al.* 1996). Therefore the pathogenic significance of food isolates should be ascertained before the food is condemned as a health risk. The bio-serotypes responsible for human disease are frequently isolated from pigs and occasionally pork products (Schiemann 1989).

The minimum reported growth temperature for *Y. enterocolitica* is −1.3°C and the bacterium grows relatively well at chill temperatures (Walker *et al.* 1990b). As with *L. monocytogenes*, reducing the storage temperature has an increasingly dramatic effect on the growth of *Y. enterocolitica* (Fig. 7.2). Furthermore, storage at refrigeration temperatures will interact with other preservation factors present in foods to prevent growth of this bacterium. Factors affecting the growth of *Y. enterocolitica* have been discussed by Walker and Stringer (1990). *Y. enterocolitica* is a heat-sensitive bacterium and will be readily eliminated from foods by heating (Lovett *et al.* 1982). It has, however, been reported from cooked meats, seafoods and pasteurised dairy products, which indicates that post-process contamination had occurred. Greater attention is required for the environmental control of *Y. enterocolitica* in food manufacturing establishments. As yet, relatively little is published on this aspect.

Aeromonas hydrophila

The role of *A. hydrophila* as an agent of food-borne disease is still a matter of controversy as no fully documented outbreaks have been reported. This bacterium however, does possess many of the characteristics of other pathogenic bacteria (Cahill 1990). As with *Y. enterocolitica*, the number of reported cases of *A. hydrophila* gastroenteritis in England and Wales has risen during the 1980s (Anon. 1991a). The reasons for this are as described previously. Incidents of food-borne disease implicating *A. hydrophila* have included oysters and prawns (Todd *et al.* 1989) – both chilled foods. Within the genus *Aeromonas*, some of the other major motile species (i.e. *A. hydrophila*, *A. sobria* and *A. caviae)* may be considered to be pathogenic (Stelma 1989). All three of these species have been isolated from a variety of chilled foods (Abeyta and Wekell 1988, Fricker and Tompsett 1989).

The minimum reported growth temperature for *A. hydrophila* is −0.1 to 1.2°C (four strains tested) and so growth will occur at chill temperatures (Walker and

Stringer 1987). As with the previous psychrotrophic pathogens, temperature control is important and temperature abuse will greatly increase the rate of growth. Relatively little is published about the heat resistance of *A. hydrophila* but the bacterium is considered to be heat-sensitive and so may be readily eliminated from foods (Palumbo and Buchanan 1988). The effects of other factors (e.g. pH, salt, preservatives etc.) on the growth of *A. hydrophila* have been reviewed by Palumbo and Buchanan (1988). Little has been published on the presence of *A. hydrophila* in the processing environment, but it is likely that it will be isolated particularly from wet areas.

Bacillus cereus

The role of *B. cereus* as a spoilage bacterium of chilled foods is well recognised (Griffiths and Phillips 1990). Many such strains may grow at temperatures as low as 1°C (Coghill and Juffs 1979). This bacterium may also cause food-borne disease but the number of reported cases is generally low (Border and Norton 1997). The minimum reported growth temperatures of these strains is usually 10–15°C (Goepfert *et al.* 1972, Johnson 1984) although some isolates from outbreaks which involved vegetable pie, pasteurised milk and cod were able to grow and produce toxins at 4°C (van Netten *et al.* 1990, Jaquette and Beuchat 1998). In addition, psychrotrophic, presumptively enterotoxigenic strains were frequently isolated from pasteurised milks and some cook-chill meats (van Netten *et al.* 1990). If temperature abuse of the product occurred, the time until the toxin was detected was reduced by 50% when the temperature was raised from 4 to 7°C.

Bacillus cereus may be of particular significance in foods which have been heated or pasteurised, as the heat treatment may have eliminated other competitor microorganisms. During subsequent chilled storage, spores which may survive the heat treatment may germinate and grow. Although little published information is available, the heat resistance of psychrotrophic *B. cereus* (and other related species) is generally lower than that of the mesophilic strains (Reinheimer and Bargagna 1989). Other *Bacillus* species (i.e. *B. subtilis* and *B. licheniformis*) may also cause human disease (Kramer and Gilbert 1989). Although psychrotrophic strains of these have been isolated from milk, their association with human disease is at present unclear.

Clostridium botulinum

Human botulism is caused by the ingestion of a neurotoxin, and, based on the antigenic analysis of this, seven types can be distinguished (named A–G) (Hauschild 1989). Traditionally, food-borne disease was caused by types A and B. It is now well recognised that types E and F may also cause disease following the ingestion of preformed toxin. The strains responsible for disease can be divided into two main groups. Firstly, types A and some strains of B and F are proteolytic and so often cause putrefaction of foods if substantial growth occurs (Hauschild 1989). Secondly, types E and others of B and F are non-proteolytic and so the consequences of growth in foods will be less pronounced (Hauschild,

Table 7.3 Comparison of proteolytic and non-proteolytic strains of *Clostridium botulinum* (Betts 1992, Hauschild 1989)

	Cl. botulinum	
	Proteolytic	Non-proteolytic
Minimum temperature	10–12°C	3.3–5.0°C
Minimum pH	4.6	5.0
Maximum salt	10%	5–6.5%
Minimum a_w	0.93	0.95–0.97
D value at 100°C for spores	25 min.	< 0.1 min

1989). The minimum growth temperature of the mesophilic proteolytic strains is considered to be 10°C and so these are of limited significance with chilled foods. In 1961, Schmidt *et al.* reported that type E *Cl. botulinum* was able to grow and produce toxin in a beef stew after incubation at 3.3°C for 32 days. It is now recognised that non-proteolytic strains of types B and F are also capable of growth and toxin production at 5°C or less (Ecklund *et al.* 1967, Simunovic *et al.* 1985). Therefore, these non-proteolytic strain may grow in chilled foods.

The growth of non-proteolytic *Cl. botulinum* is of particular concern in 'sous-vide' processing. This consists of packing foods under vacuum in air-impermeable sealed bags which are then heat processed and stored chilled for extended periods. Whilst the time and temperature of cooking is specific to the food type, it will destroy vegetative microbial cells but may not be sufficient to destroy bacterial spores. These may subsequently germinate and grow in the absence of air during refrigerated storage (Betts 1992).

It should be noted that the heat resistance of the psychrotrophic non-proteolytic strains is considerably lower than that of the mesophilic proteolytic strains (Table 7.3). The risk of botulism from 'sous-vide' products can be minimised by the use of an appropriate time–temperature profile during heating, adequately controlled chilled storage and/or alterations in the product formulation to prevent growth (Betts 1992, Betts 1996). The minimum pH and a_w values for growth also differ between the proteolytic and non-proteolytic strains (Table 7.3). Overall, the non-proteolytic strains are less resistant to low pH and a_w values (Hauschild 1989).

Microorganisms capable of initiating growth at temperatures of 5–10°C

There are a number of other pathogenic bacteria which, although unable to grow at temperatures below 5°C, may grow if temperature abuse occurs. These include *Salmonella* species, *Escherichia coli* and *Staphylococcus aureus*, with generally accepted minimum growth temperatures of 5.1, 7.1 and 7.7°C respectively (Alcock 1987, Angelotti *et al.* 1961) At temperatures up to 10°C, the growth rate of these bacteria is generally slow (Matches and Liston 1968). These bacteria do, however, cause food-borne disease, frequently implicating chilled foods. Psychrotrophic strains of salmonellae have very occasionally been reported and this may be of more concern with regard to the public health issues of chilled foods (d'Aoust 1991). Several types of *E. coli* are well recognised as agents of food-borne disease. At present the type of most concern is *E. coli* 0157:H7 and other verocytotoxigenic *E. coli* (VTEC) which may produce severe haemorrhagic colitis (Kaper and O'Brien 1998). Limited growth of some strains may occur at 5–10ºC (Alcock 1987, Kauppi *et al.* 1998). This organism has been reviewed by Bell and Kyriakides (1998a). Whilst *Staph. aureus* may grow at temperatures as low at 7.7ºC, disease is caused by the ingestion of a preformed toxin. The minimum temperature for toxin production is greater than for growth and has been reported to be 14.3ºC (Alcock 1987).

Overall, the bacterial species above do not grow at temperatures below 5ºC, but may survive at these temperatures. Often, pathogens and spoilage bacteria will survive adverse conditions (e.g. low pH or high salt) better at refrigeration temperatures compared with higher temperatures (Faith *et al.* 1998). Therefore, if the infectious dose of the bacterium is low and/or growth of the pathogen has already occurred (e.g. during slow cooling), growth during chilled storage may not be a prerequisite for disease.

Microorganisms capable of initiating growth at temperatures greater than 10ºC

These species include mesophilic *Cl. botulinum*, mesophilic *B. cereus* and other *Bacillus* species, *Cl. perfringens* and *Campylobacter* species. In general, these will not grow below 10ºC and growth is limited at temperatures between 10 and 15ºC (Walker and Stringer 1990). Of particular concern in this bacterial group are the *Campylobacter* species which comprise the most commonly reported cause of gastrointestinal disease in the UK (Border and Norton 1997). Although many of the reported cases are sporadic, outbreaks have frequently implicated the consumption of raw milk and undercooked chicken (Skirrow 1990). This bacterial group is unusual as the minimum temperature for growth is 25–30ºC and so it will not grow on most foods. The infectious dose of the microorganism is very low and so growth may not be necessary for disease to occur (Butzler and Oosterom 1991).

Whilst disease caused by the mesophilic spore-forming bacteria has implicated chilled foods, this is usually as a consequence of poor temperature control during cooling after cooking (Gould and Russell 1991, Shaw 1998). These bacteria may grow extremely rapidly during a long slow cooling regime after cooking and then persist during chilled storage.

7.8 Temperature control

With chilled foods, good temperature control is essential, not only to maintain the microbiological safety and quality of foods, but also to minimise changes in the biochemical and physical properties of the food. The temperatures of storage of chilled foods may vary greatly during manufacture, distribution, retail sale and in the home. Consequently, during the life of a chilled food, considerable opportunities exist for temperature abuse to occur. The greater the abuse of temperature, then the greater the potential for microbial growth to occur. This may result in a product becoming unsafe and/or a loss in product quality. Temperature control is the key issue with regard to chilled foods and an integral part of the preservation system. In many of the stages in the food chain after primary chilling, the refrigeration equipment is designed to maintain the product temperature. It may not be able rapidly to reduce the temperature of foods that have been abused at higher temperature.

7.9 Predictive microbiology

As discussed previously, chill temperatures will not prevent microbial growth completely and additional preservative factors, such as reduced pH and water activity, may be required to extend the time period before significant microbial growth occurs. Traditionally, the effect of combinations of preservation systems on target organisms would have been tested using laboratory studies (often called challenge tests). Whilst challenge tests have an important role, they tend to be expensive, time consuming and results obtained are limited to the specific conditions tested. Should any of these change, the test needs to be repeated. However, the chilled foods market is very dynamic and there is a great demand for continual development of new products (Stringer and Dennis 2000). These need to be developed and marketed rapidly.

Predictive microbiology is a tool which can provide rapid reliable answers concerning the likely growth of specific organisms under defined conditions, including conditions not previously examined. Models can be used to predict the probability of growth, the time until growth occurs or the growth rate of microorganisms. The use of predictive models to describe the microbial kinetics is not new and reference to these techniques can be found in publications dating from the 1920s (Esty and Meyer 1922). Microbiological modelling has been reviewed by Gould (1989a) and McMeekin *et al.* (1993).

The development of a microbiological model generally uses the following stages:

- careful selection and appropriate preparation of the target microorganism
- inoculation of the target microorganism into a growth medium (microbiological media or food) with defined characteristics
- storage of the medium under controlled conditions
- sampling of the medium for the target microorganism at relevant intervals
- construction of a model to describe the target microorganism's response
- validation of the model's predictions – preferably in food to ensure the predictions are meaningful
- refinement or further enhancement of the model.

The types of models which have been used vary greatly and include the Arrhenius equation, non-linear Arrhenius (Schoolfield) models, Bêlehrádek-type (Ratkowsky or square root) models, polynomial models, mechanistic models (all reviewed by McMeekin *et al.* 1993) and a dynamic modelling approach (Baranyi and Roberts 1994).

7.9.1 Food pathogens

Over the past decade, there has been considerable work done on predictive modelling of a wide range of pathogenic bacteria, e.g. kinetic growth models have been published for *Salmonella* (Gibson *et al.* 1988), *L. monocytogenes* (Farber *et al.* 1996), *Cl. botulinum* (Graham *et al.* 1996). In order to make such models accessible to food manufacturers, there is a requirement for them to be packaged as user-friendly software.

There are two main systems currently available for predicting the growth of food pathogens. In the UK, the Food MicroModel system is the largest and most comprehensive system. It was developed from a Ministry of Agriculture Fisheries and Food sponsored research programme and the software is available for purchase from Leatherhead Food Research Association (Leatherhead, UK). There is an extensive range of pathogen models in the system including those shown in Table 7.4.

The models in the Food MicroModel system were produced from data obtained in laboratory growth media and validated by comparing predictions from the model with data obtained from the literature or obtained from inoculated food studies.

Another comprehensive modelling programme has been produced in the USA by the United States Department of Agriculture (USDA). It is called the Pathogen Modeling Program and was designed by Dr Robert L. Buchanan and Dr Richard Whiting. The models in this system include those shown in Table 7.5.

This programme is available free of charge and can be obtained from the internet (http://www.arserrc.gov). The models in this programme have been produced from extensive growth data in laboratory media, but have not been validated in foods.

Table 7.4 Some pathogen models

Growth models	Thermal death models
Aeromonas hydrophila	*Cl. botulinum*
Bacillus cereus	*E. coli* O157:H7
Clostridium botulinum	*L. monocytogenes*
Clostridium perfringens	*Salmonella*
Escherichia coli O157:H7	*Y. enterocolitica*
Listeria monocytogenes	
Salmonella	
Staphylococcus aureus	
Yersinia enterocolitica	

Table 7.5 Some models in the Pathogen Modeling Program (USDA)

Growth models	Survival models
A. hydrophila	*E. coli* O157:H7
B. cereus	*L. monocytogenes*
E. coli O157:H7	*Salmonella*
Salmonella spp.	*S. aureus*
Shigella flexneri	
S. aureus	
Y. enterocolitica	

7.9.2 Food spoilage

With regard to the modelling of food spoilage organisms, there are few systems available although many individual models have been published. Work in Tasmania has developed *Pseudomonas* predictor models applicable to milk and raw meats (McMeekin and Ross 1996). Campden and Chorleywood Food Research Association (CCFRA) has developed a collection of models which can be used to assess spoilage rates or likely stability of foods, including chilled foods. This collection of models is called Forecast and is available to potential users via an enquiry service (+44 (0) 1386 842000) which runs the model on behalf of clients after a detailed consultation with respect to their needs. The consultancy aspect of this approach also allows subsequent expert interpretation and consideration of model validation status. Table 7.6 shows the range of models currently available within Forecast. All models within the Forecast system have been produced from data obtained in laboratory media and have been validated in relevant foods using literature data or inoculated challenge test studies.

Limited models on spoilage organisms are available in the Food MicroModel programme previously mentioned and these include: *Brochothrix thermosphacta, Saccharomyces cerevisiae, Lactobacillus plantarum, Zygosaccharomyces bailii*. In addition to bacteria and yeasts, models have also been developed for mould growth (Valík *et al.* 1999). Furthermore, Membré and Kubaczka (1998) have applied similar models to product degradation (i.e. pectin breakdown) rather than just microbial growth.

Table 7.6 Current options for CCFRA Forecast models

Model	pH	Salt (% w/v)	Temperature °C
Bacillus spp.	4.0–7.0	0.5–10.0	5–25
Pseudomonas spp.	5.5–7.0	0.0–4.0	0–15
Enterobacteriaceae	4.0–7.0	0.5–10	0–30
Yeasts (chilled)	2.5–6.3	0.5–10	1–22
Lactic acid bacteria	2.9–5.8	0.5–10	2–30

7.9.3 Practical application of models

Figure 7.3 shows how a model has been put to practical use by comparing predicted values of numbers against predetermined standards for termination of shelf-life. Many other potential applications exist, for example:

- What level of microorganisms will be present under different temperatures of storage?
- How much salt is needed to restrict microbial numbers to a pre-set level after one week storage at 8ºC?
- What will be the effect of increasing the product pH from 5.0 to 5.4?

Several authors have reported deficiencies or inaccuracies in model predictions (Dalgaard and Jørgensen 1998, Hygtiä *et al.* 1999) in that they predict faster growth than that observed in foods. However, many of the models, particularly those for pathogens, are designed to be 'fail safe' and foods may contain additional antimicrobial factors not present in the model, which may inhibit or prevent the predicted growth. Consequently, it is important to determine that any models used contain the important preservation factors relevant to the study and that the model has been validated in appropriate foods. Most of the models developed have been based on single organisms or groups of organisms in pure cultures and may not therefore take into account any effects of

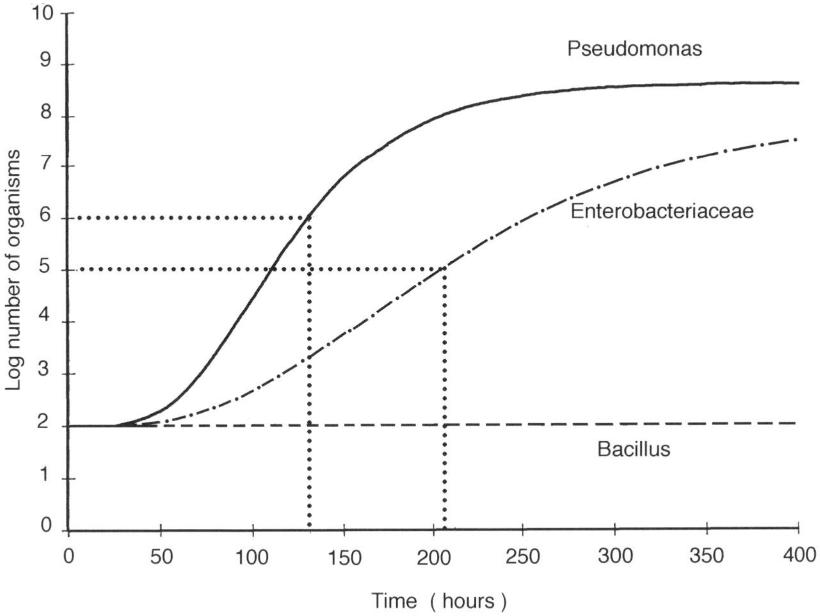

Fig. 7.3. A graphical representation of predictions made using CCFRA forecast conditions: pH 6.0, salt 3% w/v, temperature of storage 6ºC. The user's tolerance for enterobactericeae, *Pseudomonas* spp. and *Bacillus* spp. are clearly shown in relation to the predicted shelf-life.

microbial interaction and competition likely to be seen in foods. Pin and Baranyi (1998) have used modelling techniques to examine the interactions between spoilage bacteria. In the wrong hands, the information from predictive models may be misused and may have serious consequences. It is important that the right questions be asked in order to obtain useful information.

There are many advantages to the use of predictive models in the development and manufacture of chilled foods. They can help to focus resources during product development to assess the microbiological safety and stability of hundreds of different ingredient combinations before stepping into the development kitchen. Predictive models can be used as decision-making tools to allow productive focusing of effort in process and product development and risk and hazard assessment. They can be of great value in complex HACCP studies if used correctly. They should be followed up with targeted practical trials and challenge tests. Used in this way, predictive models can be powerful tools for industrial food microbiologist. Recently, several workers have proposed the development of predictive models with computational neural networks (Hajmeer *et al.* 1997) and their incorporation in decision support systems for microbiological quality and safety (Wijtzes *et al.* 1998). Predictive models also have a role to play in education and training, in that they allow demonstration of microbial behaviour and risk without the need for expensive laboratory exercises.

It should be stressed that microbiological models will never completely remove the requirements for microbiological expertise or to conduct microbiological challenge tests and shelf-life studies, but can be very useful for an indication of the safety and stability of chilled products and ingredients.

7.10 Conclusions

Chilled foods comprise a diverse and complex group of commodities which contain a large number of ingredients. The composition and number of microorganisms present is affected by the indigenous microflora, microorganisms contaminating before and after processing, the growth rates and abilities of the microorganisms, the spoilage abilities of the microorganisms, the intrinsic properties of the food, the effects of processing and packaging, and the time and temperatures of storage. Consequently, the microbial safety and spoilage of chilled foods is very complex, but certain general principles may be applied.

1. The microbiological status of all raw materials should be known and only materials of good quality used.
2. All stages of processing should be defined, monitored and controlled to ensure their correct operation. This is of particular significance in foods which rely on a combination of factors to ensure microbial stability.
3. The temperatures and times of chill storage should be controlled during all stages, from raw materials through retail sale and preferably to the home.

The lower the temperature throughout the process, the slower the rate of growth.

4. Attention must be given to the hygiene of the entire process to ensure that microbial contamination is minimised.

These objectives may be best achieved through the application of a quality system including Hazard Analysis Critical Control Points (HACCP) (Leaper 1997) which may be powerfully integrated with other systems, including risk analysis (Jouve *et al.* 1998). The use of appropriate and validated models may greatly help in the decision-making processes of HACCP and risk analysis. Finally, greater education of all involved in food manufacture, distribution and retail sale and better education of the consumer in areas of hygiene and temperature control will be of great benefit.

7.11 References

ABEYTA C and WEKELL M M, (1988) Potential sources of *Aeromonas hydrophila. J. Food Safety* **9** 11–22.

ADVISORY COMMITTEE ON THE MICROBIOLOGICAL SAFETY OF FOOD (ACMSF), (1995) Report on Verocytotoxin-producing *Escherichia coli.* HMSO, London.

ALCOCK S J, (1984) Growth characteristics of food-poisoning organisms at suboptimal temperatures. II Salmonellae, *Campden Food Preservation Research Association Technical Memorandum No 364.*

ALCOCK S J, (1987) Growth characteristics of food-poisoning organisms at suboptimal temperatures, *Campden Food Preservation Research Association Technical Memorandum No. 440.*

ANGELOTTI R, FOTER M J and LEWIS K H, (1961) Time–temperature effects of salmonellae and staphylococci in foods, *Am. J. Pub. Health*, **36** 559–63.

ANON, (1989) *Guidelines for Cook-Chill and Cook-Freeze Catering Systems*, HMSO, London.

ANON, (1991a) *The Microbiological Safety of Food, Part II* HMSO, London.

ANON, (1991b) *Principles and Practices for the Safe Processing of Foods*, Butterworth-Heinemann, Oxford.

BARANYI J and ROBERTS T A, (1994) A dynamic approach to predicting bacterial growth in food, *International Journal of Food Microbiology*, **23** 277–94.

BAYNE H G and MICHENER H D, (1979) Heat resistance of *Byssochlamys* ascospores *Appl. Environ. Microbiol.*, **37** 449–53.

BELL C and KYRIAKIDES A, (1998a) *E. coli*: a practical approach to the organism and its control in foods, Blackie Academic and Professional, London.

BELL C and KYRIAKIDES A, (1998b) *Listeria*: a practical approach to the organism and its control in foods, Blackie Academic and Professional, London.

BETTS G D, (1992) The microbiological safety of sous-vide processing, *Campden and Chorleywood Food Research Association Technical Manual No. 39.*

BETTS G D, (1996) A code of practice for the manufacture of vacuum and modified atmosphere packaging chilled foods, *Campden and Chorleywood Food Research Association, CCFRA Guideline No. 11.*

BLACK R E, JACKSON R L, TSAI T, MEDVESKY M, SHAYEGANI M, FEELEY J C, MACLEOD K I E, and WAKELEE A W, (1978) Epidemic *Yersinia enterocolitica* infection due to contaminated chocolate milk, *New Engl. J. Med.*, **298** 7679.

BORCH E, KANT-MUERMANS M-L, and BLIXT, Y. (1996) Bacterial spoilage of meat and cured meat products. *International Journal of Food Microbiology* **33** 103–20.

BORDER P and NORTON M, (1997) Safer eating: microbiological food poisoning and its prevention. The Parliamentary Office of Science and Technology, London.

BRADSHAW J G, PEELER J T and TWEDT R M, (1991) Thermal resistance of *Listeria* spp. in milk, *J. Food Prot.*, **54** 12–4.

BUTZLER J P and OOSTEROM J, (1991) *Campylobacter*: pathogenicity and significance in foods, *Int. J. Food Microbiol.*, **12** 1–8.

CAHILL M M (1990) Virulence factors in motile *Aeromonas* species: a review, *J. Appl. Bacteriol.*, **69** 1–16.

COGHILL D and JUFFS H S, (1979) Incidence of psychrotrophic spore-forming bacteria in pasteurised milk and cream products and effect of temperature on their growth, *Australian J. Dairy Technol.*, **3** 150–3.

CONNER D E and KOTROLA J S, (1995) Growth and survival of *Escherichia coli* O157:H7 under acidic conditions, *Applied and Environmental Microbiology*, **61** 382–5.

COUSIN M A, (1982) Presence and activity of psychrotrophic microorganisms in milk and dairy products: a review, *J. Food Prot.*, **45**, 172–207.

COX L J, KLEISS T, CORDIER J L, CORDELLANA C, KONKEL P, PEDRAZZINI C, BEUMER R and SIEBENGA A, (1989) *Listeria* spp. in food processing, non-food processing and domestic environments, *Food Microbiol*, **6** 49–61.

DAINTY R H, (1996) Chemical/biochemical detection of spoilage, *International Journal of Food Microbiology*, **33** 19–34.

DALGAARD P and JØRGENSEN L V, (1998) Predicted and observed growth of *Listeria monocytogenes* in seafood challenge tests and naturally contaminated cold smoked salmon, *International Journal of Food Microbiology*, **40** 105–15.

DAY B P F, (2000) Chilled food packaging. In: Stringer, M. F. and Dennis, C. (ed.), *Chilled Foods: a comprehensive guide*, 2nd edn. Woodhead Publishing Ltd., Cambridge, pp. 137–50.

D'AOUST J Y, (1991) Psychrotrophy and foodborne *Salmonella, Int. J. Food Microbiol.*, **13** 207–16.

DENG Y, RYU J H and BEUCHAT L R, (2000) Tolerance of acid adopted and non-adopted *Escherichia coli* O157:H7 cells to reduced pH as affected by type of acidulant. *Journal of Applied Microbiology*, **86** 203–10.

DOYLE M P, (1990) Pathogenic *Escherichia coli*, *Yersinia enterocolitica* and *Vibrio parahaemolyticus, The Lancet*, **336** 1111–15.

ECKLUND M W, WIELER D L and POYSKY F T, (1967) Outbreak and toxin production of non-proteolytic type B *Clostridium botulinum* at 3.3 to 5.6°C, *J. Bacteriol*, **93** 1461–2.

EDDY B P, (1960) The use and meaning of the term 'psychrophilic', *J. Appl. Bacteriol.*, **23** 189–90.

ESTY J R and MEYER K F, (1992) The heat resistance of spores of *Cl. botulinum* and allied anaerobes, *J. Infect. Dis.*, **31** 650–63.

FAITH N G, WIERZBA R K, IHNOT A M, ROERING A M, LORANG T D, KASPER, C W and LUCHANSKY J B (1998) Survival of *Escherichia coli* O157 in full and reduced fat pepperoni after manufacture of sticks, storage of slices at 4° or 21°C under air vacuum and baking of slices on frozen pizza at 135, 191 and 246°C, *Journal of Food Protection*, **61** 383–9.

FARBER J M, CAI Y and ROSS W H, (1996) Predictive modelling of the growth of *Listeria monocytogenes* in CO_2 environments, *International Journal of Food Microbiology*, **32** 133–44.

FILTENBORG O, FRISVAD J C and THRANE U, (1996) Moulds in food spoilage. *International Journal of Food Microbiology*, **33** 85–102.

FRICKER C R and TOMPSETT S, (1989) *Aeromonas* spp. in foods: a significant cause of food poisoning, *Int. J. Food Microbiol.*, **9** 17–23.

GARDNER G A, (1981) *Brochothrix thermosphacta* (*Microbacterium thermosphactum*) in the spoilage of meats: a review. In: Roberts, T. A. *et al.* (eds) *Psychrotrophic Microorganisms in Spoilage and Pathogenicity*, Academic Press, London, pp. 139–73.

GARDNER G A, (1983) Microbial spoilage of cured meats. In: Roberts, T. A. and Skinner, F. A. (eds) *Food Microbiology: Advances and Prospects*, Academic Press, London pp. 179–202.

GAZE J E, (1992) Food pasteurisation treatments, *Campden Food & Drink Research Association Technical Manual No. 27.*

GAZE J E, BROWN G D, GASKELL D E and BANKS J G, (1989) Heat resistance of *Listeria monocytogenes* in homogenates of chicken, beef steak and carrot, *Food Microbiol.*, **6** 251–59.

GEORGE S M, LUND B M and BROCKLEHURST T F, (1988) The effect of pH and temperature on initiation of growth of *Listeria monocytogenes*, *Letters in Appl. Microbiol.*, **6** 153–6.

GIBSON A M, BRATCHELL N and ROBERTS T A, (1988) Predicting microbial growth: growth responses of salmonellae in a laboratory medium as affected by pH, sodium chloride and storage temperature, *Int. J. Food Microbiol.*, **6** 155–78.

GILL C D and MOLIN G, (1991) Modified atmospheres and vacuum packaging. In: Russell, N. J. and Gould, G. W. (eds), *Food Preservatives*, Blackie and Son Ltd., Glasgow, pp. 172–99.

GILL C O, (1983) Meat spoilage and evaluation of the potential storage life of fresh meat, *J. Food Prot.*, **46**, 444–52.

GILMOUR A and WALKER S J, (1988) Isolation and identification of *Yersinia enterocolitica* and *Yersinia enterocolitica*-like bacteria *J. Appl. Bacteriol.* Suppl., **65** 213S–236S.

GLASS K A and DOYLE M P, (1991) Relationship between water activity of fresh pasta and toxin production by proteolytic *Clostridium botulinum.*, *J. Food Prot.*, **54** 162–5.

GOEPFERT J M, SPIRA W M and KIM H U, (1972) *Bacillus cereus* food poisoning: a review. *J. Milk Food Technol.* **35**, 213–27.

GOULD G, (1989a). Predictive modelling of microbial growth and survival in foods, *Food Sci. Technol. Today*, **3** 89–92.

GOULD G W, (1989b) Heat-induced injury and inactivation. In: Gould, G. W. (ed.), *Mechanisms of Action of Food Preservation Procedures*, Elsevier Appl. Sci. London, pp. 11–42.

GOULD G W, (1996) Methods for preservation and extension of shelf-life, *International Journal of Food Microbiology*, **33** 51–64.

GOULD G W and JONES M V, (1989) Combination and synergistic effects. In: Gould, G W (ed.), *Mechanisms of Action of Food Preservation Procedures*, Elsevier Appl. Sci., London, pp. 400–21.

GOULD G W and RUSSELL N J, (1991) Major food-poisoning and food-spoilage microorganisms. In: Russell, N. J. and Gould, G. W. (eds), *Food Preservatives*, Blackie and Son Ltd., Glasgow, pp. 1–21.

GRAHAM A F, MASON D R and PECK M W, (1996) Predictive model of the effect of temperature, pH and sodium chloride on growth from spores of non-proteolytic *Clostridium botulinum, International Journal of Food Microbiology*, **31** 69–85.

GRAM L and HUSS H H, (1996) Microbiological spoilage of fish and fish products, *International Journal of Food Microbiology*, **53** 121–38.

GREENWOOD M H and HOOPER W L, (1989) Improved methods for the isolation of *Yersinia* species from milk and foods *Food Microbiol.*, **6** 99–104.

GRIFFITHS M W and PHILLIPS J D, (1990) Incidence, source and some properties of psychrotrophic *Bacillus* spp. found in raw and pasteurized milk, *J. Soc. Dairy Technol*, **43** 62–6.

GUTIÉRREZ R, GARÉIA T, GONZALEZ I, SANZ B, HERNÁNDEZ P E and MARTIN R, (1997) A quantitative PCR-ELISA for the rapid enumeration of bacteria in refrigerated raw milk, *Journal of Applied Microbiology*, **83** 518–23.

HAJMEER M N, BASHEER I A and NAJJAR Y M, (1997) Computational neural networks for predictive microbiology II. Application to microbial growth, *International Journal of Food Microbiology*, **34** 51–66.

HAUSCHILD A H W, (1989) *Clostridium botulinum.* In: Doyle, M. P. (ed.), *Foodborne Bacterial Pathogens*, Marcel Dekker, New York, pp. 111–89.

HERBERT R A, (1989) Microbial growth at low temperatures. In: Gould, G. W. (ed.), *Mechanisms of Action of Food Preservation Procedures*, Elsevier Appl. Sci., London, pp. 71–96.

HOLAH J T, (1999) Effective microbiological sampling of food processing environments, *Campden and Chorleywood Food Research Association, Guideline No. 20.*

HOLAH J T, TAYLOR J and HOLDER J S, (1993) The spread of *Listeria* by cleaning systems, *Campden Food & Drink Research Association Technical*

Memorandum No. 673.

HUIS IN'T VELD J H T, (1996) Microbial and biochemical spoilage of foods: an overview, *International Journal of Food Microbiology* **33**, 1–18.

HYGTIÄ E, HIELM S, MOKKILA M, KINNUNEN A and KORKEALA H, (1999) Predicted and observed growth and toxigenesis by *Clostridium botulinum* type E in vacuum-packaged fishery product challenge tests, *International Journal of Food Microbiology*, **47** 161–9.

JAQUETTE C B and BEUCHAT L R, (1998) Combined effects of pH, nisin and temperature on growth and survival of psychrotrophic *Bacillus cereus, Journal of Food Protection*, **61** 563–70.

JAY J M, (1978) *Modern Food Microbiology*, 2nd edn. D. van Nostrand Co., New York.

JOHNSON K M, (1984) *Bacillus cereus* foodborne illness – an update, *J. Food Prot.*, **47** 145–53.

JOUVE J L, STRINGER M F and BAIRD-PARKER A C, (1998) *Food Safety Management Tools*, ILSI – Europe, Brussels.

KABARA J J and EKLUND T, (1991) Organic acids and their esters. In: Russell N J and Gould G W (eds), *Food Preservatives*, Blackie and Son Ltd., Glasgow, pp. 44–71.

KAPER B and O'BRIEN A D, (1998) *Escherichia coli* O157:H7 and other Shiga toxin-producing *E. coli* strains, ASM Press, Washington.

KAUPPI K L, O'SULLIVAN D J and TATINI S R, (1998) Influence of nitrogen source on low temperature growth of verocytotoxigenic *Escherichia coli, Food Microbiology*, **15** 355–64.

KRAMER J M and GILBERT R J, (1989) *Bacillus cereus* and other *Bacillus* species. In: Doyle, M P (ed.), *Foodborne Bacterial Pathogens*, Marcel Dekker, New York, pp. 21–69.

LEAPER S, (1997) HACCP: a practical guide (2nd edn), *Campden and Chorleywood Food Research Association Technical Manual, No. 38.*

LOGUE C M, SHERIDAN G, WAUTERS G, MCDOWELL D A and BLAIR I S, (1996) *Yersinia* spp and numbers, with particular reference to *Y.enterocolitica* occurring on Irish meat and meat products, and the influence of alkali treatment on their isolation, *International Journal of Food Microbiology*, **33** 257–74.

LOVETT L, BRADSHAW J G and PEELER J T, (1982) Thermal inactivation of *Yersinia enterocolitica* in milk *Appl. Environ. Microbiol.*, **44** 517–19.

LÜCKE F K and EARNSHAW R G, (1991) Starter cultures. In: Russell, N J and Gould, G W (eds), *Food Preservatives*, Blackie and Son Ltd., Glasgow, pp. 215–34.

LUND B M, (1990) The prevention of foodborne listeriosis, *Br. Food J.*, **92** 13–22.

MCLAUCHLIN J A, (1987) A review: *Listeria monocytogenes*, recent advances in the taxonomy and epidemiology of listeriosis in humans, *J. Appl. Bacteriol.*, **63**, 1–2.

MCMEEKIN T A and ROSS T, (1996) Shelf-life prediction: status and future possibilities, *International Journal of Food Microbiology*, **31** 65–84.

MCMEEKIN T A, OLLEY J N, ROSS T and RATKOWSKY D A, (1993) *Predictive microbiology: theory and application*, Research Studies Press, Somerset.

MATCHES J R and LISTON J, (1968) Low temperature growth of *Salmonella*, *J. Food Sci.*, **33** 641–5.

MEMBRÉ J M and KUBACZKA M, (1998) Degradation of pectin compounds during pasteurized vegetable juice spoilage by *Chryseomonas luteola*: a predictive microbiology approach, *International of Food Microbiology*, **42** 159–66.

MICHENER H D and ELLIOTT R P, (1964) Minimum growth temperatures for food poisoning, fecal indicator and psychrophilic microorganisms, *Adv. in Food Res.*, **13** 349–96.

MITSCHERLICH E and MARTH E H, (1984) *Microbial Survival in the Environment*, Springer-Verlag, Berlin.

MORITA R Y, (1973) Psychrophilic bacteria, *Bacteriol. Rev.*, **39** 144–67.

MURRAY E G D, WEBB R A and SWAN M B R, (1926) A disease of rabbits characterised by a large mononuclear leucocytosis caused by a hitherto undescribed bacillus Bacterium *monocytogenes* (n. sp), *J. Pathol. Bacteriol*, **29** 407–39.

NEILL S D, (1974) A study of the microflora of raw milk stored at low temperature, *PhD Thesis*, The Queen's University of Belfast, Northern Ireland.

PALUMBO S A and BUCHANAN R L, (1988) Factors affecting growth or survival of *Aeromonas hydrophila* in foods, *J. Food Safety*, **9** 37–51.

PIN C and BARANYI J, (1998) Predictive models as means to quantify the interactions of spoilage organisms, *International Journal of Food Microbiology*, **41** 59–72.

PITT J I and HOCKING A D (1985) Spoilage of fresh and perishable foods. In: *Fungi and Food Spoilage*, Academic Press, Sydney, pp. 365–82.

PRESCOTT S C and GEER L P, (1936) Observations on food poisoning organisms under refrigeration conditions, *Refrigeration Engineering*, **32** 211–2, 282–3.

RALOVICH B S, (1987) Epidemiology and significance of listeriosis in the European countries. In: Schonberg, A. (ed.), *Listeriosis: Joint WHO/ROI Consultation on Prevention and Control*, Vet. Med. Hefte, Berlin, pp. 51–5.

REINHEIMER J A and BARGAGNA M L, (1989). Response of psychrotrophic strains of *Bacillus* to different heat treatments, *Microbiol-Aliments-Nutr.*, **5** 117–22.

RIDELL J and KORKEALA H, (1997) Minimum growth temperatures of *Hafnia alvei* and other *Enterobacteriaceae* isolated from refrigerated meat determined with a temperature gradient incubator. *International Journal of Food Microbiology* **35**, 287–92.

RUSSELL N J and GOULD G W, (1991) *Food preservatives*, Blackie, Glasgow.

SCHIEMANN D A, (1989) *Yersinia enterocolitica* and *Yersinia pseudotuberculosis*. In: Doyle, M. P. (ed.), *Foodborne Bacterial Pathogens*, Marcel Dekker, New York, pp. 601–72.

SCHLIEFSTEIN J I and COLEMAN M B, (1939) An unidentified microorganism resembling *B. lignieri* and *Past. pseudotuberculosis*, and pathogenic for man, *New York State J. Med.*, **39** 1749–53.

SCHMIDT C F, LECHOWICH R V and FOLINAZZO J F, (1961) Growth and toxin production by Type E *Clostridium botulinum* below 40°F, *J. Food Sci.*, **26** 626–34.

SCHUCHAT A, SWAMINATHAN B and BROOME C V, (1991) Epidemiology of human listeriosis, *Clin. Microbiol. Rev.*, **4** 169–83.

SHAW R, (1998) Identification and prevention of hazards associated with slow cooling of hams and other large cooked meats and meat products, *Campden and Chorleywood Food Research Association Review No. 8.*

SIMUNOVIC J, OBLINGER J L and ADAMS J P, (1985) Potential for growth of non-proteolytic types of *Clostridium botulinum* in pasteurized and restructured meat products: a review, *J. Food Prot.*, **48** 265–76.

SKIRROW M B, (1990) *Campylobacter*, *The Lancet*, **336** 921–3.

SPERBER W H, (1983) Influence of water activity on foodborne bacteria – a review, *J. Food Prot.*, **46** 142–50.

STELMA G N, (1989) *Aeromonas hydrophila*, In: Doyle, M. P. (ed.) *Foodborne Bacterial Pathogens*, Marcel Dekker, New York, pp. 1–19.

STRINGER M F and DENNIS C, (2000) The market for chilled foods, In *Chilled Foods: a comprehensive guide*, 2nd edn. Woodhead Publishing, Cambridge.

TACKET C O, NAVAIN J P, SATTIN R, LOFGREN J R, KONIGSBERG C, RENDTORFF R C, RAUSA A, DAVIS B R and COHEN M L, (1984) A multistate outbreak of infections caused by *Yersinia enterocolitica* transmitted by pasteurised milk, *J. American Med. Assoc.*, **51** 483–6.

TACKET C O, BALLARD L, HARRIS N, ALLARD L, NOLAN C, QUAN T and COHEN M L, (1985) An outbreak of *Yersinia enterocolitica* infections caused by contaminated tofu, *American J. Epidemiol.*, **121** 705–11.

TERPLAN G, SCHOEN R, SPRINGMEYER W, DEGLE I and BECKER H, (1987) Investigations on incidence, origin and behaviour of *Listeria* in cheese. In: Schönberg, A. (ed.), *Listeriosis – Joint WHO/ROI Consultation on Prevention and Control.*, Vet. Med. Hefte, Berlin, pp. 98–105.

TODD L S, HARDY J C, STRINGER M F and BARTHOLOMEW B A, (1989) Toxin production by strains of *Aeromonas hydrophila* grown in laboratory media and prawn purée, *Int. J. Food Microbiol.*, **9** 145–56.

VALÍK L, BARANYI J and GÖRNER F, (1999) Predicting fungal growth: the effect of water activity on *Penecillium roquefortii, International Journal of Food Microbiology*, **47** 141–46.

VAN NETTEN R, VAN DE MOOSDIJK A, VAN HOENSEL P and MOSSEL D A A, (1990) Psychrotrophic strains of *Bacillus cereus* producing enterotoxin. *J. Appl. Bacteriol.*, **69** 73–9.

VENKITANARAYANEN K S, FAUSTMAN C, CRIVELLO J F, KHAN M I, HOAGLAND T A and BERRY B W, (1997) Rapid estimation of spoilage bacterial load in aerobically stored meat by a quantitative polymerase chain reaction, *Journal of Applied Microbiology*, **82** 359–64.

WALKER S J, (1988) Major spoilage microorganisms in milk and dairy products, *J. Soc. Dairy Technol.*, **41** 91–2.

WALKER S J and STRINGER M F, (1987) Growth of *Listeria monocytogenes* and *Aeromonas hydrophila* at chill temperatures, *Campden Food and Drink Research Association Technical Memorandum No 462*.

WALKER S J and STRINGER M F, (1990) Microbiology of chilled foods. In: Gormley, T. R (ed.), *Chilled Foods – The State of the Art*. Elsevier Appl. Sci., Barking, pp. 269–304.

WALKER S J, (1990) *Listeria monocytogenes*: an emerging pathogen. In: Turner, A. (ed.), *Food Technology International Europe*, Sterling Publications, London, pp. 237–40.

WALKER S J, ARCHER P and BANKS J G, (1990a) Growth of *Listeria monocytogenes* at refrigeration temperatures, *J. Appl. Bacteriol.*, **68** 157–62.

WALKER S J, ARCHER P and BANKS J G, (1990b) Growth of *Yersinia enterocolitica* at chill temperatures in milk and other media *Milchwissenschaft*, **45** 503–6.

WHITFIELD F B, (1998) Microbiology of food taints. *International Journal of Food Science and Technology*, **33** 31–51.

WIJTZES T, VAN'T RIET K, HUIS IN'T VELD J H J and ZWIETERING M H, (1998) A decision support system for the prediction of microbial food safety and quality, *International Journal of Food Microbiology*, **42** 79–90.

8

Conventional and rapid analytical microbiology

R. P. Betts, Campden and Chorleywood Food Research Association

8.1 Introduction

The detection and enumeration of microorganisms either in foods or on food contact surfaces forms an integral part of any quality control or quality assurance plan. Microbiological tests done on foods can be divided into two types: (a) quantitative or enumerative, in which a group of microorganisms in the sample are counted and the result is expressed as the number of the organisms present per unit weight of sample; or (b) qualitative or presence/absence, in which the requirement is simply to detect whether a particular organism is present or absent in a known weight of sample.

The basis of methods used for the testing of microorganisms in foods is very well established, and relies on the incorporation of a food sample into a nutrient medium in which microorganisms can replicate thus resulting in a visual indication of growth. Such methods are simple, adaptable, convenient and generally inexpensive. However, they have two drawbacks: firstly, the tests rely on the growth of organisms in media, which can take many days and result in a long test elapse time; and secondly, the methods are manually oriented and are thus labour intensive.

Over recent years, there has been considerable research into rapid and automated microbiological methods. The aim of this work has been to reduce the test elapse time by using methods other than growth to detect and/or count microorganisms and to decrease the level of manual input into tests by automating methods as much as possible. These rapid and automated methods have gained some acceptance within the food industry and could form an important quality control tool in the chilled foods area. Positive release of chilled foods on the results of a rapid method could increase the shelf-life of a product by one or two days compared with a conventional microbiological technique. In addition, the availability of very rapid microbiological test methods indicates a potential for on-line control and the use of such systems in Hazard Analysis Critical Control Point (HACCP) procedures.

8.2 Sampling

Although this chapter deals with the methodologies employed to test foods, it is important for the microbiologist to consider sampling. No matter how good a method is, if the sample has not been taken correctly and is not representative of the batch of food that it has been taken from, then the test result is meaningless. It is useful to devise a sampling plan in which results are interpreted from a number of analyses, rather than a single result. It is now common for microbiologists to use two or three class sampling plans, in which the number of individual samples to be tested from one batch are specified, together with the microbiological limits that apply. These type of sampling plans are fully described in Anon. 1986.

Once a sampling plan has been devised then a representative portion must be taken for analysis. In order to do this the microbiologist must understand the food product and its microbiology in some detail. Many chilled products will not be homogeneous mixtures but will be made up of layers or sections, a good example would be a prepared sandwich. It must be decided if the microbiological result is needed for the whole sandwich (i.e. bread and filling), or just the bread, or just the filling, indeed in some cases one part of a mixed filling may need to be tested, when this has been decided then the sample for analyses can be taken, using appropriate aseptic technique and sterile sampling implements (Kyriakides *et al.*, 1996). The sampling procedure having been developed, the microbiologist will have confidence that samples taken are representative of the foods being tested and test methods can be used with confidence.

8.3 Conventional microbiological techniques

As outlined in the introduction, conventional microbiological techniques are based on the established method of incorporating food samples into nutrient media and incubating for a period of time to allow the microorganisms to grow. The detection or counting method is then a simple visual assessment of growth. These methods are thus technically simple and relatively inexpensive, requiring no complex instrumentation. The methods are however very adaptable, allowing the enumeration of different groups of microorganisms.

Before testing, the food sample must be converted into a liquid form in order to allow mixing with the growth medium. This is usually done by accurately weighing the sample into a sterile container and adding a known volume of sterile diluent (the sample to diluent ratio is usually 1:10); this mixture is then homogenised using a homogeniser (e.g. stomacher or pulsifier) that breaks the sample apart, releasing any organisms into the diluent. The correct choice of diluent is important. If the organisms in the sample are stressed by incorrect pH or low osmotic strength, then they could be injured or killed, thus affecting the final result obtained from the microbiological test. The diluent must be well buffered at a pH suitable for the food being tested and be osmotically balanced. When testing some foods (e.g. dried products) which may contain highly stressed microorganisms, then a suitable recovery period may be required before the test commences, in order to ensure cells are not killed during the initial phase of the test procedure (Davis and Jones 1997).

8.3.1 Conventional quantitative procedures

The enumeration of organisms in samples is generally done by using plate count, or most probable number (MPN) methods. The former are the most widely used, whilst the latter tend to be used only for certain organisms (e.g. *Escherichia coli*) or groups (e.g. coliforms).

Plate count method

The plate count method is based on the deposition of the sample, in or on an agar layer in a Petri dish. Individual organisms or small groups of organisms will occupy a discrete site in the agar, and on incubation will grow to form discrete colonies that are counted visually. Various types of agar media can be used in this form to enumerate different types of microorganisms. The use of a non-selective nutrient medium that is incubated at 30ºC aerobically will result in a total viable count or mesophilic aerobic count. By changing the conditions of incubation to anaerobic, a total anaerobe count will be obtained. Altering the incubation temperature will result in changes in the type of organism capable of growth, thus showing some of the flexibility in the conventional agar approach. If there is a requirement to enumerate a specific type of organism from the sample, then in most cases the composition of the medium will need to be adjusted to allow only that particular organism to grow. There are three approaches used in media design that allow a specific medium to be produced: the elective, selective and differential procedures.

Elective procedures refer to the inclusion in the medium of reagents, or the use of growth conditions, that encourage the development of the target organisms, but do not inhibit the growth of other microorganisms. Such reagents may be sugars, amino acids or other growth factors. Selective procedures refer to the inclusion of reagents or the use of growth conditions that inhibit the development of non-target microorganisms. It should be noted that, in many cases, selective agents will also have a negative effect on the growth of the target microorganism, but this will be less great than the effect on non-target cells. Examples of selective procedures would be the inclusion of antibiotics in a medium or the use of anaerobic growth conditions. Finally, differential procedures allow organisms to be distinguished from each other by the reactions that their colonies cause in the medium. An example would be the inclusion of a pH indicator in a medium to differentiate acid-producing organisms. In most cases, media will utilise a multiple approach system, containing elective, selective and differential components in order to ensure that the user can identify and count the target organism.

The number of types of agar currently available are far too numerous to list. For details of these, the manuals of media manufacturing companies (e.g. Oxoid, LabM, Difco, Merck) should be consulted.

MPN method

The second enumerative procedure mentioned earlier was the MPN method. This procedure allows the estimation of the number of viable organisms in a sample based on probability statistics. The estimate is obtained by preparing decimal (tenfold) dilutions of a sample, and transferring sub-samples of each dilution to (usually) three tubes of a broth medium. These tubes are incubated, and those that show any growth (turbidity) are recorded and compared to a standard table of results (Anon. 1986) that indicate the contamination level of the product.

As indicated earlier, this method is used only for particular types of test and tends to be more labour and materials intensive than plate count methods. In addition, the confidence limits are large even if many replicates are studied at each dilution level. Thus the method tends to be less accurate than plate counting methods.

8.3.2 Conventional qualitative procedures

Qualitative procedures are used when a count of the number of organisms in a sample is not required and only their presence or absence needs to be determined. Generally such methods are used to test for potentially pathogenic microorganisms such as *Salmonella* spp., *Listeria* spp., *Yersinia* spp. and *Campylobacter* spp. The technique requires an accurately weighed sample (usually 25g) to be homogenised in a primary enrichment broth and incubated for a stated time at a known temperature. In some cases, a sample of the primary enrichment may require transfer to a secondary enrichment broth and further incubation. The final enrichment is usually then streaked out onto a selective agar plate that allows the growth of the organisms under test. The long enrichment procedure is used because the sample may contain very low levels of the test organism in the presence of high numbers of background microorganisms. Also, in processed foods the target organisms themselves may be in an injured state. Thus the enrichment methods allow the resuscitation of injured cells followed by their selective growth in the presence of high numbers of competing organisms.

The organism under test is usually indistinguishable in a broth culture, so the broth must be streaked onto a selective/differential agar plate. The microorganisms can then be identified by their colonial appearance. The formation of colonies on the agar that are typical of the microorganism under test are described as presumptive colonies. In order to confirm that the colonies are composed of the test organism, further biochemical and serological tests are usually performed on pure cultures of the organism. This usually requires colonies from primary isolation plates being restreaked to ensure purity. The purified colonies are then tested biochemically by culturing in media that will indicate whether the organism produces particular enzymes or utilises certain sugars.

At present a number of companies market miniaturised biochemical test systems that allow rapid or automated biochemical tests to be quickly and easily set up by microbiologists. Serological tests are done on pure cultures of some isolated organisms, e.g. *Salmonella* using commercially available antisera.

8.4 Rapid and automated methods

The general interest in alternative microbiological methods has been stimulated in part by the increased output of food production sites. This has resulted in

1. Greater numbers of samples being stored prior to positive release – a reduction in analysis time would reduce storage and warehousing costs.
2. A greater sample throughput being required in laboratories – the only way that this can be achieved is by increased laboratory size and staff levels, or by using more rapid and automated methods.
3. A requirement for a longer shelf-life in the chilled foods sector – a reduction in analysis time could expedite product release thus increasing the shelf-life of the product.
4. The increased application of HACCP procedures – rapid methods can be used in HACCP verification procedures.

There are a number of different techniques referred to as rapid methods and most have little in common either with each other or with the conventional procedures that they replace. The methods can generally be divided into quantitative and qualitative tests, the former giving a measurement of the number of organisms in a sample, the latter indicating only presence or absence. Laboratories considering the use of rapid methods for routine testing must carefully consider their own requirements before purchasing such a system. Every new method will be unique, giving a slightly different result, in a different timescale with varying levels of automation and sample throughput. In addition, some methods may work poorly with certain types of food or may not be able to detect the specific organism or group that is required. All of these points must be considered before a method is adopted by a laboratory. It is also of importance to ensure that staff using new methods are aware of the principles of operation of the techniques and thus have the ability to troubleshoot if the method clearly shows erroneous results.

8.4.1 Electrical methods

The enumeration of microorganisms in solution can be achieved by one of two electrical methods, one measuring particle numbers and size, the other monitoring metabolic activity.

Particle counting

The counting and sizing of particles can be done with the 'Coulter' principle, using instruments such as the Coulter Counter (Coulter Electrics, Luton). The method is based on passing a current between two electrodes placed on either side of a small aperture. As particles or cells suspended in an electrolyte are drawn through the aperture they displace their own volume of electrolyte solution, causing a drop in d.c. conductance that is dependent on cell size. These changes in conductance are detected by the instrument and can be presented as a series of voltage pulses, the height of each pulse being proportional to the volume of the particle, and the number of pulses equivalent to the number of particles.

The technique has been used extensively in research laboratories for experiments that require the determination of cell sizes or distribution. It has found use in the area of clinical microbiology where screening for bacteria is required (Alexander *et al.* 1981). In food microbiology however, little use has been made of the method. There are reports of the detection of cell numbers in milk (Dijkman *et al.*, 1969) and yeast estimation in beer (MaCrae 1964), but little other work has been published. Any use of particle counting for food microbiology would probably be restricted to non-viscous liquid samples or particle-free fluids, since very small amounts of sample debris could cause significant interference, and cause aperture blockage.

Metabolic activity

Stewart (1899) first reported the use of electrical measurement to monitor microbial growth. This author used conductivity measurements to monitor the putrefaction of blood, and concluded that the electrical changes were caused by ions formed by the bacterial decomposition of blood constituents. After this initial report a number of workers examined the use of electrical measurement to monitor the growth of microorganisms. Most of the work was successful; however, the technique was not widely adopted until reliable instrumentation capable of monitoring the electrical changes in microbial cultures became available.

There are currently four instruments commercially available for the detection of organisms by electrical measurement. The Malthus System (IDG, Bury, UK) based on the work of Richards *et al.* (1978) monitors conductance changes occurring in growth media as does the Rabit System (Don Whitley Scientific, Yorkshire), whilst the Bactometer (bioMerieux, Basingstoke, UK), and the Batrac (SyLab, Purkersdorf, Austria) (Bankes 1991) can monitor both conductance and capacitance signals. All of the instruments have similar basic components: (a) an incubator system to hold samples at a constant temperature during the test; (b) a monitoring unit that measures the conductance and/or capacitance of every cell at regular frequent intervals (usually every 6 minutes); and (c) a computer-based data handling system that presents the results in usable format.

The detection of microbial growth using electrical systems is based on the measurement of ionic changes occurring in media, caused by the metabolism of microorganisms. The changes caused by microbial metabolism and the detailed electrochemistry that is involved in these systems has been previously described in some depth (Eden and Eden 1984, Easter and Gibson 1989, Bolton and Gibson 1994). The principle underlying the system is that as bacteria grow and metabolise in a medium, the conductivity of that medium will increase. The electrical changes caused by low numbers of bacteria are impossible to detect using currently available instrumentation, approximately 106 organisms/ml must be present before a detectable change is registered. This is known as the *threshold of detection*, and the time taken to reach this point is the *detection time*.

In order to use electrical systems to enumerate organisms in foods, the sample must initially be homogenised. The growth well or tube of the instrument containing medium is inoculated with the homogenised sample and connected to the monitoring unit within the incubation chamber or bath. The electrical properties of the growth medium are recorded throughout the incubation period. The sample container is usually in the form of a glass or plastic tube or cell, in which a pair of electrodes is sited. The tube is filled with a suitable microbial growth medium, and a homogenised food sample is added. The electrical changes occurring in the growth medium during microbial metabolism are monitored via the electrodes and recorded by the instrument.

As microorganisms grow and metabolise they create new end-products in the medium. In general, uncharged or weekly charged substrates are transformed into highly charged end-products (Eden and Eden 1984), and thus the conductance of the medium increases. The growth of some organisms such as yeasts does not result in large increases in conductance. This is possibly due to the fact that these organisms do not produce ionised metabolites and this can result in a decrease in conductivity during growth.

When an impedance instrument is in use, the electrical resistance of the growth medium is recorded automatically at regular intervals (e.g. 6 minutes) throughout the incubation period. When a change in the electrical parameter being monitored is detected, then the elapsed time since the test was started is calculated by a computer; this is usually displayed as the detection time. The complete curve of electrical parameter changes with time (Fig. 8.1) is similar to a bacterial growth curve, being sigmoidal and having three stages: (a) the inactive stage, where any electrical changes are below the threshold limit of detection of the instrument; (b) the active stage, where rapid electrical changes occur; and (c) the stationary or decline stage, that occurs at the end of the active stage and indicates a deceleration in electrical changes.

The electrical response curve should not be interpreted as being similar to a microbial growth curve. It is accepted (Easter and Gibson 1989) that the lag and

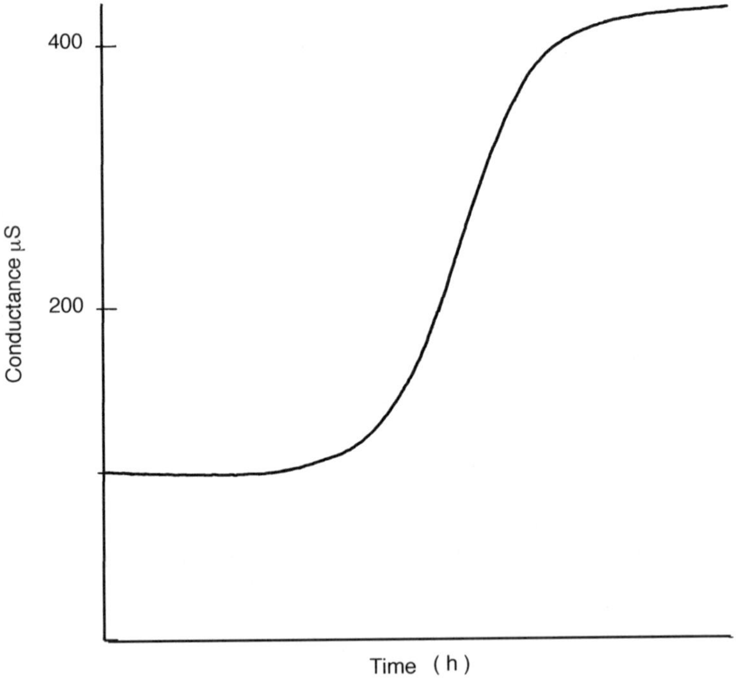

Fig. 8.1 A conductance curve generated by the growth of bacteria in a suitable medium.

logarithmic phases of microbial growth occur in the inactive and active stages of the electrical response curve, up to and beyond the detection threshold of the instrument. The logarithmic and stationary phases of bacterial growth occur during the active and decline stages of electrical response curves.

In order to use detection time data generated from electrical instruments to assess the microbiological quality of a food sample, calibrations must be done. The calibration consists of testing samples using both a conventional plating test and an electrical test. The results are presented graphically with the conventional result on the *y*-axis and the detection time on the *x*-axis (Fig. 8.2). The result is a negative line with data covering 4 to 5 log cycles of organisms and a correlation coefficient greater than 0.85 (Easter and Gibson, 1989), Calibrations must be done for every sample type to be tested using electrical methods; different samples will contain varying types of microbial flora with differing rates of growth. This can greatly affect electrical detection time and lead to incorrect results unless correct calibrations have been done.

So far, the use of electrical instruments for total microbial assessment has been described. These systems, however, are based on the use of a growth medium and it is thus possible, using media engineering, to develop methods for the enumeration or detection of specific organisms or groups of organisms. Many examples of the use of electrical measurement for the detection/ enumeration of specific organisms have been published; these include:

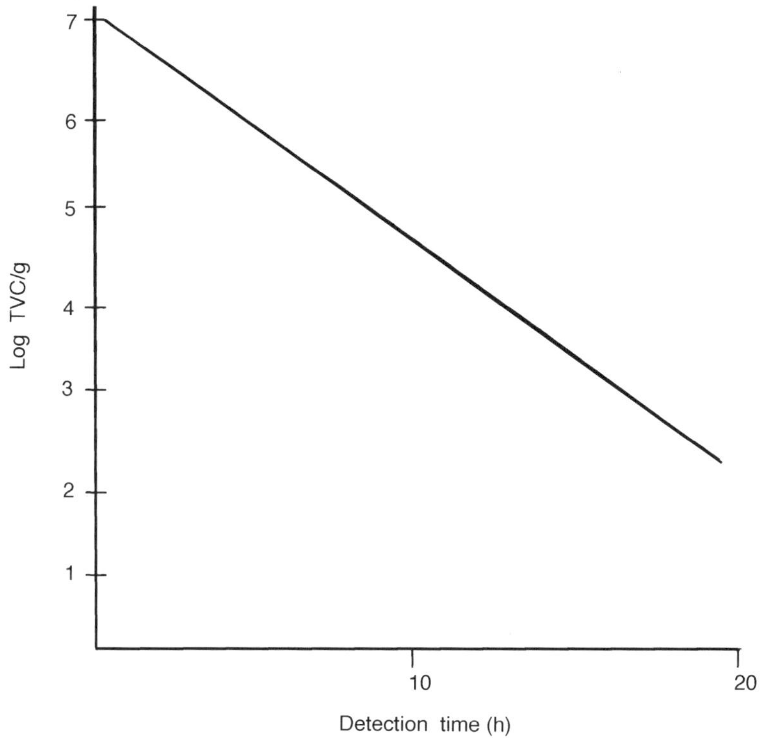

Fig. 8.2 Calibration curve showing changes in conductance detection time with bacterial total viable count (TVC).

Enterobacteriaceae (Cousins and Marlatt 1990, Petitt 1989), *Pseudomonas* (Banks *et al.* 1989), *Yersinia enterocolitica* (Walker 1989) and yeasts (Connolly *et al.*, 1988), *E.coli* (Druggan *et al.* 1993), *Campylobacter* (Bolton and Powell 1993). In the future, the number of types of organism capable of being detected will undoubtedly increase. Considerable research is currently being done on media for the detection of *Listeria*, and media for other organisms will follow.

Most of the electrical methods described above involve the use of direct measurement, i.e. the electrical changes are monitored by electrodes immersed in the culture medium. Some authors have indicated the potential for indirect conductance measurement (Owens *et al.* 1989) for the detection of microorganisms. This method involves the growth medium being in a separate compartment to the electrode within the culture cell. The liquid surrounding the electrode is a gas absorbent, e.g. potassium hydroxide for carbon dioxide. The growth medium is inoculated with the sample and, as the microorganisms grow, gas is released. This is absorbed by the liquid surrounding the electrode, causing a change in conductivity, which can be detected.

This technique may solve the problem caused by microorganisms that produce only small conductance changes in conventional direct conductance cells. These organisms, e.g. many yeast species, are very difficult to detect using conventional direct conductance methods, but detection is made easy by the use of indirect conductance monitoring (Betts 1993). The increased use of indirect methods in the future could considerably enhance the ability of electrical systems to detect microorganisms that produce little electrical change in direct systems, thus increasing the number of applications of the technique within the food industry.

8.4.2 Adenosine triphosphate (ATP) bioluminescence

The non-biological synthesis of ATP in the extracellular environment has been demonstrated (Ponnamperuma *et al.*, 1963), but it is universally accepted that such sources of ATP are very rare (Huernnekens and Whiteley 1960). ATP is a high-energy compound found in all living cells (Huernnekens and Whiteley 1960), and it is an essential component in the initial biochemical steps of substrate utilisation and in the synthesis of cell material.

McElroy (1947) first demonstrated that the emission of light in the bioluminescent reaction of the firefly, *Photinus pyralis*, was stimulated by ATP. The procedure for the determination of ATP concentrations utilising crude firefly extracts was described by McElroy and Streffier (1949) and has since been used in many fields as a sensitive and accurate measure of ATP. The light-yielding reaction is catalysed by the enzyme luciferase, this being the enzyme found in fireflies causing luminescence. Luciferase takes part in the following reaction:

1. Luciferase + Luciferin + ATP → Mg^{2+} Luciferase – Luciferin – AMP + PP

   The complex is then oxidised:

2. Luciferase – Luciferin – AMP + O_2 → (Luciferase – Luciferin – AMP = O) + H_2O

   The oxidised complex is in an excited stage, and as it returns to its ground stage a photon of light is released:

3. Luciferase – Luciferin – AMP = 0 → (Luciferase – Luciferin – AMP = 0) + Light

The light-yielding reaction is efficient, producing a single photon of light for every luciferin molecule oxidised and thus every ATP molecule used (Seliger and McElroy 1960).

Levin *et al.* (1964) first described the use of the firefly bioluminescence assay of ATP for detecting the presence of viable microorganisms. Since this initial report considerable work has been done on the detection of viable organisms in environmental samples using a bioluminescence technique (Stalker 1984). As all viable organisms contain ATP, it could be considered simple to use a bioluminescence method to rapidly enumerate microorganisms. Research, however, has shown that the amount of ATP in different microbial cells varies depending on species, nutrient level, stress level and stage of growth (Stannard

1989, Stalker 1984). Thus, when using bioluminescence it is important to consider:

1. the type of microorganism being analysed; generally, vegetative bacteria will contain 1 fg of ATP/cell (Karl 1980), yeasts will contain ten times this value (Stannard 1989), whilst spores will contain no ATP (Sharpe *et al.* 1970);
2. whether the cells have been subjected to stress, such as nutrient depletion, chilling or pH change. In these cases a short resuscitation may be required prior to testing;
3. whether the cells are in a relatively ATP-free environment, such as a growth medium, or are contained within a complex matrix, like food, that will have very high background ATP levels.

When testing food samples one of the greatest problems is that noted in 3 above. All foods will contain ATP and the levels present in the food will generally be much higher than those found in microorganisms within the food. Data from Sharpe *et al.* (1970) indicated that the ratio of food ATP to bacterial ATP ranges from 40000:1 in ice-cream to 15:1 in milk. Thus, to be able to use ATP analysis as a rapid test for foodborne microorganisms, methods for the separation of microbial ATP were developed. The techniques that have been investigated fall into two categories: either to physically separate microorganisms from other sources of ATP, or to use specific extractants to remove and destroy non-microbial ATP. Filtration methods have been successfully used to separate microorganisms from drinks (LaRocco *et al.* 1985, Littel and LaRocco 1986) and brewery samples (Hysert *et al.* 1976). These methods are, however, difficult to apply to particulate-containing solutions as filters rapidly become blocked. A potential way around this problem has been investigated by some workers and utilise a double filtration system/scheme (Littel *et al.* 1986), the first filter removing food debris but allowing microorganisms through, the second filter trapping microorganisms prior to lysis and bioluminescent analysis. Other workers (Baumgart *et al.* 1980, Stannard and Wood 1983) have utilised ion exchange resins to trap selectively either food debris or microorganisms before bioluminescent tests were done.

The use of selective chemical extraction to separate microbial and non-microbial ATP has been extensively tested for both milk (Bossuyt 1981) and meat (Billte and Reuter 1985) and found to be successful. In general, this technique involves the lysis of somatic (food) cells followed by destruction of the released ATP with an apyrase (ATPase) enzyme. A more powerful extraction reagent can then be used to lyse microbial cells, which can then be tested with luciferase, thus enabling the detection of microbial ATP only.

There are a number of commercially available instruments aimed specifically at the detection of microbial ATP; Lumac (Netherlands), Foss Electric (Denmark), Bio Orbit (Finland) and Biotrace (UK) all produce systems, including separation methods, specifically designed to detect microorganisms in foods. Generally, all of the systems perform well and have similar specifications, including a minimum detection threshold of 10^4 bacteria (10^3 yeasts) and analysis times of under one hour.

In addition to testing food samples for total viable microorganisms, there have been a number of reports concerning potential alternative uses of ATP bioluminescence within the food industry. The application of ATP analysis to rapid hygiene testing has been considered (Holah 1989), both as a method of rapidly assessing microbiological contamination, and as a procedure for measuring total surface cleanliness. It is in the latter area that ATP measurement can give a unique result. As described earlier, almost all foods contain very high levels of ATP, thus food debris left on a production line could be detected in minutes using a bioluminescence method, allowing a very rapid check of hygienic status to be done. The use of ATP bioluminescence to monitor surface hygiene has now been widely adopted by industry. The availability of relatively inexpensive, portable, easy to use luminometers has now enabled numerous food producers to implement rapid hygiene testing procedures that are ideal for HACCP monitoring applications where surface hygiene is a critical control point. Reports suggest (Griffiths 1995) that all companies surveyed who regularly use ATP hygiene monitoring techniques note improvements in cleanliness after initiation of the procedure. Such ATP based test systems can be applied to most types of food processing plant, food service and retail establishments and even assessing the cleanliness of transportation vehicles such as tankers.

One area that ATP bioluminescence has not yet been able to address has been the detection of specific microorganisms. It may be possible to use selective enrichment media for particular microorganisms in order to allow selective growth prior to ATP analysis. This approach would, however, considerably increase analysis time and some false high counts would be expected. The use of specific lysis agents that release ATP only from the cells being analysed have been investigated (Stannard 1989) and shown to be successful. The number of these specific reagents is, however, small and thus the method is of only limited use. Perhaps the most promising method developed for the detection of specific organisms is the use of genetically engineered bacteriophages (Ulitzur and Kuhn, 1987, Ulitzur *et al.* 1989, Schutzbank *et al.* 1989).

Bacteriophages are viruses that infect bacteria. Screening of bacteriophages has shown that some are very specific, infecting only a particular type of bacteria. Workers have shown it is possible to add into the bacteriophage the genetic information that causes the production of bacterial luciferase. Thus, when a bacteriophage infects its specific host bacterium, the latter produces luciferase and becomes luminescent. This method requires careful selection of the bacteriophage in order to ensure false positive or false negative results do not occur; it does however indicate that, in the future, luminescence-based methods could be used for the rapid detection of specific microorganisms (Stewart 1990).

In conclusion, the use of ATP bioluminescence in the food industry has been developed to a stage at which it can be reliably used as a rapid test for viable microorganisms, as long as an effective separation technique for microbial ATP is used. Its potential use in rapid hygiene testing has been realised and the technique is being used within the industry. Work has also shown that luminescence can allow the rapid detection of specific microorganisms but such a system would need to be commercialised before widespread use within the food industry.

8.4.3 Microscopy methods

Microscopy is a well established and simple technique for the enumeration of microorganisms. One of the first descriptions of its use was for rapidly counting bacteria in films of milk stained with the dye methylene blue (Breed and Brew 1916). One of the main advantages of microscope methods is the speed with which individual analyses can be done; however, this must be balanced against the high manual workload and the potential for operator fatigue caused by constant microscopic counting.

The use of fluorescent stains, instead of conventional coloured compounds, allows cells to be more easily counted and thus these stains have been the subject of considerable research. Microbial ecologists first made use of such compounds to visualise and count microorganisms in natural waters (Francisco *et al.* 1973, Jones and Simon 1975). Hobbies *et al.* (1977) first described the use of Nuclepore polycarbonate membrane filters to capture microorganisms before fluorescent staining, whilst enumeration was considered in depth by Pettipher *et al.* (1980), the method developed by the latter author being known as the direct epifluorescent filter technique (DEFT).

The DEFT is a labour-intensive manual procedure and this has led to research into automated fluorescence microscope methods that offer both automated sample preparation and high sample throughput. The first fully automated instrument based on fluorescence microscopy was the Bactoscan (Foss Electric, Denmark), which was developed to count bacteria in milk and urine. Milk samples placed in the instrument are chemically treated to lyse somatic cells and dissolve casein micelles. Bacteria are then separated by continuous centrifugation in a dextran/sucrose gradient. Microorganisms recovered from the gradient are incubated with a protease to remove residual protein, then stained with acridline orange and applied as a thin film to a disc rotating under a microscope. The fluorescent light from the microscope image is converted into electrical impulses and recorded. The Bactoscan has been used widely for raw milk testing in continental Europe, and correlations with conventional methods have reportedly been good (Kaereby and Asmussen 1989). The technique does, however, have a poor sensitivity (approximately 5×10^4cells/ml) and this negates its use on samples with lower bacterial counts.

An instrument-based fluorescence counting method, in which samples were spread onto a thin plastic tape, was developed for the food industry. The instrument (Autotrak) deposited samples onto the tape, which was then passed through staining and washing solutions, before travelling under a fluorescence microscope. The light pulses from the stained microorganisms were then enumerated by a photomultiplier unit. Tests on food samples using this instrument (Betts and Bankes 1988) indicated that the debris from food samples interfered with the staining and counting procedure and gave results that were significantly higher than corresponding total viable counts.

Perhaps the most recent development in fluorescence microscope techniques to be used within the food industry for rapid counting is flow cytometry. In this technique the stained sample is passed under a fluorescence microscope system as a liquid in a flow cell. Light pulses caused by the light hitting a stained particle are transported to a photomultiplier unit and counted. This technique is automated, rapid and potentially very versatile. Of the microscope methods discussed here, the DEFT has perhaps the widest usage, whilst flow cytometry could offer significant advantages in the future. These two procedures will therefore be discussed in more detail.

DEFT

The DEFT was developed for rapidly counting the numbers of bacteria in raw milk samples (Pettipher *et al.* 1980, Pettipher and Rodrigues 1982). The method is based on the pretreatment of a milk sample in the presence of a proteolytic enzyme and surfactant at 50ºC, followed by a membrane filtration step that captures the microorganisms. The pretreatment is designed to lyse somatic cells and solubilise fats that would otherwise block the membrane filter. After filtration the membrane is strained with the fluorescent nucleic acid binding dye, acridine orange, then rinsed and mounted on a microscope slide. The membrane is then viewed with an epifluorescent microscope. This illuminates the membrane with ultraviolet light, causing the stain to emit visible light that can be seen through the microscope. As the stain binds to nucleic acids it is concentrated within microbial cells by binding to DNA and RNA molecules; thus any organisms on the membrane can be easily visualised and counted. The complete pretreatment and counting procedure can take as little as 30 minutes.

Although the DEFT was able to give a very rapid count, it was very labour intensive, as all of the pre-treatment and counting was done manually. This led to a very poor daily sample throughput for the method. The development of semi-automated counting methods based on image analysis (Pettipher and Rodridgues 1982) overcame some of the problems of manual counting and thus allowed the technique to be more user friendly.

The early work on the uses of DEFT for enumerating cells in raw milk was followed with examinations of other types of foods. It was quickly recognised that the good correlations between DEFT count and conventional total viable counts that were obtained with raw milk samples did not occur when heat-treated milks were examined (Pettipher and Rodrigues 1981). Originally this was considered to be due to heat-inducing staining changes occurring in Gram-positive cocci (Pettipher and Rodrigues 1981); however, more recent work (Back and Kroll 1991) has shown similar changes occur in both Gram-positives and Gram-negatives. Similar staining phenomena have also been observed with heat-treated yeasts (Rodrigues and Kroll 1986) and in irradiated foods (Betts *et al.* 1988). Thus the use of DEFT as a rapid indication of total viable count are mainly confined to raw foods.

The types of food with which DEFT can be used has been expanded since the early work with raw milk. Reports have covered the use of the method with frozen meats and vegetables (Rodrigues and Kroll 1989), raw meats (Shaw *et al.* 1987), alcoholic beverages (Cootes and Johnson 1980, Shaw 1989), tomato paste (Pettipher *et al.* 1985), confectionery (Pettipher 1987), dried foods (Oppong and Snudden 1988) and hygiene testing (Holah *et al.* 1988). In addition, some workers (Rodrigues and Kroll 1988) have suggested that the method could be modified to detect and count specific groups of organisms.

In conclusion, the DEFT is a very rapid method for the enumeration of total viable microorganisms in raw foods and has been used with success within the industry. The problems of the method are a lack of specificity and an inability to give a good estimate of viable microbial numbers in processed foods. The former could be solved by the use of short selective growth stages or fluorescently labelled antibodies; however, these solutions would have time and cost implications. The problem with processed foods can be eliminated only if alternating straining systems that mark viable cells are examined; preliminary work (Betts *et al.* 1989) has shown this approach to be successful, and the production and commercialisation of fluorescent viability stains could advance the technique. At present the high manual input and low sample throughput of DEFT procedures has limited the use of the procedure in the food industry.

Flow cytometry

Flow cytometry is a technique based on the rapid measurement of cells as they flow in a liquid stream past a sensing point (Carter and Meyer 1990). The cells under investigation are inoculated into the centre of a stream of fluid (known as the sheath fluid). This constrains them to pass individually past the sensor and enables measurements to be made on each particle in turn, rather than average values for the whole population. The sensing point consists of a beam of light (either ultraviolet or laser) that is aimed at the sample flow and one or more detectors that measure light scatter or fluorescence as the particles pass under the light beam. The increasing use of flow cytometry in research laboratories has largely been due to the development of the reliable instrumentation and the numerous staining systems. The stains that can be used with flow cytometers allow a variety of measurements to be made. Fluorescent probes based on enzyme activity, nucleic acid content, membrane potential and pH all have been examined, whilst the use of antibody-conjugated fluorescent dyes confers specificity to the system.

Flow cytometers have been used to study a range of eukaryotic and prokaryotic microorganisms. Work with eukaryotes has included the examination of pathogenic amoeba (Muldrow *et al.* 1982) and yeast cultures (Hutter and Eipel 1979), whilst bacterial studies have included the growth of *Escherichia coli* (Steen *et al.* 1982), enumeration of cells in bacterial cultures (Pinder *et al.* 1990) and the detection of *Legionella* spp. in cooling tower waters (Tyndall *et al.* 1985).

Flow cytometric methods for the food industry have been developed and have been reviewed by Veckert *et al.* (1995). Donnelly and Baigent (1986) explored the use of fluorescently labelled antibodies to detect *Listeria monocytogenes* in milk, and obtained encouraging results. The method used by these authors relied on the selective enrichment of the organisms for 24 hours, followed by staining with fluorescein isothiocyanate labelled polyvalent *Listeria* antibodies. The stained cells were then passed through a flow cytometer, and the *L. monocytogens* detected. The author suggested that the system could be used with other types of food. A similar approach was used by McCelland and Pinder (1994) to detect *Salmonella typhimurium* in dairy products.

Patchett *et al.* (1991) investigated the use of a Skatron Argus flow cytometer to enumerate bacteria in pure cultures and foods. The results obtained with pure cultures showed that flow cytometer counts correlated well with plate counts down to 10^3 cells/g. With foods, however, conflicting results were obtained. Application of the technique to meat samples gave a good correlation with plate counts and enabled enumeration down to 10^5 cells/g. Results for milk and paté were poorer, the sensitivity of the system for paté being 10^6 cells/ml, whilst cells inoculated into milk were not detected at levels in excess of 10^7 ml. The poor sensitivity of this flow cytometer with foods was thought to be due to interference of the counting system caused by food debris and it was suggested that the application of separation methods to partition microbial cells from food debris would overcome the problem.

Perhaps the most successful application of flow cytometric methods to food products has been the use of a Chemunex Chemflow system to detect contaminating yeast in dairy and fruit products (Bankes *et al.* 1991). The procedure used with this system calls for an incubation of the product for 16–20 hours followed by centrifugation to separate and concentrate the cells. The stain is then added, and a sample is passed through the flow cytometer for analysis. An evaluation of the system by Pettipher (1991), using soft drinks inoculated with yeasts, showed that it was reliable and user friendly. The results obtained indicated that cytometer counts correlated well with DEFT counts, however, the author did not report how the system compared to plate counts.

Investigations of the Chemflow system by Bankes *et al.* (1991) utilised a range of dairy and fruit-based products inoculated with yeast. Results indicated that yeast levels as low as 1 cell/25g could be detected in 24 hours in dairy products. In fruit juices a similar sensitivity was reported: however, a 48 hour-period was required to ensure that this was achieved. The system was found to be robust and easy to use. The Chemflow system has now been adapted to detect bacterial cells as well as yeasts, and applications are available for fermentor biomass and enumeration of total flora in vegetables. The Chemflow system has been fully evaluated in a factory environmental (Dumain *et al.* 1990) testing fermented dairy products. These authors report a very good correlation between cytometer count and plate count (r = 0.98), results being obtained in 24 hours, thus providing a time saving of three days over classical methods.

In conclusion, flow cytometry can provide a rapid and sensitive method for the rapid enumeration of microorganisms. The success of the system depends on the development and use of (a) suitable staining systems, and (b) protocols for the separation of microorganisms from food debris that would otherwise interfere with the detection system. In the future a flow cytometer fitted with a number of light detection systems could allow the analysis of samples for many parameters at once, thus considerably simplifying testing regimes.

Solid phase cytometry
A relatively new cytometric technique has been developed by Chemunex (Maisons-Alfort, France) based on solid phase cytometry. In this procedure samples are passed through a membrane filter which captures contaminating microorganisms. A stain is then applied to the filter to fluorescently mark metabolically active microbial cells. After staining, the membrane is then transferred to a Chemscan RDI instrument, which scans the whole membrane with a laser, counting fluorescing cells. The complete procedure takes around 90 minutes to perform and can detect single cells in the filtered sample. The Chemscan RDI solid phase cytometry system is an extremely powerful tool for rapidly counting low levels of organisms. It is ideally suited to the analysis of waters or other clear filterable fluids, and using specific labelling techniques could be used to detect particular organisms of interest. Foods containing particulate materials could, however, be problematic as organisms would need to be separated from the food material before filtration and analysis.

8.4.4 Immunological methods

Antibodies and antigens
Immunological methods are based on the specific binding reaction that occurs between an antibody and the antigen to which it is directed. *Antibodies* are protein molecules that are produced by animal white blood cells, in response to contact with a substance causing an immune response. The area to which an antibody attaches on a target molecule is known as the *antigen*. Antigens used in immunochemical methods are of two types. The first occurs when the analyte is of low molecular weight and thus does not stimulate an immune response on its own; these substances are described as haptens and must be bound to a larger carrier molecule to elicit an immune response and cause antibody production. The second type of antigen is immunogenic and is able to elicit an immune response on its own.

Two types of antibody can be employed in immunological tests. These are known as *monoclonal* and *polyclonal* antibodies. *Polyclonal* antibodies are produced if large molecules such as proteins or whole bacterial cells are used to stimulate an immune response in an animal. The many antigenic sites result in numerous different antibodies being produced to the molecule or cell. *Monoclonal* antibodies are produced by tissue culture techniques and are derived from a single white blood cell; thus they are directed towards a single antigenic site. The binding of an antigen is highly specific. Immunological methods can therefore be used to detect particular specific microorganisms or proteins (e.g. toxins). In many cases, when using these methods a label is attached to the antibody, so that binding can be visualised more easily when it occurs.

Labels

The labels that can be used with antibodies are of many types and include radiolabels, fluorescent agents, luminescent chemicals and enzymes; in addition agglutination reactions can be used to detect the binding of antibody to antigen.

Radioisotopes have been extensively used as labels, mainly because of the great sensitivity that can be achieved with these systems. They do however have some disadvantages, the main one being the hazardous nature of the reagents. This would negate their use in anything other than specialist laboratories, and certainly their use within the food industry would be questionned.

Fluorescent labels have been widely used to study microorganisms. The most frequently used reagent has been fluorescein. However, others such as rhodamine and umbelliferone have also been utilised. The simplest use of fluorescent antibodies is in microscopic assays. Recent advances in this approach have been the use of flow cytometry for multiparameter flow analysis of stained preparations, and the development of enzyme-linked immunofluorescent assays (ELIFA), some of which have been automated.

Luminescent labels have been investigated as an alternative to the potentially hazardous radiolabels (Kricka and Whitehead 1984). The labels can be either chemiluminescent or bioluminescent, and have the advantage over radiolabels that they are easy to handle and measure using simple equipment, whilst maintaining a similar sensitivity (Rose and Stringer 1989). A number of research papers have reported the successful use of immunoluminometric assays (Lohneis *et al.* 1987); however, none have yet been commercialised.

Antibodies have been used for the detection of antigens in precipitation and agglutination reactions. These assays tend to be more difficult to quantify than other forms of immunoassay and usually have only a qualitative application. The assays are quick and easy to perform and require little in the way of equipment.

A number of agglutination reactions have been commercialised by manufacturers and have been successfully used within the food industry. These methods have tended to be used for the confirmation of microbial identity, rather than for the detection of the target organisms. They offer a relatively fast test time, are easy to use and usually require no specialist equipment, thus making ideal test systems for use in routine testing laboratories.

Several latex agglutination test kits are available for the conformation of *Salmonella* from foods. These include the Oxoid *Salmonella* Latex Kit (Oxoid) designed to be used with the Oxoid Rapid *Salmonella* Test Kit (Holbrook *et al.* 1989); the Micro Screen *Salmonella* Latex Slide Agglutination Test (Mercia Diagnostics Ltd.); the Wellcolex Colour *Salmonella* Test (Wellcome Diagnostics) (Hadfield *et al.* 1987a, b); and the Spectate *Salmonella* test (Rhone

Poulenc Diagnostics Ltd.) (Clark *et al.* 1989). The latter two kits use mixtures of coloured latex particles that allow not only detection but also serogrouping of *Salmonella*. Latex agglutination test kits are also available for *Campylobacter* (Microscreen, Mercia Diagnostics), *Staphyloccocus aureus* (Staphaurex, Wellcome Diagnostics), *Shigella* (Wellcolex Colour *Shigella* Test, Wellcome Diagnostics) and *Escherichia coli* 0157:H7 (Oxoid). Agglutination kits have also been developed for the detection of microbial toxins, e.g. Oxoid Staphylococcal Enterotoxin Reverse Passive Latex Agglutination Test (Rose *et al.* 1989, Bankes and Rose 1989).

Enzyme immunoassays have been extensively investigated as rapid detection methods for foodborne microorganisms. They have the advantage of specificity conferred by the use of a specific antibody, coupled with coloured or fluorescent end-points that are easy to detect either visually or with a spectrophotometer or fluorimeter. Most commercially available enzyme immunoassays use an antibody sandwich method in order initially to capture and then to detect specific microbial cells or toxins. The kits are supplied with two types of antibody: capture antibody and conjugated antibody. The capture antibody is attached to a solid support surface such as a microtitre plate well. An enriched food sample can be added to the well and the antigens from any target cells present will bind to the antibodies. The well is washed out, removing food debris and unbound microorganisms. The enzyme conjugated antibody can then be added to the well. This will bind to the target cell forming an antibody sandwich. Unbound antibodies can be washed from the well and the enzyme substrate added. The substrate will be converted by any enzyme present from a colourless form, into a coloured product. A typical microplate enzyme immunoassay takes between two and three hours to perform and will indicate the presumptive presence of the target bacterial cells. Thus positive samples should always be confirmed by biochemical or serological methods.

There are a number of commercially available enzyme immunoassay test kits for the detection of *Listeria*, *Salmonella*, *Escherichia coli* 0157, *Staphylococcal enterotoxins*, and *Bacillus diarrhoeal* toxin, from food samples. The sensitivity of these systems is approximately 10^6 cells/ml, so that a suitable enrichment procedure must be used before analysis using the assay. Thus, results can be obtained in two to three days, rather than the three to five days required for a conventional test procedure.

Over recent years a number of highly automated immunoassays have been developed, these add to the benefit of the rapid test result, by reducing the level of manual input required to do the test. Automation of enzyme immunoassays has taken a number of forms; a number of manufacturers market instruments which simply automate standard microplate ELISAs. These instruments hold reagent bottles and use a robotic pipetting arm, which dispenses the different reagents required in the correct sequence. Automated washing and reading completes the assay with little manual input needed. At least two manufacturers have designed immunoassay kits around an automated instrument, to produce very novel systems.

The Vidas system (bioMerieux, Basingstoke) uses a test strip, containing all of the reagents necessary to do an ELISA test, the first well of the strip is inoculated with an enriched food sample, and placed into the Vidas instrument, together with a pipette tip internally coated with capture antibody. The instrument then uses the pipette tip to transfer the test sample into the other cells in the strip containing various reagents needed to carry out the ELISA test. All of the transfers are completely automatic, as is the reading of the final test result. Vidas ELISA tests are available for a range of organisms including, *Salmonella, Listeria, Listeria monocytogenes, E.coli* 0157, *Campylobacter* and staplylococcal enterotoxin. Evaluations of a number of these methods have been done (Blackburn *et al* 1994, Bobbitt and Betts 1993) and indicated that results were at least equivalent to conventional test methods.

The EIAFOSS (Foss Electric, Denmark) is another fully automated ELISA system, in this case the instrument transfers all of the reagents into sample containing tubes, in which all of the reactions occur. The EIAFOSS procedure is novel as it uses antibody coated magnetic beads as a solid phase. During the assay these beads are immobilised using a magnet mounted below the sample tube. EIAFOSS test kits are available for *Salmonella, Listeria, E.coli* 0157 *and Campylobacter*, and evaluations have indicated that these methods operate well (Jones and Betts 1994).

The newest immunoassay procedure that has been developed into a commercial format is arguably the simplest to use. Immunochromatography operates on a dipstick, composed of an absorbent filter material which contains coloured particles coated with antibodies to a specific organism. The particles are on the base of the dipstick and when dipped into a microbiological enrichment broth, they move up the filter as the liquid is moved by capillary action. At a defined point along the filter material lies a line of immobilised specific antibodies. In the presence of the target organism, binding of that organism to the coloured particles will occur. This cell/particle conjugate moves up the filter dipstick by capillary action until it meets the immobilised antibodies where it will stick. The build up of coloured particles results in a clearly visible coloured line, indicating a positive test result.

A number of commercial kits are based around this procedure including the Oxoid *Listeria* Rapid Test (Jones *et al.* 1995a) and the Celsis Lumac Pathstik (Jones *et al.* 1995b.) have been developed and appear to give good results. The immunochromatography techniques require an enrichment in the same way as other immunoassays; they do not however require any equipment or instrumentation, and once the dipstick is inoculated, need only minutes to indicate a positive or negative result.

Immunoassay conclusions

In conclusion, immunological methods have been extensively researched and developed. There are now a range of systems that allow the rapid detection of the specific organism to which they are directed. Numerous evaluations of commercially available immuno-based methods have indicated that the results generally correlate well with conventional microbiological methods. Enzyme immunoassays in particular appear to offer a simple way of reducing analysis times by one or two days; automation or miniaturisation of these kits has reduced the amount of person time required to do the test and simplified the manual procedures considerably.

The main problem with the immunological systems is their low sensitivity. The minimum number of organisms required in an enzyme immunoassay system to obtain a positive result is approximately 10^5/ml. As the food microbiologist will want to analyse for the presence or absence of a single target organism in 25g of food, an enrichment phase is always necessary. The inclusion of enrichment will always add 24–48 hours to the total analysis time.

8.4.5 Nucleic acid hybridisation

Nucleic acids

The specific characteristics of any organism depend on the particular sequence of the nucleic acids contained in its genome. The nucleic acids themselves are made up of a chain of units each consisting of a sugar (deoxyribose or ribose, depending on whether the nucleic acid is DNA or RNA), a phosphorus-containing group and one of four organic purine or pyrimidine bases. DNA is constructed from two of these chains arranged in a double helix and held together by bonds between the organic bases. The bases specifically bind adenine to thymine and guanine to cytocine. It is the sequence of bases that make different organisms unique.

The development of nucleic acid probes

Nucleic acid probes are small segments of single-stranded nucleic acid that can be used to detect specific genetic sequences in test samples. Probes can be developed against DNA or RNA sequences. The attraction of the use of gene probes in the problem of microbial detection is that a probe consisting of only 20 nucleotide sequences is unique and can be used to identify an organism accurately (Gutteridge and Arnott 1989).

In order to be able to detect the binding of a nucleic acid probe to DNA or RNA from a target organism, it must be attached to a label of some sort that can easily be detected. Early work was done with radioisotope labels such as phosphorus (^{32}P) that could be detected by autoradiography or scintillation counting. Radiolabels, however, have inherent handling, safety and disposal problems that make them unsuitable for use in food laboratories doing routine testing. Thus the acceptance of widespread use of nucleic acid probes required the development of alternative labels.

A considerable amount of work has been done on the labelling of probes with an avidin-biotin link system. This is based on a very high binding specificity between avidin and biotin. The probe sequence of nucleic acid is labelled with biotin and reacted with target DNA. Avidin is then added, linked to a suitable detector, e.g. avidin-alkaline phosphatase, and binding is detected by the formation of a coloured product from a colourless substrate. These alternative labelling systems proved that non-radiolabelled probes could be used for the detection of microorganisms. However, the system was much less sensitive than isotopic procedures, requiring as much as a 100-fold increase in cell numbers for detection to occur, compared to isotope labels.

In order to develop non-isotopic probes with a sensitivity approaching that of isotope labels, it was necessary to consider alternative probe targets within cells. Probes directed toward cell DNA attach to only a few sites on the chromosome of the target cell. By considering areas of cell nucleic acid that are present in relatively high copy number in each cell and directing probes toward these sites, it is possible to increase the sensitivity of non-isotopic probes considerably. Work on increasing probe sensitivity centred on the use of RNA as a target. RNA is a single-stranded nucleic acid that is present in a number of forms in cells. In one form it is found within parts of the cell protein synthesis system called ribosomes. Such RNA is known as ribosomal RNA (rRNA), and is present in very high copy numbers within cells. By directing nucleic acid probes to ribosomal RNA it is possible to increase the sensitivity of the assay system considerably.

Probes for organisms in food

Nucleic acid hybridisation procedures for the detection of pathogenic bacteria in foods have been described for *Salmonella* spp. (Fitts 1985 Curiale *et al.* 1986), *Listeria* spp. (Klinger *et al.* 1988, Klinger and Johnson 1988), *Yersinia enterocolitica* (Hill *et al.* 1983b, Jagow and Hill 1986), *Listeria monocytogenes* (Datta *et al.* 1988), enterotoxigenic *Escherichia coli* (Hill *et al.* 1983a, 1986), *Vibrio vulnificus* (Morris *et al.* 1987), enterotoxigenic *Staphylococcus aureus* (Notermans *et al.* 1988), *Clostridium perfringens* (Wernars and Notermans 1990) and *Clostridium botulinum* (Wernars and Notermans 1990).

The first commercially available nucleic-probe-based assay system for food analysis was introduced by Gene Trak Systems (Framingham, MA, USA) in 1985 (Fitts 1985). This test used *Salmonella*-specific DNA probes directed against chromosomal DNA to detect *Salmonella* in enriched food samples. The format of the test involved hybridisation between target DNA bound to a membrane filter and phosphorous 32-labelled probes. The total analysis time for the test was 40–44 hours of sample enrichment in non-selective and selective media, followed by the hybridisation procedure lasting 4–5 hours. Thus the total analysis time was approximately 48 hour. The *Salmonella* test was evaluated in collaborative studies in the USA and appeared to be at least equivalent to standard culture methods (Flowers *et al.* 1987). Gene Trak also produced a hybridisation assay for *Listeria* spp., based on a similar format (Klinger and Johnson 1988).

The Gene Trak probe kits gained acceptance within the United States and a number of laboratories began using them. In Europe, however, there was a reluctance among food laboratories to use radioisotopes within the laboratory. In addition, ^{32}P has a short half-life, which caused difficulties when transporting kits to distant sites. In 1988 Gene Trak began marketing non-isotopically labelled probes for *Salmonella, Listeria* and *Escherichia coli.* The detection system for the probes was colorimetric. In order to overcome the reduction in sensitivity caused by the use of non-isotopic labels, the target nucleic acid within the cell was ribosomal RNA. This nucleic acid is present in an estimated 500 to 20,000 copies per cell.

The colorimetric hybridisation assay is based on a liquid hybridisation reaction between the target rRNA and two separate DNA oligonucleotide probes (the capture probe and the reporter probe) that are specific for the organism of interest. The capture probe molecules are extended enzymatically with a polymer of approximately 100 deoxyadenosine monophosphate residues. The reporter probe molecules are labelled chemically with the hapten fluorescein.

Following a suitable enrichment of the food under investigation, a test sample is transferred to a tube and the organisms lysed, releasing rRNA targets. The capture and detector probes are then added and hybridisation is allowed to proceed. If target rRNA is present in the sample, hybridisation takes place between the probes and the target 16s rRNA. The solution containing the target probe complex is then brought into contact with a solid support dipstick, containing bound deoxythymidine homopolymer, under conditions that will allow hybridisation between the poly-deoxyadenosine polymer of the capture probe and the poly-deoxythymidine on the dipstick. Unhybridised nucleic acids and cellular debris are then washed away, leaving the capture DNA-RNA complex attached to the surface of the dipstick. The bound fluoresceinated reporter probe is detected by the addition of an antifluorescein antibody conjugated to the enzyme horseradish peroxidase. Subsequent addition of a chromogenic substrate for the enzyme results in colour development that can be measured spectrophotometrically.

Results of the colorimetric assays (Mozola *et al.* 1991) have indicated a good comparison between the probe methods and conventional cultural procedures for both *Salmonella* and *Listeria.* The sensitivity of the kits appeared to be between 10^5 and 10^6 target organisms/ml, and thus the enrichment procedure is a critical step in the methodology. Since the introduction of the three kits previously mentioned, Gene Trak have begun to market systems for *Staphylococcus aureus, Campylobacter* spp. and *Yersinia enterocolitica.*

Commercially available nucleic acid probes for the confirmation of *Campylobacter, Staphylococcus aureus* and *Listeria* are available from Gen-probe (Gen Probe Inc., San Diego, USA). These kits are based on a single-stranded DNA probe that is complementary to the ribosomal RNA of the target organism. After the ribosomal RNA is released from the organism, the labelled DNA probe combines with it to form a stable DNA:RNA hybrid. The hybridised probe can be detected by its luminescence.

The assay method used is termed a hybridisation protection assay and is based on the use of a chemiluminescent acridinium ester. This ester reacts with hydrogen peroxide under basic conditions to produce light that can be measured in a luminometer. The acridinium esters are covalently attached to the synthetic DNA probes through an alkylamine arm. The assay format is based on differential chemical hydrolysis of the ester bond. Hydrolysis of the bond renders the acridinium permanently non-chemiluminescent. When the DNA probe which the ester is attached, hybridises to the target RNA, the acridinium is protected from hydrolysis and can thus be rendered luminescent. The test kits for *Campylobacter* and *Listeria* utilise a freeze-dried probe reagent. The *Campylobacter* probe reacts with *C. jejuni, C. coli* and *C. pylori*; the *Listeria* probe reacts with *L. monocytogenes.* In both cases a full cultural enrichment protocol is necessary prior to using the probe for confirmation testing.

An evaluation of the *L. monocytogenes* probe kit (Bobbit and Betts 1991) indicated that it was totally specific for the target organism. The sensitivity required approximately 10^6 *L. monocytogenes* to be present in order for a positive response to be obtained. The kit appeared to offer a fast reliable culture confirmation test and had the potential to be used directly on enrichment broth, thus reducing test times even further.

Probes – the future

The development and use of probes in the food industry has advanced little in recent years. The kits that are currently available show great promise but are not as widely used as immunoassays. Microbiologists must always consider the usefulness of analysing the genetic information within cells, for example to detect the presence of genes coding for toxins could be detected, even when not expressed, and screening methods could be devised for pathogenicity plasmids, such as that in *Yersinia enterocolitica.* It may be however, that the advances in molecular biology mean that the best way to test for such information is by using nucleic acid amplification methods such as the Polymerase Chain Reaction (PCR).

Nucleic acid amplification techniques

In recent years, several genetic amplification techniques have been developed and refined. The methods usually rely on the biochemical amplification of cellular nucleic acid and can result in a 10^7-fold amplification in two to three hours. The very rapid increase in target that can be gained with nucleic acid amplification methods, makes them ideal candidates for development of very rapid microbial detection systems. A number of amplification methods have been developed and applied to the detection of microorganisms:

- Polymerase Chain Reaction (PCR) and variations, including nested PCR, reverse transcriptase (RT) PCR and multiplex PCR.
- Q Beta Replicase.
- Ligase Amplification Reaction (LAR).
- Transcript Amplification System (TAS), also known as Self Sustained Sequence Replication (3SR) or Nucleic Acid Sequence Based Amplification (NASBA).

Of these amplification methods only PCR has been commercialised as a kit-based procedure for the detection of food-borne microorganisms. Much research has been done with NASBA and there are a number of research papers outlining its use for detecting food pathogens but as yet, no commercially available kits are on the market.

Polymerase chain reaction (PCR)

PCR is a method used for the repeated *in vitro* enzymic synthesis of specific DNA sequences. The method uses two short oligonucleotide primers that hybridise to opposite strands of a DNA molecule and flank the region of interest in the target DNA. PCR proceeds via series of repeated cycles, involving DNA denaturation, primer annealing and primer extension by the action of DNA polymerase. The three stages of each cycle are controlled by changing the temperature of the reaction, as each stage will occur only at particular defined temperatures. These temperature changes are accomplished by using a specialised instrument known as a thermocycler. The products of primer extension from one cycle, act as templates for the next cycle, thus the number of target DNA copies doubles at every cycle.

Reverse transcriptase – PCR

This involves the use of an RNA target for the PCR reaction. The PCR must work on a DNA molecule; thus initially reverse transcriptase is used to produce copy DNA (cDNA). The latter is then used in a conventional PCR reaction. The RT-PCR reaction is particularly applicable to certain microbiological tests. Some food-borne viruses contain RNA as their genetic material; thus RT-PCR must be used if amplification and thus detection of these viruses is necessary. A second use of RT-PCR is in the detection of viable microorganisms. One of the problems associated with PCR is its great sensitivity and ability to amplify very low concentrations of a target nucleic acid. Thus, if using PCR to detect the presence or absence of a certain microorganism in a food, PCR could 'detect' the organism, even if it had been previously rendered inactive by a suitable food process. This could result in a false positive detection. A way to overcome this problem is to use an RT-PCR targeted against cellular messenger RNA, which is only produced by active cells and once produced has a short half-life. Thus a detection of specific mRNA by an RT-PCR procedure is indicative of the presence of a viable microorganism.

NASBA

NASBA is a multi-enzyme, multicycle amplification procedure, requiring more enzymes and reagents than standard PCR. It does however, have the advantage of being isothermal, therefore all stages of the reaction occur at a single temperature and a thermocycler is not required. Various research papers have been published which use NASBA to detect foodborne pathogeses (e.g. Uyttendaele *et al.* 1996), however, the procedure has yet to be commercialised.

Commercial PCR-based kits

Currently there are three manufacturers producing kits based on PCR for the detection of food-borne microorganisms. BAX (Qualicon, USA) utilises tabletted reagents and a conventional thermocycler, geL electrophoresis-based approach. Positive samples are visualised as bands on an electrophoresis gel. BAX kits are available for *Salmonella* (Bennett *et al.* 1998), *Listeria* Genus *Listeria monocytogenes, E. coli* 0157:H7; the tests for *Salmonella* and *E.coli* 0157 have been through an Association of Official Analytical Chemists Research Institute (AOACRI) testing procedure and have gained AOACRI Performance Tested Status.

The second of the commercially available PCR kits is the Probelia kit (Sanofi, France); this uses conventional PCR followed by an immunoassay and colorimetric detection system. Kits are available for *Salmonella* and *Listeria.*

The final commercial PCR system is the TaqMan system (Perkin Elmer, USA). This uses a novel probe system incorporating a TaqMan Label. This is non-fluorescent in its native form, but once the probe is bound between the primers of the PCR reaction, it can be acted upon by the DNA polymerase enzyme used in PCR to yield a fluorescent end product. This fluorescence is detected by a specific fluorescence detection system. TaqMan kits are available for *Salmonella* and under development for *Listeria* and *E. coli* 0157. Perhaps one of the most interesting future aspects of TaqMan is its potential to quantify an analyte. Currently PCR-based systems are all based on presence/absence determinations, TaqMan procedures and instrumentation can give information on actual numbers. Therefore the potential for using PCR for rapidly counting microorganisms could now be achieved.

Separation and concentration of microorganisms from foods

In recent years there has been considerable interest in the potential for separating microorganisms from food materials and subsequently concentrating them to yield a higher number per unit volume. The reason for this interest is that many of the currently available rapid test methods have a defined sensitivity, examples are: 10^4/ml for ATP luminescence, 10^6/ml for electrical measurement, 10^5-10^6/ml for immunoassay and DNA probes and approximately 10^3/ml for current PCR based kits. These sensitivity levels mean that a growth period is usually required before the rapid method can be applied and this growth period may significantly increase the total test time.

One way in which this problem can be addressed is by separating and concentrating microorganisms from the foods, in order to present them to the analytical procedure in a higher concentration. An additional advantage being that the microbial cells may be removed from the food matrix, which in some cases may contain materials which interfere with the test itself. A simple example of the use of concentration, is in the analysis of clear fluids (water, clear soft drinks, wines, beers, etc.). Here contamination levels are usually very low, thus large volumes are membrane filtered to concentrate the microorganisms onto a small area. These captured organisms can then be analysed. A thorough review of separation concentration methods has been given by Betts (1994). They broadly fall into five categories:

1. filtration
2. centrifugation
3. phase separation
4. electrophoresis
5. immuno-methods.

Of these categories only one has reached commercialisation for use in solid foods, these are the immuno-methods. Immunomagnetic separation relies on coating small magnetic particles with specific antibodies for a known cell. The coated particles can be added into a food suspension or enrichment, and if present, target cells will attach to the antibodies on the particles.

Application of a magnetic field retains the particles and attached cells allowing food debris and excess liquid to be poured away, thus separating the cells from the food matrix and concentrating them. This type of system has been commercialised by Dynal (Norway), LabM (England), and Denka (Japan), an automated system incorporating the procedure is produced by Foss Electric (EIAFOSS). The various companies produce kits for *Salmonella, Listeria, E. coli* 0157, other verocytotoxin producing *E. coli* and *Campylobacter*. Immunomagnetic separation systems for detecting the presence of *E. coli* 0157 have been very widely used and become accepted standard reference methods in many parts of the world.

Identification and characterisation of microorganisms

Once an organism has been isolated from a food product it is often necessary to identify it, this is particularly relevant if the organism is considered to be a pathogen. Traditionally, identification methods have involved biochemical or immunological analyses of purified organisms. With the major advances now taken in molecular biology, it is now possible to identify organisms by reference to their DNA structure. The sensitivity of DNA based methods will in fact allow identification to a level below that of species (generally referred to as characterisation or sub-typing). Sub-typing is a powerful new tool that can be used by food microbiologists not just to name an organism, but also to find out its origin. Therefore it is possible in some cases to isolate an organism in a finished product, and then through a structured series of tests find whether its origin was a particular raw material, the environment within a production area or a poorly cleaned piece of equipment.

A number of DNA-based analysis techniques have been developed that allow sub-typing, many of these have been reviewed by Betts *et al.* (1995). There is, however, only one technique that has been fully automated, and made available to food microbiologist on a large scale, and that is Ribotyping through use of the Qualicon RiboPrinter (Qualicon, USA). This fully automated instrument accepts isolated purified colonies of bacteria, and produces DNA band images (RiboPrint patterns), that are automatically compared to a database to allow identification and characterisation. The technique has successfully been used within the food industry to identify contaminants, indicate the sources and routes of contamination and check for culture authenticity (Betts 1998).

8.5 Microbiological methods – the future

Conventional microbiological methods have remained little changed for many decades. Microbiologists generally continue to use lengthy enrichment and agar-growth-based methods to enumerate, detect and identify organisms in samples. As the technology of food production and distribution has developed, there has been an increasing requirement to obtain microbiological results in shorter time periods.

The rapid growth of the chilled foods market, producing relatively short shelf-life products, has led this move into rapid and automated methods, as the use of such systems allows: (a) testing of raw materials before use; (b) monitoring of the hygiene of the production line in real time; and (c) testing of final products over a reduced time period. All of these points will lead to better quality food products with an increased shelf-life.

All of the methods considered in this chapter are currently in use in Quality Control laboratories within the food industry. Some (e.g. electric methods) have been developed, established and used for a considerable time period, whilst others (e.g. Polymerase Chain Reaction), are a much more recent development. The future of all of these methods is good; they are now being accepted as standard and routine, rather than novel. Some users are beginning to see the benefits of linking different rapid methods together to gain an even greater test rapidity, e.g. using an enzyme immunoassay to detect the presence of *Listeria* spp., then using a species-specific nucleic acid probe to confirm the presence or absence of *L. monocytogenes.*

One of the problems of many of the rapid methods is a lack of sensitivity. This does in many cases mean that lengthy enrichments are required prior to using rapid methods. Research on methods for the separation and concentration of microorganism from food samples would enable microorganisms to be removed from the background of food debris and concentrated, thus removing the need for long incubation procedures. The developments in DNA-based methods for both detection and identification/characterisation have given new tools to the food microbiologist, there is no doubt that these developments will continue in the future giving significant analytical possibilities that are currently difficult to imagine.

8.6 References and further reading

ALEXANDER M K, KHAN M S and DOW C S, (1981) Rapid screening for bacteriuria using a particle counter, pulse-height analyser and computer. *Journal of Clinical Pathology* **34**, 194–8.

ANON, (1986) *Microorganisms in Foods 2. Sampling for microbiological analysis: Principles and specific applications*, ICMSF. Blackwell Scientific, Oxford.

BACK J P and KROLL R G, (1991) The differential fluorescence of bacteria stained with acridline orange and the effect of heat, *Journal of Applied Bacteriology*, **71** 51–8.

BANKES P, (1991) An evaluation of the Bactrac 4100. *Campden Food & Drink Research Association Technical Memorandum 628*, Campden Food & Drink Research Association, UK.

BANKES P and ROSE S A, (1989) Rapid detection of Staphylococcal enterotoxins in foods with a modification of the reversed passive latex agglutination assay, *Journal of Applied Bacteriology*, **67** 395–9.

BANKES P, ROWE D and BETTS R P, (1991) The rapid detection of yeast spoilage using the Chemflow system. *Campden Food & Drink Research Association Technical Memorandum 621*, Campden Food & Drink Research Association, UK.

BANKS J G, ROSSITER L M and CLARK A E, (1989) Selective detection of *Pseudomonas* in foods by a conductance technique. In: *Rapid Methods and Automation in Microbiology and Immunology* Florence 1987. Balows, A., Tilton, R.C. and Turano, A. (eds) Brixia Academic Press, Brescia, pp. 725–7.

BAUMGART J, FRICKLE K and KUY C, (1980) Quick determination of surface bacterial content of fresh meat using a bioluminescence method to determine adenosine triphosphate, *Fleischwirtschaft*, **60** 266–70.

BENNETT A R, GREENWOOD D, TENNANT C, BANKS J G and BETTS R P, (1998). Rapid and Definitive detection of *Salmonella* in foods by PCR, *Letters in Applied Microbiology*, **26** (6) 437–41.

BETTS R P, (1993) Rapid electrical methods for the detection and enumeration of food spoilage yeasts, *International Biodeterioration and Biodegradation*, **32** 19–32.

BETTS R P, (1994) The separation and rapid detection of microorganisms In: *Rapid Methods and Automation in Microbiology and Immunology* Spencer R.C., Wright E.P. and Newsom S.W.B., (eds) pp. 107–120, Intercept Press, Hampshire, England.

BETTS R P, (1998) Foodborne bacteria in the spotlight, *Laboratory News* September p. A6–7.

BETTS R P and BANKES P, (1988) An evaluation of the Autotrak – A rapid method for the enumeration of microorganisms. *Campden Food and Drink Research Association Technical Memorandum 491*, Campden and Chorleywood Food Research Association, UK.

BETTS R P, BANKES P, FARR L and STRINGER M F, (1988) The detection of irradiated foods using the Direct Epifluorescent Filter Technique, *Journal of Applied Bacteriology*, **64** 329–35.

BETTS R P, BANKES P and BANKS J G, (1989) Rapid enumeration of viable microorganisms by staining and direct microscopy. *Letters in Applied*

Microbiology, **9** 199–202.

BETTS R P, BANKES P and GREEN J, (1991) An evaluation of the Tecra enzyme immunoassay for the detection of *Listeria* in foods. In: *Food Safety and Quality Assurance: Applications of Immunoassays Systems.* Morgan M R A, Smith C J and Williams P A (eds) Elsevier, London, pp. 283–98.

BETTS R P, STRINGER M, BANKS J G and DENNIS C, (1995) Molecular Methods in Food Microbiology. *Food Australia*, **47** (7) 319–22.

BILLTE M and REUTER G, (1985) The bioluminescence technique as a rapid method for the determination of the microflora of meat, *International Journal of Food Microbiology*, **2** 371–81.

BLACKBURN C DE W and STANNARD C J, (1989) Immunological detection methods for *Salmonella* in foods. In: *Rapid Microbiological Methods for Foods, Beverages and Pharmaceuticals. Society for Applied Bacteriology Technical Series 25*, Stannard, C.L., Pettit, S.B. and Skinner, F.A. (eds) Blackwell Scientific, Oxford.

BLACKBURN C DE W, CURTIS L M, HUMPHESON L and PETITT S, (1994) Evaluation of the Vitek Immunodiagnostic Assay System (VIDAS) for the detection of *Salmonella* in foods, *Letters in Applied Microbiology*, **19** (1) 32.

BOBBITT J A and BETTS R P, (1991) Evaluation of Accuprobe Culture Confirmation Test for *Listeria monocytogenes. Campden Food & Drink Research Association Technical Memorandum 630*, Campden Food & Drink Research Association, UK.

BOBBITT J A and BETTS R P, (1993) Evaluation of the VIDAS *Listeria* system for the detection of *Listeria* in foods. *Campden Food and Drink Research Association Technical Memorandum 674*, Campden and Chorleywood Food Research Association, UK.

BOLTON F J and GIBSON D M, (1994) Automated electrical techniques in microbiological analysis. In *Rapid Analysis Techniques in Food Microbiology*, ed. Patel P., Blackie Academic, London.

BOLTON F J and POWELL S J, (1993) Rapid methods for the detection of *Salmonella* and *Campylobacter* in meat products, *European Food and Drink Review*, Autumn 73–81.

BOSSUYT R, (1981) Bacteriological quality determination of raw milk by an ATP assay technique, *Milchwissenschaft*, **36** 257–60.

BREED R S, and BREW J D, (1916) Counting bacteria by means of the microscope. *Technical Bulletin 49.* New York Agricultural Experimental Station, Albany, New York.

CANDLISH A, (1992) Rapid testing methods for *Salmonella* using immunodiagnostic kits. Presented at: Complying with Food Legislation through Immunoassay, February 1992, Glasgow.

CARTER N P and MEYER E W, (1990) Introduction to the principles of flow cytometry. In: *Flow cytometry, a practical approach*, Ormerod, M.G. (ed.) IRL Press, Oxford, pp. 1–28.

CLARK C, CANDLISH A A G and STEELL W, (1989) Detection of *Salmonella* in foods using a novel coloured latex test. *Food and Agricultural Immunology*,

1 3–9.

CLAYDEN J A, ALCOCK S J and STRINGER M F, (1987) Enzyme linked immunoabsorbant assays for the detection of *Salmonella* in foods. In: *Immunological Techniques in Microbiology. Society for Applied Bacteriology Technical Series 24* Grange, J.M., Fox, A. and Morgan, N.L. (eds). Blackwell Scientific, Oxford, pp. 217–30.

CONNOLLY P, LEWIS S J and CORRY J E L, (1988) A medium for the detection of yeasts using a conductimetric method, *International Journal of Food Microbiology*, **7** 31–40.

COOMBES P, (1990) Detecting *Salmonella* by conductance. *Food Technology International Europe*, No. 4229, 241–6.

COOTES R L and JOHNSON R, (1980) A fluorescent staining technique for determination of viable and non-viable yeast and bacteria in wines, *Food Technology in Australia*, **32** 522–24.

COUSINS D L and MARLATT F, (1990) An evaluation of a conductance method for the enumeration of enterobacteriaceae in milk, *Journal of Food Protection*, **53** 568–70.

CURIALE M S, FLOWERS R S, MOZOLA M A and SMITH A E, (1986) A commercial DNA probe based diagnostic for the detection of *Salmonella* in food samples: In: *DNA Probes: Applications in Genetic and Infectious Disease and Cancer*, Lerman, L. S. (ed.) Cold Spring Harbour Laboratory, New York, pp. 143–8.

CURIALE M S, KLATT M L, ROBINSON B J and BECK L T, (1990a) Comparison of colorimetric monoclonal enzyme immunoassay screening methods for detection of *Salmonella* in foods, *Journal of the Association of Official Analytical Chemists*, **73** 43–50.

CURIALE M S, MCIVER D, WEATHERSBY S and PLANER C, (1990b) Detection of *Salmonella* and other enterobacteriaceae by commercial deoxyribonucleic acid hybridization and enzyme immunoassay kits, *Journal of Food Protection*, **53** 1037–46.

DATTA A R, WENTZ B A, SHOOK D and TRUCKSESS M W, (1988) Synthetic oligonucleotide probes for detection of *Listeria monocytogenes, Applied and Environmental Microbiology*, **54** 2933–7.

DAVIS S C and JONES K L, (1997) A study of the recovery of microorganisms from flour. *Campden and Chorleywood Food Research Association R&D Report No. 51*, Campden and Chorleywood Food Research Association, UK.

DIJKMAN, A.J., SCHIPPER, C.J., BOOY, C.J. and POSTHUMUS, G. (1969) The estimation of the number of cells in farm milk, *Netherlands Milk and Dairy Journal*, **23** 168–81.

DONNELLY C W and BAIGENT G T, (1986) Method for flow cytometric detection of *Listeria monocytogenes* in milk, *Applied and Environmental Microbiology*, **52** 689–95.

DRUGGAN P, FORSYTHE S J and SILLEY P, (1993) Indirect impedance for microbial screening in the food and beverage industries. In: *New Techniques in Food and Beverage Microbiology*, (eds R G Kroll and A Gilmour) SAB Technical Series 31, London, Blackwell.

DUMAIN P P, DESNOUVEAUZ R, BLOC'H L, LECONTE C, FUHRMANN B, DE COLOMBEL E, PLESSIS M C and VALERY S, (1990) Use of flow cytometry for yeast and mould detection in process control of fermented milk products: The Chemflow System – A Factory Study, *Biotech Forum Europe*, **7** (3) 224–9.

EASTER M C and GIBSON D M, (1985) Rapid and automated detection of *Salmonella* by electrical measurement, *Journal of Hygiene, Cambridge*, **94** 245–62.

EASTER M C and GIBSON D M, (1989) Detection of microorganism by electrical measurements. In: *Rapid Methods in Food Microbiology. Progress in Industrial Microbiology Vol. 26* Adams, M.R. and Hope, C.F.A. (eds) Elsevier pp. 57–100.

EDEN R and EDEN G, (1984) *Impedance Microbiology*, Research Studies Press. Letchworth, Herts, England.

FITTS R, (1985) Development of a DNA-DNA hybridization test for the presence of *Salmonella* in foods, *Food Technology*, **39** 95–102.

FLOWERS R S, KLATT M J and KEELAN S L, (1988) Visual immunoassay for the detection of *Salmonella* in foods. A collaborative study. *Journal of the Association of Official Analytical Chemists*, **71** 973–80.

FLOWERS R S, MOZOLA M A, CURIALE M S, GABIS D A and SILLIKER J H, (1987) Comparative study of DNA hybridization method and the conventional cultural procedure for detection of *Salmonella* in foods, *Journal of Food Science*, **52** 842–5.

FRANCISCO D D, MAH R A and RABIN A C, (1973) Acridine orange epifluorescent technique for counting bacteria in natural waters, *Transactions of the American Microscopy Society*, **92** 416–21.

GRIFFITHS M W, (1995) Bioluminescence and the food industry. *Journal of Rapid Methods and Automation in Microbiology*, **4** 65–75.

GUTTERIDGE C S and ARNOTT M L, (1989) Rapid Methods: An Over the Horizon View. In: *Rapid Methods in Microbiology. Progress in Industrial Microbiology Vol. 26*, Adams, M.R. and Hope, C.F.A. (eds) Elsevier, pp. 297–319.

HADFIELD S G, JOVY N F and MCILLMURRAY M B, (1987b) The application of a novel coloured latex test to the detection of *Salmonella.* In: *Immunological Techniques in Microbiology. Society for Applied Bacteriology Technical Series 24.* Grange, J.M., Fox, A. and Morgan, N.L. (eds) Blackwell Scientific, Oxford, pp. 145–52.

HADFIELD S G, LANE A and MCILLMURRAY M B, (1987a) A novel coloured latex test for the detection and identification of more than one antigen, *Journal of Immunological Methods*, **97** 153–8.

HILL W E, MADDEN J M, MCCARDELL B A, SHAH D B, JAGOW J A, PAYNE W L and BOUTIN B K, (1983a) Foodborne entertoxigenic *Esch coli:* detection and enumeration by DNA colony hybridisation *Applied and Environmental Microbiology*, **46** 636–41.

HILL W E, PAYNE W L and AULISIO C C G, (1983b) Detection and enumeration of virulent *Yersinia enterocolitica* in food by colony hybridisation, *Applied and Environmental Microbiology*, **46** 636–41.

HILL W E, WENTZ B A, JAGOW J A, PAYNE W L and ZON G, (1986) DNA colony hybridisation method using synthetic oligonucleotides to detect enterotoxigenic *Esch. Coli:* a collaborative study, *Journal of the Association of Official Analytical Chemists*, **69** 531–6.

HOBBIES J F, DALEY R J and JASPER S, (1977) Use of nucleoport filters for counting bacteria by fluorescence microscopy, *Applied and Environmental Microbiology*, **33** 1225–8.

HOLAH J T, (1989) Monitoring the Hygienic Status of Surfaces. In: *Hygiene – The issues for the 90'*, Symposium Proceedings, Campden and Chorleywood Food Research Association, UK.

HOLAH J T, BETTS R P and THORPE R H, (1988) The use of direct epifluorescent microscopy (DEM) and the direct epifluorescent filter technique (DEFT) to assess microbial populations on food contact surfaces, *Journal of Applied Bacteriology*, **65** 215–21.

HOLBROOK R, ANDERSON J M, BAIRD-PARKER A C and STUCHBURY S H, (1989) Comparative evaluation of the Oxoid *Salmonella* Rapid Test with three other rapid *Salmonella* methods, *Letters in Applied Microbiology*, **9** 161–4.

HUERNNEKENS F M and WHITELEY H R, (1960) In: Florkin, M. and Mason, H.S. (eds) *Comparative Biochemistry*, Volume 1. Academic Press, New York, p. 129.

HUGHES D, SUTHERLAND P S, KELLY G and DAVEY G R, (1987) Comparison of the Tecra *Salmonella* tests against cultural methods for the detection of *Salmonella* in foods, *Food Technology in Australia*, **39** 446–54.

HUTTER K J and EIPEL H E, (1979) Rapid determination of the purity of yeast cultures by immunofluorescence and flow cytometry, *Journal of the Institute of Brewing*, **85** 21–2.

HYSERT D W, LOVECSES F and MORRISON N M, (1976) A firefly bioluminescence ATP assay method for rapid detection and enumeration of brewery micro-organisms, *Journal of the American Society of Brewing Chemistry*, **34** 145–50.

JAGOW J and HILL W E, (1986) Enumeration by DNA dolony hybridisation of virulent *Yersinia enterocolitica* colonies in artificially contaminated food, *Applied and Environmental Microbiology*, **51** 411–43.

JONES J G and SIMON B M, (1975) An investigation of errors in direct counts of aquatic bacteria by epifluorescence microscopy with reference to a new method of dyeing membrane filters, *Journal of Applied Bacteriology*, **39** 1–13.

JONES K L and BETTS R P, (1994) The EIAFOSS system for rapid screening of *Salmonella* from foods. *Campden Food & Drink Research Association Technical Memorandum 709*, Campden and Chorleywood Food Research Association, UK.

JONES K L, MACPHEE S, TURNER A and BETTS R P, (1995a) An evaluation of the Oxoid *Listeria* Rapid Test incorporating clearview for the detection of

Listeria from foods. *Campden and Chorleywood Food Research Association R&D Report No 19*, Campden & Chorleywood Food Research Association, UK.

JONES K L, MACPHEE S, TURNER A and BETTS R P, (1995b) An evaluation of Pathstick for the detection of *Salmonella* from foods. *Campden & Chorleywood Food Research Association R&D Report No 11*, Campden and Chorleywood Food Research Association, UK.

KAEREBY F and ASMUSSEN B, (1989) Bactoscan – Five years of experiences. In: *Rapid methods and Automation in Microbiology and Immunology*, Balows, A., Tilton, R.C. and Turano, A. (eds) Florence 1987. Brixia Academic Press, Brescia, pp. 739–45.

KARL D M, (1980) Cellular nucleotide measurements and applications in microbial ecology, *Microbiological Review*, **44** 739–96.

KLINGER J D and JOHNSON A R, (1988) A rapid nucleic acid hybridisation assay for *Listeria* in foods, *Food Technology*, **42** 66–70.

KLINGER J D, JOHNSON A, CROAN D, FLYNN P, WHIPPIE K, KIMBALL M, LAWNE J and CURIALE M, (1988) Comparative studies of a nucleic acid hybridisation assay for *Listeria* in foods, *Journal of the Association of Official Analytical Chemists*, **73** 669–73.

KRICKA L J and WHITEHEAD T P, (1984) Luminescent immunoassays: new labels for an established technique, *Diagnostic Medicine*, May, 1–8.

KYRIAKUDES A, BELL C and JONES K L, (eds) (1996) A Code of Practice for Microbiology Laboratories Handling Food Samples. *Campden & Chorleywood Food Research Association Guideline No. 9*, Campden and Chorleywood Food Research Association, UK.

LAROCCO K A, GALLIGAN P, LITTLE K J and SPURGASH A, (1985) A rapid bioluminescent ATP method for determining yeast concentration in a carbonated beverage, *Food Technology*, **39** 49–52.

LEVIN G V, GLENDENNING J R, CHAPELLE E W, HEIM A H and ROCECK E, (1964) A rapid method for the detection of microorganisms by ATP assay: its possible application in cancer and virus studies, *Bioscience*, **14** 37–8.

LITTEL K J and LAROCCO K A, (1986) ATP screening method for presumptive detection of microbiologically contaminated carbonated beverages, *Journal of Food Science*, **51** 474–6.

LITTEL K J, PIKELIS S and SPURGASH A, (1986) Bioluminescent ATP assay for rapid estimation of microbial numbers in fresh meat, *Journal of Food Protection*, **49** 18–22.

LOHNEIS M, JASCHKE K H and TERPLAN G, (1987) An immunoluminomtric assay (ILMA) for the detection of staphylococcal enterotoxins, *International Journal of Food Microbiology*, **5** 117–27.

MCCLELLAND R G and PINDER A C, (1994) Detection of *Salmonella typhimurium* in dairy products with flow cytometry and monoclorical antibodies *Applied and Environmental microbiology*, **60** (12), 4255.

MACRAE R M, (1964) Rapid Yeast Estimations, in: *Proceedings of European Brewing Convention Brussels 1963*. Elsevier Publishing, Amsterdam, pp.

510–12.

MCELROY W D, (1947) The energy source for bioluminescence in an isolated system, *Proceedings of the National Academy of Science, USA*, **33** 342–5.

MCELROY W D. and STREFFIER B L, (1949) Factors influencing the response of the bioluminescent reaction to adenosine triphosphate, *Archives of Biochemistry*, **22** 420–33.

MATTINGLY J A, BUTMAN B T, PLANK M, DURHAM R J and ROBINSON B J (1988) Rapid monoclonal antibody based enzyme-linked immunosorbant assay for the detection of *Listeria* in food products, *Journal of the Association of Official Analytical Chemists*, **71** 679–81.

MORRIS J G, WRIGHT A C, ROBERTS D M, WOOD P K, SIMPSON L M and OLIVER J D, (1987) Identification of environmental *Bibrio vulnificus* isolates with a DNA probe for the cytotoxin-hemolysin gene, *Applied and Environmental Microbiology*, **53** 193–5.

MOZOLA M, HALBERT D, CHAN S, HSU H Y, JOHNSON A, KING W, WILSON S, BETTS R P, BANKES P and BANKS J G, (1991) Detection of foodborne bacterial pathogens using a colorimetric DNA hybridisation method. In: *Genetic Manipulation Techniques and Applications. Society for Applied Bacteriology Technical Series No. 28* Grange, J. M., Fox, A. and Morgan, N. L. (eds) Blackwell Scientific, Oxford, pp. 203–16.

MULDROW L L, TYNDALL R L and FLIERMANS C B, (1982) Application of flow cytometry to studies of free-living amoebae, *Applied and Environmental Microbiology*, **44** (6) 1258–69.

NEAVES P, WADDELL M J and PRENTICE G A, (1988) A medium for the detection of Lancefield group D cocci in skimmed milk powder by measurement of conductivity changes, *Journal of Applied Bacteriology*, **65**, 437–48.

NOTERMANS S, HEUVELMAN K J and WERNARS K, (1988) Synthetic enterotoxin B, DNA probes for the detection of enterotoxigenic *Staphylococcus aureus*, *Applied and Environmental Microbiology*, **54** 531–3.

OPPONG D and SNUDDEN B H, (1988) Comparison of acridine orange staining using fluorescence microscopy with traditional methods for microbiological examination of selected dry food products, *Journal of Food Protection*, **51** 485–8.

OWENS J D, THOMAS D S, THOMPSON P S and TIMMERMAN J W, (1989) Indirect conductimetry: A novel approach to the conductimetric enumeration of microbial populations, *Letters in Applied Microbiology*, **9** 245–50.

PATCHETT R A, BACK J P, PINDER A C and KROLL R G, (1991) Enumeration of bacteria in pure cultures and foods using a commercial flow cytometer *Food Microbiology*, **8** 119–25.

PETTIT S B, (1989) A conductance screen for enterobacteriaceae in foods, In: *Rapid Microbiological Methods for Foods Beverages and Pharmaceuticals*, Stannard, C J, Petitt, S B, Skinner, F A (eds) Blackwell Scientific, Oxford, pp. 131–42.

PETTIPHER G L, (1987) Detection of low numbers of osmophilic yeasts in crème fondant within 24 h using a pre-incubated DEFT count, *Letters in Applied*

Microbiology, **4** 95–8.

PETTIPHER G L, (1991) Preliminary evaluation of flow cytometry for the detection of yeasts in soft drinks, *Letters in Applied Microbiology*, **12** 109–12.

PETTIPHER G L and RODRIGUES U M, (1981) Rapid enumeration of bacteria in heat treatment milk and milk products using a membrane filtration-epifluorescence microscopy technique, *Journal of Applied Bacteriology*, **50** 157–66.

PETTIPHER G L and RODRIGUES U M, (1982) Semi-automated counting of bacteria and somatic cells in milk using epifluorescence microscopy and television image analysis, *Journal of Applied Bacteriology*, **53** 323–9.

PETTIPHER G L, MANSELL R, MCKINNON C H and COUSINS C M, (1980) Rapid membrane filtration-epifluorescence microscopy technique for the direct enumeration of bacteria in raw milk, *Applied and Environmental Microbiology*, **39** 423–9.

PETTIPHER G L, WILLIAMS R A and GUTTERIDGE C S, (1985) An evaluation of possible alternative methods to the Howard Mould Count, *Letters in Applied Microbiology*, **1** 49–51.

PHILLIPS A P and MARTIN K L, (1988) Limitations of flow cytometry for the specific detection of bacteria in mixed populations, *Journal of Immunological Methods*, **106** 109–17.

PINDER A C, PURDY P W, POULTER S A G and CLARK D C, (1990) Validation of flow cytometry for rapid enumeration of bacterial concentrations in pure cultures, *Journal of Applied Bacteriology*, **69** 92–100.

PONNAMPERUMA C, SAGAN C and MARINER R, (1963) Synthesis of adenosine triphosphate under possible primitive earth conditions, *Nature*, **199** 222–6.

RICHARDS J C S, JASON A C, HOBBS G, GIBSON D M and CHRISTIE R H, (1978) Electronic measurement of bacterial growth, *Journal of Physics, E: Scientific Instruments*, **11** 560–8.

RODRIGUES U M and KROLL R G, (1986) Use of the direct epifluorescent filter technique for the enumeration of yeasts, *Journal of Applied Bacteriology*, **61** 139–44.

RODRIGUES U M and KROLL R G, (1988) Rapid selective enumeration of bacteria in foods using a microcolony epifluorescence microscopy technique. *Journal of Applied Bacteriology*, **64** 65–78.

RODRIGUES U M and KROLL R G, (1989) Microcolony epifluorescence microscopy for selective enumeration of injured bacteria in frozen and heat treated foods, *Applied and Environmental Microbiology*, **55** 778–87.

ROSE S A and STRINGER M F, (1989) Immunological Methods. In: *Rapid Methods in Food Microbiology. Progress in Industrial Microbiology Vol. 26*, Adams, M.R. and Hope, C F A (eds) Elsevier, pp. 121–67.

ROSE S A, BANKES P and STRINGER M F, (1989) Detection of staphylococcal enterotoxins in dairy products by the reversed passive latex agglutination (Setrpla) kit, *International Journal of Food Microbiology*, **8** 65–72.

SCHUTZBANK T E, TEVERE V and CUPO A, (1989) The use of bioluminescent bacteriophages for the detection and identification of *Salmonella* in foods.

In: *Rapid methods and Automation in Microbiology and Immunology*, Florence 1987. Balows, A, Tilcon, R C and Turano, A (eds) Brixia Academic Press, Brescia, pp. 241–51.

SELIGER H and MCELROY M D, (1960) Spectral emission and quantum yield of firefly bioluminescence, *Archives of Biochemistry and Biophysics*, **88** 136–41.

SHARPE A N, WOODROW M N and JACKSON A K, (1970) Adenosine triphosphate levels in foods contaminated by bacteria, *Journal of Applied Bacteriology*, **33** 758–67.

SHAW B G, HARDING C D, HUDSON W H and FARR L, (1987) Rapid estimation of micobial numbers in meat and poultry by the direct epiflourescent filter technique, *Journal of Food Protection*, **50** 652–7.

SHAW S., (1989) Rapid methods and quality assurance in the brewing industry. In: *Rapid Methods in Food Microbiology, Progress in Industrial Microbiology* Vol. 26. Adams, M.R. and Hope, C F A (eds) Elsevier, pp. 273–96.

SKARSTAD K, STEEN H B and BOYE E, (1983) Cell cycle parameters of growing *Escherichia coli.* B/r studied by flow cytometry, *Journal of Bacteriology*, **154** 652–6.

SKJERVE E and OLSVIK O, (1991) Immunomagnetic separation of *Salmonella* from foods, *International Journal of Food Microbiology*, **14** 11–18.

STALKER R M, (1984) Bioluminescent ATP analysis for the rapid enumeration of bacteria-principles, practice, potential areas of inaccuracy and prospects for the Food Industry. *Campden Food and Drink Research Association Technical Bulletin 57,* Campden & Chorleywood Food Research Association, UK.

STANNARD C J, (1989) ATP Estimation. In: *Rapid Methods in Food Microbiology. Progress in Industrial Microbiology* Vol. 26. Adams, M.R. and Hope, C.F.A. (eds) Elsevier.

STANNARD C J, and WOOD, J.M. (1983) The rapid estimation of microbial contamination of raw meat by measurement of adenosine triphosphate (ATP), *Journal of Applied Bacteriology*, **55** 429–38.

STEEN H B and BOYE E, (1980) *Escherichia coli* growth studied by dual parameter flow cytophotometry, *Journal of Bacteriology*, **145** 145–50.

STEEN H B, BOYE E, SKARSTAD K, BLOOM B, GODAL T and MUSTAFA S (1982) Applications of flow cytometry on bacteria: Cell cycle kinetics, drug effects and quantification of antibody binding, *Cytometry*, **2** (4) 249–56.

STEWART G N, (1899) The change produced by the growth of bacteria in the molecular concentration and electrical conductivity of culture media, *Journal of Experimental Medicine*, **4** 23S–24S.

STEWART G S A B, (1990) A review: *In vivo* bioluminescence: New Potentials for Microbiology, *Letters in Applied Microbiology*, **10** 1–9.

TYNDALL R L, HAND R E Jr, MANN R C, EVAN C and JERNIGAN R, (1985) Application of flow cytometry to detection and characterisation of *Legionella* spp., *Applied and Environmental Microbiology*, **49** 852–7.

ULITZUR S and KUHN J, (1987) Introduction of lux genes into bacteria, a new approach for specific determination of bacteria and their antibiotic susceptibility. In: *Proceedings of Xth International Symposium on Bioluminescence and Chemoluminescence*, Freiburg, Germany 1986. Scholmerich, J., Andreesen, A., Kapp, A., Earnst, M., Woods, E.G. (eds) J. Wiley. pp. 463–72.

ULITZUR S, SUISSA M and KUHN J C, (1989) A new approach for the specific determination of bacteria and their antimicrobial susceptibilities. In: *Rapid methods and automation in microbiology and immunology* Balows, A., Tilton, R.C. and Turano, A. (eds). Florence 1987. Brixia Academic Press, Brescia, pp. 235–40.

UYTTENDAELE M, SCHUKKINK A, VAN GEMEN B and DEBEVERE J, (1996) Comparison of the nucleic acid amplifications system NASBA and agar isolation for the detection of pathogenic campylobacters in naturally contaminated poultry, *Journal of Food Protection*, **59** (7) 683–9.

VANDERLINDE P B and GRAU F H, (1991) Detection of *Listeria* spp. in meat and environmental samples by an enzyme-linked immunosorbant assay (ELISA), *Journal of Food Protection*, **54** 230–1.

VECKERT J, BREEUWER P, ABEE T, STEPHENS P and NEBE VON CARON G, (1995) Flow Cytometry Applications in physiological study and detection of foodborne organisms, *International Journal of Food Microbiology*, **28** (2) 317–26.

VERMUNT A E M, FRANKEN A A J M, and BEUMER R R, (1992) Isolation of *Salmonellas* by immunomagnetic separation, *Journal of Applied Bacteriology*, **72** 112–8.

WALKER S J, (1989) Development of an impedimetric medium for the detection of *Yersinia enterocolitica* in pasteurised milk. In: *Modern Microbiological Methods for Dairy Products*, Proceedings of a seminar organised by the International Dairy Federation. IDF, pp. 288–91.

WALKER S J, ARCHER P and APPLEYARD J, (1990) Comparison of *Listeria*-Tek ELISA kit with cultural procedures for the detection of *Listeria* species in foods, *Food Microbiology*, **7** 335–42.

WERNARS K and NOTERMANS S, (1990) Gene probes for detection of foodborne pathogens. In: *Gene probes for bacteria* Macario, A.J.L. and de Macario, E.C. (eds). Academic Press, London, pp. 355–88.

WOLCOTT M J, (1991) DNA-based rapid methods for the detection of foodborne pathogens, *Journal of Food Protection*, **54** 387–401.

9

Non-microbiological factors affecting quality and safety

H. M. Brown and M. N. Hall, Campden and Chorleywood Food Research Association

9.1 Introduction

As the chilled foods market has expanded and become more competitive, so have the demands for diversity, quality and longer shelf-life. Meeting these demands in a responsible, safe and cost-effective manner requires the application of an understanding of the factors that affect product safety and quality. Many problems can be avoided by applying this knowledge to a formalised HACCP approach to identify critical control points relating to quality as well as safety and to make realistic predictions of shelf-life. Considering these issues early in the product development process offers the best chance of providing a product that meets the consumer's expectations and delivers the desired market opportunities to the company. Food is probably the most chemically complex substance that most people encounter. There are over half a million naturally occurring compounds in fresh plant food and more are formed as a result of processing, cooking and storage. They are responsible for the appearance, flavour, texture and nutritional value of the food (quality), and for its physiological effects when consumed (safety).

Non-microbiological factors that affect quality and safety of chilled foods can be broadly divided into chemical, biochemical and physico-chemical factors. Each of these is dependent on properties of the food (e.g. pH, water activity) and the conditions in which the food is held (e.g. temperature, gaseous atmosphere). Attention to the selection of raw materials in order to achieve high quality is paramount, since subsequent processing cannot compensate for poor-quality raw materials, particularly for chilled foods in which the perception of 'freshness' is one of the most important criteria for its purchase.

The effects of chemical, biochemical and physio-chemical factors are rarely mutually exclusive but these categories provide a convenient framework for discussion. The effects of these factors are not always detrimental and in some instances they are essential for the development of the desired characteristics of a product. In this chapter, some of the characteristics of chemical, biochemical and physico-chemical reactions are described, along with examples that are of significance to chilled foods.

9.2 Characteristics of chemical reactions

Chemical reactions will proceed if reactants are available, if they are in a suitable form and if the activation energy threshold of the reaction is exceeded. The presence of inorganic catalysts reduces the activation energy threshold and causes reactions to proceed that would otherwise not have done so. The reaction rate is dependent on the concentration of the reactants and on the temperature. Increases in temperature speed up the random movement of reactant molecules, increasing the probability of their coming into contact. A general assumption is that for every 10°C rise in temperature the rate of reaction doubles.

9.3 Chemical reactions of significance in chilled foods

9.3.1 Lipid oxidation

Lipid oxidation is one of the major causes of deterioration in the quality of meat and meat products. Cooked meats and poultry rapidly develop a characteristic oxidized flavour, termed 'warmed-over' flavour (WOF) by Tims and Watts (1958). The flavour is best described as that associated with reheated meat and has been described as such by sensory assessors during free profiling of precooked meat, reheated after chill storage (Churchill *et al.* 1988, Lyon 1987). Further descriptors have been defined for WOF in pork (Byrne *et al.* 1999a) and chicken meat (Byrne *et al.* 1999b) and have resulted in the development of sensory vocabularies containing 16 and 18 terms respectively. In cooked meats held at chill storage temperatures, this stale, oxidized flavour becomes apparent within a short time (48 hours) which contrasts with the slower onset of rancidity during frozen storage (weeks) (Pearson and Gray 1983). Although WOF has generally been recognized as affecting only cooked meat, there is evidence that it develops just as rapidly in raw meat that has been ground and exposed to the air (Greene 1969, Sato and Hegarty 1971) and in restructured fresh meat products as a consequence of disruption of the tissue membranes and exposure to oxygen (Gray and Pearson 1987). Nevertheless, the significance of the development of this flavour to food processors has increased with the advent and expansion of markets for cooked chilled ready meals such as 'TV dinners', airline catering, and fast food outlets. The consumer expectation in these situations is for 'freshly prepared' flavours. The continued development and success of fast food facilities and precooked chilled meals will depend to some extent on the ability of processors to overcome the development of WOF.

Lipid oxidation has long been considered to be the primary cause of WOF, supported by studies correlating increases in WOF determined sensorily (Love 1988) with measurements of the thiobarbituric acid (TBA) number (an indicator of lipid oxidation) (Igene *et al.* 1979, Igene *et al.* 1985, Smith *et al.* 1987), and identification of the volatile compounds extracted from the headspace above meat samples (St Angelo *et al.* 1987, Ang and Lyon 1990, Churchill *et al.* 1990). As with other examples of oxidative rancidity, the process of lipid oxidation results in the formation of many different compounds, some of which are more significant than others to the undesirable odour and flavour associated with rancidity. This gives rise to a less than perfect relationship between measured chemical markers and sensory assessment of rancidity.

The reactivity of food lipids is influenced by the degree of unsaturation of constituent fatty acids, their availability and the presence of activators or inhibitors. The composition of fats in meat reflects a number of factors, including the diet of the animal and the type of fat. Lipids are most abundant as either storage depot (adipose) fats or in cell membranes as phospholipids. During cooking, the unsaturated phospholipids, as opposed to the storage triglycerides, are rendered more susceptible to oxidation by disruption and dehydration of cell membranes. The higher degree of unsaturation of fatty acids in the phospholipids contributes to their more rapid rate of oxidation (Igene *et al.* 1981). The role of phospholipids in the formation of WOF (Igene and Pearson 1979) and TBA reactive substances (Roozen 1987, Pikul and Kummerow 1991) has been demonstrated.

Autoxidation of lipids is generally accepted to involve a free radical chain reaction (Fig. 9.1), which is initiated when a labile hydrogen atom is abstracted from a site on the lipid (RH) with the production of lipid radicals ($R^{\bullet}$) (initiation). Reaction with oxygen yields peroxyl radicals ($ROO^{\bullet}$) and this is followed by abstraction of another hydrogen from a lipid molecule. A hydroperoxide (ROOH) and another free radical ($R^{\bullet}$) which is capable of perpetuating the chain reaction, are formed (propagation). Decomposition of the hydroperoxides involves further free radical mechanisms and the formation of non-radical products including volatile aroma compounds.

Despite much research effort, the mechanism of initiation leading to the formation of the lipid (alkyl or allyl) radical ($R^{\bullet}$) in meat is still an area of confusion and debate. The involvement of iron has been established (Minotti and Aust 1987), but beyond this various mechanisms have been suggested but not supported by conclusive evidence (Ashgar *et al.* 1988).

The rate of formation of free radicals is increased by the presence of metal catalysts. In the case of warmed-over flavour development in cooked meats, both free ferrous ions and haemoproteins, including metmyoglobin in the presence of hydrogen peroxide (Asghar *et al.* 1988) have been shown to have a prooxidant effect. The availability of free iron is known to increase as a result of cooking (Igene *et al.* 1979) as haemoproteins are broken down and release free

Initiation $RH \longrightarrow R^{\bullet}$

Propagation $R^{\bullet} + O_2 \longrightarrow ROO^{\bullet}$

$ROO^{\bullet} + RH \longrightarrow ROOH + R^{\bullet}$

Fig. 9.1. Free radical chain reaction.

iron. The amount released is dependent on the rate of heating and the final temperature, and therefore on the method of heating. Slow heating releases more free iron than fast heating – roasting or braising of meat releases more than does microwave heating (Shricker and Miller 1983).

Procedures for the prevention of WOF were reviewed by Pearson and Gray (1983). The method used is very often restricted by the requirements of the final product. Phenolic antioxidants such as BHT and BHA are of little value in intact meat cuts (Watts 1961), whereas they may be more suited to comminuted meat products since a more even distribution of the antioxidant can be achieved. Overheating or retorting of meat to produce compounds that have antioxidant activity (Maillard Reaction products) may be suitable for canned products but tends to result in a product with characteristics contrary to the 'fresh' perception that is a necessary part of many chilled foods. Alternatively, these compounds can be added to meat, but they are then restricted by the same considerations that apply to artificial antioxidants. Reduction of WOF has also been achieved by use of vitamin E. Kerry *et al.* (1999) demonstrated that addition of alpha-tocopherol to cooked pig meat reduced lipid oxidation and WOF. Difficulties in achieving adequate distribution of the antioxidant in the meat could be overcome by the incorporation of vitamin E supplements to the feed of the animals. Addition of alpha-tocopherol acetate to the diet of rabbits (Lopez-Bote *et al.* 1997) and broiler chicks (O'Neill *et al.* 1998) has been shown to be reflected by an increase in the muscle tissue and result in reduced WOF development. Investigations of natural antioxidants present in vegetables have shown some benefits for using extracts from green peppers, onions and potato peelings (Pratt and Watts 1964) and herbs and spices, particularly rosemary, sage, marjoram (Hermann *et al.* 1981) and clove (Jayathilakan *et al.* 1997). Reports of the effectiveness of rosemary oleoresin as an antioxidant in precooked meats are conflicting although Murphy *et al.* (1998) found rosemary oleoresin and sodium tripolyphosphate to be effective in the prevention of WOF in precooked roast beef slices. Precooked pork balls processed with rosemary stored at 4°C for 48 hours did not develop oxidized flavours as the controls did (Korczack *et al.* 1988), whereas restructured beef steaks processed with oleoresin rosemary stored at refrigerated temperatures showed no significant improvement in comparison to the controls (Stoick *et al.* 1991).

The addition of nitrite between 50 and 200 ppm is an effective inhibitor of the development of WOF (Sato and Hegarty 1971, Cho and Rhee 1997). Nitrite and

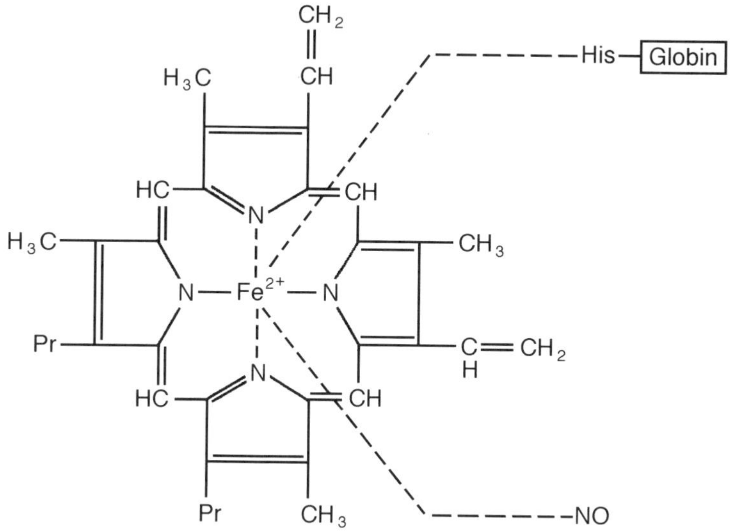

Fig. 9.2. Nitrosylmyoglobin. Myoglobin with the nitrite ligand.

haemoproteins form nitrosylmyochrome and nitrosylhaemochrome complexes in which the iron is stabilised by the linking of nitric oxide to the porphyrin ring (Fig. 9.2); however, the pink coloration of the meat may be undesirable; causes of pinking in uncured cooked meat is further considered later in this chapter. The effectiveness of pyrophosphate, tripolyphosphate and hexametaphosphate, which chelate metal ions, particularly prooxidative ferrous ions, was demonstrated by Tims and Watts (1958) in pork. It has since been verified for ground beef (Sato and Hegarty 1971), for restructured beef steaks (Mann *et al.* 1989), and for battered and breaded chicken (Brotsky 1976). Phosphates in combination with ascorbic acid may exert a synergistic effect, such that cooked ground pork was protected against lipid oxidation for up to 35 days at 4°C (Shahidi *et al.* 1986).

An alternative approach is to protect the meat from oxidation. This can be achieved by creating an oxygen barrier, using a sauce or a gravy that can be in place at the time of cooking and during subsequent storage. This principle has been demonstrated by comparing the shelf-life of frozen meat to that of the same meats cooked without gravy coverings (Dalhoff and Jul 1965). Cooked pork covered with gravy could be stored at −18°C for more than 100 weeks, whereas pork stored without gravy was unacceptable after 22 weeks.

Modified-atmosphere packaging to reduce WOF has been applied to precooked turkey and pork and pork products. Although those stored in nitrogen and carbon dioxide atmospheres were less 'oxidized' than those in air, vacuum packaging was the most effective (Nolan *et al.* 1989 and Juncher *et al* (1998). Shaw (1997) reviewed the potential benefits of the use of MAP for cook-chill ready meals. Protection against oxidation at the time of cooking is also beneficial. Cooking and subsequent storage of chicken breasts in a nitrogen

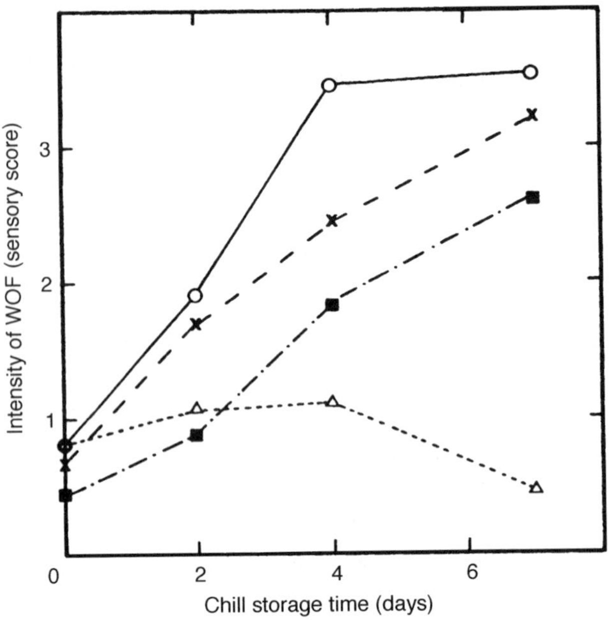

Fig. 9.3. Effect of cooking and storage atmospheres on WOF development in chicken breasts. (○—○) cooked and stored in air; (×—×) cooked in nitrogen, stored in air; (■—■) cooked in air, stored in nitrogen; ([△]—[△]) cooked and stored in nitrogen.

atmosphere reduced the TBA values and sensory scores for WOF intensity as compared with those cooked in air and stored in either nitrogen or air (Fig. 9.3).

Autooxidation or oxidative rancidity is by no means confined to meat and meat products. Dairy products and fatty fish are also highly susceptible. Migration of copper into cream on churning can initiate the oxidative sequence of reactions causing rapid flavour impairment. Buttermilk has a high proportion of unsaturated phospholipids, particularly phosphatidylethanolamine, that can bind metal ions in a prooxidative fashion, and the presence of a metal–phospholipid complex at an oil-water interface facilitates lipid hydroperoxide formation.

Fish fats contain a high proportion of n−3 polyunsaturated fatty acids, which are vulnerable to oxidation by atmospheric oxygen leading to deteriorative changes. Despite this, rancid flavours only appear to affect the acceptability of fattier species such as trout, sardine, herring and mackerel; and even then, trout and gutted mackerel oxidize at temperatures above 0°C whereas herring remains relatively unaffected. Castell (1971) has suggested that in fish, oxidized lipids become bound in lipid-protein complexes rather than forming carbonyl compounds associated with rancid flavours. The lipid–protein complexes also contribute to the toughened texture of poorly stored fish. Competing demands for available oxygen from microorganisms and enzymes, which differ between species, may also influence whether oxygen is available for autooxidation. In trout, reports of lipoxygenase activity in the skin tissue have suggested the potential to initiate lipid oxidation by providing a source of initiating radicals (German and Kinsella 1985). A complicating factor in the assessment of the significance of oxidation to the quality of fish is that many products distributed chilled have previously been frozen, particularly for example, herring, to spread seasonal availability.

9.3.2 Pink discoloration in meat products

Discoloration in foods is a common problem which can take many forms and be associated with a wide range of chemical reactions: biochemical or enzymic browning is considered later in this chapter. Pink discoloration in cooked meats is a long-standing and all too common problem affecting manufacturing, retailing food service and domestic sectors and is often interpreted as undercooking. The problem is particularly evident with sliced meats, reformed roast products, pasties and casseroles. Various causes of pinking have been identified and these are indicated in Table 9.1 on the basis of the pigment type thought to be involved. Maga (1994) has reviewed the causes and factors affecting pink discoloration in cooked white meats.

Myoglobin is a monomeric globular haem-protein found in all vertebrates which together with haemoglobin give rise to the red colour of meats. The amount of myoglobin varies from species to species, tissue to tissue and is affected by a wide range of environmental factors. As indicated in Table 9.1 myoglobin can be present in several forms, some of which can impart a red or pink residual colour to the meat even after cooking. Recent work has indicated that over 80% of instances of pinking are due to nitrosomyoglobin arising from nitrate contamination and its subsequent bacterial reduction to nitrite (Brown *et al.* 1998).

9.4 Characteristics of biochemical reactions

Biochemical reactions are catalysed by specialized proteins called enzymes. They are highly specific and efficient catalysts, lowering the activation threshold so that the rate of reaction of thermodynamically possible reactions is

Table 9.1 Pigment types and causes giving rise to pink coloration in meat products (Brown *et al.* 1998)

Pigment type	Cause of pink discoloration
Oxymyoglobin	Low temperature cooking
Nitrosomyoglobin	Nitrite contamination directly or from reduced nitrate; nitrogen oxides in ovens
Carboxymyoglobin	Carbon monoxide in ovens; gamma-irradiation
Reduced denatured myoglobin	High pH, slow cooking, high salt and availability of reducing agents

dramatically increased. The specificity of enzymes for a particular substrate is indicated in the name, usually by attachment of the suffix '-ase' to the name of the substrate on which it acts: for example, lipase acts on lipids, protease on proteins. The catalytic activity of enzymes is highly dependent on the conformational structure of the protein, and many of the characteristics of enzyme-catalysed reactions result from the influence of the localized environment. Heat, extremes of acidity of alkalinity, and high ionic strength may denature the enzyme, causing impairment or loss of activity. Enzyme inhibitors and activators that bind either reversibly or irreversibly may act by causing changes in conformational structure or acting directly at the active site.

The temperature at which denaturation takes place is often a reflection of the environmental conditions that the enzyme naturally operates in. For most enzymes from warm-blooded animals, denaturation begins around 45°C, and by 55°C rapid denaturation destroys the catalytic function of the enzyme protein; enzymes from fruit and vegetables are generally denatured at higher temperatures (70–80°C); and some microbial enzymes, e.g. lipases and proteases, can withstand temperatures in excess of 100°C (Cogan 1977).

In the living cell, enzymes catalyse a vast array of reactions that taken together constitute metabolism. In the cellular environment, control and coordination of enzyme activity is achieved by means of feedback mechanisms and compartmentalisation. Disruption which occurs at the time of slaughter or harvest may necessitate steps being taken to prevent the subsequent action of enzymes (blanching of vegetables is a good example); or the activity of enzymes may be enhanced if they improve product quality, as in the case of 'conditioning' of meats, where protease activity is used to break down muscle fibres to develop full flavour and tenderness.

The rate of enzyme-catalysed reactions increases with substrate concentration but only up to a limit (maximal activity) at which the enzyme is saturated with substrate. Further increases in substrate concentration do not increase the rate of reaction. The rate of reaction increases with temperature in the same way as chemical reactions up to an optimum temperature for activity. At temperatures above this, denaturation of the enzyme protein takes place and activity is lost. At chill storage temperatures, the activity of enzymes in most foods is low, but there are notable exceptions. Enzymes in cold-blooded species may be adapted to be active at cold temperatures. In cod, lipase activity at 0°C shows a marked lag phase before maximal activity is achieved and the rate of activity decreases to 0°C and increases to a maximum at −4°C.

Enzymes from different sources, although catalysing conversion of the same substrates to the same reaction products, may have different characteristics in terms of rate of reaction, or pH or temperature optima, depending upon their origin. In a chilled pasta salad composed of cooked pasta, onion, red and green peppers, cucumber, sweetcorn, mushrooms and vinaigrette dressing, shelf-life was limited by browning of either the sweetcorn or the mushrooms depending on the holding temperature (Gibbs and Williams 1990). Holding the salad at storage temperatures between 2°C and 15°C showed that the temperature characteristics

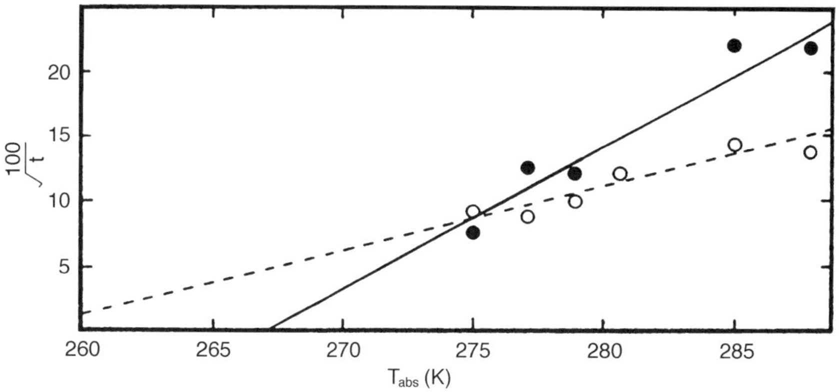

Fig. 9.4. Organoleptic changes in chill-stored pasta salad in vinaigrette. Temperature dependence of the rates of browning of sweetcorn (●—●) and mushrooms (○—○) (Gibbs and Williams 1990).

of the browning reaction, likely to be catalysed by the enzyme polyphenoloxidase, were quite different in the mushrooms and sweetcorn (Fig. 9.4). In mushrooms, the rate of browning reaction appeared to be less temperature-sensitive than was the reaction in sweetcorn, such that at higher temperatures the shelf-life of the salad was limited by browning of the sweetcorn, and at lower temperatures by browning of the mushrooms. To prevent such changes or to predict the shelf-life as a function of temperature, the subtleties of the reactions causing the changes in visual appearance need to be known.

Enzymes in food may be endogenous, that is, they are present naturally in the tissues of the plant or animal that comprises the food. Many hundreds of enzymes fall into this category, though not all will have a significant effect on food quality. Exogenous enzymes in food may be added by the manufacturer to perform a specific function, such as papain for the tenderization of meat, proteases for cheese ripening, or naringinase for the debittering of citrus juices particularly grapefruit juice. Enzymes may be present as a result of 'contamination' by migration from one food to another when they are in contact; an example would be the migration of lipases from unblanched peppers in a pizza topping to the cheese where, if the appropriate triacylglycerols are available, lipolysis will result in soapy flavours. Alternatively, there may be 'contamination' by extracellular enzymes from microorganisms such as lipases and proteases, where the organism may be destroyed by heat processing but the enzyme which is resistant to the heat treatment remains.

9.5 Biochemical reactions of significance in chilled foods

9.5.1 Enzymic browning

In fruits and vegetables, enzymic browning occurs due to damage such as bruising and preparation procedures of cutting, peeling and slicing. The yellowish brown through to black pigments that are formed can appear very rapidly and are unappetizing. In the intact tissue the enzymes responsible, generically referred to as 'phenolases', are separated from the substrate. However, when they are brought into contact as a result of damage, naturally occurring phenolic compounds are enzymically oxidized to form yellowish quinone compounds (Vámos-Vigyázó 1981). A sequence of polymerization reactions follow, giving rise to brown products such as melanins.

The extent of browning is dependent on the activity and amount of the polyphenoloxidase in the specific fruit or vegetable and the availability of substrates which may be catechol, tyrosine or dopamine amongst others, but there is always a requirement for oxygen. A number of approaches have been taken to prevent or retard enzymic browning. Reduction of the available oxygen concentration has been achieved via various approaches: vacuum packaging which retarded enzymic browning in potato strips (O'Beirne and Ballantyne 1987); modified atmosphere packaging, e.g. for shredded lettuce and cut carrots (McLachlan and Stark 1985); the addition of an oxygen scavenger to the pack, which retarded enzymic browning and textural changes in apricot and peach halves (Bolin and Huxsoll 1989); and restricting oxygen diffusion into tissues by immersion in water, brine or syrup solutions. In contrast, high levels of oxygen (70–100%) have also been shown to reduce ascorbic acid breakdown, lipid oxidation and enzymic browning in cut lettuce probably as a result of increasing the total antioxidant capacity of the material (Day 1998). A more direct method to prevent enzymic discoloration is to use enzyme inhibitors, though this may conflict with the 'fresh' image of the product or be restricted by legislation. Traditionally, the use of sulphite in the form of metabisulphite dips provided an effective means of preventing enzymic browning in many instances. With restrictions on the use of sulphite, alternatives have been sought. The pH optimum for phenolase activity is generally between pH 5 and 7. Reduction of the pH to less than 4 by the use of edible acids inactivates the enzyme. Citric acid and ascorbic acid dips retard browning by both a reduction in pH and complexation of copper which is essential for the enzyme to function. Levels of 10% ascorbic acid were shown to be effective for potatoes, and 0.5–1% for apples (O'Beirne 1988). Phenolases from most fruits and vegetables are readily inactivated by heat (Vámos-Vigyázó, 1981) but for salads and pre-prepared vegetables heat treatment may not be an acceptable option owing to the concomitant changes in colour and texture.

9.5.2 Glycolysis

This is a key metabolic pathway of intermediary metabolism found in almost all living organisms. Changes that take place at the time of slaughter and harvest influence the route that substrates metabolized via this pathway subsequently follow. Diversion of the pathway to produce end-products of lactic acid in meat and ethanol in vegetables have marked consequences for the subsequent quality of the food product.

Adenosine triphosphate (ATP) is consumed continuously by the living cell to maintain its structure and function. It is produced from the metabolism of glycogen via glycolysis and the Krebs citric acid cycle. At slaughter, the blood supply and therefore replenishment of oxygen to the muscles ceases, but glycolytic activity continues using the stores within muscle cells. Glycogen is metabolized to pyruvate, but, under anaerobic conditions, the Krebs citric acid cycle is no longer functional and the pyruvate is reduced by NADH to lactic acid. The supply of NADH is replenished by glycolysis allowing the conversion of glycogen to lactic acid to continue until the glycogen stores are depleted. The breakdown of each glucose unit in muscle glycogen results in the production of two molecules of lactic acid. The accumulation of lactic acid progressively lowers the pH in the muscles, this action finally ceasing when the muscle supply of glycogen is depleted and the pH is about 5.5–5.6. When ATP is no longer generated the muscle fibres go into a state of stiffness known as 'rigor'. Provided that there is an adequate supply of glycogen at the time of slaughter, the rate and extent of pH fall is dependent on the activity of key enzymes in the glycolytic pathway, competing reactions for adenosine diphosphate (ADP), and the temperature. The lower the temperature the longer the time taken to reach the pH limit, as biochemical reactions are slowed down. The rate of fall and the final pH can have a profound effect on the quality of the meat (Marsh *et al.* 1987). Lowering of muscle pH leads to protein denaturation and release of a pink proteinaceous fluid called 'drip'. Reducing the rate at which lactic acid accumulates by rapid chilling of the carcass can dramatically reduce drip loss (Taylor 1972, Swain *et al.* 1986); however, rapid chilling to temperatures below 12°C before anaerobic glycolysis has ceased produces a condition called 'cold shortening', resulting in tough meat.

Animals that were exhausted at the time of slaughter will have depleted glycogen reserves and produce less lactic acid during the attainment of rigor. Pork that has a pH greater than 6.0–6.2 at rigor is dark, firm, dry meat (DFD), and spoils microbiologically within 3–5 days owing to the high pH. Animals that were stressed at the time of slaughter to such an extent that respiration was anaerobic may attain rigor pH within one hour of slaughter. Pork which falls to pH 5.8 within 45 minutes of slaughter is pale, soft, exudative meat (PSE). It is characterized by excessive drip loss and is pale as a result of membrane leakage and protein denaturation. The shelf-life of such meat is reduced owing to enhanced microbial growth and oxidation of phospholipids.

9.5.3 Proteolysis

Activity of proteases can have both beneficial and detrimental effects depending on the situation.

Proteases in meat are important in the loss of stiffness that takes place after rigor, known as 'conditioning'. Traditionally, conditioning is allowed to occur at the slaughterhouse and should be allowed to proceed until the meat is tender and acceptable to the consumer. Ideally this takes 2–3 weeks holding at chill temperatures; but unchilled carcasses lose stiffness sooner, as proteases act faster at higher temperatures. For beef, the conditioning rate increases with temperature up to 45°C (Q_{10} 2.4), then at a slower rate to 60°C (Davey and Gilbert 1976). The role of proteases in conditioning has been reviewed (Goll *et al.* 1989, Quali and Talmant 1990). Meat proteases can be classified on the basis of preferred pH for functional activity. Proteases active at acid pH, e.g. the cathepsins, are found in small organelles, lysosomes, located at the periphery of muscle cells. The stability of lysosomes decreases with a fall in pH, allowing leakage of proteases into the cell and eventually extracellular spaces. A protease active at neutral pH and thought to be involved in conditioning is calpain I which requires free calcium ions for activity. In meat, during the onset of rigor, the lack of ATP as an energy source to pump calcium ions out of cells leads to a rise in the levels of free calcium, and conditions suitable for protease activity. The duration of rigor stiffness is dependent on the species, being about one day for beef, half a day for pork and 2–4 hours for chicken. The reasons for these differences are not fully understood. Cathepsin levels are higher in chicken and pork which condition quickly (Etherington *et al.* 1987), and in beef the myofibrillar structure is more resistant to the action of cathepsin enzymes than it is in chicken (Mikami *et al.* 1987). More precise details of the proteases responsible and the conditions that control their activity have yet to be fully understood.

In cheese making, the addition to the milk of proteases in rennin and the microbial starter culture causes the development of characteristic flavour and texture during ripening. Chymosin, an aspartyl protease in rennin, splits a single peptide bond in κ-casein, a milk protein, which results in clotting. A combination of the action of chymosin and proteases from the starter culture degrade casein to peptides. Many of these peptides can have bitter or sour flavours or no flavour at all, but intracellular proteases from the starter culture break the peptides down further to amino acids and small peptides which have flavour-enhancing properties.

Bitter flavours in dairy products may be an adverse effect of protease activity. Peptides that are composed of predominantly non-polar amino acids tend to be bitter. In fermented dairy products, conditions that favour proteolysis and the accumulation of peptide intermediates are likely to have a bitter flavour.

In fish, proteases are responsible for the condition known as 'belly burst'. Heavy feeding prior to capture enhances the concentration and activity of gut enzymes. Unless the fish is gutted or cooled soon after capture, protease activity weakens the gut wall, allowing leakage of the contents to surrounding tissues. Herring and mackerel are notably more susceptible to belly burst; herring can become unsuitable for smoking in one day. In crustacea such as lobster and prawns the process is even more rapid, with gut enzymes attacking the flesh within hours of death. Rapid chilling and processing after catching is required.

9.5.4 Lipolysis

The hydrolysis of triacylglycerols at an oil-water interface is catalysed by lipase (Fig. 9.5). The specificity of lipases varies, some being able to attack esters at all

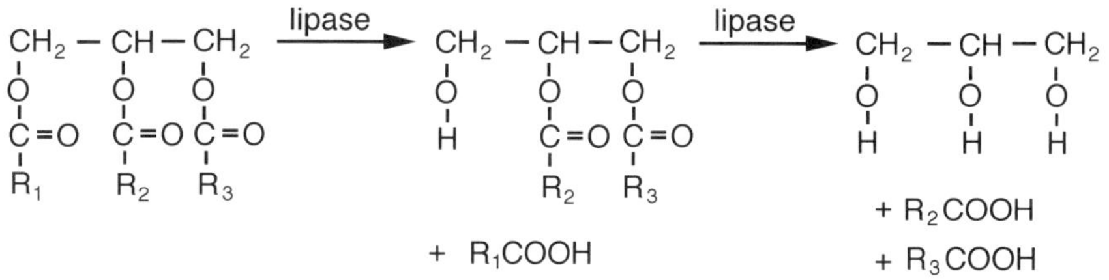

Fig. 9.5. Action of lipase on triacylglycerol.

three positions in the triacylglycerol whilst others are restricted to positions 1 and 3.

The activity of lipases of either endogenous or microbial origin is responsible for changes in functional properties of some dairy products such as a reduction in the skimming properties of skim milk and the churning capacity of cream, but particularly for the soapy and rancid flavours of foodstuffs. Long-chain fatty acids are usually associated with soapy flavours, and short-chain fatty acids with unpleasant rancid flavours; for example, the odour of valeric acid is described as being like 'sweaty feet', and hexanoic acid as 'goat-like'. The flavour threshold of these compounds is generally low, e.g. 14 ppm for hexanoic acid, so even a very little lipolytic activity can have a marked effect on quality.

In milk, the release of as little as 1-1.5% of the fatty acids from triacylglycerols can cause it to be unpalatable (Table 9.2). Endogenous milk lipases are most likely to be responsible if lipolysis occurs before the milk has been heat-treated and if the total viable count is less than 10^6 per ml. Flavour changes due to endogenous lipases in milk are a rare occurrence. Endogenous lipases are denatured by pasteurization, but extracellular microbial lipases released by psychrotrophic bacteria such as *Pseudomonas* spp are heat-stable, withstanding pasteurization and, in some cases, HTST treatments. As psychrotrophic organisms are able to grow at 2–4°C, the preferred holding temperature for milk or cream in bulk storage tanks, significant levels of lipase may be reached. Heat-resistant lipases may take weeks to have an effect on

Table 9.2 Free fatty acid concentrations in dairy products and rancid flavour threshold values (Allen 1989)

Product	Free fatty acid values (meq/g fat)	
	Normal	Likely to cause problems
Milk powder	0.3–1.0	1.5–2.0
Ice cream	0.5–1.2	1.7–2.1
Butter	0.5–1.0	2.0
Cheese:		
Cheddar	1.2	2.9
Brie	1.2	–
Blue	40.0	–

product quality and are usually of greater significance to the quality of ambient and long shelf-life products.

In cheese-making, hydrolysis due to lipase activity in the rennet may be needed to develop the required flavour (Peppler and Reed 1987). Almost all strongly flavoured cheeses, such as Stilton, Roquefort, Gorgonzola and Parmesan, depend on free fatty acids for their flavour. With the advent of microbial proteases as rennet substitutes there is a need to add lipases with the appropriate specificity to achieve the precise mixture of fatty acids responsible for the desired flavour. The difficulties associated with achieving the appropriate specificity and amounts of enzyme needed have been demonstrated with Cheddar cheeses. Differences in fatty acid levels that give a normal Cheddar and a rancid Cheddar can be reached despite extremely small differences in the amount of lipase (Law and Wigmore 1985).

9.6 Characteristics of physico-chemical reactions

Physico-chemical reactions that affect the quality of chilled foods occur as a result of physical changes to the product or the chemical or biochemical reactions that follow. Thus migration of components either by diffusion or osmosis and light absorption by natural or artificial pigments, fall within this category.

9.7 Physico-chemical reactions of significance in chilled foods

9.7.1 Migration

In mayonnaise-based salads, such as coleslaw and potato-based salads, the major quality changes observed are sensory changes related to the distribution of oil and water between the mayonnaise and vegetable tissue (Tunaley and Brocklehurst 1982). In the case of coleslaw, a 13.5% increase in ether-extractable solids from the cabbage and a translucent appearance, indicated the uptake of oil from the mayonnaise by the cabbage within 6 hours of mixing (Tunaley *et al.* 1985). In the mayonnaise, the change in oil content was reflected by an increase in the polydispersity of the globule size. In addition, migration of water from the cabbage to the mayonnaise, owing to the difference in osmotic potential, caused the mayonnaise to become runny and 'non-coating' within the same timeframe as the cabbage becoming translucent. Investigations of differences between cabbage varieties with respect to oil absorption have shown that stored Dutch cabbage gave no change in the assessment of 'creamy-oiliness' of the mayonnaise, whereas fresh English cabbage gave a significant decrease. Other ingredients with a large difference in osmotic potential with respect to the mayonnaise, such as celery and raisins, may also present problems owing to moisture migration resulting in the formation of pools of water on the surface of the mayonnaise.

One of the most widely experienced quality changes involving the migration of water is sogginess in sandwiches. Moisture migration from the filling to the bread can be reduced by the use of fat-based spreads to provide a moisture barrier at the interface (McCarthy and Kauten 1990). In pastry- and crust-based products such as pies and pizzas, migration of moisture from fillings and toppings to the pastry and crust causes similar problems. The migration of moisture or oils may be accompanied by soluble colours; for example, in pizza toppings where cheese and salami come into contact red streaking of the cheese is seen, and in multilayered trifles migration of colour between layers can detract from the visual appearance unless an appropriate strategy for colouring is used. Migration of enzymes from one component to another, for example when sliced unblanched vegetables are placed in contact with dairy products, can lead to flavour, colour or texture problems depending on the enzymes and substrates available (Labuza 1985).

9.7.2 Evaporation

A high volume of chilled foods are sold unwrapped from delicatessen counters, particularly cooked fresh meat, fish, pâtés and cheese. The shelf-life of such products differs markedly from the wrapped equivalent – six hours versus a few days to weeks. The most common cause for this reduction in shelf-life is evaporative losses. These result in a change in appearance, to such an extent that the consumer will select products which have been loaded into the cabinet most recently in preference to those which have been held in the display cabinet. The practical display-life of unwrapped meat products is determined by surface colour changes that may make the product seem unattractive. Changes in appearance are related to weight loss due to evaporation (Table 9.3). The direct cost of evaporative loss from unwrapped foods in chilled display cabinets was estimated to be in excess of £5 million per annum in 1986 (Swain and James 1986). In stores where the rate of turnover of product is high, the average weight loss will be greater because of the continual exposure of freshly wetted surfaces to the air stream.

Weight losses from the surface of unwrapped foods are dependent on the rate of evaporation of moisture from the surface and the rate of diffusion of moisture from within the product. Temperature, relative humidity and air velocity are the most influential factors affecting weight loss. Weight loss during storage of fruit and vegetables is mainly due to transpiration. Most have an equilibrium humidity of 97–98% and will lose water if kept at humidities less than this. For practical reasons, the recommended range for storage humidities is 80–100% (Sharp 1986). The rate of water loss is dependent on the difference between the water vapour pressure exerted by the produce and the water vapour pressure in the air, and air speed over the product. Loss of as little as 5% moisture by weight causes fruit and vegetables to shrivel or wilt. As the temperature of air increases, the amount of water required to saturate it increases (approximately doubling for each 10°C rise in temperature). If placed in a sealed container, foods will lose or

Table 9.3 Evaporative weight loss from, and the corresponding appearance of, sliced beef topside after 6 hours' display (James 1985)

Evaporative loss (g/cm^2)	Change in appearance
Up to 0.01	Red, attractive and still wet; may lose some brightness
0.015–0.025	Surface becoming drier; still attractive but darker
0.025–0.035	Distinct obvious darkening; becoming dry and leathery
0.05	Dry, blackening
0.05–0.10	Black

gain water until the humidity inside the container reaches a value characteristic of that food at that temperature. If the temperature is increased and the water vapour in the atmosphere remains constant, then the humidity of the air will fall. Minimizing temperature fluctuations is crucial for the prevention of moisture loss in this situation.

9.7.3 Chill injury

Although low-temperature storage of fruit and vegetables is considered to be the most effective method for preserving the quality of perishable horticultural products, for chill-sensitive crops it may be more harmful than beneficial. Most fruit and vegetables of tropical and subtropical origin are injured by exposure to low but not freezing temperatures (10–15°C) (Couey 1982). Some temperate fruit and vegetables are also susceptible to injury, but at lower threshold temperatures (below 5°C to 10°C) (Bramlage 1982).

Chill injury is indicated by a range of different symptoms that adversely affect quality. Pitting, a general collapse of the tissue, is induced by dehydration and low temperatures. It is most evident in mangoes, avocados, grapefruit and limes, in which the outermost covering is harder and thicker than that underneath. Surface discoloration is common in fruits with thin soft peels, such as bell peppers, aubergines and tomatoes. Uneven or incomplete ripening is induced in tomatoes, melons and bananas. Most frequently internal breakdown and a weakening of the tissues makes the fruit or vegetable susceptible to decay by post-harvest plant pathogens. Chill injury may occur within a short space of time if temperatures are considerably below the critical level. In some cases, symptoms may only develop and become detectable after removal from cold storage and on holding at warmer temperatures, making it difficult to determine immediately after exposure to low temperatures whether chill injury has occurred.

Changes in physical structures occurring at the time of chill injury have been described; however, their association with the development of symptoms of chill injury has not been established in the majority of cases. Changes in membrane lipid structure and composition (Whitaker 1991), alterations of the cytoskeletal structure of cells and conformational changes in some regulatory enzymes and structural proteins leading to loss of compartmentalization within cells have been reported. Resulting changes in plant physiology include loss of membrane integrity, leakage of solutes, stimulation of ethylene production (Wang and Adams 1980), and bursts of respiration (Wang 1982).

Approaches to alleviate chill injury are highly dependent on the fruit or vegetable in question (Jackman *et al.* 1988). The most obvious is to avoid exposure of chill-sensitive fruit and vegetables to low temperatures. However, as already stated, chilling provides a means of reducing respiration rate, evaporation and transpiration and therefore extends storage-life. Temperature treatments – such as pre-storage conditioning at temperatures just above the threshold (acclimation) (suitable for cucumber and bananas); intermittent warming during storage (suitable for apples and stone fruits); or holding at ambient temperatures for a short time prior to chill storage – are effective in some cases. Controlled-atmosphere storage has been shown to be beneficial in a limited number of cases, e.g. avocados (Spalding and Reeder 1975), peaches (Anderson 1982) and okra (Ilker and Morris 1975), but it is considered to aggravate chill injury by imposing the additional stresses of low oxygen and high carbon dioxide levels on the produce (Wade 1979).

Chemical treatments have been shown to be effective on some fruit and vegetables. On the basis that changes in membrane structure lead to chill injury, treatments leading to an alteration or protection of components of cell membranes have been used. Treatment of tomato seedlings with ethanolamine increased the levels of unsaturated fatty acids incorporated into membrane phospholipids; this reduced damage to cellular components during chilling (Ilker *et al.* 1976). Free radical scavengers or antioxidants such as ethoxyquin and sodium benzoate, diphenylamine and butylated hydroxytoluene have been shown to be effective on cucumbers, bell peppers (Wang and Baker 1979) and apples (Huelin and Coggiola 1970). Coating of fruits in waxes or oils (provided they are approved for food use) prior to chilling are effective by preventing moisture loss and reducing oxygen available for oxidation. Incorporation of the fungicides benomyl or thiabendazole (TBZ) into this type of coating has been shown to have further advantages for peaches and nectarines (Schiffman-Nadel *et al.* 1975). The ultimate goal for alleviating chill injury is to select, breed or genetically engineer fruit and vegetable crops to prevent chill sensitivity. Plant breeding and selection have had varying degrees of success. A better understanding of the mechanisms responsible for chill injury should provide the insight required for targeted genetic engineering programmes to overcome this problem, though the varying causes of chill injury are unlikely to be overcome by universal solutions.

9.7.4 Syneresis

The weeping or slow spontaneous movement and separation of liquid from a colloidal semi-solid mass is termed syneresis. It occurs as a result of physico-chemical changes in carbohydrates or proteins which influence their ability to hold water.

As a food ingredient, starch fulfils a number of essential functions – thickens, gels, stabilizes emulsions, controls moisture migration, and influences texture. An inherent limitation of native starches and flour is a lack of stability at low temperatures and at fluctuating temperatures. At low temperatures they become prone to weeping or syneresis.

Native starch is a complex carbohydrate composed of the homopolymers, amylose and amylopectin. Amylose is a linear chain molecule composed of 1,4-linked α-D-glucopyranose building blocks. Amylopectin has a backbone structure like amylose but, in addition, 1,6-linkages give it a branched structure that confers a greater water-holding capacity than amylose. The ratio of amylopectin to amylose therefore alters the properties and texture of a starch. For example, wheat flour, a traditional thickener used in gravies and sauces, provides desirable flavour and opacity, but has no low-temperature stability and chilling results in syneresis. When the starch grains are swollen, the linear amylose molecules tend to leach out into solution and reassociate into aggregates aligned by hydrogen bonding. The reassociated amylose tends to expel water resulting in opacity and syneresis. Cooling or freezing causes the overall structure to shrink, greatly accelerating the rate at which syneresis occurs.

Problems of syneresis often occur as a result of improper selection of starch. Incorporation of stabilized waxy maize-based starches into products that are to be chilled resists retrogradation and syneresis. Alternatively, stabilized starches are available that have been modified specifically with monofunctional blocking groups to prevent associations between leached amylose molecules and thereby prevent syneresis. Use of a modified starch in conjunction with a wheat flour provides stability in the final product.

Syneresis in milk is known as 'wheying off', the point at which the curds and whey separate. It is obviously desirable for cheese making, but not in milk-based products such as yoghurts. Homogenization of milk for yoghurt production decreases syneresis by increasing the hydrophilicity and water-binding capacity by enhancing casein and fat globule membrane interactions and other protein-protein interactions (Tamime and Deeth 1980). Heat processing for yoghurt manufacture (85°C for 30 minutes, or 90–95°C for 5–10 minutes) is unique in dairy processing. It is believed to bring about important changes in the physico-chemical structure of the proteins, which minimizes syneresis and results in maximal firmness of the yoghurt coagulum.

9.7.5 Staling

The market for sandwiches containing a wide variety of fillings that need chill storage has grown considerably. However, the staling of bread is one of the few reactions that has a negative temperature coefficient (McWeeney 1968); that is, bread stales more rapidly at reduced temperatures (Meisner 1953). The term 'staling' in relation to bread is used to describe an increase in crumb firmness and crumb-texture hardness, loss of crust crispness and increased toughness, and disappearance of the fresh bread flavour and emergence of a stale bread flavour. Despite extensive research into the mechanism of staling, most researchers are only prepared to agree that firmness changes are attributed to physico-chemical reactions of the starch component, mainly due to its amylopectin fraction, and some include involvement of flour proteins.

The shelf-life of commercial bread is considered to be two days (Maga 1975), which will be reduced by holding at chill temperatures. The use of modified atmosphere packaging, particularly carbon dioxide, is believed to slow the rate of staling of bread (Avital *et al.* 1990).

9.8 Non-microbiological safety issues of significance in chilled foods

Non-microbiological safety issues associated with chilled foods are rarely a result of, or exacerbated by, chilled storage temperature. Some arise as a consequence of the ingredient combinations or minimal processing that subsequent chill storage enables. In most instances, judicious selection of raw materials and a carefully tailored monitoring programme, based on an assessment of the risks posed by individual ingredients and the final product, contributes to the assurance of product safety. If possible, it is always preferable for shelf-life to be limited by changes in quality rather than safety because changes in quality can usually be discerned by the smell, taste or appearance of the product, but such changes cannot be relied upon to indicate when safety limits the shelf-life.

9.8.1 Natural toxicants

There is a tendency to associate 'natural' with a wholesome and healthy image, yet in some cases there is an awareness that some naturally occurring chemical compounds in food may contribute to human illness. Such an example is greening of potatoes, which is commonly associated with the potential to cause harm. Glycoalkaloids, the group of toxic compounds that can be found in potatoes stored under stress conditions, accumulate just beneath the peel and at eye regions, so peeling reduces potential human exposure. Cooking is not thought to reduce glycoalkaloid concentrations (Bushway and Ponnampalam 1981). However, as a result of dietary advice to increase the intake of fibre, an increasing number of potato products incorporate or retain the skin, e.g. chilled, filled, baked potatoes and potato skins; such products could present a higher risk. It has been generally agreed that tubers for human consumption should not exceed 20 mg glycoalkaloid per 100 g fresh tuber weight. Monitoring of the levels of glycoalkaloids is advised, particularly in new cultivars and after changes in storage and processing procedures, to ensure that they do not exceed recommended limits.

Pulses and grain legumes have long been known to contain highly toxic lectins (haemagglutinins) which agglutinate red blood cells. Haemagglutinins have been detected in a wide range of leguminous seeds including lentils, soyabeans, lima or butter beans, and red kidney beans (Liener 1974). During the last decade a number of incidents of food poisoning have been associated with red kidney beans, and in one case with butter beans (Bender and Reaidi 1982, Rodhouse *et al.* 1990). A tendency to partially cook pulses or to eat them raw, particularly red kidney beans in salads, led to numerous cases of gastrointestinal disturbances. Soaking of beans for at least 5 hours leaches out lectins and boiling in fresh water for at least 10 minutes heat-inactivates any that remain, preventing the possibility of food poisoning.

The inclusion of nuts, figs and dates in exotic salads such as hosaf carries the associated risk of contamination by mycotoxins (fungal toxins). Mycotoxins are contaminants rather than natural toxicants, being secondary metabolites of the fungal species e.g. *Aspergillus*, *Penicillium*, and *Fusarium*. These fungal species grow on a wide variety of substrates, most notably cereals and ground nuts and other high carbohydrate seeds (e.g. figs) under environmental conditions ranging from tropical to domestic refrigeration temperatures. Unfortunately, mycotoxin production is associated with storage conditions designed to prevent fungal growth. Mycotoxins are chemically very diverse (they include groups such as aflatoxins, ochratoxins and trichothecenes), ranging in molecular complexity and toxicity (some are extremely toxic and others are carcinogenic). Control of mycotoxin contamination has focused on treatments to prevent mould growth or mycotoxin production during storage (Moss and Frank 1987), and on the development of improved analytical methods for their detection. Improved quality of raw materials and post harvest treatments, coupled with improved storage and distribution conditions, reduces the incidence of contamination. In keeping with the principles of HACCP awareness of the possibility of mycotoxin contamination should be accompanied by the implementation of a suitable monitoring programme, based on an assessment of the potential risks involved, and written into the raw material specifications.

9.8.2 Phycotoxins

Toxic compounds produced by algae (phycotoxins) enter the food chain via seafood, usually either shellfish (shellfish toxins) or finfish (ciguatoxins). The growing awareness of the beneficial dietary effects associated with eating fish and seafood products and the availability of the chill chain to distribute these products has resulted in an increase in their geographic availability and consumption (Przybyla Wilkes 1991). Importation of seafoods means more exotic forms of phycotoxin are now potentially found on a global scale (Scoging 1991).

Four different forms of shellfish poisoning are recognized: Paralytic Shellfish Poisons (PSP), Diarrhetic Shellfish Poisons (DSP), Amnesic Shellfish Poisons (ASP) and Neurotoxic Shellfish Poisons (NSP). Shellfish, particularly bivalve molluscs, e.g. mussels, clams and oysters, accumulate these toxins and are unharmed by them. Subsequent consumption of shellfish by humans produces immediate and severe effects, depending on the type of toxin involved. Accumulation of toxins by shellfish coincides with high levels of particular algal species in coastal waters, so-called 'algal blooms'. These result from increased availability of nutrients and light in surface waters associated with seasonal climate and hydrographic changes. In the UK, extensive monitoring is undertaken by the Ministry of Agriculture, Fisheries and Food during high-risk periods. Coastal waters, shellfish and some crustacea are analysed for PSP toxins. Prohibition orders on the collection of shellfish are put in place when toxins accumulate to levels which are regarded as unsafe for human consumption (West *et al.* 1985). This is currently believed to be the most effective control method, as the toxins which the shellfish accumulate are reduced, but not eliminated, by cooking (Krogh 1987) or by holding shellfish in purification tanks.

PSP is linked with algal species which occur in waters where ambient temperatures are around 15–17ºC. Initial symptoms, seen within 30 minutes of consumption, are tingling and numbness in the mouth and fingertips which spreads throughout the body, causing impaired muscle coordination and, in severe cases, paralysis. The major toxin is saxitoxin, though 18 other toxic derivatives have been identified which are either natural algal toxins or metabolized derivatives found in shellfish.

DSP intoxications are common in Japan, but outbreaks have also been recorded in France, Italy and The Netherlands. Symptoms occurring within 30 minutes of consumption are vomiting, abdominal pain and diarrhoea. The major toxic components are okadaic acid and dinophysic toxins found in mussels, clams and scallops. Denaturation of these toxins only occurs after processing at 100ºC for 163 minutes; therefore monitoring and prohibition orders are the only real safeguard.

ASP is believed to be caused by a toxic amino acid, domoic acid, produced by a diatom occurring in USA, Japanese and Canadian coastal waters. Symptoms include nausea, diarrhoea and confusion/disorientation headaches and, in severe cases, memory loss.

NSP intoxications have been mainly associated with the consumption of oysters, clams and other bivalve molluscs in North America. Symptoms occur within 3 hours of consumption and include gastrointestinal disturbances, numbness of the mouth, muscular aches and dizziness. The dinoflagellate responsible for NSP, *Ptychodiscus brevis*, is notorious for the massive fish kills that occur every 3–4 years off the west coast of Florida.

The lack of availability of analytical standards has hampered the development of suitable chemical methods for the determination of these toxins. Most monitoring programmes rely on the use of a mouse bioassay to detect levels of toxicants. Restriction of harvesting of shellfish at those times of the year when algal blooms occur is currently the safest method of prevention.

Ciguatera toxins are the largest global public health non-microbial problem associated with seafood. Most incidents occur in the USA, danger areas being the Pacific, Caribbean and Indian Oceans. To date, three incidents have been recorded in the UK (Scoging 1991). Finfish that harbour the toxin include the barracuda, red snapper, grouper, amberjack, surgeon fish and sea bass. Ciguatoxin is a neuromuscular toxin that affects the membrane potential of neural cells. Symptoms vary widely with the dose ingested but include vomiting, abdominal plain, dizziness, blurred vision, and reversal of the sensations of hot and cold. Onset is usually within a few hours of consumption and the effects can persist for several months. The toxins are heatstable and unaffected by processing methods. The appearance of the fish gives no indication of the toxin. Development of a dipstick immunoassay to detect ciguatoxins has facilitated sampling in the field (Hokama *et al.* 1989).

9.8.3 Scombroid fish poisoning

Scombroid fish poisoning occurs throughout the world, though most incidents are recorded in the USA, Japan and the UK. In the USA, scombrotoxicosis was the cause of 29% of food-poisoning incidents caused by chemical agents between 1973 and 1987 (Hughes and Potter 1991), and in the UK, 348 suspected incidents were reported between 1976 and 1986 (Bartholomew *et al.* 1987). *Scombridae* and *Scomberesocidae* families (tuna, mackerel, saury, bonito and seerfish), but incidents have also been associated with non-scombroid fish (sardines, herring, pilchards, anchovies and marlin) (Bartholomew *et al.* 1987, Morrow *et al.* 1991).

Scombrotoxicosis is characterized by the rapid onset (within a few minutes to 2–3 hours of eating the fish) of symptoms which can include flushing, headache, cardiac palpitations, dizziness, itching, burning of the mouth and throat, rapid and weak pulses, rashes on the face and neck, swelling of the face and tongue, abdominal cramp, nausea, vomiting and diarrhoea. The similarity of these symptoms with those related to food allergy has often resulted in misdiagnosis.

Histamine has been considered to be the cause of scombrotoxic poisoning for a number of reasons. Analysis of the fish remaining 'on the plate' usually reveals it contains high levels of histamine; metabolites of histamine have been detected in the urine of victims; symptoms resemble those of known histamine responses; and administration of antihistamine drugs reduces the severity of symptoms. Scoging (1998) of the Food Hygiene Laboratory (Public Health Laboratory Service) proposed guidelines with respect to histamine levels and the potential for illness. Histamine is a spoilage product resulting from decarboxylation of the amino acid L-histidine which is abundant in scombroid fish flesh. Formation of histamine requires the enzyme histidine decarboxylase, which is produced by the normal bacterial microflora of fish skin, gut and gills. If fish is stored above 4°C, these organisms proliferate and levels of histamine in the flesh increase. Prevention of scombroidfish poisoning would therefore appear to be highly dependent on good handling practices – rapid chilling of the catch, and adequate chilling of the fish prior to preparation for eating.

However, in medically supervised feeding studies, deliberately spoiled mackerel and mackerel with added histamine, fed to volunteers, failed to reproduce scornbrotoxic symptoms (Clifford *et al.* 1989). These workers suggested that histamine alone is unlikely to be the causative agent. Other amines, such as cadaverine, have been suggested as potentiators or as synergists to histamine (Bjeldanes *et al.* 1978). Further feeding studies, using mackerel implicated in a scombroid-fish poisoning outbreak which reproduced symptoms in volunteers, showed the potency of the mackerel was not related to the histamine dose (Ijomah *et al.* 1991), or the content of other amines (cadaverine, putrescine, spermidine, spermine, tyramine), or any relationship between the levels of these amines (Clifford *et al.* 1991). Vomiting and diarrhoea were abolished by administration of antihistamine drugs. It has been suggested that histamine, released by the human body as a part of the natural defence mechanism, is responsible for the observed symptoms, that dietary histamine has a minor role in scombroid-fish poisoning, and that, as yet, the agent in fish which is responsible for triggering the release of histamine by the body is unidentified.

9.8.4 Allergens

Food allergy, as opposed to food intolerance, is an immunological reaction to some component of the food. This component or antigen can stimulate the body to release specific immunoglobulin E (IgE) antibodies that give rise to anaphylaxis. Such a reaction can range from a trivial event such as a sneeze to a life-threatening incident. Many food types have been associated with allergic reactions but perhaps best known are those involving milk, soya, shellfish and nuts, particularly peanuts. These issues are not specific to chilled foods but with the increase in formulated chilled food products and the severity of the potential hazards associated with the use of peanuts mention is appropriate here. Studies have indicated that peanut allergy has been reported by 0.5% of the adult population in the UK (Emmett *et al.* 1999) and that those with sensitivity to peanuts commonly show reaction to other nut types, i.e. Hazel nut and Brazil nut (Pumphrey *et al.* 1999). The allergenicity of peanut residues is heat stable and Ara h 1, a major peanut allergen, has been shown to retain its IgE binding characteristics despite significant structural denaturation (Koppelman *et al.* 1999). The possible carry over of allergenic material from product to product or production line to production line, therefore, necessitates stringent hygiene practices with associated quality assurance measures. Wherever possible, products containing peanut residues should be prepared and processed in separate areas, away from products that consumers do not expect to contain peanut residues. HACCP procedures should be used to identify all potential sources of cross contamination. Where there is potential for cross contamination, product scheduling and appropriate cleaning regimes are essential.

9.8.5 Products of lipid oxidation

Lipid oxidation products are of great significance to the sensory properties of food, but, in addition, attention has been given to the health risks that they may pose, and to their role in reduction of nutrient availability via free radical production and destruction of fat-soluble vitamins A and E.

Lipid hydroperoxides and their decomposition products may bind and polymerize proteins, and cause damage to membranes and biological components, thus affecting vital cell functions (Halliwell and Gutteridge 1986, Frankel 1984). Lipid peroxides and oxidized cholesterol may be involved in tumour promotion and in atherosclerosis. Malonaldehyde, a secondary product of lipid oxidation, has been implicated as a catalyst in the formation of N-nitrosamines and as a mutagen (Pearson *et al.* 1983, Jurdi-Haldernan *et al.* 1987, Sanders 1987). The significance to human health of eating foods which contain high levels of lipid hydroperoxides and their decomposition products is still to be established, particularly as the rate of formation of lipid peroxides *in vivo* is much greater than that arising from dietary intake. Nevertheless, whilst possible health risks associated with lipid oxidation products remain controversial, high levels of lipid peroxides are undesirable in the diet. Pre-cooked meats have been identified, amongst other products, as an area which requires further research to improve methods for retarding the development of rancidity (Addis and Warner 1991).

9.9 Conclusion

The objective of this chapter has been to illustrate, by example, the way in which many non-microbiological factors interact to influence the quality and safety of chilled foods. The contribution that an understanding of food chemistry can make towards optimization or prevention of these interactions is evident. Further understanding is required for expansion and continued success in the production of safe, high quality chilled foods which achieve the desired shelf-life.

9.10 References

ADDIS P B and WARNER G J, (1991) The potential health aspects of lipid oxidation products in food. In: Aruoma, O I. and Halliwell, B (eds), *Free Radicals and Food Additives*, Taylor and Francis, London, pp. 77–119.

ALLEN J C, (1989) Rancidity in dairy products. In: Allen, J C. and Hamilton, R J. (eds) *Rancidity in foods* 2nd edn, Elsevier Applied Science, London, pp. 199–210.

ANG C Y W and LYON B G, (1990) Evaluations of warmed-over flavor during chill storage of cooked broiler breast, thigh and skin by chemical, instrumental and sensory methods, *J. Food Sci.*, **55** (3) 644–8, 673.

ANDERSON R E, (1982) Long-term storage of peaches and nectarines intermittently warmed during controlled atmosphere storage, *J. Am. Soc. Hort. Sci.*, **107** (2) 214–16.

ASGHAR A A, GRAY J I, BUCKLEY D J, PEARSON A M and BOOREN A M, (1988)

Perspectives on warmed-over flavor, *Food Technol.*, **42** (6) 102–8.

AVITAL Y, MARMHEIM C H and MILTZ J, (1990) Effect of carbon dioxide atmosphere on staling and water relations in bread, *J. Food Sci.*, **55** (2) 413–16, 461.

BARTHOLOMEW B, BERRY P, RODHOUSE J, GILBERT R and MURRAY C K, (1987) Scombrotoxic fish poisoning in Britain; features of over 250 suspected incidents from 1976 to 1986, *Epidemiol Infect*, **99** 775–82.

BENDER A E and REAIDI G B, (1982) Toxicity of kidney beans (*Phaseolus vulgaris*) with particular reference to lectins, *J. Plant Foods*, **4** (1) 15–22.

BJELDANES L F, SCHUTZ D E and MORRIS M M, (1978) On the aetiology of scombroid poisoning: cadaverine potentiation of histamine toxicity in the guinea pig, *Food and Cosmetics Toxicology*, **16** 157–9.

BOLIN H R and HUSOLL C C, (1989) Storage stability of minimally processed fruit, *J. Food Proc. Pres*, **13**(4) 281–92.

BRAMLAGE W J, (1982) Chilling injury of crops of temperate origin,. *HortSci*, **17** (2) 165–8.

BROTSKY E, (1976) Automatic injection of chicken parts with polyphosphate, *Poultry Sci.*, **55** (2) 653–60.

BROWN H M, OSBORN H and LEDWARD D, (1998) Undesirable pink colouration in cooked meat products. *RSS No. 80*, CCFRA Research Summary Sheets, Campden and Chorleywood Food Research Association, Chipping Campden, Gloucestershire GL55 6LD.

BUSHWAY R J and PONNAMPALAM R, (1981) α-Chaconine and α-solanine content of potato products and their stability during several modes of cooking, *J. Agric. Food Chem.*, **29** (4) 814–17.

BYRNE D V, BAK L S, BREDIE W L P, BERTELSEN G and MARTENS M, (1999a) Development of a sensory vocabulary for warmed-over flavour. I. In porcine meat, *J. Sens. Std.*, **14** (1) 47–65.

BYRNE D V, BAK L S, BREDIE W L P, BERTELSEN G and MARTENS M, (1999b) Development of a sensory vocabulary for warmed-over flavour. II. In chicken meat, *J. Sens. Std.*, **14** (1) 67–78.

CASTELL C H, (1971) Metal-catalysed lipid oxidation and changes of proteins in fish, *J. Am. Chem. Soc*, **48** (11) 645–9.

CHO S H and RHEE K S, (1997) Lipid oxidation in mutton: species related and warmed-over flavours, *Journal of Food Lipids*, **4** (4) 283–93.

CHURCHILL H M, GRIFFITHS N G and WILLIAMS B M, (1988) Warmed-over flavour in meat, *Campden Food & Drink Research Association Technical Memorandum No. 489.*

CHURCHILL H M, GRIFFITHS N G and WILLIAMS B M, (1990) The effect of reheating on warmed-over flavours in chicken and baked potatoes, *Campden Food & Drink Research Association Technical Memorandum No. 591.*

CLIFFORD M N, WALKER R and WRIGHT J, (1989) Studies with volunteers on the role of histamine in suspected scombrotoxicosis, *J. Sci. Food Agric.*, **47** 365–75.

CLIFFORD M N, WALKER R, WRIGHT L, MURRAY C K and HARDY R, (1991) Is there a role for amines other than histamines in the aetiology of scombrotoxicosis? *Food Additives and Contaminants*, **8** (5) 641–52.

COGAN T M, (1977) A review of heat resistant lipases and proteinases and the quality of dairy products, *Ir. J. Food Sci. Technol*, **1** 95–105.

COUEY H M, (1982) Chilling injury of crops of tropical and subtropical origin, *HortSci.*, **17** (2) 162–5.

DALHOFF E and JUL M, (1965) In: Penzer, W T. (ed.) *Progress in Refrigeration Science and Technology*, Vol. 1. AVI Publishing Co., Westport, CT, p. 57.

DAVEY C L and GILBERT K V, (1976) The temperature coefficient of beef ageing, *J. Sci. Food. Agric.*, **27** (3) 244–50.

DAY B P F, (1998) Novel MAP – a brand new approach, *Food Manufacture*, **73** (11), 24–6.

EMMETT S E, ANGUS F J, FRY J S and LEE P N, (1999) Perceived prevalence of peanut allergy in Great Britain and its association with other atopic conditions and with peanut allergy in other household members, *Allergy*, **54** (4) 380–5.

ETHERINGTON D J, TAYLOR M A J and DRANSFIELD E, (1987) Conditioning of meat from different species: relationship between tenderizing and the levels of cathepsin B, cathepsin L, calpain I, calpain II and beta-glucuronidase, *Meat Sci.*, **20** (1) 1–18.

FRANKEL E N, (1984) Recent advances in the chemistry of the rancidity of fats. In: Bailey, A J. (ed.) *Recent Advances in the Chemistry of Meat*, The Royal Society of Chemistry, Special Publication, No. 47, pp. 87–18.

GERMAN J B and KINSELLA J E, (1985) Lipid oxidation in fish tissues. Enzymatic initiation via lipoxygenase, *J. Agric. Food Chem.*, **33** (4) 680–3.

GIBBS P A and WILLIAMS A P, (1990) Using mathematical models for shelf life prediction. In: Turner, A. (ed.), *Food Technology International Europe*, Stirling Publications International Ltd.

GOLL D E, KLEESE W C and SZPACENKO A, (1989) Skeletal muscle proteases and protein turnover. In: Campion, D R, Hausman, G J and Martin, R J (eds), *Animal Growth and Regulation*, Plenum, New York, pp. 141–82.

GRAY J L and PEARSON A M, (1987) Rancidity and warmed-over flavor. In: Pearson A M and Dutson, T R (eds), *Advances in Meat Research. Vol. 3. Restructured Meat and Poultry Products*, Van Nostrand Reinhold, New York, pp. 221–69.

GREENE, B E, (1969) Lipid oxidation and pigment changes in raw beef, *J. Food Sci.*, **34** (2) 110–13.

HALLIWELL B and GUTTERIDGE J M C, (1986) *Free radicals in biology and medicine*, Clarendon Press, Oxford.

HERMANN K, SCHUTTE M, MULLER H and BISMER R, (1981) Ueber die antioxidative Wirkung von Gerwürzen, *Deutsche Lebensmittel-Rundschau*, **77** 134.

HOKAMA Y, HONDA S A A, KOBAYASHI M N, NAKAGAWA L K, ASAHINA A Y and MIYAHARA J T, (1989) Monoclonal antibody (Mab) in detection of ciguatoxin (CTX) and related polyethers by the stick-enzyme immunoassay (S-EIA) in fish tissues associated with ciguatera poisoning. In: Natori, S.,

Hashimoto, K. and Ueno, Y. (eds), *Mycotoxins and Phycotoxins '88*. Elsevier Science Publishers, Amsterdam, pp. 303–10.

HUELIN F E and COGGIOLA I M, (1970) Superficial scald, a functional disorder of stored apples. V. Oxidation of alpha-farnesene and its inhibition by diphenylamine, *J. Sci. Food Agric.*, **21** (1) 44–8.

HUGHES M J and POTTER M E, (1991) Scombroid-fish poisoning: from pathogenesis to prevention, *The New England Journal of Medicine*, **324** 766–8.

IGENE J O and PEARSON A M, (1979) Role of phospholipids and triglycerides in warmed-over flavour development in meat model systems, *J. Food Sci.*, **44** (5) 1285–90.

IGENE J O, KING J A, PEARSON A M and GRAY J I, (1979) Influence of heme pigments, nitrite, and non-heme iron on development of warmed-over flavor (WOF) in cooked meat, *J. Agric. Food Chem.*, **27** (4) 838–42.

IGENE J O, PEARSON A M and GRAY J I, (1981) Effects of length of frozen storage, cooking and holding temperatures upon component phospholipids and the fatty acid composition of meat triglycerides and phospholipids, *Food Chem*, **7** (4) 289–303.

IGENE J O, YAMAUCHI K, PEARSON A M, GRAY J I and AUST S D, (1985) Evaluation of 2-thiobarbituric acid reactive substances (TBRS) in relation to warmed-over flavor (WOF) development in cooked chicken, *J. Agric. Food Chem.*, **33** (3) 364–7.

IJOMAH P, CLIFFORD M N, WALKER R, WRIGHT L, HARDY R and MURRAY C K, (1991) The importance of endogenous histamine relative to dietary histamine in the aetiology of scombrotoxicosis, *Food Additives and Contaminants*, **8** (4) 531–2.

ILKER Y and MORRIS L L, (1975) Alleviation of chilling injury of okra. *HortSci.* **10**, 324.

ILKER Y, WARING A J, LYONS J M and BREIDENBACH R W, (1976) The cytological responses of tomato-seedling cotyledons to chilling and the influence of membrane modifications upon these responses, *Protoplasma*, **90** 229–52.

JACKMAN R L, YADA R Y, MARANGONI A, PARKIN K L and STANLEY D W, (1988) Chilling injury. A review of quality aspects, *J. Food Qual.*, **11** (4) 253–78.

JAMES S J, (1985) Display conditions from the product's point of view. Institute of Food Research Bristol Laboratory, Teach-in on retail display cabinets.

JAYATHILAKAN K, VASUNDHARA T S and KUMUDAVALLY K V, (1997) Effects of spices and Maillard reaction products on rancidity development in precooked refrigerated meat, *Journal of Food Science and Technology, India*, **34** (2) 128–31.

JUNCHER D, HANSEN T B, ERIKSEN H, SKOVGAARD I M, KNOCHEL S and BERTELSEN G, (1998) Oxidative and sensory changes during bulk and retail storage of hot-filled turkey casserole, *Zeitschrift fuer Lebensmittel Untersuchung und Forschung A*, **206** (6) 378–81.

JURDI-HALDERNAN D, MCNEIL J H and YARED D M, (1987) Antioxidant activity of onion and garlic juices in stored cooked ground lamb, *J. Food Protect*, **50** (5) 411–13.

KERRY J P, BUCKLEY D J, MORRISSEY P A, O'SULLIVAN K and LYNCH P B, (1999) Endogenous and exogenous alpha-tocopherol supplementation: effects on lipid stability (TBARS) and warmed-over flavour (WOF) in porcine *M. longissimus dorsi* roasts held in aerobic and vacuum packs, *Food Research International,* **31** (3) 211–16.

KOPPELMAN S J, BRUIJNZEEL-KOOMEN C A F M, HESSING M and JONGH, H H J DE (1999) Heat-induced conformational changes in Ara h 1, a major peanut allergen, do not affect its allergenic properties, *Journal of Biological Chemistry*, **274** (8) 4770–4.

KORCZACK L, FLACZYK E and PAZOLA Z, (1988) Effects of spices on stability of meat products kept in cold storage, *Fleischwirtschaft*, **68** 64–6.

KROGH P, (1987) Scientific Report on paralytic shellfish poisons in Europe. Document VI/3964/87-EN rev. 1, Commission of the European Communities, Directorate General for Agriculture, VI/B/II.2, Brussels.

LABUZA T P, (1985) An integrated approach to food chemistry: illustrative cases. In: Fennema, O. R. (ed.) *Food Chemistry*. Marcel Dekker, New York, pp. 913–38.

LAW B A and WIGMORE A S, (1985) Effect of commercial lipolytic enzymes on flavour development in Cheddar cheese, *J. Soc. Dairy Technol.*, **38** (3) 86–8.

LIENER I E, (1974) Phytohaemagglutinins: their nutritional significance, *J. Agric. Food Chem.*, **22** (1) 17–22.

LOPEZ-BOTE C, REY A, RUIZ J, ISABEL B and SANZ-ARIAS R, (1997) Effects of feeding diets high in monounsaturated fatty acids and alpha tocopherol acetate to rabbits on resulting carcass fatty acid profile and lipid oxidation, *Animal Science*, **64** (1) 177–86.

LOVE J (1988) Sensory analysis of warmed-over flavour in meat, *Food Technol.*, **42**(6) 140–3.

LYON B G, (1987) Development of chicken flavour descriptive attribute terms aided by multivariate statistics, *J. Sensory Studies*, **2** (1) 55–67.

MCCARTHY M J and KAUTEN R J, (1990) Magnetic resonance imaging applications in food research. *Trends in Food Science & Technology* December, 134–9.

MCLACHLAN A and STARK R, (1985) Modified atmosphere packaging of selected prepared vegetables, *Campden Food & Drink Research Association Technical Memorandum No. 412.*

MCWEENY D J, (1968) Reactions in food systems: negative temperature coefficients and other abnormal temperature effects, *J. Food Technol.*, **3** (1) 15–30.

MAGA J A, (1975) Bread staling *CRC Crit. Rev. Food Technol.*, **5** (4) 443–86.

MAGA J A, (1994) *Pink discolouration in cooked white meat. Food Reviews Int.*, **10** (3) 273–86.

MANN T F, REAGAN J O, LILLARD D A, CAMPION D R, LYON C E and MILLER M F, (1989) Effects of phosphate in combination with nitrite or Maillard reaction products upon warmed-over flavour in precooked, restructured beef chuck roasts, *J. Food Sci.*, **54** (6) 1431–3, 1437.

MARSH B B, RINGKOB T R, RUSSELL R L, SWARTZ D R and PAGEL L A, (1987) Effects of early-*post mortem* glycolytic rate on beef tenderness, *Meat Sci.*, **21** (4) 241–8.

MEISNER J A, (1953) Importance of temperature and humidity in the transportation and storage of bread, *Baker's Dig.*, **27** 109.

MIKAMI M, WHITING A H, TAYLOR M A L, MACIEWICZ R A and ETHERINGTON D J, (1987) Degradation of myofibrils from rabbit, chicken and beef by cathepsin L and lysosomal lystates *Meat Sci.* **21** (2) 81–97.

MINOTTI G and AUST S D, (1987) The requirement for iron (III) in the initiation of lipid peroxidation by iron (II) and hydrogen peroxide, *J. Biol. Chem.*, **262**, 1098.

MORROW J D, MARGOLIES G R, ROWLAND J and ROBERTS L J, (1991) Evidence that histamine is the causative toxin of scombroid-fish poisoning, *The New England Journal of Medicine*, **324**, 716–20.

MOSS M O and FRANK M, (1987) Prevention: effects of biocides and other agents on mycotoxin production. In: Watson, D H. (ed.) *Natural Toxicants in Food Progress and Prospects*, Ellis Horwood, Chichester, pp. 231–52.

MURPHY A, KERRY J P, BUCKLEY J and GRAY I, (1998) The antioxidative properties of rosemary oleoresin and inhibition of off-flavours in precooked roast beef slices, *J. Sci. Food. Agric.*, **77** (2) 235–43.

NOLAN N L, BOWERS J A and KROPF D H, (1989) Lipid oxidation and sensory analysis of cooked pork and turkey stored under modified atmospheres, *J. Food Sci.*, **54** (4) 846–9.

O'BEIRNE D, (1988) Modified atmosphere packaging of ready-to-use potato strips and apple slices. Abstracts 18th Annual Food Science and Technology Research Conference, *Ir. J. Food Sci. Technol*, **12** (1) 94–5.

O'BEIRNE D and BALLANTYNE A, (1987) Some effects of modified atmosphere packaging and vacuum packaging in combination with antioxidants on quality and storage life of chilled potato strips, *Int. J. Food Sci. Technol*, **22** (5) 515–23.

O'NEILL L M, GALVIN K, MORRISSEY P A and BUCKLEY D J, (1998) Comparison of the effects of dietary olive oil, tallow and vitamin E on the quality of broiler meat and meat products, *British Poultry Science*, **39** (3) 365–71.

PEARSON A M and GRAY J T, (1983) Mechanisms responsible for warmed-over flavor in cooked meat. In: Waller, G R. and Feather, M S. (eds) *The Maillard Reactions in Foods and Nutrition*, American Chemical Society, Washington DC, p. 287.

PEARSON A M, GRAY J T, WOLDZAK A M and HORENSTEIN N A, (1983) Safety implications of oxidized lipids in muscle foods, *Food Technol.*, **37** (7) 121–9.

PEPLER H J and REED G, (1987) Enzymes in food and feed processing. In: Rehm, H J. and Reed, G. (eds) *Biotechnology*, VCH, Weinheim.

PIKUL J and KUMMEROW F A, (1991) Thiobarbituric acid reactive substance formation as affected by distribution of polyenoic fatty acids in individual phospholipids, *J. Agric. Food Chem.*, **39** 451–7.

PRATT D E and WATTS B M, (1964) Antioxidant activity of vegetable extracts. I. Flavone aglycones, *J. Food Sci.*, **29** 27–33.

PRZYBYLA WILKES A, (1991) Hooking onto seafood, *Food Product Design*, August, 51–5.

PUMPHREY R S H, WILSON P B, FARAGHER E B and EDWARDS S R, (1999) Specific immunoglobulin E to peanut, Hazel nut and Brazil nut in 731 patients: similar patterns found at all ages, *Clinical and Experimental Allergy*, **29** (9) 1256–9.

QUALI A and TALMANT A, (1990) Calpains and calpastatin distribution in bovine, porcine and ovine skeletal muscles, *Meat Sci.*, **28** (4) 331–48.

RODHOUSE J C, HAUGH C A, ROBERTS D and GILBERT R J, (1990) Red kidney bean poisoning in the UK: an analysis of 50 suspected incidents between 1976 and 1989, *Epidemiol. Infect.*, **105**, 485–91.

ROOZEN J P, (1987) Effects of types I, II, and III antioxidants on phospholipid oxidation in a meat model for warmed over flavour, *Food Chem.*, **24** 167–85.

ST ANGELO, A L, VERCELLOTI J R, LEGENDRE M G, VINNETT C H, KUAN J W, JAMES JR C and DUPUY H P, (1987) Chemical and instrumental analyses of warmed-over flavour in beef, *J. Food Sci.*, **52** (5) 1163–8.

SANDERS T, (1987) Toxicological considerations in oxidative rancidity of animal fats, *Food Sci. Technol. Today*, **1** 162–4.

SATO K and HEGARTY G R, (1971) Warmed-over flavor in cooked meats, *J. Food Sci.*, **36** (7) 1098–102.

SCHIFFMANN-NADEL M, CHALUTZ E, WAKS J and DAGAN M, (1975) Reduction of chilling injury in grapefruit by thiabendazole and benomyl during long-term storage, *J. Am. Soc. Hort. Sci.*, **100** 270–2.

SCHRICKER B R and MILLER D D, (1983) Effects of cooking and chemical treatment on heme and nonheme iron in meat, *J. Food Sci.*, **48** (4) 1340–3, 1349.

SCOGING A C, (1991) Illness associated with seafood, *Communicable Disease Report*, **1** (11) R117–22.

SCOGING A C, (1998) Scombrotoxic (histamine) fish poisoning in the United Kingdom: 1987 to 1996, *Communicable Disease and Public Health*, **1**(3) 204–5.

SHAHIDI F, RUBIN L J, DIOSADY L L, KASSUM N, FONG J C and WOOD D F, (1986) Effect of sequestering agents on lipid oxidation in cooked meats, *Food Chem.*, **21** (2) 145–52.

SHARP A K, (1986) Humidity: measurement and control during the storage and transport of fruits and vegetables, *CSIRO Food Res. Quart.*, **46** (4) 79–85.

SHAW R, (1997) Cook-chill ready meals: opportunities for MAP., *Food Review*, **24** (4) 23, 25, 27, 29, 31.

SMITH D M, SALIH A M and MORGAN R G, (1987) Heat treatment effects on warmed-over flavor in chicken breast meat. *J. Food Sci.*, **52** (4) 842–5.

SPALDING D H and REEDER, W F, (1975) Low-oxygen high carbon dioxide controlled atmosphere storage for control of anthracnose and chilling injury of avocados, *Phytopath.*, **65**(4) 458–60.

STOICK S M, GRAY J I, BOOREN A M and BUCKLEY D J, (1991) Oxidative stability of restructured beef steaks processed with oleoresin rosemary, tertiary butylhydroquinone, and sodium tripolyphosphate, *J. Food Sci.*, **56** (3) 596–600.

SWAINE M V L and JAMES S J, (1986) Evaporative weight loss from unwrapped meat and food products in chilled display cabinet. In: *Meat Chilling '86. International Institute of Refrigeration Commission C2, Bristol.*

SWAIN M V L, GIGIEL A J and JAMES S J, (1986) Carbon dioxide chilling of hot boned meat. In: *Meat Chilling '86. International Institute of Refrigeration Commission C2, Bristol* pp. 195–202.

TAMIME A Y and DEETH H C, (1980) Yoghurt: technology and biochemistry, *J. Food Protect.*, **43** (12) 939–77.

TAYLOR A A, (1972) Influence of carcass chilling rate on drip in meat. In: *Proceedings of the Meat Research Institute Symposium No. 2, Meat Chilling Why and How?* Bristol, pp. 5.1–5.8.

TIMS M J and WATTS B M, (1958) Protection of cooked meats with phosphates, *Food Technol.*, **12** 240–3.

TUNALEY A and BROCKLEHURST T F, (1982) A study on the shelf life of coleslaw, *Chilled Foods*, **1** (6) 12–13.

TUNALEY A, BROWNSEY G and BROCKLEHURST T F, (1985) Changes in mayonnaise-based salads during storage, *Lebensm. Wiss. u. Technol.*, **18** (4) 220–4.

VÁMOS-VIGYÁZÓ L, (1981) Polyphenol-oxidase and peroxidase in fruits and vegetables, *CRC Crit. Rev. Food Sci. Nutri.*, **15** (1) 49–127.

WADE N L, (1979) Physiology of cool-storage disorders of fruit and vegetables. In: Lyons, J. M., Graham, D. and Raison, J K. (eds) *Low Temperature Stress in Crop Plants*, Academic Press, New York, pp. 81–96.

WANG C Y, (1982) Physiological and biochemical responses of plants to chilling stress, *Hort Sci*, **17** (2) 173–86.

WANG C Y and ADAMS D O, (1980) Ethylene production by chilled cucumbers (*Cucumis sativus* L), *Plant Physiol.*, **66** (5) 841–3.

WANG C Y and BAKER J E, (1979) Effects of two free radical scavengers and intermittent warming on chill injury and polar lipid composition of cucumber and sweet pepper fruits, *Plant Cell Physiol.*, **20**, 243–51.

WATTS B M, (1961) The role of lipid oxidation in lean tissues in flavor deterioration of meat and fish. In: *Proceedings Flavor Chemistry Symposium, Campbell Soup. Co., Camden, NJ* p. 83.

WEST P A, WOOD P C and JACOB M, (1985) Control of food poisoning risks associated with shellfish, *Journal of the Royal Society of Health*, **1** 15–21.

WHITAKER B D, (1991) Changes in lipids of tomato fruit stored at chilling and non-chilling temperatures, *Phytochemistry*, **30** (3) 757–61.

Part IV

Safety and quality issues

10

Shelf-life determination and challenge testing

G. Betts and L. Everis, Campden and Chorleywood Food Research Association

10.1 Introduction

All food products are susceptible to deterioration in their quality during storage. Chilled foods in particular are highly perishable and the time during which the quality is maintained at a consumer acceptable standard can be termed the shelf-life. The definition of shelf-life has been given by several authors as the time between the production and packaging of the product and the point it becomes unacceptable under defined environmental conditions (Ellis 1994) or the time at which it is considered unsuitable for consumption (Singh 1994). The end of a product's shelf-life will be due to deleterious changes to quality caused by biological, chemical, biochemical and physiochemical means, or by food safety concerns due to the growth of food pathogens which may not necessarily cause any changes in product quality.

There are few reference books available which give lists of shelf-lives for chilled foods as the shelf-life of each specific product is unique and based on the particular recipe, raw ingredients and manufacturing and storage conditions used. If there are any changes to these, then the shelf-life will be liable to change (see Section 10.2). Whilst there is some guidance available in the literature for chilled foods (Ellis 1994) and MAP foods (Day 1992) the shelf-life of products should be defined scientifically during product development following the procedures outlined in this chapter.

The rationale for arriving at a particular shelf-life will undoubtedly encompass safety, quality and commercial decisions. It is unlikely that all of these will be in agreement and the safety of the product must always assume the highest priority. There are, however, many commercial and marketing pressures to consider which will put some constraints on whether the shelf-life obtained from microbiological evaluation is acceptable from a commercial viewpoint. For example, how does the shelf-life compare with that of similar competitors' products? Does the shelf-life provide sufficient time for the sale of a significant proportion of the product within the shelf-life, thereby minimising 'end-of-shelf-life' stock disposal?

Is the shelf-life long enough to suit weekly shopping, which is the way that most chilled foods are purchased (Evans *et al* 1991). If there is a commercially viable minimum shelf-life, then this needs to be considered at the product development stage and the recipe altered accordingly.

Another constraint for chilled food manufacturers is the rapid expansion of the chilled foods market. Over the past ten years there has been considerable development in the number and types of products available (Dennis and Stringer 2000). There are approximately 7,458 new products per year, of which 3,616 are chilled (CCFRA 1999). The chilled foods market in 1997 was £5.1 billion (Anon. 1998). Such an active market requires rapid development of new product formats and ingredient combinations with short launch times. Traditionally, the safety and quality of new products would have been evaluated solely by the use of laboratory studies which are time consuming and expensive. Predictive mathematical modelling techniques are now available and are gaining increasing use in the development of new products. Their use in shelf-life determination will be discussed later in Section 10.3.

In addition to commercial pressures for extensive and rapid product development there is a consumer pressure for fresh tasting products with less salt and preservatives which require minimum preparation (Gibson and Hocking 1997). These requirements have the potential to increase the growth of food spoilage organisms and pathogens and thus decrease the likely shelf-life attainable under chilled storage conditions. Such product changes mean that new combinations of ingredients and preservative factors need to be used to maximise shelf-life. This will also be discussed throughout this chapter. Determination of the shelf-life of a product is decided by a combination of safety requirements, quality and marketing issues and customer demands. Arriving at the correct shelf-life is essential for product success. This chapter illustrates how this can be achieved.

10.2 Factors affecting shelf-life

10.2.1 Product considerations

Before a new product can be developed there are a number of fundamental considerations to be made which will affect the shelf-life likely to be achieved.

Product description

The first step to take is to decide what essential product characteristics are required; for example, is it a dairy product or a tomato-based product? Is it to be a homogeneous sauce or will it contain particulate matter? The generic product type will give an initial indication of the microorganisms likely to be of concern to the product and thus the shelf-life likely to be achieved. For example, non-acidic dairy products are more susceptible to rapid growth of microorganisms and are likely to have a shorter shelf-life than, for example, acidic products which are more inhibitory to growth.

For pasteurised chilled products, the heat treatment required may also be different based on these essential product characteristics (Gaze and Betts 1992) and the heat processing requirements can be decided at this stage.

Product packaging

The desired product packaging format needs to be considered. If the product is to be pasteurised in-pack, then provided that there is no post-process contamination, there should be fewer vegetative spoilage organisms or pathogens in the product and the attainable shelf-life should be relatively long. Bacterial spores will remain present in the product but these grow relatively slowly at chill temperatures (see Chapter 7). If the product is assembled after cooking, then there is a greater risk of contamination with microorganisms even in good hygienic conditions. The shelf-life of these products is likely to be shorter than those for in-pack pasteurised products.

The gaseous atmosphere of the packaging needs to be considered. If the product is packed under normal atmospheric conditions then aerobic spoilage organisms such as *Pseudomonas* species can grow and rapidly spoil chilled products held at < 5°C. If, however, modified atmosphere conditions are used which exclude oxygen, then spoilage will be by facultative or strict anaerobes which grow more slowly under good chill conditions (see Chapter 7).

Preservatives

Is the product to be marketed as preservative free or low salt? Removing such ingredients will allow more rapid growth of microorganisms and thus reduce the shelf-life unless additional preservative factors are included, e.g. addition of lemon juice. The effect of removing traditional preservatives should be considered at this early stage and a list of alternatives investigated.

Shelf-life constraints

It may be the case that there is a minimum shelf-life that must be achieved to make the product commercially viable, e.g. ten days to allow storage over a weekend period. This must be defined at the beginning of the shelf-life determination in order that the product formulation and packaging specifications chosen are likely to achieve the minimum desired shelf-life. Often products have a perceived maximum shelf-life by the consumer which is shorter than that which can be achieved in reality. For example, Evans (1998) reviewed the mean actual and perceived shelf-lives of a range of chilled products and found that for pâté the actual storage life was over 10 days whilst the perceived storage life was only 4 days.

Having formulated the specific product and packaging characteristics, it is likely that a target shelf-life will be derived based on past experience. This should then be confirmed by a scientific approach outlined in Section 10.4. Once the shelf-life has been determined there are a number of factors which will affect shelf-life as described below. These also need to be considered at the product development stage to ensure that they are under control during routine production of the product.

10.2.2 Raw materials

The raw materials used in the preparation of a product will influence the biochemistry and microbiology of the finished product. In order to achieve a consistent shelf-life, the quality of the raw materials needs to be standardised and the attributes most likely to affect product shelf-life should be laid down in specifications. Variations in the quality of raw ingredients can lead to variations in the final product which may affect product shelf-life. Variations in raw material can occur for a number of reasons: natural variation, variety change, a change of supplier, seasonal availability or pre-processing applied to raw materials. The manufacture of coleslaw provides an example of where shelf-life can be influenced by the seasonal availability of freshly harvested cabbage which has a low yeast count, whereas cabbage from cold storage has a higher yeast count. Use of cabbage from cold stores results in coleslaw with a markedly shorter shelf-life owing to the higher starting levels of yeast introduced via the raw ingredients.

If an ingredient for a raw product does not meet an agreed specification, e.g. for levels of microorganisms, it is still possible to use the ingredient for a different purpose, e.g. to be added to a product before cooking, provided there is no compromise to food safety. The likely consequences of using higher levels of organisms can be evaluated using predictive models (Section 10.3). Tolerance limits for those ingredients that exert a key preservative effect in the final product, such as the percentage of salt, need to be established during the development of the product or in challenge testing (see Section 10.6) and be stated in the product process and formulation specifications. Any variability in the levels of these ingredients due to inaccuracy in weighing ingredients during routine production will affect the shelf-life achieved. Ingredients which are crucial to product safety or stability during the assigned shelf-life should be identified using product hazard analysis (Leaper 1997) and the levels of these ingredients must be controlled during routine production. For example, for chilled MAP foods a salt level of 3.5% in the aqueous phase can be a key controlling factor for these foods and the salt level must be monitored for each batch of product manufactured (Betts 1996).

Product formulation can be used to overcome natural variability of critical factors in raw materials and thereby reduce the variation of the final product. The pH is one of the most important factors affecting the degree of heat processing required to achieve sterilisation. In tomatoes there is a variability of acidity between cultivars. Product formulation can be used to overcome this variability either by blending high and low acidity cultivars or by the addition of permitted organic acids. Again, it is crucial that where pH is used as a key preservative factor, the pH of each batch of product should be monitored to ensure that the target level is achieved.

10.2.3 Assembly of product

In complex and multi-component products, contact between components may result in migration of flavourings, colourings, moisture or oil from one component to another. This may limit shelf-life due to quality changes, e.g. in multi-layered trifles where the visual appearance is impaired by the migration of colouring components from one layer to another, and in fruit pies where migration of moisture from the filling to the pastry crust leads to a loss of texture. Alternatively, migration brings together substrates for chemical reactions, and it is the products of these reactions which influence product shelf-life. Pizza toppings containing unblanched green peppers may be susceptible to rancid off-flavours as a result of the enzyme lipoxygenase in the green peppers coming into contact with fatty acid substrates in the pizza base or in the cheese.

The way in which multi-component products are assembled can have major effects on the microbiological safety of the product. If components which are stable due to low A_w or low pH are placed in contact with components which are inherently unstable due to a high A_w or high pH, there will be a layer formed between the two components where the A_w and pH is now suitable for microbial growth. Any microorganisms in the stable component could grow in this layer. Migration or contact between components can be limited to extend shelf-life and needs to be considered during product development. Where migration and contact are detrimental to product quality, the use of edible films or packaging in separate compartments could be considered.

10.2.4 Processing

Processing encompasses a wide range of treatments that may be given to food from simply chopping or washing, (e.g. ready-to-eat chilled salads), to heat processing, acidification, addition of preservatives, fermentation or salting. Processing exerts a considerable effect on the microflora, chemical, biochemical and sensory properties of a food. In some cases, the purpose of food processing may be primarily intended to achieve the desired characteristics of the product, e.g. fermented salads, and can result in changes that extend the shelf-life by reducing pH. In other cases, processing may be specifically selected to influence shelf-life limiting factors, e.g. heat processing to inactivate microorganisms and deleterious food enzymes. If a process is used to achieve product shelf-life, there should be an awareness that small changes in the processing conditions can have a marked effect on shelf-life. Any stages of product manufacture which are essential for the safe shelf-life of the product should be identified during a hazard analysis and suitably controlled during routine production.

10.2.5 Hygiene

Poor hygienic control during preparation, processing and packaging can result in a high level of organisms being introduced into the product; for example, poor cleaning of meat slicing equipment results in an increase in the microbiological count. This may have adverse effects on the safety and quality of the product, which would affect shelf-life. The shelf-life for a particular product can only reflect the range of variables that were included in the testing procedures. Consistent hygienic control during subsequent production runs is essential if product with the assigned shelf-life is to be produced (see Chapters 13 and 14).

10.2.6 Packaging

Packaging prevents contamination by microorganisms, protects against physical damage and can be used to isolate the product from adverse environmental factors such as light, atmospheric oxygen or humidity. Packaging materials can be used to filter out light of specific wavelengths to prevent or reduce photocatalysed reactions that would result in oxidation or nutrient degradation. Products sensitive to oxidation by atmospheric oxygen can be protected by selection of packaging materials with the appropriate barrier properties to oxygen ingress and, similarly, selection of barrier properties with respect to moisture can be used to prevent the product from drying out or retain the humidity within the pack.

Modified atmosphere packaging extends the shelf-life of meats and vegetables either by directly affecting the critical quality factor (as in the case of fresh red meat, where the bright red colour is maintained by a high oxygen atmosphere), or by influencing the rate at which reactions leading to adverse quality changes proceed. In addition, active packaging, e.g. $O_2/CO_2/H_2O$ absorbers, or ethanol emitters may extend shelf-life (see Chapter 6).

Where the atmosphere within the package is modified, i.e. is not the normal atmospheric composition, then the effects on microorganisms should be considered. Excluding oxygen from packs will prevent the normal growth of spoilage organisms of air-packed chilled foods, i.e. *Pseudomonas* spp. Therefore, foods may become unsafe due to growth of anaerobic pathogens such as *C. botulinum* (Anon. 1992), but these products may appear to be organoleptically stable (Betts 1996).

10.2.7 Storage and distribution

The conditions applied to the final product during storage and distribution can have a marked effect on shelf-life. Temperature, lighting and humidity influence which microorganisms grow, which biochemical reactions and physical changes take place and the rate they occur. To make an assessment of the shelf-life of the final product, the conditions that the product is likely to encounter and the effects that they will have on the product need to be known or determined. Limited numbers of time-temperature surveys of chilled products during storage, transportation, retailing and handling in the home have been performed, some of which have been summarised by Bøgh-Søresen and Olsson (1990).

Currently, the UK Food Safety Temperature Regulations (Food Safety (Temperature Control) Regulations 1995 SI No. 2200) allow a maximum of 8°C during distribution and retail display of chilled products. It is important that this is considered when the shelf-life of products is determined, as any assessments done at lower temperatures will be affected by use of a higher temperature (see Section 10.4).

10.2.8 Consumer handling

Consumer handling of chilled products can affect the quality and the safety of the product. Factors such as the time taken to carry the product home, consumer perceptions of chilled foods and domestic storage conditions need to be taken into consideration when setting up the time and temperature regime to be used in storage trials. This is perhaps the part of the chill chain that is most variable and over which the manufacturer has least influence and control. A survey of consumer handling of chilled foods in the UK (Evans 1998) indicated that most consumers shopped at least once a week for quantities of chilled foods. In the majority of cases, transport to the home was by car or foot (83%), taking an average of 43 minutes to get food from the retail store into the home refrigerator. The temperature of the foods generally ranged from 4–20°C. Domestic refrigerator temperatures were found to have an overall mean temperature of 6°C with a range of −1°C to +11°C. On average, only 30% of refrigerators were operating below 5°C. Of the refrigerators included in the survey, 7.3% were running at average temperatures of greater than 9°C, though positional temperature differences, particularly in fridge-freezers and larder freezers, indicate this figure to be higher if product is stored in the top of these refrigerators. Evans (1998) has also provided additional data on consumer practices.

Whilst such information gives an indication of the temperatures and times that are likely to be needed in shelf-life trials to simulate consumer handling, there is still the decision to be made with respect to what is a reasonable worst case to use. If the 'worst case' temperatures and holding times that have been recorded were used in shelf-life tests, few of the products currently in the market-place would achieve the target shelf-life. The manufacturer has to estimate where 'reasonable' abuse ends and 'unreasonable' abuse begins.

Consumer handling of products may not be as intended or envisaged by the manufacturer. Many chilled products are purchased on the basis of the 'fresh image', but then frozen at home. Opening and partial use of vacuum or modified atmosphere packaged products invalidates the shelf-life information with respect to the remaining product. A survey of consumer perceptions of the shelf-life of chilled foods has indicated that whilst consumers believe that most chilled foods should be stored for two days or less, in reality the same householders were storing certain chilled foods for considerably longer periods of time (Evans *et al.* 1991).

10.2.9 Legislative requirements

Legislative responsibilities for the chilled food manufacturer are described fully in Chapter 2. The main legislative restrictions for chilled foods in relation to shelf-life are the distribution and storage temperatures which can be used. In the UK, chilled foods can be stored at temperatures up to 8°C and this should be taken into consideration when defining the shelf-life during product development. It may be possible that a product would be able to be stored at lower temperatures throughout much of its shelf-life during retail distribution and storage, however, with the maximum temperature specification of 8°C, it is possible that on occasion a batch of the product would be stored at 8°C throughout its shelf-life and therefore it should be able to withstand this time and temperature regime whilst maintaining product quality and safety.

In addition, if the chilled product is to be exported to other EU countries, there will be different chilled temperature restrictions. These will also need to be considered during product development. There are many different requirements for chilled products throughout the EU and a working document, first draft of a proposal for a European parliament and council regulation on the hygiene of foodstuffs, is being circulated which would harmonise temperature regulations throughout the EU (Anon. 1997).

With respect to MAP chilled foods, there are guidelines which restrict the shelf-life of these products to ten days or less at chill temperatures of ≤ 8°C, unless specific controlling factors are in place to minimise the potential for growth of psychrotrophic *C. botulinum* (Anon. 1992, Betts 1996).

Any deviations from these guidelines should be made only after scientific evidence that the alternative preservation systems in the products will prevent the growth of, or toxin production by, *C. botulinum*.

10.2.10 Effects of intrinsic/extrinsic factors

The factors discussed above, namely the type and source of ingredients and the subsequent processing and packaging, will influence the types and levels of microorganisms that will be present, and the chemical and biochemical reactions that can occur, in the final product. The ability of organisms to grow or cause problems, or for chemical reactions to proceed in the final product, will be dependent on the properties of the final product, i.e. pH, A_w (known as intrinsic factors), and on the external factors that the final product encounters, such as temperature (known as extrinsic factors). Intrinsic factors include:

- water activity (A_w) (available water)
- pH/total acidity
- type of acid
- preservatives, including salt and spices
- nutrients
- natural microflora
- redox potential (Eh)

- available oxygen
- natural biochemistry (enzymes, chemical reactants).

Extrinsic factors include:

- heat treatment (processing, cooking or reheating of the food prior to consumption)
- headspace gas composition
- temperature throughout storage and distribution
- relative humidity (Rh)
- light (UV and IR).

Table 10.1 gives examples of minimum growth conditions for pathogenic and spoilage organisms of concern to chilled foods. It must be stressed that such

Table 10.1 Minimum growth conditions for microorganisms which may be associated with chilled food

Type of microorganism	Minimum pH for growth	Minimum A_w for growth	Anaerobic growth[a]	Minimum growth temp[b] (°C)
Pathogens [(c)]				
Salmonella	4.0	0.94	Yes	7
Staphylococcus aureus	4.0 (4.5 for toxin)	0.83 (0.90 for toxin)	Yes	6 (10 for toxin)
Bacillus cereus (psychrotrophic)	4.4	0.91	Yes	< 4
Clostridium botulinum proteolytic A, B, F	4.6	0.93	Yes	10
Non-proteolytic B, E, F	5.0	0.97	Yes	3.3
Listeria monocytogenes	4.3	0.92	Yes	0
Escherichia coli	4.4	0.95	Yes	7.0
Vibrio parahaemolyticus	4.8	0.94	Yes	5
Yersinia enterocolitica	4.2	0.96	Yes	–2
E. coli O157	4.5	0.95	Yes	–6.5
Spoilage organisms [(d)]				
Pseudomonas	5.5	0.97	No	< 0
Enterobacter aerogenes	4.4	0.94	Yes	2
Lactic acid bacteria	3.8	0.94	Yes	4
Micrococci	5.6	0.9	No	4
Yeasts	1–5	0.8	Yes	–5
Moulds	< 2.0	0.6	No	< 0

Notes: The table lists various species and indicates approximate growth and survival limits with the various factors acting alone. Interactions between factors are likely to considerably alter these values.

[a] For example, in vacuum pack.
[b] Minimum growth temperatures are for growth in typical neutral pH, high water activity, chilled foods.
[c] Data for pathogens taken from Anon. 1997.
[d] Data for spoilage organisms taken from Brown 1991.

values are typical values only and are dependent on the particular strain of organisms used in the study and the storage conditions used. Such data should only be used as an indication of the microorganisms likely to grow in a product and should be supported by thorough scientific evaluation.

10.3 Modelling shelf-life

As discussed previously, one of the most important factors in determination of product shelf-life is the potential for the growth of microorganisms. There is therefore a requirement to test new chilled products to assess the potential for growth of pathogens or spoilage organisms and thus define the period of time during which the product is considered acceptable for its intended use. Traditionally, the only objective approach to this has been extensive laboratory studies of pilot scale production of the product (see Section 10.4). This approach is still the best way to do final testing of the product, but it is very time consuming and costly during early product development stages. With the vast number of new products reaching the shelves every week, an alternative approach is required to aid shelf-life determination. A technique gaining increasing application in chilled food production is predictive models which can be powerful tools at all stages of product development and manufacture.

Chapter 7 describes the types of models available for microorganisms; their use in shelf-life determination is elaborated upon here. Three examples of the use of models are given below.

1. *During product conception/development.* To give an initial indication of major problems, e.g. to take an extreme situation, if a neutral pH product with a 20-day shelf-life is envisaged but pathogen models show that *L. monocytogenes*, *B. cereus* and *Salmonella* could all grow within two days then it will be apparent at the earliest point that the product will not be feasible.
2. *During Product Formulation.* For example, having decided on a product type which is likely to be safe and stable for the desired shelf-life, then models can help in final product recipe formulation. There may be three different combinations of salt and pH which will give the same shelf-life at a given temperature, e.g:

pH	salt % (w/v)
6.4	0.8
6.2	0.7
6.0	0.4

However, one of these may have a better taste in terms of consumer preference. The effect of these different recipes on microorganisms can be assessed without the requirement for laboratory trials at this stage.

3. In Setting Specifications. There are very few microbiological specifications for the chilled food manufacturer; however, two very useful publications on final product specifications are PHLS (1996) and IFST (1999) which can help in deciding microbiological specifications. These two documents take different approaches. For example, the PHLS document gives levels of different microorganisms which are acceptable at the end of the shelf-life, i.e. at point of consumption. If a chilled foods manufacturer wanted to use these specifications to ensure end-of-life microbiological levels were acceptable, then predictive models can be used to give an indication of the levels of microorganisms which could be present at the beginning of life for any product formulation to achieve the end-of-life specifications. If, for example, the recommended level of TVC or Enterobacteriaceae at end-of-life was 10^6 cfu per gram, then predictive models could be used as shown in the examples below.

Example 1

Enterobacteriaceae pH 5.5, salt 5.25 % (w/v), 15°C

Initial level per gram	Intended shelf-life	Predicted level @ 5 days	Time to 10^6 (days)
10	5 days	1.39×10^4	13.5
100	5 days	2.39×10^5	9
10^3	5 days	2.75×10^6	6
10^4	5 days	1.5×10^7	3

As can be seen from the above, in order to achieve a shelf-life of five days, with an end specification of no more than 10^6 cfu/g, the initial level of Enterobacteriaceae should be no more than 10^3 cfu/g.

The IFST document gives maximum levels of microorganisms which should be present after manufacture of product. Predictive models can be used to give an indication of the likely increase in these levels during different chilled storage regimes.

Example 2

Enterobacteriaceae pH 5.5, salt 2% (w/v)

Temperature	Initial level (manufacture specification)	Time to 10^6(days) (end specification)
3°C	100	33
5°C	100	17
8°C	100	7
10°C	100	4

As can be seen, if after manufacture the level is 100 cfu/g, the temperature of storage can have a huge effect on the likely shelf-life, e.g. at 3°C it would be 33 days compared to just seven days at 8°C.

Most predictive model systems will give information which can be used to help in defining the shelf-life of a product. Data will be available as one or all of the following:

- lag time, i.e. time before an increase in numbers occurs
- generation time, i.e. time taken for each cell to double
- growth curve which will show the lag time, exponential (or fastest) growth phase and stationary phase of growth
- time taken to reach a target level of cells, or conversely
- numbers of cells present after a pre-determined time.

Currently, most microbial predictive models are based on the growth of one species or genus of organism or, in some cases, a group of related organisms, e.g. lactic acid bacteria, Enterobacteriaceae. Where more than one organism is a cause for concern, predictions need to be obtained from the relevant models and interpreted by suitably trained personnel.

Example 3
This shows how different spoilage models can be used to assess the likely shelf-life of a chilled meat product.

- **Organisms of concern**
  Lactic acid bacteria
  Enterobacteriaceae

- **End-of-life determined when:**
  Lactics @ 10^6 per gram
  Enterobacteriaceae @ 10^6 per gram

- **Product formulation**
  pH = 5.5
  A_w = 0.983

  % Salt = 3

Organism	Initial level (cfu/g)	Time to 10^6 days	Shelf-life days
Lactic acid bacteria	100	9	9
Enterobacteriaceae	100	7	7

As can be seen, the Enterobacteriaceae will be limiting to the shelf-life because the Enterobacteriaceae reach a level of 10^6 cfu/g within seven days whereas the lactics do not reach this level until the ninth day.

10.4 Determination of product shelf-life

In order to assess the shelf-life of a new or existing product it is usual to conduct microbiological analyses of the final product for typical spoilage organisms and pathogens during storage at specified times and temperatures. As this may require the use of contract services it can be expensive and it is therefore important that the shelf-life trials are done at the right stage during product development.

A useful approach to shelf-life determination has been described by Brown (1991), where a three-staged approach is taken, i.e.

Phase I	Pilot Scale
Phase II	Pre-production
Phase III	Full Production.

10.4.1 Shelf-life considerations at pilot scale

At this stage, it is likely that the product is a new idea or marketing concept and will need to be produced on a small scale in the development kitchen in order to assess the likelihood of product success in terms of flavour, colour, texture and eating quality. Many of the details about the product are unlikely to be known, so only a preliminary consideration of product shelf-life is possible (Fig. 10.1).

Initially, the product can be described on a theoretical basis by listing the likely ingredients, processing and packaging properties, and storage requirements. This listing can be used to evaluate the critical properties of the product with respect to shelf-life and to highlight likely changes in quality, possibilities of microbiological spoilage and any potential food safety problems. Ingredients identified as likely to contain food-poisoning organisms and spoilage bacteria can be monitored at all stages in the preparation of the product. Changes in the numbers of organisms due to each particular operation can then be noted and used when assessing risks associated with the product, and should suggest, or help in assessing, the most important changes to monitor during storage trials. For example, cooking may reduce the numbers of organisms, whereas a holding period or the introduction of a particular ingredient may increase the numbers. It is important even at the pilot plant scale to consider the likely hazards associated with the product using an approach such as Hazard Analysis of Critical Control Points (e.g. Leaper 1997).

Using the data collected in the paper exercise on the ingredients, processing and packaging, an assessment should be made of the potential for contamination with food-poisoning organisms and their growth during storage. If there is a possibility of the product containing infectious pathogens (e.g. *Listeria* or *Salmonella*) or toxin producers such as *Staphylococcus aureus*, *Bacillus cereus* or the Clostridia, then their potential for harm will need to be evaluated in later shelf-life trials. Initial indications of the likely growth of spoilage organisms or pathogens can be gained from predictive models (Section 10.3).

Shelf-life assessments made on product produced at the kitchen scale are limited to some extent, as it is inevitable that, during the development of a chilled product, the product and processes will change several times. Moreover, the conditions are unlikely to equate to those of a 'real' production situation. The

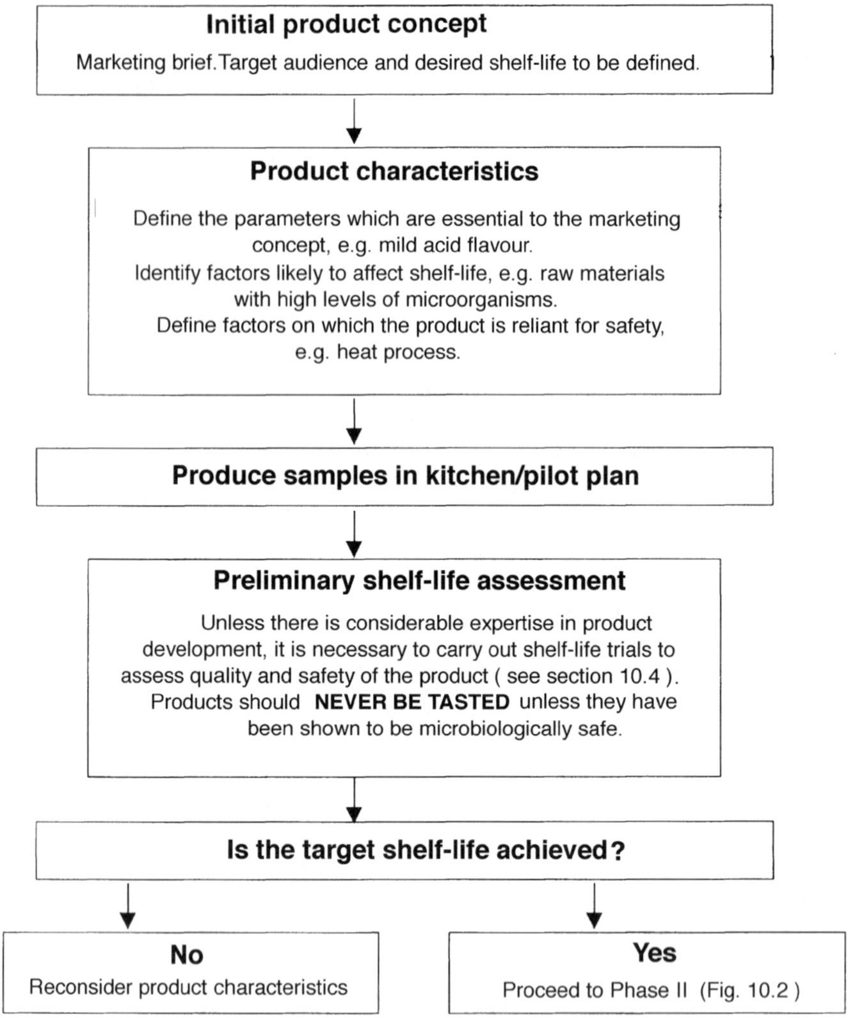
Initial product concept
Marketing brief. Target audience and desired shelf-life to be defined.
Product characteristics
Define the parameters which are essential to the marketing concept, e.g. mild acid flavour.
Identify factors likely to affect shelf-life, e.g. raw materials with high levels of microorganisms.
Define factors on which the product is reliant for safety, e.g. heat process.
Produce samples in kitchen/pilot plan
Preliminary shelf-life assessment
Unless there is considerable expertise in product development, it is necessary to carry out shelf-life trials to assess quality and safety of the product (see section 10.4).
Products should NEVER BE TASTED unless they have been shown to be microbiologically safe.
Is the target shelf-life achieved?
No
Reconsider product characteristics
Yes
Proceed to Phase II (Fig. 10.2)

Fig. 10.1. Pilot-scale shelf-life determination (adapted from Brown 1991).

quality of the raw materials and standards of handling and hygiene will differ from full scale production and the size of the 'batch' will be considerably smaller, resulting in less variability in the final product. Despite these limitations, data indicative of product shelf-life can be gained from observations of products made on the kitchen scale, realistically portioned, packed and stored at 8°C (unless knowledge to the contrary of the storage conditions for the product is available) for at least their target shelf-life, or a shorter period if the changes become acceptable.

The material produced on a kitchen or pilot scale should not be subject to extensive sensory or microbiological examination, but each of the changes contributing to quality should be identified and described. However, before the product can be tasted it must be shown to be microbiologically safe. The overall objective is to identify and describe product acceptability and the extent of sensory change during storage. The type of observation needed is that which will be useful to describe the characteristics of the product that lead to the product becoming unacceptable to the consumer. In many instances there is likely to be a particular quality change, a 'critical quality parameter', that limits shelf-life. By combining the background data with that collected from actual storage trials of the product, it should be possible to gain an overall impression of whether the product is likely to meet its target shelf-life.

At this stage of testing, the assessors will normally have been closely involved with development of the product. If the product does not achieve the quality desired, they should be able to decide whether changes in processing, packaging or formulation can be used to achieve the desired quality, and whether they require re-testing on the kitchen scale or whether such changes could be adopted for the pre-production run. If the quality changes are unacceptable then the product or processing concept will need to be reconsidered with a view to overcoming these problems. Again, the use of predictive models may help to decide whether changes in product composition are likely to affect microbial shelf-life.

10.4.2 Shelf-life considerations at pre-production scale (Fig. 10.2).

The objectives of pre-production are to scale up production, to provide and confirm product, process and formulation specifications, and to ensure that the product produced is viable (meets the marketing brief, including the target shelf-life). Pre-production runs are undertaken on either pilot plant equipment or, preferably, by the production of batches on full scale equipment. When the development of the process is complete, details should be documented in accordance with the requirements of any quality system in place, e.g. HACCP.

The product specification will need to address raw materials, product formulation, assembly of the product, processing, hygiene and cleaning, and packaging. Deviation from these specifications may have an impact on product shelf-life, therefore shelf-life is only applicable to product produced under these defined conditions. Any modifications to any aspect of production should be recorded. Shelf-life testing should be most extensive at the pre-production stage. To set up shelf-life tests, the storage conditions, a sampling protocol, and the analyses to be performed need to be defined.

Storage conditions

Storage conditions should be chosen carefully to match or represent real conditions to which the product is likely to be exposed during its life. The key storage condition for chilled food is the time–temperature regime. Both time and temperature must be specified, covering the temperature range encountered and the times spent in distribution, retail cabinets and in consumer use, including

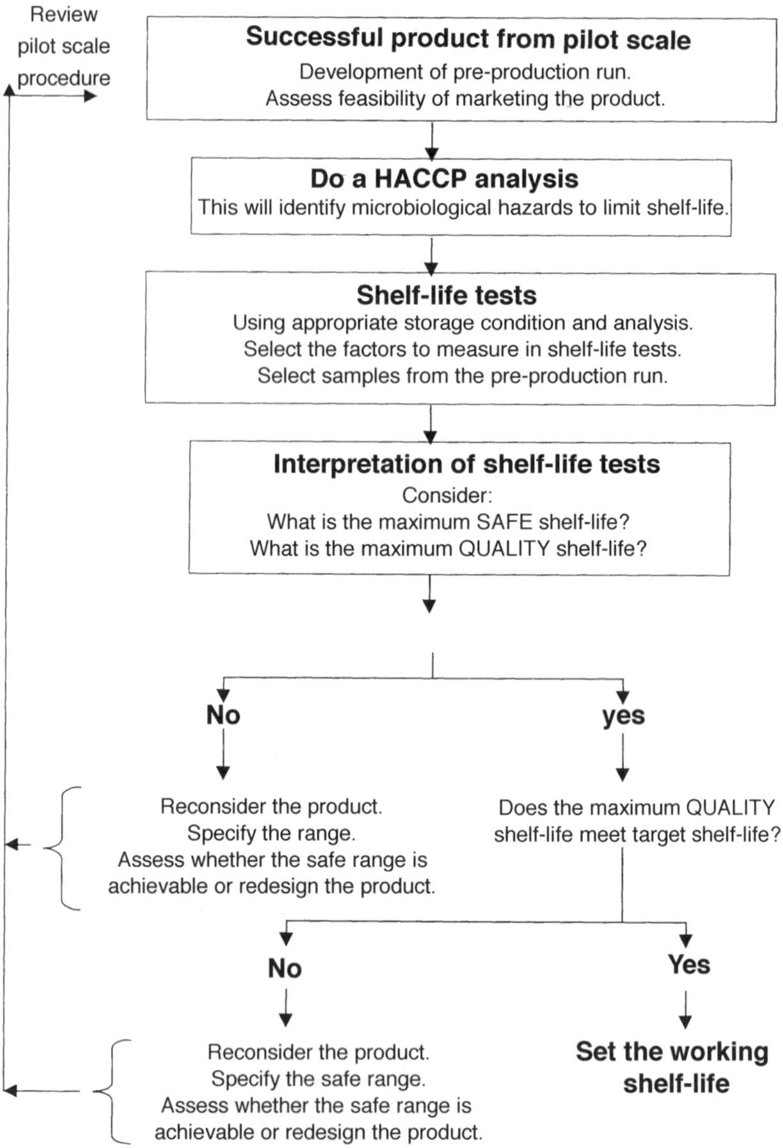
Review
pilot scale
procedure
Successful product from pilot scale
Development of pre-production run.
Assess feasibility of marketing the product.
Do a HACCP analysis
This will identify microbiological hazards to limit shelf-life.
Shelf-life tests
Using appropriate storage condition and analysis.
Select the factors to measure in shelf-life tests.
Select samples from the pre-production run.
Interpretation of shelf-life tests
Consider:
What is the maximum SAFE shelf-life?
What is the maximum QUALITY shelf-life?
No
yes
Reconsider the product.
Specify the range.
Assess whether the safe range is
achievable or redesign the product.
Does the maximum QUALITY
shelf-life meet target shelf-life?
No
Yes
Reconsider the product.
Specify the safe range.
Assess whether the safe range is
achievable or redesign the product.
Set the working
shelf-life

Fig. 10.2 Pre-production runs (adapted from Brown 1991).

transport from store to home and storage in domestic refrigerators. The sequence in which holding times and temperatures are applied requires careful consideration. A product purchased 'early' in its shelf-life may spend a larger proportion of the total shelf-life in a domestic refrigerator at temperatures which are generally higher than legislation permits in retail outlets (Rose *et al.* 1990). Similarly, the impact of conditions encountered during transportation from store to home may differ depending upon where it occurs in the shelf-life.

It is therefore difficult to determine every combination of times and temperatures likely to be experienced by a product and a decision must be taken on what combinations will be representative of normal conditions. It has been suggested, on the basis of experience within the food industry, (Brown 1991), that holding the product at not less than 8°C with a hold at 22°C for four hours at the earliest possible sale time, provides a reasonable default from which to begin. As experience of the product and product performance increases, standard shelf-life testing conditions can then be reviewed.

It is important to remember that the shelf-life of a product will be true only for the batch of product and the temperatures tested. If the product achieves the described shelf-life when tested at 8°C, it will undoubtedly achieve a longer shelf-life at lower temperatures, e.g. 5°C. Conversely, if shelf-life trials were done at 5°C and the product was held at higher temperature, it is likely that the shelf-life will be shorter than anticipated.

A final consideration before deciding time and temperatures is whether they meet customer requirements. Many retailers of chilled product have built up specific shelf-life testing requirements on the basis of knowledge of their own distribution and storage facilities. It is advisable to ensure that any tests considered will meet these requirements before starting shelf-life studies.

Sampling times

Once the time and temperature regime has been established, sufficient product must be held under these conditions to allow several units of product to be assessed on more than one sampling occasion. The extent and frequency of sampling is very dependent on previous experience of the product or similar products, and the stability of the product. It is undesirable to use fewer than three replicates for each sampling time but preferable to use five as this will allow greater statistical analysis of the data. The number and frequency of sampling occasions is dependent on the target shelf-life of the product.

Sampling at the beginning of shelf-life (day 0), at the target shelf-life, and on at least three occasions in between is suggested. For a short shelf-life product (2–5 days) this may result in samples being taken daily, and less frequently for a longer shelf-life product. It is advisable to take additional samples beyond the target shelf-life to monitor the margin of safety and/or quality. Key parameters to monitor in shelf-life tests should have been identified as a result of the observations made at the kitchen or pilot scale stage, and through HACCP assessments. To determine the extent of change with time, sensory assessments (Chapter 12), microbiological analyses and chemical analyses (Chapters 8 and 9) will be needed.

In normal production, there will be some variability of most key parameters. It is important that the relevant extremes of these key parameters are included in shelf-life tests to ensure that the shelf-life is applicable to all the extremes of product that may be produced. To ensure such product is available it may be appropriate to make a pre-production batch that represents the extreme situation for shelf-life testing. In some cases, either the presence of a microorganism potentially critical to the shelf-life of a product cannot be guaranteed, or the level of contamination with it is likely to be subject to considerable variability. Shelf-life determination may then be best approached by deliberately adding the organism to the product in challenge testing studies. If such an approach is adopted, an awareness of the limitations is essential for interpretation of these tests (see Section 10.5). Shelf-life tests should ideally be repeated on more than one pre-production run to test day-to-day variation and to provide a level of confidence and experience of the product. The number of sampling occasions may be reduced based on data from the first trials.

A route for interpreting the information collected in shelf-life trials is illustrated in Figure 10.2. Safety should always be of prime importance; therefore, the first limit to establish is the maximum safe shelf-life of the product. The main reasons for chilled foods becoming unsafe are the growth of pathogens or toxins produced by microorganisms if present. Controls and monitoring procedures used in routine HACCP analysis should minimise the likelihood of the presence of pathogens.

Secondly, the shelf-life tests should enable the maximum quality shelf-life to be defined. Establishing the criteria of importance, and defining the lowest acceptable standard is a matter of manufacturer policy. Many companies use a mix of 'brand image', price, market share and customer complaint level to set and confirm the final end of shelf-life quality standards. Changes in appearance, texture and flavour will occur as a result of the chemical or biochemical reactions, changes in physical structure, and the growth of spoilage microorganisms. These may be measured in terms of microbial count, the value of a specific quality-related factor such as the peroxide value for oxidised fats, or by sensory evaluation. Changes in quality will reach a point where the product no longer achieves the standards laid out in the marketing brief. This time interval defines the quality shelf-life.

Once the shelf-life has been determined in terms of safety and quality, these can be compared with the target shelf-life. Ideally, both should exceed the target shelf-life. The shelf-life should be set as the safety or quality limit, whichever is the shorter, though it is always preferable that shelf-life is limited by quality rather than safety, as in most cases changes in quality can be discerned by smell, taste or appearance, whereas such changes cannot be relied upon to indicate safety limits. Results from challenge tests should also be reviewed in order to assess the safety of the product. As scaling up from pre-production to full scale production may cause some changes, it is suggested that the maximum shelf-life is reduced to provide an additional safety margin. If the safe shelf-life is less than the target, then either the product needs to be reconsidered or the viability of marketing the product with a shorter shelf-life needs to be assessed.

10.4.3 Shelf-life at full production scale (Fig. 10.3)

The objective of full-scale production is to produce product for sale. Shelf-life testing at this stage is to confirm the determinations made at the pre-production

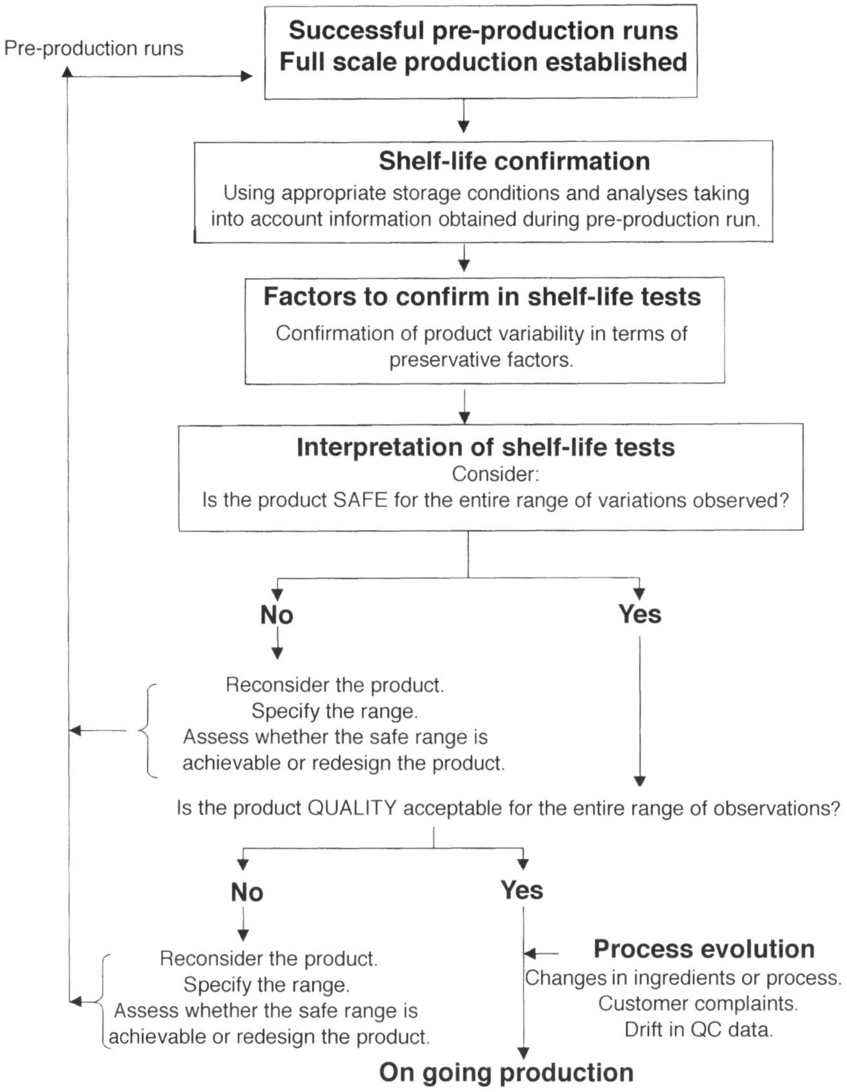
Pre-production runs
Successful pre-production runs
Full scale production established
Shelf-life confirmation
Using appropriate storage conditions and analyses taking into account information obtained during pre-production run.
Factors to confirm in shelf-life tests
Confirmation of product variability in terms of preservative factors.
Interpretation of shelf-life tests
Consider:
Is the product SAFE for the entire range of variations observed?
No
Yes
Reconsider the product.
Specify the range.
Assess whether the safe range is achievable or redesign the product.
Is the product QUALITY acceptable for the entire range of observations?
No
Yes
Reconsider the product.
Specify the range.
Assess whether the safe range is achievable or redesign the product.
Process evolution
Changes in ingredients or process.
Customer complaints.
Drift in QC data.
On going production

Fig. 10.3 Full-scale production flow chart (adapted from Brown 1991).

stage. The same key parameters should be monitored in shelf-life tests, but special care should be taken to test the full range of variation that is produced, particularly if this is greater than at the pre-production stage. If the shelf-life differs from that of pre-production runs, then it is necessary to reconsider both pre-production and full production runs to identify the cause.

Shelf-life tests should, ideally, be performed on product produced during the first three full production runs. As experience of the product increases and feedback via consumer complaints becomes available, the shelf-life can be adjusted accordingly. It is inevitable that the ingredients, the process or distribution of the product will change in time. Although each small change may, by itself, seem irrelevant to the shelf-life of the product, taken together these changes may have a marked effect. It is important that those responsible for shelf-life testing be informed of even apparently small changes, and that periodically the whole process be reviewed and confirmatory tests performed if necessary.

10.5 Maximising shelf-life

There are a number of steps that can be taken to increase the shelf-life of a product whilst maintaining product safety. These are described below.

10.5.1 Product formulation changes

Minor changes in the composition of a product may be sufficient to prevent or delay the growth of spoilage organisms and thus increase the period of time before it is unacceptable for use. This is particularly the case where the factor being changed, e.g. salt, is limiting to growth. There is currently a trend to reduce the levels of added salt in ready-to-eat products. Where this is the case, a minor change in the acidity of the product may achieve a similar effect in increasing shelf-life. The example below illustrates this point. With no salt present, this product would be limited to a shelf-life of just five days, but with 2% salt the shelf-life can be increased.

Example
Enterobacteriaceae

pH 5.50
Temperature 8°C

Salt (% $^{w}/_{v}$)	Count @ 5 days
5	1.1×10^2
4	2.89×10^2
3	5.78×10^3
2	6.8×10^5
1	3.1×10^7
0.5	8.8×10^7

10.5.2 Use of new technologies

The most effective way to inactivate pathogens, spoilage organisms and enzymes is to apply a heat process. However, this may cause various changes to the flavour of a product and may detract from the fresh taste. There is the potential for using alternative cold-processing systems for products which will inactivate microorganisms and in some cases enzymes, but will not cause heat damage to the product. Examples of such technologies include the use of high pressure and ultrasound (see Section 10.7) (Mermelstein, 1998).

10.5.3 Storage temperatures

This is one of the most effective ways for all chilled foods manufacturers to maximise the shelf-life of a product. If final products and raw materials could be stored at 2°C instead of 5°C and retail storage was done at 5°C instead of 8°C this could markedly increase the shelf-life. This is illustrated in Table 10.2 for a ready-meal product stored at various on-site and retail temperatures.

10.5.4 Use of modified atmosphere

Chapter 7 discusses different packaging formats available for chilled products. With regard to the microbiological shelf-life, the use of increased levels of carbon dioxide and exclusion of oxygen can markedly increase shelf-life (Betts 1996). It must be stressed, however, that the use of MAP for chilled products may restrict the shelf-life to ten days unless it can be shown to be able to prevent the growth of *C. botulinum* (see Section 10.2).

10.6 Challenge testing

There is often some confusion surrounding the differences between shelf-life determination and challenge testing. During the development of new chilled products there are two different aspects which need to be considered:

Table 10.2 Levels of *Pseudomonas* (Ps) present after each storage period from an initial level of 10 cfu/g

On-site storage for 2 days		Retail display for 6 days		Likely
Temp	Level of Ps after 2 days (cfu/g)	Temp	Final level of Ps (cfu/g)	shelf-life (days)
2°C	16	3°C	7.8×10^5	9.0
		8°C	7.3×10^8	5.5
3°C	21	3°C	1.0×10^6	9.0
		8°C	7.5×10^8	5.5
5°C	54	3°C	2.3×10^6	8.5
		8°C	8.7×10^8	5.0

Notes: Product characteristics: Chilled ready-meal; Salt – 0.8% w/v; pH – 6.2; Target life – 8 days; End of life – *Pseudomonas* @ 10^7/g (starting level 10/g)

1. Is the product safe and stable during normal production and storage conditions and for how long, i.e. what is its shelf-life?
2. Is the product likely to be safe and stable during its shelf-life if it became contaminated with food pathogens and spoilage organisms, i.e. are the product formulation and storage conditions inherently safe with respect to the inoculated organisms?

In the first case stated above, only the microorganisms that are naturally present in the batch of product will be present and able to grow during storage. Ideally, under good manufacturing conditions and using a HACCP approach to product manufacture, there will be minimal chance of food pathogens, e.g. *Salmonella*, being present in the product. Therefore, during shelf-life determination there may be an absence of food pathogens and the product may be considered to be safe during the assigned shelf-life. However, it is possible that on occasions during routine production, the product may be contaminated with food pathogens or different spoilage organisms not present in the batch used for shelf-life determination. If this occurred, then the manufacturer would need to know if the product was likely to remain safe and stable during storage. This 'what if' scenario is the rationale behind the use of challenge testing trials. With the requirement under the Food Safety (UK) Regulations to show 'due diligence' with respect to the safety of foods, many food manufacturers are doing inoculated challenge tests on new and existing products.

Microbiological challenge testing is the laboratory simulation of what can happen to a product during manufacture, distribution and subsequent storage. This involves inoculation of the product with relevant microorganisms and holding of the product under a range of controlled environmental conditions to assess the risk of food poisoning or to establish product stability in the case of food spoilage organisms. Notermans *et al.* (1993) describes four stages to challenge testing as (1.) an appropriate experimental design; (2.) an inoculation procedure; (3.) a test procedure; and (4.) interpretation of results. Important aspects of these stages are described further below.

1. Design of challenge test

There are three main ways in which a challenge test can be applied to new or existing chilled products:

1. to determine the safety of the product
2. to determine the potential for spoilage of the product
3. to assess the stability of new recipes or reformulated products (Anon. 1987).

Safety is the prime consideration when producing food for commercial sale and it must be ensured that the product represents a minimum hazard to the consumer.

Microbiological challenge testing should be done if the microorganism of concern is suspected to be present in low numbers (Notermans and in't Veld

1994) or has the potential to be contained in raw material or introduced into the product at some stage during production and distribution.

With respect to assessing the safety of a chilled product with regard to food pathogens, the organisms likely to be a hazard for the product and necessitate challenge testing, should be identified during routine hazard analysis, e.g. HACCP. Use of this approach to identify organisms of concern is described by Notermans and in't Veld, (1994).

Having identified which organisms are to be tested, it is necessary to consider what particular aspects of the product and storage conditions are to be challenged. There are a variety of factors involved in the overall preservation of a food; these are the same intrinsic factors or extrinsic factors which are considered when determining shelf-life. In challenge testing, variation of each of these factors in turn enables those factors effective in preserving a food to be determined. Challenge tests can therefore be done to evaluate different formulations of the food. They can also be done to assess the effect of storing the product under a variety of controlled storage conditions. These should include conditions to simulate abuses which may be encountered during distribution and consumer handling after purchase. Processing and packaging conditions may be key elements in the shelf-life of chilled foods, and must not be ignored in challenge tests.

Another important aspect to consider during design of a challenge test is the number of samples to be evaluated at each sampling point to allow sufficient data to be obtained for statistical analysis. In addition, the number of sampling times needs to be defined. As a starting point, the regime used for shelf-life determination can be used, i.e. there should be a minimum of five sampling times, one at the beginning, one at the end and three spaced throughout the total time period. On each occasion, there should be a minimum of three and ideally five or even ten samples analysed in order to gain confidence in the reproducibility of the results. It is not unusual for levels of microorganisms to stay constant, and even decrease over a period of hours, days or weeks before beginning to increase (Curiale 1998). Sufficient sampling times should be done throughout the trial to ensure that any initial decrease in microbial levels is not taken to be the end point of the trial and so subsequent growth following this is missed.

Each experiment needs to be adequately controlled to ensure that the results obtained are meaningful. These should include positive controls in which the organisms under test are known to grow, (i.e. in laboratory media), as well as negative controls in which growth is not expected to occur. Suitable controls may be a standard product of known shelf-life, uninoculated samples, products stored under standard extrinsic conditions. Controls provide a measure against which change may be judged but also serve as indicators of reproducibility between experiments.

Challenge testing is neither a quick nor a simple matter. It is usually used if other methods of ensuring safety or stability need to be supplemented or are thought inadequate in the circumstances of the particular foodstuff. At the outset, it is important to define clearly the reasons for undertaking the challenge test and the aims of the experiments. It must then be decided whether a challenge test will meet those objectives. Once it has been determined that a challenge test is to be undertaken, it is necessary to determine the most effective way of satisfying the objectives. This involves consideration of the type of product to be tested, the factors believed to contribute to its stability and the conditions to which the product might reasonably be expected to be subjected to during its normal life. Experiments should then be carefully designed to elucidate the preservative mechanisms operating in the product and the tolerances of these factors without causing failure of stability or safety.

2. Inoculation procedures

The choice of organisms for use in challenge tests is very important as they must provide a realistic challenge to the product. If the organisms chosen were particularly resistant to a preservative used in the product, then they may grow whereas other resistant strains would not. Conversely, if the organisms used are very sensitive to the antimicrobials used then they would fail to grow and the product would appear to be safe or stable when in fact it may have failed the challenge test if more realistic organisms were chosen.

The cultures chosen should ideally have been isolated from a food source similar to the product under test, or conditioned by growing them in a product sample, or laboratory media with similar characteristics (Notermans and in't Veld, 1994). Cultures from recognised culture collections, e.g. National Collection of Industrial and Marine Bacteria, are preferable as they allow full traceability for comparison between different challenge tests but should be checked to ensure that they behave in a similar manner to freshly isolated strains. It is preferable to use a cocktail of two or more strains of each microorganism in order to provide a greater challenge to the product.

Other important considerations are the size, e.g. 0.1ml and method of inoculation used. In terms of inoculum size, a minimal amount should be used so that the characteristics of the product, such as A_w, are not affected. Anon. (1987) details inoculation procedures for liquid, dry and intermediate moisture products. The number of organisms per millilitre must be realistic. The levels must be high enough to be easily detected, e.g. minimum level of 100 cells per gram of product, however, they should not be so high that they easily overcome the preservation capability of the product. Conducting a challenge test with an inoculum size of 10^6 cells per gram is unrealistic and is likely to lead to product failure.

3. Interpretation of results

Before the experiment begins it is important to define the criteria of acceptability of the product. The criteria used will vary depending on whether the system under consideration is intrinsically stable or unstable, and whether safety or spoilage is being assessed. For intrinsically stable systems the end-point of a challenge test will be defined by microbial growth or undesirable organoleptic changes after a given period of time. For intrinsically unstable systems there may be a variety of end-points, including growth to a specified number of organisms per unit weight of food, change of lag phase or generation time, or organoleptic spoilage. In the case of safety testing, the end-point may be the start of growth pathogens, the growth of pathogens to a specified number, the number of organisms per unit weight of food or toxin production.

If the specified end-point is reached earlier than anticipated then the product should be reformulated or the processing and storage conditions adjusted accordingly.

If the product survives the challenge test, i.e. the test criteria are not reached during the desired shelf-life, then the product can be considered suitable in relation to the specific conditions tested.

If there are any changes, however small, to the product formulation, processing, storage, distribution or retail conditions then the results of the challenge test can no longer be considered to be reliable. A worked example of how challenge testing procedures can be applied to a chilled pasta product is discussed by Anon. (1987).

10.7 Future trends

This chapter has outlined current thinking with respect to shelf-life determination with particular emphasis on microbial safety and quality. Although this process may be considered an extensive approach, as confidence in setting shelf-life increases, this will enable new approaches to be assessed and utilised.

One of the major areas to assist in shelf-life determination is predictive microbiological modelling. In particular, the development of food sector specific models will increase in the future. Models have been developed for fish products available from the Ministry of Fisheries, Technical University at Denmark (Seafood Spoilage Predictor on the internet – http://www.dfu.min.dk/micro/ssp/help/usingssp.htm) and are currently in progress for cured meat products. Such models will enable an evaluation to be made of the interactive effects of a mixed microbial flora typically found in these products on food spoilage. There will be an increasing use of alternative technologies to improve the quality and/or shelf-life of chilled foods.

Currently, high pressure is already used for chilled pâtés, orange juice and ethnic dishes such as salsa (Mermelstein 1998). As such technology becomes more widely used, the equipment costs will be reduced and the usage may increase. The chilled food market seems set to increase sales over the next few years and accurate, effective shelf-life determination will be one of the keys to its success.

10.8 References

ANON, (1987). Microbiological Challenge Testing. Campden Food & Drink Research Association Technical Manual No. 20.

ANON, (1992) Advisory Committee on the Microbiological Safety of Food (ACMSF). Report on Vacuum Packaging and Associated Processes. HMSO London.

ANON, (1997) Harmonisation of Safety Criteria for Minimally Processed Foods. Inventory Report. FAIR Concerted Action FAIR CT96-1020.

ANON, (1998) Chilled Foods Market Report Plus. Keynote Publications. 1998.

BETTS G D, (1996) A Code of Practice for the Manufacture of Vacuum and Modified Atmosphere Packaged Chilled Foods. CCFRA Guideline No. 11. CCFRA, Chipping, Campden, UK.

BØGH-SORENSEN L and OLSSON P, (1990) The chill chain. In: Gormley, T. R. (ed.) Chilled Foods. The state of the art. Elsevier Applied Science, pp. 245.

BROWN H M, (1991) Evaluation of the Shelf Life for Chilled Foods. Campden Food & Drink Research Association Technical Manual No. 28.

CCFRA, (1999) Product Intelligence Dept.

CURIALE M S, (1998) Limiting growth: microbial shelf-life testing. *Food Product Design* **7** (11) 72–83.

DAY B F P D, (1992) Guidelines for the Good Manufacturing and Handling of Modified Atmosphere Packed Food Products. CCFRA Technical Manual No. 34, Chipping Campden, UK.

DENNIS C and STRINGER M, (2000) Introduction: the chilled foods market. In: Chilled Foods: a comprehensive guide, 2nd Edition. Eds. M Stringer and C Dennis. Woodhead Publishing Ltd., Cambridge, pp. 1–16.

ELLIS M J, (1994) The methodology of shelf-life determination. In: Shelf-life Evaluation of Foods. Eds. C.M.D Man and A.A. Jones. Blackie Academic and Professional.

EVANS J A, STANTON J I, RUSSELL S L and JAMES S J, (1991) Consumer handling of chilled foods: A survey of time and temperature conditions. MAFF Publications, London, pp. 102.

EVANS J A, (1998) Consumer perceptions and practice in the handling of chilled foods. In: Sous vide and cook-chill processing for the food industry. Ed. S. Ghazala. Aspen Publishers Inc., Maryland.

GAZE J E and BETTS G D, (1992) Food Pasteurisation Treatments. CCFRA Technical Manual No. 27. CCFRA, Chipping Campden, UK.

GIBSON A M and HOCKING A D, (1997) Advances in the predictive modelling of fungal growth in food *Trends in Food Science and Technology*, **8** 353–8.

IFST, (1999) Development and use of microbiological criteria for foods. IFST Booklet. ISBN 0 905367 **16** 2.

LEAPER S, (1997) HACCP: a practical guide (second edition). CCFRA Technical Manual No. 38, Chipping Campden, UK.

MERMELSTEIN N H, (1998) High pressure processing begins, *Food Technology*, **2** (6) 104–6.

NOTERMANS S and IN'T VELD P, WIJTZES T and MEAD G C, (1993) A user's guide to microbial challenge testing for ensuring the safety and stability of food products, *Food Microbiology*, **10** 145–57.

NOTERMANS S and IN'T VELD P, (1994) Microbiological challenge testing for ensuring safety of food products, International Journal of Food Microbiology, **24** 33–39.

PHLS, (1996) Microbiological guidelines for some ready to eat foods sampled at the point of sale, *PHLS Microbiology Digest*, **13** (1) 41–43.

ROSE S A, STEADMAN S and BRUNSKILL R, (1990) A temperature survey of domestic refrigerators. CCFRA Technical Memoranda 577.

SINGH, (1994) Scientific principles of shelf-life evaluation. In: Shelf-life Evaluation of Foods. Eds. C M D Man and A A Jones. Blackie Academic and Professional.

11

Microbiological hazards and safe process design

M. H. Brown, Unilever Research, Sharnbrook

11.1 Introduction

Many different ingredients and raw materials are processed to make chilled foods. At harvest or slaughter these materials may have a wide range of microbes in or on them. Some of them carry the micro-organisms that cause their eventual spoilage (e.g. bacilli or Lactic acid bacteria) whilst others pick them up during harvesting or processing. Many food poisoning bacteria occur naturally with farm animals and agricultural produce (e.g. *Salmonella*, *E.coli* O157 and *Campylobacter*) and hence can contaminate meat and poultry, milk and vegetable products. The numbers and types present will vary from one ingredient to another and often product safety at the point of consumption will depend on manufacturing, consumer use and the presence or numbers of pathogens in the raw material and eventually in the manufactured product. In order to ensure safe products with a reliable shelf-life, the manufacturer must identify which food poisoning and spoilage bacteria are likely to be associated with particular raw materials and products (e.g. by microbiological surveys). Therefore it is essential to design food processing procedures according to principles that ensure that the hazard of food poisoning is controlled. This is especially important in the prepared foods and cook-chill sectors where safety relies on the control of many features of the manufacturing process (ICMSF 1988, Kennedy 1997). The appropriate means of control should be incorporated into the product and process design and implemented in the manufacturing operation. Often the means of control exist at several stages along the supply chain. For example Gill *et al.* (1997) have suggested that the overall hygienic quality of beef hamburger patties could be improved only if hygienic quality beef (i.e. lowest possible levels of contamination with pathogens) was used for manufacture and there was better management of retail outlets with regard to patty storage and cooking. Good manufacturing practice guides are available for many sectors of the chilled food industry (e.g. IFST Guide to Food and Drink Good Manufacturing Practice: IFST 1998, UK Chilled Food Association Guidelines: CFA 1997). These guides outline responsibilities in relation to the manufacture of safe products; adherence to their principles will ensure that the product remains wholesome and safe under the expected conditions of use.

Product and process design will always be a compromise between the demands for safety and quality on the one hand, and cost and operational limitations on the other. Heat is the main means of ensuring product safety and the elimination of spoilage bacteria. The heating that can be applied may sometimes be limited by quality changes in the product. Usually, minimum cooking processes, either in-factory or in-home, will be designed to kill specific bacteria such as infectious pathogens or those causing spoilage. The skill of the product designer is to balance these competing demands for quality and safety and decide where an acceptable balance lies. Even so, usually more than one process step contributes to quality and safety, for example refrigerated storage is used to retard or prevent the growth of vegetative cells and spores that have survived factory heating. Hence the safety of chilled foods which have no inherent preservative properties, depends almost exclusively on suitable refrigeration temperatures being maintained throughout the supply chain including, for example, the defrosting of frozen ingredients and loading of refrigerated vehicles. Where preservation is used, for example, reduced pH/ increased acidity or vacuum packing, chilling will also contribute to the effectiveness of the preservation system and introduces the need for additional controls during processing.

The techniques of risk assessment, either formal or more commonly informal, may be used to guide the manufacturer in achieving a predictable and acceptable balance between the sale of raw or undecontaminated components, cooking and the chances of pathogen survival. Successful process design must consider not only contaminants likely to be carried by the raw materials, but also the shelf-life of the food and its anticipated storage conditions with distributors, retailers or customers, CFDRA (1990). In this sense, the customer is an integral part of the safety chain and some additional level of risk attributable to consumer mishandling or mis-use is always accepted by a manufacturer when he designs products whose safety and high quality shelf-life relies on customer use (e.g. cooking or chilled storage). Brackett (1992) has pointed out that chilled foods contain few, or no, antimicrobial additives to prevent growth of pathogenic micro-organisms and are susceptible to the effects of inadequate refrigeration that may allow pathogen growth. He also highlights related issues such as over reliance on shelf-life as a measure of quality and the need to consider the needs of sensitive groups (such as immunocompromised consumers) in the product design. If the product design relies on the customer carrying out a killing step to free the product of pathogens, such as salmonellae, it is important that helpful, accurate and validated heating or cooking instructions are provided by the manufacturer and that use of these instructions results in high product quality. Good control of heat processing and hygiene in the factory and the home or food service outlet are essential for product safety. The prevention of product re-contamination or cross-contamination after heating plays an even more critical role when products are sold as ready-to-eat.

It is essential that foods relying on chilled storage for their safety are stored at or below the specified temperature(s) (from $-1°$ to $+8°C$) during manufacture, distribution and storage. Storage at higher temperatures can allow the growth of any hazardous micro-organisms that may be present. Inappropriate processing in conjunction with temperature or time abuse during storage will certainly lead to the growth of spoilage micro-organisms and premature loss of quality. Labuza and Bin-Fu (1995) have proposed the use of time/temperature integrators (TTI) for monitoring the conditions and the extent of temperature abuse in the distribution chain. In conjunction with predictive microbial kinetics the impact of storage temperature on the safe shelf-life of meat and poultry products can be estimated. The risks associated with any particular products can be investigated either by practical trials (such as challenge testing) or by the use of mathematical modelling.

The use of predictive models for microbial killing by heat (interchange of time and temperature to calculate process lethality based on D and z values) or the extent of microbial growth can improve supply chain management. In the UK, Food MicroModel (FMM: www.lfra.co.uk) and in the US, the Pathogen Modelling Program (www.arserrc.gov/mfs/regform.htm) are computer-based predictive microbiology databases applicable to chilled products. Panisello and Quantick, (1998) used FMM to make predictions on the growth of pathogens in response to variations in the pH and salt content of a product and specifically the effect of lowering the pH of pâté. Zwietering and Hasting (1997) have taken this concept a stage further and developed a modelling approach to predict the effects of processing on microbial growth during food production, storage and distribution. Their process models were based on mass and energy balances together with simple microbial growth and death kinetics and were evaluated using a meat product line and a burger processing line. Such models can predict the contribution of each individual process stage to the microbial level in a product.

Zwietering *et al.* (1991) and Zwietering *et al.* (1994a, b) have modelled the impact of temperature and time and shifts in temperature during processing on the growth of *Lactobacillus plantarum*. Such predictive models can, in principle, be used for suggesting the conditions needed to control microbial growth or indicate the extent of the microbial 'lag' phase during processing and distribution where temperature fluctuations may be common and could allow growth. Impe *et al.* (1992) have also built similar models describing the behaviour of bacterial populations during processing in terms of both time and temperature, but have extended their models to cover inactivation at temperatures above the maximum temperature for growth.

Adair and Briggs (1993) have proposed the development of expert systems, based on predictive models to assess the microbiological safety of chilled foods.

Such systems could be used to interpret microbiological, processing, formulation and usage data to predict the microbiological safety of foods. However to be realistic, the models are only as good as the data input and at present there is both uncertainty and variability associated with the data available. Betts (1997) has also discussed the practical application of microbial growth models to the determination of shelf-life of chilled foods and points out the usefulness of models in speeding up product development and the importance of validating the output of models in real products. Modelling technology can offer advantages in terms of time and cost, but is still in its infancy (Pin and Baranyi 1998). Its usefulness is limited, as there is variation not only in the microbial types present in raw materials and products but also in their activities and interactions altering growth or survival rates or the production of metabolites recognised by customers as spoilage.

There are, not surprisingly, major differences between manufacturers in the degree of time or temperature abuse they design their product to withstand and hence the risks they are prepared to accept on behalf of their customers. This can result in major differences in the processes, ingredients and packaging used and the shelf-lives given to apparently similar products.

11.2 Definitions

Definitions are given below, firstly in order to avoid misunderstanding and secondly to introduce general comments and guidance for the design of processes which control microbiological risks adequately. They are discussed in the following groups: raw materials; Chilled foods; Safety and quality control; Processes.

11.2.1 Raw materials

Undecontaminated materials

These include any food components of the final product, that have not been decontaminated so that they are effectively free of bacteria prejudicing or reducing the microbiological safety or shelf-life of the finished product. Such starting materials should be handled in the factory so that numbers of contaminants are not increased and they cannot contaminate any other components that have already been decontaminated. For example, the layout of processing areas should be designed on the forward flow principle to prevent cross contamination; uncooked material should not be handled by personnel also handling finished product (except with the appropriate hygiene controls and separation), or allowed to enter high care areas (see below). If it is anticipated that these materials may contain pathogenic microbes, the severity of the risks should be assessed. Their handling, processing and usage should be controlled accordingly to prevent cross contamination or the manufacture of products which may be accidentally harmful to customers (see below).

Decontaminated materials
These materials will have been treated, usually with heat, to reduce their microbial load. If they are intended for direct incorporation into ready-to-eat products then the heat treatment used in their preparation should be sufficient to ensure the safety of the product (i.e. predictable absence of pathogens) depending on whether it is of short or long shelf-life (see 'Safe process design' below). Suitable precautions must be taken to prevent their recontamination after treatment and during handling in the factory. Hence primary packaging should be removed from decontaminated materials only in high-hygiene areas.

11.2.2 Chilled foods

This broad group covers all foods which rely on chilled storage (originally defined as from -1° to $+8^{\circ}C$ (Anon. 1982) but see below) as a component of their preservation system. It may therefore include foods made entirely from raw or uncooked ingredients. Some such foods may require cooking prior to consumption in order to make them edible, e.g. raw fish and meat products, and it is accepted that such foods may unavoidably contain pathogenic micro-organisms from time to time.

Prepared chilled or ready-to-eat foods
These chilled foods may contain raw or uncooked ingredients (Risk Classes 1 and 2, see 'Risk classes' below and Table 11.1), such as salad or cheese components. But their preparation by the manufacturer is such that the food is either obviously ready-to-eat or only requires re-heating, rather than full cooking, prior to use. The manufacturer should do his best to ensure that such foods are free of hazardous pathogens or hazardous levels of pathogens at the end of their shelf-life, and ingredients should be sourced with this objective in view. A scheme for the layout of process lines used in their manufacture is given in Figs 11.1 and 11.2.

Cooked ready-to-eat foods
Such foods (Risk Classes 3 and 4, see below, 'Risk classes' and Table 1) are made entirely from cooked ingredients and therefore should be freed of infectious pathogens during processing. Cooking procedures during production should be designed to ensure this and handling procedures after cooking, including cooling, should be designed to prevent recontamination of the product or its components, such as primary packaging materials. Often, the appearance of such foods makes it obvious to the customer that no heating, or mild re-heat, is all that are required before eating. Heating requirements should be made clear by any instructions. Typical process line layouts are shown in Figs 11.3 and 11.4.

REPFEDS
For the wider range of in-pack pasteurised foods, Mossel *et al.* (1987) and Notermans *et al.* (1990) have proposed the more informative name: 'refrigerated pasteurised foods of extended durability' (or 'REPFED'), which includes sous-

Table 11.1 Risk classes of chilled foods

Risk class[a]	Typical shelf-life	Critical hazard	Relative risk	Required minimum heat treatment	Required manufacturing class[b]		
					MA	HA	HCA
1	1 week	Infectious pathogens	High	Customer cook (minimum 70°C, 2 min.)	✓	(✓)	
2	1–2 weeks	Infectious pathogens	Low	Pasteurization by manufacturer (minimum 70°C, 2 min.)	✓	✓	✓
3	>2 weeks	Infectious pathogens and spore-formers	Low	Pasteurization by manufacturer (minimum 90°C, 10 min.)	✓		✓
4	>2 weeks	Spore-formers	Low	Pasteurization by manufacturer (minimum 90°C, 10 min.)	✓	✓	

Notes
[a] Class 1: Raw chill-stable foods, e.g. meat, fish etc.; Class 2: Products made from a mixture of cooked and low-risk raw components; Class 3: Products cooked or baked and assembled or primary packaged in a high-care area; Class 4: Products cooked in-pack.
[b] MA: Manufacturing area; HA: Hygienic area; HCA: High-care area.

vide and other foods with preservation and pasteurisation combinations that ensure long shelf-lives under chill conditions. These products are processed to free them of spoilage bacteria and pathogens capable of growth at chill temperatures, and hence allow very long shelf-lives (42 days or so). Therefore processing, handling and packaging must specifically ensure that they are free of infectious pathogens and the spore-forming pathogens capable of growing under chilled conditions. There is still a lack of knowledge on realistic safety boundaries and the risks associated with these products, with respect to the most severe hazard non-proteolytic *Clostridium botulinum* (Peck 1997). The determinants of effectiveness of complex combination preservation systems that rely on mild heating and chilled storage are not fully understood.

11.2.3 Safety and quality control

Good manufacturing practice

Good manufacturing practice (GMP) covers the boundaries and fundamental principles, procedures and means needed to design an environment suitable for the production of food of acceptable quality. Good hygienic practice (GHP) describes the basic hygienic measures that establishments should meet and

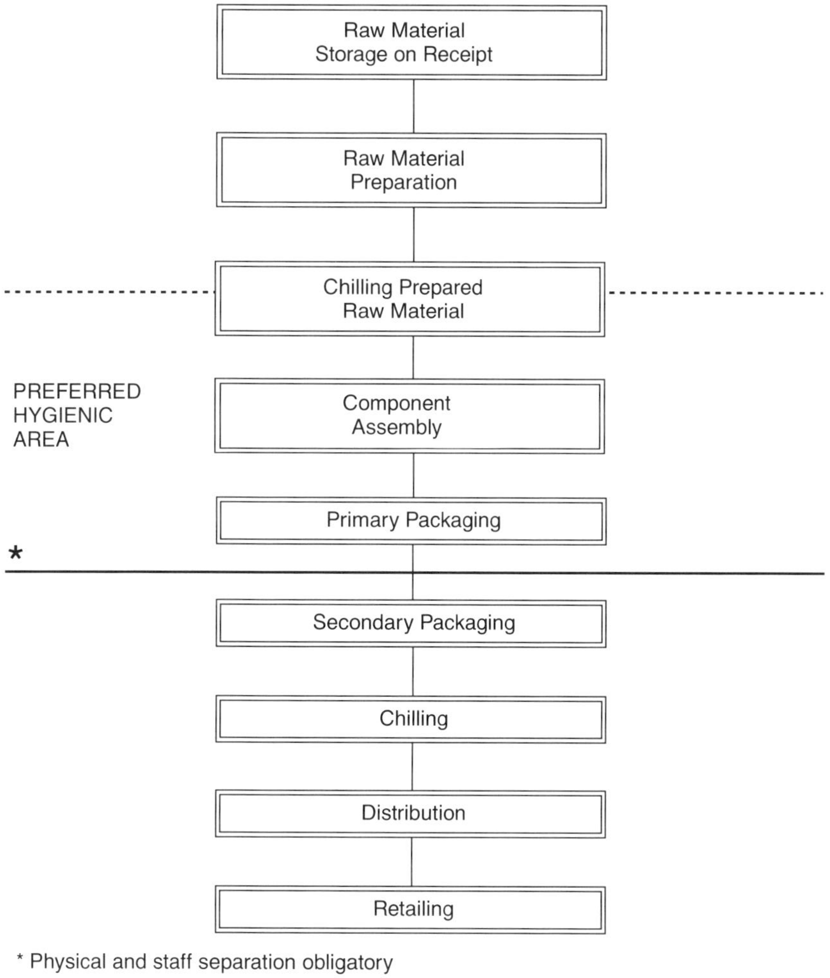

Fig. 11.1 Typical flow diagram for the production of chilled foods prepared from only raw components. (Class 1)

which are pre-requisites to other approaches, in particular HACCP. GMP codes and the hygiene requirements they contain are the relevant boundary conditions for the hygienic manufacture of foods and should always be applied. Governments (see Anon. 1984, 1986), the Codex Alimentarius Committee on Food Hygiene (FAO/WHO) and the food industry, often acting in collaboration with food inspection and control authorities and other groups have developed GMP/GHP requirements (Jouve *et al.* 1998). Generally GHP/GMP requirements cover the following:

- the hygienic design and construction of food manufacturing premises
- the hygienic design, construction and proper use of machinery

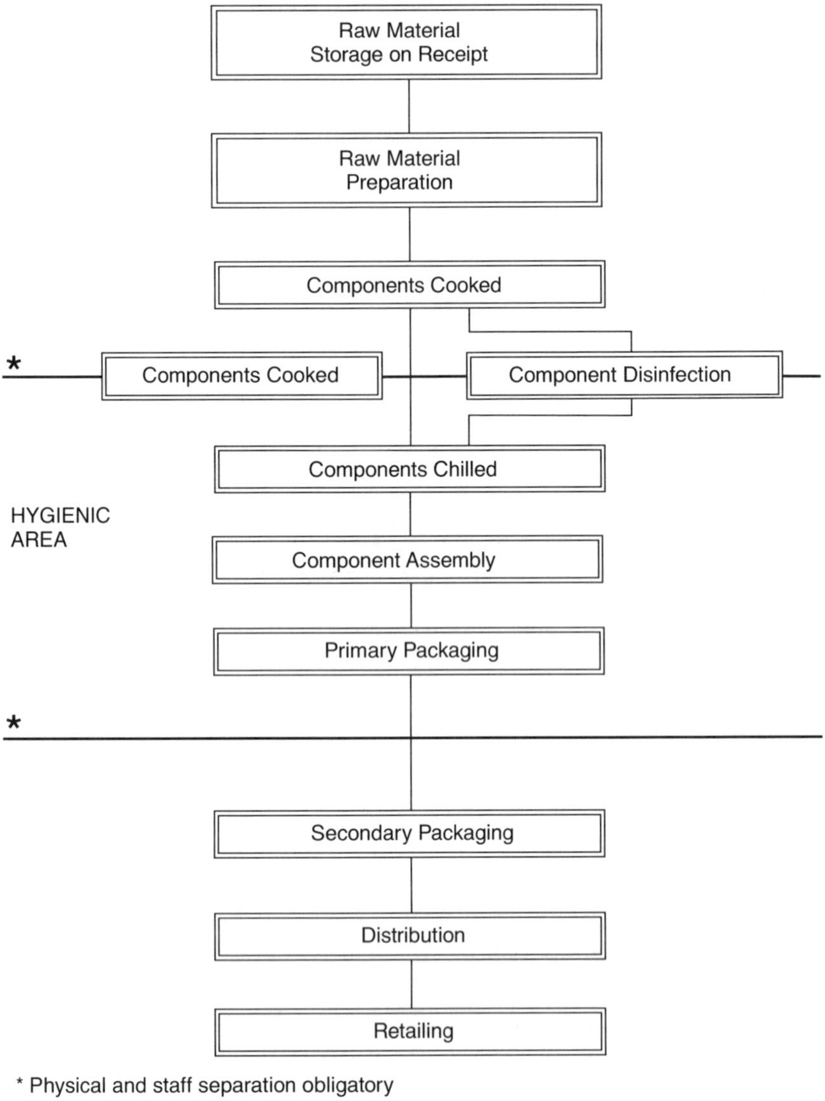

Fig. 11.2 Typical flow diagram for the production of chilled foods prepared from both cooked and raw components. (Class 2)

- cleaning and disinfection procedures (including pest control)
- general hygienic and safety practices in food processing including
  - the microbiological quality of raw materials
  - the hygienic operation of each process step
  - the hygiene of personnel and their training in hygiene and the safety of food.

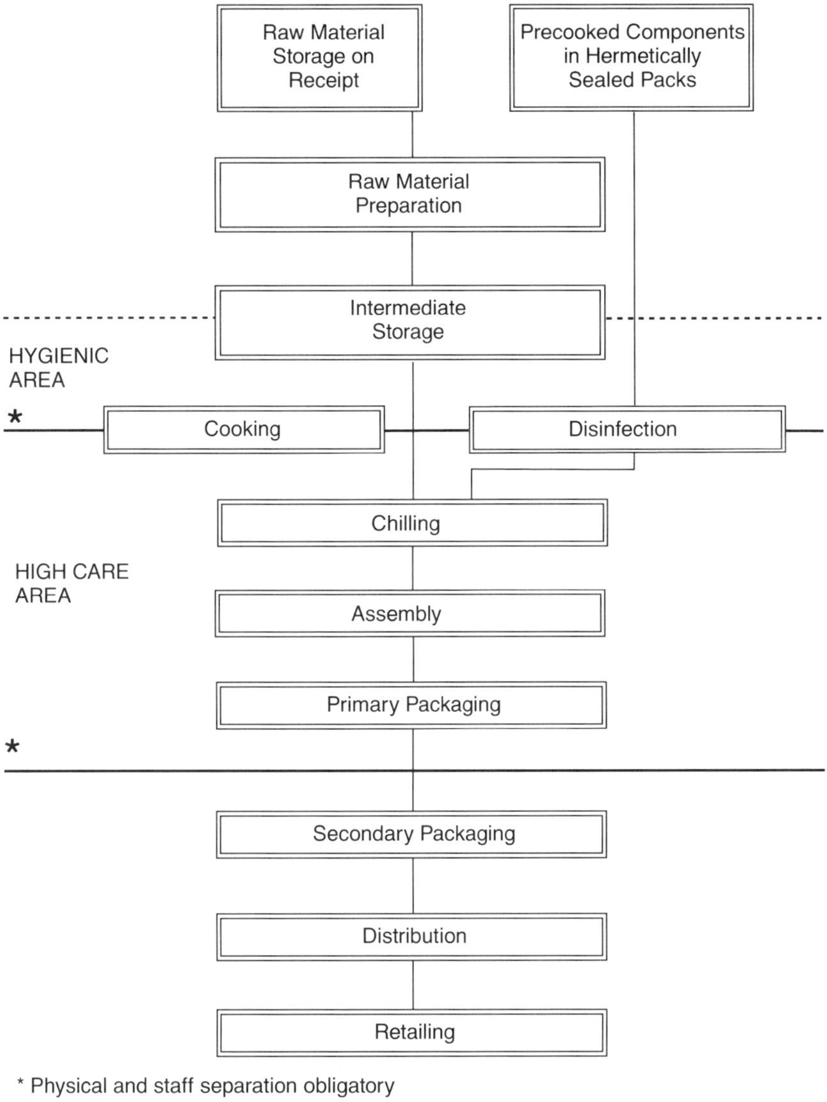

Fig. 11.3 Typical flow diagram for the production of pre-cooked chilled meals from cooked components. (Class 3)

HACCP

The Hazard Analysis Critical Point Control System (HACCP) is a food safety management system using the approach of identifying hazards and controlling the critical points in food handling and processing to prevent food safety problems. It is a system or approach that can be used to assure food safety in all scales and types of food manufacture and is an important element in the overall management of food quality and safety. The widespread introduction of

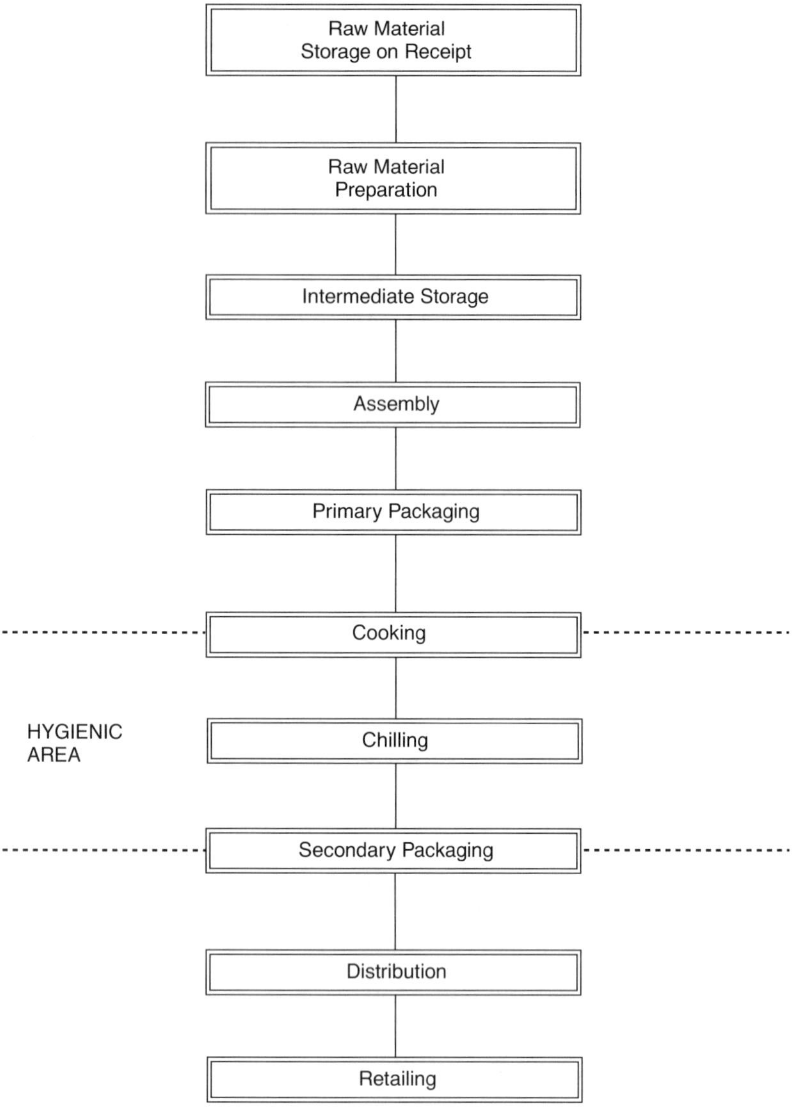

Fig 11.4 Typical flow diagram for the production of chilled foods cooked in their own packaging prior to distribution. (Class 4)

HACCP has promoted a shift in emphasis from end-product inspection and testing to preventive control of hazards at all stages of food production, but especially at the critical control points (CCPs). As such, it is a management technique ideally suited to the manufacture of chilled foods, where many elements of the process contribute to safety and shelf-life, the shelf-life is restricted and any delay to await the results of microbiological testing uses-up shelf-life. HACCP involves:

- the identification of realistic (microbiological) hazards, such as pathogenic agents and the conditions leading to their presence, growth or survival (HACCP is also used for the control of chemical and physical hazards)
- the identification of specific requirements for the control of hazards and identification of process stages where this is achieved
- procedures and equipment to measure and document the efficacy of the controls that are an integral part of the HACCP system
- the documentation of limits and the actions required when these are exceeded.

For steps in the manufacturing process that are not recognised as CCPs, the use of GMP/GHP provides assurance that suitable control and standards are being applied. The identification and analysis of hazards within the HACCP programme will provide information to interpret GMP/GHP requirements and direct training, calibration etc. for specific products or processes. The Microbiology and Food Safety Committee of the National Food Processors Association (NFPA 1993) has considered HACCP systems for chilled foods produced at a central location and distributed chilled to retail establishments. They used chicken salad as a model for proposing critical control points and give practical advice on HACCP planning; development of a production flow diagram, hazard identification, establishing critical limits, monitoring requirements; and verification procedures to ensure the HACCP system is working effectively. There are also USDA recommendations and outline HACCP flow diagrams for chilled food processes, cook-in-package and cook-then-package Snyder (1992).

Risk analysis

Ensuring the microbiological safety and wholesomeness of food requires the identification of realistic hazards and their means of control (risk assessment). The ability of a food producer to assess the impact of process, product and market changes on the level of risk and the type of hazard are important to the assurance of consistent standards of food safety. The effects of changes on risk and hazards need to be identified and can include the development of new products and processes, the use of different raw material sources, or the targeting of new customer groups, such as children. Food producers have always assessed these risks using either empirical or experiential approaches. As causal links have been established between food-borne illness and the presence, or activities (toxigenisis), of food-poisoning micro-organisms, so the control of specified microbial hazards has progressively become the means of ensuring food safety. These practical approaches have now developed into formal systems with well defined procedures and are known as Microbiological Risk Assessment (MRA). They are described in Microbiological Risk Assessment; an interim report (ACDP, 1996) or by the Codex scheme (Figure 1, Codex Alimentarius, 1996).

The overall aim of risk analysis is to reduce risk by

- identifying realistic microbiological hazards and characterising them according to severity
- examining the impact of raw material contamination, processing and use on the level of risk
- and communicating clearly and consistently, via the output of the study, the level of risk to the consumer.

When risk assessment is put together with risk communication (distribution of information on a risk and on the decisions taken to combat a risk) and used to promote sound risk management (actions to eliminate or minimise risk), a risk analysis is produced (ACDP 1996).

Stages in a risk assessment
Clear formulation of a problem is an essential prelude to risk assessment. A process or ingredient change, the emergence of a new pathogen or a change in public concern may trigger a risk assessment over a hazard and this can lead to the review of control, factory layout or sourcing options or the revision of cooking instructions.

The first stage of the assessment is to identify the hazard (hazard identification). For example, concern may be over the presence of salmonella in a product, as ingesting products containing infective cells may cause salmonellosis. The chances of causing harm are governed by many factors specific to the hazard (its virulence, incidence and concentration) and its prevalence in raw materials.

Exposure assessment describes the likely exposure of customers to the hazard, based on the size of the portion consumed and the impact of prior manufacturing etc. on the quantity of the infectious agent (i.e. *Salmonellae*) present (and infectious) at consumption. For a cooked product, exposure will depend on the numbers of salmonella entering the heating process, the heating characteristics of the product and the heating conditions used either in-factory or in-home, these combine to determine the number of pathogens surviving at consumption. If the heat sensitivity of salmonella and the product's heat treatment are known, then numbers likely to survive can be estimated. For many chilled foods (e.g. burgers or flash-fried poultry products), microbiological safety is not necessarily guaranteed by manufacturing processes, but can be a joint responsibility with the customer (Notermans *et al.* 1996). This makes the consumer part of the process for ensuring that the end product is safe, and therefore an assessment of their effect an essential part of a risk assessment.

Quantification of the risks of infection after product consumption is known as hazard characterisation. It links the sensitivity of consumers to infection (i.e. usually making use of expert opinion or knowledge of the dose-response within populations) with the concentration of the agent in the portion. The output of these three stages is a risk characterisation, which describes for a certain consumer the risks of (Salmonella) infection associated with the consumption of a particular product, sourced and manufactured under specified conditions.

To facilitate communication of decisions on risk and their basis, information must be accessible to the management, customers and staff. Where risk decisions or conclusions are communicated effectively, risk management practices can be readily implemented, consistent standards applied and dangerous changes may be stopped. Implementation of an effective process for communication and understanding on a consistent, scientific, and yet practical basis is an unsolved problem. Risk assessment has been reviewed (Jaykus 1996) and applied to specific problems, listeriosis (Miller *et al.* 1997), the role of indicators (Rutherford *et al.* 1995) and links with HACCP (Elliot *et al.* 1995).

Precautionary principle

Generally, the actions taken to protect public health are based on sound science. However, from time to time decisions have to be taken in an area of scientific uncertainty, for example if the prevalence or severity of a new pathogen is unknown. Any decisions made using the precautionary principle should control the perceived health risks without resorting to excessively restrictive control measures and should be proportionate to the severity of the food safety problem. An example is the measures taken to control the presence of *E.coli* O157: H7 in vegetables (by pasteurisation) or salad crops (by disinfection during washing) or by the proposal of Good Agricultural Practices (De Roever 1998), when prevalence of the pathogen is unknown and the illness caused severe. The resulting actions need to have been derived in an understandable and justifiable way and any assumptions and uncertainties need to be clear. Most of all the reduction in risk achieved must be acceptable to all the parties involved. Decisions arising in this way should be regarded as temporary awaiting further information that will allow a more reliable risk assessment and lead to the appropriate control measures, as described above.

11.2.4 Processes

Cooking

Cooking indicates that a process step delivers or has delivered sufficient heat to cause all parts of a food to reach the required sensory quality and should have caused a significant reduction in the numbers of any infectious pathogens that may be present. A 6-log reduction in numbers of infectious pathogens, or 70ºC for two minutes, is usually considered to be a diligent minimum target (see 'Safe process design' below). Some cooking stages may deliver considerably more heat than this, for example those involving prolonged periods of boiling. If the cooking stage is used as a pasteurization step, then recontamination must be prevented. It is important that cooking specifications distinguish between heat treatments (i.e. the conditions within a food) and the process conditions needed to provide the heat treatment. For example, for a given heat treatment, process conditions will vary according to the diffusion of heat through the product, its dimensions and the transfer of heat to or through its surface.

Pasteurisation

Pasteurisation is a processing stage involving heat. It is designed to bring about consistently a predictable reduction in the numbers of specified types of micro-organism in a food or ingredient. The safety and shelf-life requirements should determine the minimum severity of any pasteurisation stage, and the processes used may not necessarily heat ingredients sufficiently to give the required sensory quality or destroy all the micro-organisms present. The effectiveness of a particular heat treatment may be altered by various factors, including the preservation system employed in a particular food. For example, the heat resistance of bacteria and their spores is generally increased by low water activity, but decreased by low pH. Gaze and Betts (1992) have produced an overview of types of pasteurisation process and their microbiological targets. They also use the example of a pre-cooked chilled product to provide advice on process design and on manufacturing control points based on published heat resistance and growth data. Minimum pasteurisation processes should be targeted at foodborne pathogens, but in practice most are more severe, being targeted at the more heat-resistant bacteria causing spoilage in the product (see Gaze and Betts 1992 and CFDRA 1992).

P-value

Pasteurisation values (*P*-values) specify the effectiveness of a pasteurisation heat treatment. They are used to indicate the equivalent heat treatment corresponding to a specified heating time at a stated reference temperature.

z-value

z-value is an empirical value, quoted in temperature (C or F degrees), and used for calculating the increase or decrease in temperature that is needed to alter by a factor of 10 the rate of inactivation of a particular micro-organism. It assumes that the kinetics of microbial death at constant temperature is exponential (i.e. 'log-linear'). Although z is fundamental to the calculation of sterilisation process equivalence, it should be used with extreme caution for pasteurisation processes, as the death kinetics of many types of micro-organisms are not log-linear; especially where vegetative micro-organisms are concerned and when heating rates are low. For instance, 'heat adaptation' may occur. This may even raise the resistance of the micro-organism during the heating process (Mackey and Derrick 1987). 'Shoulders' and 'tails' on survivor curves are more commonly seen (Gould 1989). In practice, the validity of the z concept is therefore particularly limited at low temperatures, such as those involved in pasteurisation. Consequently, where there are other factors such as preservatives interacting with the heat treatment, or if processes are designed to cause large log reductions (i.e. in excess of ten thousand-fold), then tailing of the survivor curves may become particularly important. Actual or challenge trials should be undertaken to establish confidently safe processes.

Re-heating

The customer usually does re-heating and it is a procedure intended by the manufacturer only to ensure the optimum culinary quality of a product.

Depending on the product design, such a process may or may not provide adequate heating for safety; this is especially true where products are designed for microwave reheating. A re-heating process should, therefore, only be recommended to the customer if the food has effectively been freed of hazardous contaminants during processing and has remained so during any further processing, packaging, distribution and shelf-life. The chemical and physical characteristics of a ready meal, including saltiness, type of tray, and geometry and layout of components, affect microwave heating uniformity and rate. Ryynanen and Ohlsson (1996) heated four-component chilled ready meals in a domestic microwave oven and found that arrangement and geometry of components and type of tray mainly affected heating uniformity. Where microwaves are being used for cooking their effectiveness should be validated; a suitable method uses alginate beads containing micro-organisms with a known heat resistance (Holyoak *et al.* 1993).

Cooling

Cooling reduces the product temperature after factory cooking. Its aim should be to ensure that the product spends the minimum possible time in the temperature range allowing the growth of hazardous bacteria, i.e. between 55ºC and 10ºC. Cooling rates are often specified in legislation; for example, in the EEC Meat and Meat Products Directive 77/99, prepared meals must be cooled to below 10ºC within two hours of cooking. Evans *et al.* (1996) have highlighted the importance of cooling and the mandatory requirements that exist in the UK for cook-chill products (Anon. 1982) These guidelines recommend that 80mm trays should be chilled to below 10ºC in 2.5 hours, between 10 and 40mm they should be chilled to below 3ºC in 1.5 hours. Assuming that surface freezing is to be avoided and a simple, single-stage operation used, only a 10-mm deep tray can be chilled within these time limits.

Cooling of liquids or slurries may be done in line, using heat exchangers. Where batch cooling of solids and slurries is done in containers, the size of the container or the quantity of product should not be so large that rapid cooling is not possible. Product depths exceeding 10–15 cm should not be used because above this thickness conduction of heat to the surface, rather than removal of heat from the surface, will limit the rate of cooling of the bulk of the product and may allow microbial growth.

When warm or hot material is loaded into containers for air cooling, the materials of construction of containers will exert a significant effect on cooling rates, thick-walled plastic containers cooling considerably more slowly than metal ones.

The design of chillers, especially their air distribution pattern, air velocity and temperature and the way that product containers are packed into them, will control cooling rates. Racking or packing systems should allow the flows of cold air over the container surfaces so that cooling rates are maximised. Special attention should be paid to hygiene, and the control of condensation in chillers, as this is a major potential source of recontamination with *Listeria* if condensation is recirculated in aerosols over open containers, by the air flow. The importance of chilling after preparation, chilled storage and distribution have been discussed by Baird-Parker (1994) as critical control points in the manufacture of raw and cooked chilled foods. Farquhar and Symons (1992) have noted a US code of practice with recommendations for the preparation of safe chilled foods, it covers chilling, chilled storage, pre-distribution storage and handling and temperature management practices.

Chilled storage
Chilled storage should be designed to maintain the existing or specified temperature in a product. Product or ingredient containers should enter storage chills at their target temperature, because the performance of the air system (temperature and air velocity) and the way product is stacked are normally not designed to allow a significant reduction in temperature to take place.

Manufacturing area (MA)
A manufacturing area is a part of the factory that handles all types of ingredients. The process intermediates made in this area will be heat-treated before they are sold as products and will pass through hygienic or high-care areas during processing.

Hygienic area (HA)
A hygienic area is a defined processing area designed for the handling and preparation of low-risk raw materials and products containing a mixture of heat-treated and undecontaminated ingredients (Class 1). It should be designed and constructed for easy cleaning, so that high standards of hygiene can be achieved and especially to prevent bacteria, such as *Listeria*, becoming established in it and contaminating products. When it is used for the assembly of final products containing undecontaminated components such as cheese, it should not be used for the processing or preparation of any ingredients likely to carry pathogens and hence likely to increase the risks of products containing infectious pathogens. Areas conforming to this standard of hygiene should be used for the post-process handling of in-pack pasteurised products (Class 4).

High-care area (HCA)
This is a well-defined, physically separated part of a factory which is designed and operated specifically to prevent the recontamination of cooked ingredients and products after completion of the cooking process, during chilling, assembly and primary packaging. It is an integral part of the factory layout shown in Fig. 11.3 and is used for the preparation of products in Classes 2 and 3. Usually there are specific hygiene requirements covering layout, standards of construction and equipment, the training and hygiene of operatives, engineers and management and a distinct set of operational procedures (especially covering the intake and exit of food components and packaging material), all designed to limit the

chances of contamination. The usage of re-pack and re-work materials in high-care areas should be discouraged, and if this is not possible then very strict rules of segregation in time should be enforced.

Air handling
Air is a significant means of dispersing contaminants; therefore particular attention should be paid to the direction of flow of air within a manufacturing area or between areas. Within manufacturing areas the flow should be from 'clean' to 'dirty' to minimise the chances of contaminants being carried from raw to decontaminated product. The quality of the air should be related to the hygiene category of the area, see Manufacturing area (MA), Hygienic area (HA) and High-care area (HCA) requirements (CFDRA 1997).

Cleaning
Cleaning should remove food debris from process equipment, manufacturing and storage areas. Effective cleaning should remove all food debris from work surfaces, machines or an area, so that microbes cannot grow and subsequent production is not contaminated. Effective cleaning cannot be achieved unless equipment is hygienically designed and maintained. In practice, complete removal of food residues is rarely achieved by the techniques used for cleaning open plant (e.g. in slicers and dosing equipment). In factories manufacturing chilled foods, the residues after cleaning may provide growth substrates for the factory microflora, and experience has shown that many modern cleaning techniques and chemicals, when used in chilled areas, may actually select for *Listeria*. High-pressure (HPLV) cleaning if used in an uncontrolled fashion will generate aerosols that may contaminate products and equipment with food debris and bacteria. To minimise the risks of contamination, food and packaging materials should be removed from areas during cleaning. HACCP can make a valuable contribution in this area by identifying those process stages where hygiene is critical to the product quality and safety and also by checking from the process flow diagram that there is good access for cleaning within the factory layout.

Disinfection
Disinfection procedures should destroy any microbes left on cleaned surfaces and should be used at process stages where product re-contamination is a safety hazard. In practice, these procedures must often destroy or inhibit microbes remaining in the food residues invariably left after cleaning. Cleaning alone is able to achieve a satisfactory level of hygiene in hygienic or GMP areas, but high-care areas will require additional disinfection to provide extra confidence that viable bacteria are absent. Heat and a variety of chemicals are used as disinfectants. The effectiveness of disinfection will be reduced if the disinfectant is prevented from reaching the microbes by food debris; hence, thorough cleaning is always required to ensure effective disinfection. The assurance of good access of disinfectants is a primary objective of the hygienic design of

equipment and the development of cleaning schedules. The systematic monitoring of specified sites in equipment or production areas either visually or microbiologically checks the effectiveness of cleaning. The effectiveness of disinfection can be checked by swabbing or chemical means.

11.3 The microbiological hazards

The microbiological hazards of chilled foods can be roughly classified according to whether harmful microbes can infect the consumer or whether they multiply in the food and produce toxins that then cause disease soon after the food is eaten. The micro-organisms of greatest concern are listed in Table 11.2. All realistic microbiological hazards should be controlled by the product and process design. The real, not the specified, storage temperature and the length of the shelf-life must be assessed to determine whether a particular hazard is realistic in a particular product.

Infectious pathogens can be hazardous at very low levels, whereas appreciable levels or growth of toxigenic micro-organisms are needed to cause a hazard. When designing a product or process it is very risky to assume that a particular microbiological hazard will be absent, e.g. because it has not been detected in a component. Processes should be designed to control all realistic hazards.

The infectious pathogens (see Table 11.2 and Ch. 8) include *Salmonella, E.coli* O157:H7 and *Listeria monocytogenes.* They may be present in raw

Table 11.2 Food-poisoning organisms of major concern and their heat resistance and growth temperature characteristics

Minimum growth temperature	Heat resistance: Low / Medium — Vegetative cells	Heat resistance: High — Spores
Low	*Listeria monocytogenes* (INF)[a]	*Clostridium botulinum* type E, non-proteolytic B&F (TOX)
	Yersinia enterocolitica (INF)	
	Vibrio parahaemolyticus (INF)	*Bacillus cereus* (TOX)
Medium	*Aeromonas hydrophilia* (INF)	*Bacillus subtillis* (TOX)
	Salmonella species (INF)	*Bacillus licheniformis* (TOX)
		Clostridium perfringens (INF)
High	*Escherichia coli* O157 (INF)	*Clostridium botulinum* type A & proteolytic B(TOX)
	Staphylococcus aureus (TOX)[b]	
	Camplylobacter jejuni & coli (INF)	

Notes
[a]INF, infectious; [b] TOX, toxigenic.

materials such as meat, vegetables and cheese made from unpasteurised milk. All may survive for long periods in chilled products (e.g. *E.coli* O157:H7 survives for 22 days at 8ºC in crispy salad), if not eliminated by processing. All the infectious pathogens are heat sensitive and will be eliminated by the conditions used for pasteurisation (e.g. 70ºC for 2 minutes or 72ºC for 16.2 seconds). The growth of *Salmonella* and *E.coli* O157:H7 in products or in the factory environment may be controlled by refrigeration (i.e. by temperatures below about 10ºC). *E.coli* O157:H7 has a low infectious dose and causes serious illness especially in the young and the elderly, as it attaches to the wall of the intestinal tract and causes acute, bloody, diarrhoea (hemorrhagic colitis) or haemolytic uraemic syndrome (a kidney disease).

Several outbreaks of disease have been linked to chilled food and have usually had a bovine origin (e.g. undercooked ground-beef products), although unpasteurised cider and mayonnaise have been implicated. In the latter case it is thought that improper handling of bulk mayonnaise or cross-contamination with meat juices or meat products was the cause. It has been found that this *E.coli* is more tolerant of acid environments than other known strains, therefore it may survive in fermented dry sausage and yoghurt. *Salmonella enteritidis* is a potential hazard in products made from poultry and eggs, whereas the multi-drug resistant *Salmonella typhimurium* DT 104 can be found in a broad range of foods, outbreaks in the United Kingdom have been linked to poultry, meat and meat products and unpasteurised milk.

Campylobacters may cause intestinal infections leading to fever, diarrhoea and sometimes vomiting. Sources include water, milk or meat. *C. jejuni* is regularly found on retail raw poultry and outbreaks have been associated with undercooked poultry and the cross-contamination of ready-to-eat materials via the hands of kitchen staff or work areas. It does not grow below 30ºC and therefore conditions affecting survival are important, since sufficient cells must survive to form an infectious dose. Survival is better in chilled foods than at ambient or frozen temperatures. *Vibrio cholerae* may survive on refrigerated raw or cooked vegetables and cereals, if they have been sourced from tropical or warm areas where contamination is endemic. Particularly at risk are sea and other foods harvested from estuarine or inshore waters, waters subject to land-water run off or fields irrigated with sewage contaminated water. Contamination can also occur if the produce itself is cooled, washed or freshened with contaminated water. During preparation, food from this type of origin, which can include raw, pre-cooked and processed molluscs, crustaceans, fish and vegetables should be handled to minimise the chances of cross-contamination and pasteurised prior to sale.

The psychrotrophic pathogens such as *Listeria monocytogenes* (Walker and Stringer 1987), can grow at refrigeration temperatures and may readily become established on badly designed or maintained equipment and in the factory environment. They are only detected in low numbers in environmental samples associated with the primary production of food (Fenlon *et al.* 1996) and are therefore likely to be contaminants arising from manufacturing conditions.

Under otherwise optimal conditions, some strains of *L. monocytogenes* have been shown capable of slow growth at temperatures as low as −0.1°C, *Yersinia enterocolitica* at −0.9°C and *Aeromonas hydrophila* at −0.1°C (Walker 1990). For example, *L. monocytogenes* is able to grow well in components such as chill-stored, prepared vegetables and in many products lacking robust chemical preservations systems – such as chill-stored ready-meals and pâté.

The most important of the toxigenic pathogens are the cold-growing non-proteolytic strains of *Clostridium botulinum*. Their growth in pasteurised foods is a particular risk if processing has eliminated the competing flora and their growth could precede spoilage. The proteolytic strains are less hazardous because they are able to grow only at higher temperatures and, unlike the non-proteolytic strains, they normally cause spoilage that renders the product inedible. At chill temperatures the growth rate of the non-proteolytic types is slow and so requires control only in products where the designed chilled shelf-lives exceed about 10–14 days. Graham *et al.* (1996) suggest that non-proteolytic strains of *Clostridium botulinum* can grow at chill temperatures and therefore pose a potential hazard in minimally heat-processed chilled foods. They have developed models predicting growth and compared them with published data to demonstrate that they are suitable for use with fish, meat and poultry products. The models cover the combined effects of pH (5.0–7.3), salt concentration (0.1–5.0%) and temperature (4–30°C) and are based on the growth of non-proteolytic *C. botulinum* in laboratory media. Fortunately, the spores of the strains able to grow at chill temperatures are relatively heat-sensitive and therefore can be controlled by realistic cooking or pasteurisation processes (90°C × 10 min. is the process recognised as suitable for these long-life chilled foods).

Although cooking can eliminate the cold-growing types, storage temperature remains the most important control on the growth of clostridia. The United Kingdom, Advisory Committee on the Microbiological Safety of Food (1992) have discussed the potential hazards of chilled foods made using vacuum packaging and associated processes, such as *sous-vide*, and they particularly considered the risks associated with botulism. They highlighted methods to prevent and/or control the risks of botulism, including adequate heating based on the heat resistance of the spores and the restriction of shelf-life. See Smith *et al.* (1990) for details of the use of HACCP to address this problem.

At temperatures above 12–15°C the mesophilic types (producing more heat-resistant spores) are able to grow, and the processes used in the manufacture of chilled foods certainly do not inactivate their spores. *Bacillus cereus* is sometimes mentioned as a potential hazard in chilled foods, though evidence of its ability to produce harmful toxins in these foods (with the possible exception of dairy products) is equivocal. *Staphylococcus aureus* is the other toxin producer causing concern in chilled foods, though it is of importance only when the food does not contain a competing microflora and has been substantially temperature abused. Nevertheless, it is important that high-care areas and the operational procedures used in them are designed to prevent heat-processed foods becoming contaminated with *S. aureus*.

11.4 Risk classes

Customers may well consume chilled ready-to-eat foods without sufficient heating to free them of infectious pathogens. Hence the risk to the customer depends on the number and type of microbes in the foods after manufacture and any growth during distribution and storage. Therefore, the processing and hygienic principles employed in the manufacture, distribution and sale of prepared chill-stable foods should be primarily designed to control the risks of them containing infectious or toxigenic pathogens. The control of spoilage microbes should be a secondary consideration in process design, although it may often require the application of more severe heating processes or more stringent conditions of hygiene or preservation, than the control of safety. Sometimes the microbiology cannot be controlled without prejudicing the sensory quality of a food product, and there must then be a commercial decision made on the acceptable balance between a controlled loss of quality and spoilage in the marketplace. In process and product design, however, microbiological safety standards should never be compromised to improve sensory quality. If the required processing conditions cannot ensure safety against the background of realistic consumer usage, then the product should not reach the market-place (see Gould 1992, Walker and Stringer 1990).

Chilled foods may be categorised into well-defined risk classes (Table 11.1). Some prepared chilled foods (Class 1) are made entirely from raw ingredients and will obviously require cooking by the customer. Others, containing mixtures of raw and cooked components processed or packaged to ensure a satisfactory shelf-life (Class 2), may not so obviously require cooking and may contain infectious pathogens which may (e.g. *L. monocytogenes*) or may not (e.g. *Salmonella*) be able to grow during chilled storage. The manufacturer is able to control the safety of products only in the latter category (Class 2) by minimising the levels and incidence of pathogens on incoming materials (e.g. by careful choice of suppliers). To control risk, storage and processing procedures should not introduce additional contaminants or allow numbers to increase. The shelf-life and storage temperatures of such foods should be designed to ensure that only 'safe' numbers of infectious pathogens could be present if foods are stored for the full indicated shelf-life. At present there is no generally accepted estimate of the infectious dose of *Listeria* and it remains up to individual manufacturers or Trade Associations to decide on acceptable risks. *Salmonella* should be absent from these foods as the infectious dose is very low. Because *Listeria* is able to grow at chilled temperatures (for example, during storage, distribution and domestic storage), only its complete absence at the point of manufacture will ensure that ready-to-eat foods are safe, without qualification. Where it is present after manufacture, then the producer is accepting the risk associated with sensitivity of his customers to any *L. monocytogenes* that may be found at the point of consumption.

Other chilled foods may contain only cooked or otherwise decontaminated components (Class 3), or may be cooked by the manufacturer within their primary packaging (Class 4). If manufactured under well-controlled conditions such foods will be free of infectious pathogens (such as *Listeria* and *Salmonella*) and spoilage microbes, hence they will have a substantially longer shelf-life (up to 42 days or more) than those containing raw components, as they will not be subject to microbial spoilage. This substantial extension of spoilage-free shelf-life has important consequences as to which of the potential changes in their microbiology should be recognised as limiting safety during storage (see below), and hence which controls and especially pasteurisation conditions are appropriate during manufacturing.

11.5 Safe process design 1: equipment and processes

The manufacture of chilled foods is a complex process. From a microbiological point of view, processes should be designed to control the presence, growth and activity of defined types of bacteria. Some of the unit operations making up a typical process provide the opportunity to eliminate or reduce numbers of bacteria, others provide opportunities for re-contamination or growth. Product design and shelf-life (see Table 11.2) and possibly factory hygiene and layout, will determine the target bacteria for each stage of the process. At the very beginning of the supply chain, agricultural produce, farm animals and their products can act as reservoirs of food-poisoning bacteria (e.g. *Salmonella, Campylobacter and E.coli* O157). Therefore it is important that handling and processing takes account of this and reliable means of eliminating them or preventing contamination of products are put in place by the process design. The extent of precautions needed to provide effective control of a particular hazard will be proportional to the length, complexity and scale of the supply chain.

For short-shelf-life (less than 10–14 days) prepared chilled foods, the main safety risk is the presence of infectious pathogens, and processes should be designed to cause a predictable reduction in their numbers. These types of bacteria may be a hazard if they survive processing or if recontamination occurs after a process step designed to remove them. There are two main routes for re-contamination via food-handlers or cross-contamination from other foods. In designing process flows and handling procedures it must be assumed that raw foods contain low numbers of food-poisoning bacteria, therefore effective separation of material flows is necessary. Risks are further increased if the product is ready to eat and the shelf-life and storage temperatures allow growth.

Prolonging the chilled shelf-life introduces an additional hazard, arising from the growth of toxigenic bacteria, and processes need to be designed to eliminate them. This is because under normal chill conditions (i.e. below 10ºC), over two weeks or so, cold-growing strains of *Clostridium botulinum* can grow to levels where toxin production is possible. Their spores may survive in foods pasteurised using the mild processes designed to destroy infectious pathogens. More severe heat processes need to be used in the manufacture of long-life foods and these should be designed to cause a predictable reduction in the numbers of heat sensitive spores. If the preservation system of the food can prevent the outgrowth of these spores, then the heat process need be designed to control only spoilage micro-organisms. Many chilled foods do have such effective intrinsic preservation systems and are therefore safe, although having received very low heat processes.

There is still no general agreement among manufacturers or regulatory authorities on the seriousness of the risks of botulism from unpreserved chill-stored foods. However, there is ample evidence that, in spore-inoculated model systems based on ready meals, growth occurs and toxin is produced at temperatures representing those known to be found under commercial conditions (Notermans *et al.* 1990).

It is essential that if heating has not been done in the primary packaging, that unwrapped, heat-treated components intended for long-life foods are chilled, handled and assembled in a high-care-area to prevent recontamination with spores and infectious pathogens. Even if the risks of recontamination are controlled, there is a remaining risk from the survival in the components of heat-resistant bacterial spores able to grow at chilled storage temperatures. These are mostly *Bacillus* species that can eventually cause a musty form of spoilage.

11.5.1 Equipment

Many of the critical quality and safety attributes of chilled foods are controlled by the technical performance of processing plant and equipment. The lethality of these heating processes used will exert a critical effect on the chances of microbial survival and presence in the product; therefore the correct design and reliable operation of heating stages is most important. Lag periods and growth rates of any contaminants will be influenced by a variety of process factors, including cooling rates and the accuracy with which storage temperatures are held. The uniformity with which preservatives, such as curing salts and acidulants, are dosed and mixed and the effectiveness of packaging machines at producing gas-tight packs (e.g. containing an inhibitory gas mixture such as CO_2 and N_2) are also critical factors.

Food residues that have remained in a machine or in the processing area after cleaning or in a processing area for some time, are the dominant cause of microbial contamination and unless these residues are regularly removed they pose a hazard in finished products. Contamination is usually influenced to a much lesser extent by either airborne contaminants or contamination by personnel. Many codes of practice state that 'food processing equipment should be designed to be cleanable and capable of being disinfected'. However, these statements hide the real issues in designing, operating and maintaining food manufacturing equipment, which often centre on finding an acceptable balance between cost, manufacturing efficiency and hygienic design (see Chapter 15).

11.5.2 Heat processes – the use of heat to decontaminate products

Heating methods

In the manufacture of chilled foods, heat is the agent most commonly used to inactivate micro-organisms and cause beneficial textural and colour changes in products. Depending on the type of product, different types of equipment are used for heating; some examples are given below.

- Culinary steam or hot water in open vessels may directly heat mixes or particles suspended in sauces.
- Discrete pieces of meat, fish or vegetables may be cooked in trays, moulds or *sous-vide* packs heated in atmospheric ovens. Where the packs are sealed these products may be heated in water baths. Open vessels and ovens may be heated indirectly using a jacket or by direct injection, or circulation, of steam or air, or a mixture of both. Where this comes directly in contact with product, it should be of culinary quality.
- Liquid or pumpable ingredients may also be heated indirectly via jacketed equipment or using heat exchangers in pumped circuits. Normal production equipment can be used for product temperatures up to 100°C; where temperatures above this are used, or anticipated, any closed equipment should be specified and used as a pressure vessel.
- Solids such as pieces of meat or vegetable may be heated by contact heating or deep frying, for product surface temperatures exceeding 100°C, and lower centre temperatures. These depend on the initial temperature and heat transfer and heat penetration characteristics of the product.

Packaged products or ingredients may be heated in retorts or other pressurised vessels using heating media with temperatures above 100°C; quite commonly, they are also heated in water baths. It is important for packs treated in this way either to be evacuated to ensure good heat transfer or that heating processes are set to take account of the insulating effect and expansion of any headspace during heating.

Control of heating

The heating process must be designed to deliver certain minimum heat treatments as specified above. Because of the importance of this, critical control parameters and tolerances should be defined in a specification – which must be available to the operatives in charge of the process. The training of these operatives should enable them to carry out the process reliably and to monitor and record its progress. This is most effectively done by following the time/temperature conditions in the vessel or chamber or sometimes in the product itself. For each batch of product, it must be verified that the specified process and hence the target lethality have been achieved.

To make reliable use of heat for killing micro-organisms, there are some essential points that must be considered by food processors and equipment manufacturers.

- The delivery of heat to the product surface must be accurately and reliably controlled by the equipment and the way it is used and maintained, so that predictable heat transfer rates occur. Therefore a retort must always be loaded with packs positioned in the same way or the contents of a jacketed vessel must be stirred, to ensure that each product or pack is uniformly exposed to the heating medium.
- The rate of heat penetration into the product must also be known and controlled, so that target time/temperature integrals at the coldest or slowest heating points are reliably achieved. This is controlled by product formulation (for example the size of particles, the viscosity and other physical characteristics of the product), the pack size and shape and the thermal conductivity of the packaging material.
- The overall design of equipment, control systems and the provision of services, such as steam and cooling water or air, must be capable of ensuring that similar amounts of heat are delivered whenever the apparatus is run.

Equipment suitability is not the only factor dictating how successfully or reliably the designed product temperatures are achieved. Management of the process may affect the characteristics of in-process material. For example, the temperature of a component at the start of processing – whether it is frozen, thawed or warm – will dictate heating rates, and such variables should be covered by the process specification. For safe products it is essential that whatever heating technique is used can ensure that a certain minimum amount of heat is delivered to all parts of the product or pack.

Equipment performance

Many pieces of commercial heating equipment, such as ovens, are not supplied with information indicating the distribution of heat within them. In addition the heating of a product will depend on the delivery of heat to the surface of the product, product unit or pack (heat transfer) and the penetration of heat within the pack. Investigation of these features is an essential part of process or product development. For example, the uniformity of heating in a jacketed vessel will depend on mixing – which may be done by an added stirrer or may be in the hands of the operator. In an oven, or retort, heat distribution can depend on the packing density or positioning of product, creating or blocking channels between product units upsetting uniform circulation of the heating medium. At a simple level, measurement of product centre temperatures and at a more complex level, process management involves estimation of heat flow within the product in response to conditions in the equipment (Fraile and Burg, 1998a, b). It is therefore up to the user to produce his own 'map' of the distribution of heat under his particular conditions of usage, so that processes can be set using product units situated at the coldest part to ensure the consistent delivery of a minimum process. Stoforos and Taoukis (1998) have proposed a procedure for process optimisation that relies on the use of a two- or three-component time-temperature integrator for thermal process evaluation. Their proposed procedure can take account of the different *z* values associated with microbial killing and quality loss to assess the impact of particular combinations of time and temperature.

The more variable or non-uniform the delivery of heat, the higher the target heat process needs to be fixed in order to ensure that the required minimum is achieved. An essential part of process development is investigation of the range of heat treatments delivered by equipment during the predicted range of operating conditions. Lethality criteria for process design are given elsewhere in this chapter and are related to the types and number of microbes that the process is designed to kill. Hence it is important that the input level of these microbes is also controlled; if the input numbers are higher than those intended by the process designer, then survivors will be found (possibly by the customer).

Cooling

Although heat is effective at killing micro-organisms, the effectiveness of cooling processes (USDA 1988) and the hygiene of cooling equipment must be specified equally carefully (James and Bailey 1990). Even in equipment designed to achieve rapid cooling rates, the risk of product recontamination remains, either from micro-organisms endemic in the chiller or taken in and then spread by the forced air circulation, often associated with rapid cooling. The rate of cooling is also critical as it will determine the extent of dormancy of any spores surviving in the product – this will affect their readiness to germinate and grow during product storage. The extent of dormancy is particularly important in shelf-life determination when heat has been used in conjunction with chemical preservation systems, such as salt and nitrite.

11.5.3 Microbiology of heat processing

The heat process for packed products is usually delivered by hot water, by steam, by use of an autoclave, or, less commonly, by microwaves or ohmic heating. The extent of heating is selected to match the intended distribution and shelf-life of the food and the target micro-organisms for the type of raw materials and product (see minimum processes suggested for short and long shelf-life products, below). Target bacteria include the pathogenic non-spore-forming organisms, Salmonella, entero-pathogenic *E.coli*, Campylobacter *Listeria monocytogenes* and *Yersinia enterocolitica* and the spore-forming non-proteolytic strains of *Clostridium botulinum*, type E, some type B and F. Possibly some strains of *Bacillus cereus* that can grow slowly at temperatures as low as 4°C (van Netten *et al.* 1990) may be hazardous, but there is little epidemiological evidence. As raw materials are sourced globally, other pathogens may be introduced and the heat processing element of the HACCP plan should be reviewed to ensure that adequate heat processes are used.

Whilst a heat treatment of 70°C for two minutes at the coldest part of a pack will ensure at least a 10^6-fold reduction of *L. monocytogenes*, the most heat-resistant of the vegetative micro-organisms mentioned above such a treatment will have no effect on the spores of the psychrotrophic strains of *Cl. botulinum.*

Consequently, in the UK the low heat treatments of 70ºC for two minutes are recommended only for short-shelf-life products or those, e.g. in catering operations, for which 3ºC storage is certain to be maintained. (Glew 1990). In the Netherlands, a heat treatment of 72ºC for two minutes has similarly been recommended, aiming at ensuring in excess of a 10^8-fold inactivation of *L. monocytogenes* (Mossel and Struijk 1991).

Although long slow mild heating, as in the original *sous-vide* processes, may sometimes be desirable for organoleptic reasons, it is important to remember that slow heating may trigger the so-called 'heat-shock' response, during which the resistance of vegetative micro-organisms to subsequent heating increases (Mackey and Bratchell 1989). For reasons of microbiological safety, therefore, warm-up or warm-holding times during processing should be short, or increased heat resistance may be found.

Although psychrotrophic strains of *Cl. botulinum* cannot grow at or below 3ºC, the possibility of their slow growth in long-life products at temperatures just above this, demands a more severe thermal process. This process should be designed to reduce substantially, i.e. by more than 10^6-fold, the chance of survival of spores, but there is still debate about the minimum heat required. For instance, Notermans *et al.* (1990) concluded that there still remained insufficient data on the heat resistance of spores of non-proteolytic *Cl. botulinum* to ensure adequate lethality during conventional *sous-vide* processing (see below). They found that surviving spores could germinate, grow and form toxin within about three weeks at 8ºC. Pre-incubation at 3ºC shortened the subsequent time to toxin at 8ºC. They concluded that if storage below 3.3ºC cannot be guaranteed (as is often the case during retailing, and storage in the home), then storage time must be limited. However, it must be said that these products have been on the market for many years, with no recorded microbiological safety problems to date.

The comments above refer to pasteurised vacuum-packed foods that do not rely on any preservation factor other than heating, vacuum-packing and temperature control during distribution for their shelf stability and safety. Such products are normally high water activity, near-neutral pH and preservative-free. Many other pasteurised vacuum-packed products have additional intrinsic preservation designed into them that additionally enhance keepability and safety (CFDRA 1992). For example, salt- and nitrite-containing products such as hams and other cured meat products, acidified pasteurised meat sausages and a wide range of a_w-reduced traditional products, some of which are chill-stable and some even ambient-stable, e.g. the so-called SSPs ('shelf-stable products') of Leistner (1985). The processing, safety and stability requirements of these products should not be confused with those of conventional pasteurised chilled foods.

11.5.4 Pasteurisation for short shelf-life (Classes 1 and 2)

Short-shelf-life products are designed to have a shelf-life of up to 10–14 days. The heat processes they receive during manufacture should cause at least a 6-log reduction in the numbers of infectious pathogens (*Salmonella* and *Listeria*) and their handling after heating and packaging should prevent recontamination. In neutral pH, high water activity products that do not contain antimicrobial preservatives, combinations of temperatures and times equivalent to 70ºC for two minutes are more than adequate for this reduction. But practical experience has shown that longer times in this temperature range are needed for the effective control of some of the non-sporing spoilage bacteria, such as lactic acid bacteria. This combination of time and temperature is also suitable for customers to use to free the products from infectious pathogens.

11.5.5 Pasteurisation for long shelf-life (Classes 3 and 4)

Long-shelf-life products have chilled shelf-lives that are sufficiently long to allow the outgrowth of any psychrotrophic spores surviving in them. Therefore, to ensure their safety and freedom from spoilage during their shelf-life, it is essential that the heat processes used eliminate any spores capable of growth. Therefore, processes designed to ensure safety should be designed to give at least a 6-log reduction in the numbers of cold-growing strains of *Clostridium botulinum*; 90ºC for ten minutes, or an equivalent process, is generally accepted as being sufficient for safety. But a process of this severity is not sufficient to eliminate similarly the spores of all psychrotrophic *Bacillus* species. In unpreserved products some types are able to grow to levels causing spoilage within three weeks or so, at 7–10ºC – temperatures which are known to occur in the chill distribution chains in many countries (Bogh-Sorensen and Olsson 1990). These cold-growing spores often have *D* values at 90ºC of up to 11 minutes (Michels 1979).

11.5.6 Microwaves

Microwave cooking or re-heating of foods, particularly in the home, has expanded greatly in recent years. At the same time, an increasing range of chill, ambient, and frozen-stored foods designed for microwave re-heating have been developed and marketed that meet consumers' desires for improved convenience. Furthermore, it is likely that the use of microwaves to cook or pasteurise foods during processing will continue to grow in the future also. Key issues are the design and preparation of foods that have predictable microwave absorption, it is known that heating is determined by the dielectric of the food and its positioning and thickness within a product container (van Remmen *et al.* 1996). The practical problem is dosing sufficiently accurately on a commercial scale to ensure uniform and predictable heating from microwave absorption.

Whether microwave or other forms of energy generate heat there is no fundamental difference in the lethal effect on micro-organisms. There is no well-documented 'non-thermal' additional microbicidal effect from commercial or domestic microwave equipment. However, there has nevertheless been some concern about the microbiological safety of foods re-heated in domestic microwave ovens; see Sage and Ingham (1998), Tassinari and Landgraf (1997) and Heddleson *et al.* (1996) for discussions of heating variability and its consequences for microbial survival. Particularly if these foods have not been fully pre-cooked, they may have been contaminated after cooking or may contain raw ingredients. The concern has mainly been expressed following the demonstration of the presence of *Listeria monocytogenes* in a wide range of retailed foods, including some suitable for microwave re-heating. For example, a UK Public Health Laboratory Service survey found the organism to be detectable in 25 g samples of 18% of commercially available chill meals tested (Gilbert *et al.* 1989). This concern has led the Ministry of Agriculture, Fisheries and Food in the UK to review the issues thoroughly with oven manufacturers, the food industries, retailers and consumers, and to promote recommendations concerning the proper employment of microwave ovens for reheating – so that effective pasteurisation is achieved.

Although it has been suggested that the unexpected survival of micro-organisms in food heated by microwaves may result from enhanced heat resistance of the micro-organisms (e.g. *Listeria monocytogenes*; Kerr and Lacey 1989). It is now generally accepted that when survival has occurred, it has resulted solely from non-uniform heating, leading to 'cold spots' in particular parts of the product (Lund *et al.* 1989, Coote *et al.* 1991). This is a consequence of heating using microwave energy and the fact that its absorption (and hence the rate of heating) depends, far more than with conventional means of heating, on the composition and quantities of the ingredients, the geometry of the product and its pack. Measurement of the heat-induced inactivation rates of *Listeria monocytogenes* in a variety of food substrates has shown that a 10-fold reduction in numbers is achieved by heating at 70°C for 0.14–0.27 minutes (D_{70} = 0.14–0.27 minutes: Gaze *et al.* 1989). Consequently, the recommendation that *Listeria*-sensitive products should receive a minimum microwave-delivered heating throughout of 70°C for 2 minutes, as in the UK Department of Health guidelines for cook-chill foods, is intended to result in a reduction of this organism of more than 10^6-fold (Anon. 1989). Likewise, if microwaves are used during manufacturing for cooking short-shelf-life chill foods, they should be able to reliably deliver minimum amounts of heat to all parts of the product or ingredient. This is necessary to ensure a satisfactory reduction in numbers of *Listeria* and other, less heat-resistant, vegetative food-poisoning bacteria, i.e. a treatment of 70°C for two minutes, or some other time/temperature combination that delivers equivalent lethality, based on the *D*-values at 70°C of 0.14–0.27 minutes and a *z*-value of between about 6 and 7.4°C (Gaze *et al.* 1989).

11.5.7 Products cooked in their primary original packaging (sous-vide products and REPFEDs)

Although the term '*sous-vide*' strictly refers to vacuum packing, without any indication of thermal processing, it has nevertheless become an accepted name for pasteurised ingredients or foods that are vacuum-packed prior to heat processing, often in their primary packaging (Risk Class 4). *Sous-vide* processed foods include meals and meal components, soups and sauces – all of which have extended chill shelf-lives for use in catering or manufacture and, more recently, retail sale. Sous-vide products are processed at relatively low temperatures, 55+°C. The heat treatment has to be sufficient for them to remain safe and microbiologically stable at storage temperatures below 3°C (the minimum theoretical growth temperature of the cold-growing types of *Clostridium botulinum*). Depending on the severity of the heat process and the microflora of their ingredients, they may have shelf-lives of up to about six weeks, Church and Parsons (1993). An outline for a HACCP study has been published (Adams 1991)

The most comprehensive early research and evaluation of *sous-vide* processing were for catering, at the Nacka Hospital in Stockholm. Prepared foods were vacuum-packed, rapidly chilled, then stored under well-controlled refrigeration for periods of one or two months prior to reheating for consumption (Livingston 1985). This was followed by extension to a variety of more-or-less centralised catering activities in a number of European countries before the more recent extension, mostly in France, to retailed foods.

Concerns about possible microbiological safety problems have been expressed, not because of doubts about the principles underlying the *sous-vide* process, but because of the difficulty in confidently ensuring the maintenance of the required low temperature (max. 3°C) throughout long-distance distribution chains and, especially, in the home (see Betts and Gaze (1995), Juneja and Marmer (1996), Peck (1997) for a discussion of the risks of botulism and Hansen and Knochel (1996) and Ben-Embarek and Huss (1993) for the effectiveness of sous-vide processes against *Listeria monocytogenes*). Turner *et al.* (1996) for effectiveness against *Bacillus cereus* and spoilage bacteria in chicken breast). A proposal for a Code of Practice is given by Betts (1996).

11.5.8 Alternative process routes

Industrial *sous-vide* and REPFED processing procedures mostly follow the original concept, and the conventional canning route, of filling food into packs, sealing, then heating. The alternative – heating, followed by filling and sealing – unless undertaken truly aseptically, risks the introduction of micro-organisms after the heat process and prior to sealing, and is therefore principally used for short-shelf-life products. If products are hot-filled (> 80°C), extended shelf-lives may be achieved when storage is at 3°C. For this type of process, the design of the filling equipment and the type of packaging and sealing machinery are very specialised.

11.6 Safe process design 2: manufacturing areas

Manufacturing areas and production lines should be laid out on the principle of forward flow so that the chances of cross-contamination, or of products missing a process stage, are minimised.

11.6.1 Raw material and packaging delivery areas

Most factories will have designated areas for deliveries; they may be divided into different areas for the various commodities to be processed or according to their storage requirements, e.g. frozen, chilled or ambient-stable. Separation may also be governed by legislative requirements. The delivery area should allow the efficient and rapid unloading of vehicles and it should have facilities for the inspection and batch coding or maintaining batch integrity of incoming raw materials. Its organisation should allow the direct removal of these materials to storage areas; thereby allowing maintenance of storage conditions (e.g. frozen or chilled temperatures). Incoming materials should be protected from contamination and there should be facilities for the removal and disposal of secondary packaging such as cardboard boxes. If product is unpacked, clean containers may be required for product handling and storage prior to use.

Reception areas should be operated to minimise the opportunities for cross-contamination, especially when high-risk materials, which may contaminate other ingredients, are brought in or when materials for direct use in finished products (such as packaging) are handled. Areas must be capable of being effectively cleaned and should not be used for the storage of any packaging that is removed after delivery. It may be necessary to disinfect the outer packaging of materials prior to their admission to hygienic or high-hygiene areas. Ideally, materials that will be used in HCAs should be delivered to separated areas handling only low-risk ingredients and not to areas handling raw materials that are likely to contaminate them with pathogens or with excessive numbers of spoilage bacteria.

11.6.2 Storage areas

Chilled foods are made from a variety of ingredients or components made within the factory, which should all be stored so that contamination and premature spoilage are prevented. These materials differ in the storage conditions they require prior to use and also in the stage of the process at which they are made, added or combined. Therefore a chilled products factory will require a number of designated storage areas. All these areas should be controlled, most commonly for time and temperatures in the chilled (0–3°C) or frozen (below −12°C) ranges, and the rotation of stock should also be controlled. A low humidity store is required for dry ingredients and packaging materials. All temperature controlled areas should be fitted with reliable control devices, monitoring systems (to generate a record of conditions in the store) and an alarm system indicating loss of control or the failure of services. Batches of stored materials should be labelled so that their use by dates and approvals for use are clear and particular deliveries or production batches can be identified.

Layout of the store should allow access to the stored items and effective stock rotation. The operation of storage areas and the quantity of product they contain, their services and control of the means of access, such as doors, should ensure that the designed conditions (such as 2–4°C in chillers) are maintained during the working day. The services, especially supply of refrigerant, should have sufficient capacity to allow maintenance of product temperatures during high outside temperatures or peak demand for cooling. All storage areas should be easily cleaned, using either wet or dry methods, as appropriate, and racking should not be made of wood. The layout of racking and access to floors, walls and drains should allow cleaning to take place.

11.6.3 Raw material preparation and cooking areas

These areas receive ingredients from the storage areas and are used to convert them into components or ingredients by a variety of techniques, such as cutting, mixing, or cooking. Preparation and cooking may take place in a single area where the materials are to be used in cook-in-pack (Class 4) products, but when a long shelf-life and prevention of contamination are important (for Class 2 or 3 products) cooking may be done in a separate area to prevent cross-contamination.

Where the cooked product is to be taken to a hygienic area for cooling and packaging to make Class 2 or 3 ready-to-eat products, it is essential that the layout and operation of the area, and access from the preparation area, are designed to prevent recontamination after cooking. If space permits, cooking operations should be carried out in a separate area from preparation, to minimise the chances of contamination by airborne particles, dust, aerosols or personnel. Where physical separation cannot be achieved, cooked product should not be handled by personnel or with equipment that has been in contact with uncooked material. In rooms or areas where cooking is done, the cooking vessels or ovens may be used to form the barrier between 'dirty' areas, i.e. those handling uncooked material, and 'clean' areas. Air flow should be from 'clean' to 'dirty' areas, and the supply of air to extraction hoods should be designed to ensure that it does not cause contamination of cooked product. Where single-door ovens are used there is an increased risk of cross-contamination, as it is not possible to segregate raw and cooked material effectively. The entry/exit areas from these areas need to be kept clean, loading and unloading done by separate staff, and the opportunities for product contact minimised by the organisation of the area.

For short-shelf-life (Class 1) product and products containing components which have not been reliably decontaminated (Class 2), the main risk is recontamination with infectious pathogens. When the cooking stage is used to provide the 90°C/10 minute heat treatment required for long-life products (Class 3), then stringent precautions must be taken to prevent re-contamination of the cooked product with clostridial spores. These include a forward process layout, physical separation of process stages and control of airflow away from de-contaminated product. Typical process routes for short- and long-shelf-life products are shown in Figs 11.1–4. The control of condensation by the extraction of steam from cooking areas is very important and the effective ducting away of steam is essential if the recontamination of cooked product by water droplets is to be avoided. Preparation rooms or areas should be chilled, but it is often impractical to chill cooking areas, and indeed this may increase problems associated with condensation.

11.6.4 Thawing of product

Prior to use, it may be necessary to thaw frozen ingredients such as meat and fish blocks. This should be done under conditions that minimise the potential for the growth of pathogens, i.e. the maximum surface temperature of the ingredient should not be within their growth range (i.e. it should remain below 10–12ºC). If this is not possible, then thawing times should be minimised to prevent growth. Special thawing equipment such as microwave tempering units, running-water thawing baths or air thawing units may be used for microbiologically safe thawing. Such equipment should be operated according to a technically justifiable specification. Thawing of perishable products at ambient temperatures is not desirable as it can lead to the uncontrolled and unrecognised growth of hazardous micro-organisms. Where frozen ingredients are to be directly heat processed, it is important that the particles or pieces entering the cooking stage of the process are of a uniform size and controlled minimum temperature, so that cold-spots, which will be insufficiently heated, are not accidentally created.

11.6.5 Hygienic areas

Hygienic, not high-care, areas should be used for the assembly or processing of prepared foods made entirely of raw (Class 1), or of mixtures of raw and cooked (Class 2) components. Such areas should be designed and operated to prevent infectious pathogens becoming established and growing in them, by attention to a number of key requirements shown below.

Temperature

Hygienic areas should ideally be chilled to 10–12ºC or below, to limit the potential for growth of *Salmonella*. However, chilling alone will not limit the growth of *Listeria*, as these organisms can grow at chill temperatures. A very effective, if expensive, solution is to operate production areas and chillers dry, with an environmental humidity below 55–65%, which is well below the minimum water activity for the growth of *Listeria* and also low enough to ensure relatively rapid drying of surfaces. A relative humidity in this range is both comfortable for operatives and effective at drying floors, walls, ceilings and equipment. However if control of humidity is not possible, e.g. because an existing area is being used or has been upgraded, then hygienic control should be concentrated in areas known to harbour *Listeria* and these should be effectively cleaned, disinfected and dried on at least a daily basis. To do this effectively is very demanding of time, resources, and training of hygiene operatives and management.

Construction

The standards of construction of hygiene areas for storage and manufacturing should enable hygiene to be managed effectively. Floors, walls and ceilings should be constructed of materials that are impervious to water, are easy to clean, and allow routine access to services from outside the area. Floors should be sloped so that pools of water do not collect and remain on them. Drains, especially, should be designed so that food waste is not retained in them for long periods. The direction of flow should ensure that material from preparation and raw material handling areas does not flow into or through hygienic areas and any drain traps used should be capable of being cleaned to the same standards as food processing equipment. All these precautions are designed to minimise the chances of product contamination during processing and to reduce the chances of infectious pathogens being present.

Stock control

Minimum amounts of material undergoing processing should be kept in hygienic areas or in any associated chillers. Stocks of these components or materials known to contain infectious pathogens or to support the rapid growth of *Listeria*, under chilled conditions, such as prepared vegetables, should be minimised so that storage periods are as short as possible. In summary: operational procedures should be designed to minimise three things:

1. the numbers of *Listeria* or other infectious pathogens brought into the area
2. the opportunities for their growth in the area
3. the number of environments where they can survive.

It is worth remembering that during the periods between cleaning and production, when the area may not be chilled, *Listeria* and maybe other pathogens and spoilage bacteria that may be present, will continue to grow in any dilute food residues left on imperfectly cleaned equipment or in drains and chillers. The layout of the area and the operating procedures should be designed to minimise the opportunities for contamination of the product by equipment or personnel. In addition, the transfer of contaminants from one batch of material to another should be minimised by effective working practices, including a clean-as-you-go policy and a suitable product flow and a batching system providing hygiene breaks at suitable intervals. This is critical if allergenic materials are handled.

Mixed raw and cooked components

Where ready-to-eat products are to be made from mixtures of raw and cooked components, it is most important that any components known to have a high risk of containing pathogens or high numbers of spoilage bacteria are excluded from the area (and the product!) unless they have been effectively decontaminated. In practice, this often means 'cooked'. With some products, such as hard cheeses, prior processing means that they can be used safely in hygienic areas provided that their levels of surface contamination are controlled – to stop cross-contamination leading to spoilage or to a safety risk. It is important to remember that pathogens such as *L. monocytogenes* have been demonstrated to sometimes survive even the processing of spray-dried milk powder, and in cheddar, cottage and camembert cheese (see listings by Doyle, 1988). Examples of higher-risk materials, which should be excluded from such areas, are prawns and other shellfish from warm waters and untreated herbs and spices, which may carry *Salmonella*. However, it must be accepted that from time to time these products and their components will contain pathogens, and hence storage conditions and hygiene practices in manufacturing areas should be designed with this in mind.

11.6.6 High-care areas

General

High-care areas are designed for the post-cook handling, cooling and assembly of ready-to-eat products made entirely of cooked (or otherwise decontaminated) components (Class 3). They should therefore be designed and operated to prevent recontamination with food-poisoning bacteria and minimise re-contamination with spoilage bacteria. High-care areas are used for handling cooked components which will be chilled, assembled and packaged or may be further processed (e.g. sliced, cut or portioned cooked meat products, such as pate), and then packaged. The shelf-life and safety of this type of ready-to-eat product relies very largely on the prevention of recontamination, though a few of them do have, in themselves, preservation systems capable of halting the growth of *Listeria*.

High-care areas are also used for the handling of long-shelf-life (Class 3) products made from components which have been freed of hazardous bacteria (such as cold-growing strains of *Clostridium botulinum*) by the cooking process. All operations downstream from the heating process, including chilling, assembly and packaging prior to the further mild in-pack, heating that frees them of vegetative spoilage microbes must be done in a high care area. For these processes to yield safe product with a reliable long (up to 42 days) shelf-life, effective control of recontamination after cooking must be achieved. There are a number of specific requirements for high-care areas.

Physical separation

Sometimes such areas are designed and operated to a higher standard of hygiene than strictly necessary to guard against contamination by pathogens, in order to limit product recontamination with spoilage micro-organisms, such as yeasts, moulds and lactic acid bacteria. These precautions will minimise the incidence of spoilage during the products' shelf-life. Only materials, including both foodstuffs and packaging, which have been reliably decontaminated and handled to prevent recontamination, should be admitted to high-care areas. Such areas should be physically separated from all other production areas handling components that may still be contaminated and this separation should extend to the staff operating in those areas.

Surfaces

All the surfaces of the high-care area should be sealed and impervious to water and capable of being easily cleaned, disinfected and kept dry. After cleaning and disinfection, surfaces should have fewer than 10 micro-organisms per 9 cm^2 and Enterobacteriaceae should not be recoverable. The drains in high-care areas should not connect directly with areas processing, storing or handling raw materials or used for cooking; their direction of flow should be away from the high-care area.

Chillers and cooling

An integral part of the high-care area will be chillers, either designed to cool components (blast chills; James *et al.* 1987) or to maintain chill temperatures in previously cooled components. Cooling of cooked, hot products should be started as soon as practicable after cooking. As suggested before, cooling rates should be designed to prevent the growth of any surviving spore-forming bacteria. Chillers will often receive naked product and therefore both the quality of the air used for cooling and the environmental hygiene are critical (see below). As warm product will be placed in the chillers, condensation will occur and it is important that this is ducted to drain in such a way that product contamination and the retention of condensate in the area are avoided. The design, hygiene and operating temperature of the cooling elements in fan-driven evaporator units in chillers is critical in limiting the potential for product recontamination by aerosols. If the temperature of the heat exchanger coils is too high, water will condense on them, without freezing, and then be blown onto the products by the draught of the fan. Many evaporator units are designed so that the condense tray does not drain and is inaccessible for cleaning; such units can harbour *Listeria* and cause contamination in high-care areas.

Where water is used as the direct cooling medium – either as a spray or shower or in a bath for products for hermetically sealed containers – chlorination or some other disinfection procedure should be used to ensure that the product will not be recontaminated by the cooling water. Stringent cleaning and disinfection systems should be used to ensure the hygiene of circulation or recirculation systems, including heat exchangers. Packs should be dried as soon as possible after cooling and manual handling of wet packs minimised.

Air supply

The air supply to these areas should be filtered to remove particles in the 0.5–50 micron size range and the system should provide control of the air flow, from clean to dirty, by means of a slight overpressure, which prevents the ingress of untreated air. Air supply and heating and ventilation systems should be designed for easy access for inspection and cleaning.

11.6.7 Waste disposal

The efficient removal and disposal of waste from manufacturing areas is essential to maintaining the hygiene level. Suitable storage facilities and containers should be provided and the design and operation of waste disposal systems should prevent product contamination. Any equipment or utensils used for the handling of waste should not be used for the manufacture or storage of products. Such equipment must be maintained and cleaned to the same standards as the area in which it is used. If waste is not removed frequently (4–8 hourly) then separated refrigerated storage facilities should be provided.

11.7 Safe process design 3: Unit operations for decontaminated products

11.7.1 Working surfaces for manual operations

Many of the operations involved in preparation or product assembly will be carried out on tables or other flat surfaces. It is important that these surfaces are both hygienic and technically efficient, i.e. stainless steel is not suitable as a surface for cutting on, although hygienically it is excellent and easy to keep clean. For cutting, surfaces made of nylon, polypropylene or occasionally Teflon can be used. Working surfaces should enable most operations to be done effectively and also should be easy to remove for cleaning or be completely cleanable without removal. It should be possible to restore the integrity of the surface by machining or some other process, as cut or scratched surfaces are impossible to keep in a hygienic condition and are a potent source of contamination. The use of Triclosan impregnation of surfaces has been recommended to provide a degree of antimicrobial protection on surfaces and polyurethane or fluoropolymer conveyor belts, but Cowey (1997) recommends that it is effective only as an additional line of defence, not as a replacement for existing hygiene procedures.

11.7.2 Cutting and slicing

Many products designed for chilled sale, such as meats and pâtés are prepared or cooked as blocks. Automated slicing or cutting after cooking is an integral part of their conversion into consumer packs. Slicers can be potent sources of contamination because they are usually mechanically complex, providing many inaccessible and uncleanable sites that can harbour bacteria. These bacteria live in the debris continually produced by the cutting operation, which is not effectively removed by cleaning procedures used at the end of production and hence constitutes a microbiological hazard. The effectiveness of cleaning procedures is further reduced if cleaning operatives try to avoid wetting sensitive parts of machines, such as electronic controls, motors and sensors, which may be rendered inoperative by water penetration during cleaning – especially when high-pressure cleaning is used. An integral part of the hygienic design of such machines is good access for cleaning and inspection, and effective water-proofing.

An insight into the routes for microbial recontamination of product can be gained by auditing the machines for product and debris flow during operation, so that the sources and the risks of recontamination can be identified. After production, when the equipment has been cleaned, machines should be reexamined to detect areas where product is still retained, so that difficult-to-clean parts are identified, and effective cleaning schedules specially referring to them developed, implemented and monitored.

The difficulties of controlling the hygiene of such machines typify the much more widespread problem of recontamination of product by process equipment. Slicing and other forming machines handling cooked product generally operate in chilled areas; the intention of this practice is both to improve their technical performance and to slow or halt the growth of microbial contaminants by minimising temperatures in retained product debris. Temperature auditing of this type of machine will show that there are, however, many sources of heat, which are not effectively controlled by environmental cooling, e.g. motors and gearboxes. In a well-designed machine this heat is conducted away into the machine without significant temperature rise, but in other cases there will be localised hot spots in contact with food or debris where microbial growth will occur. It is not uncommon for some parts of these machines to operate at temperatures well above the design temperature of the area.

For example, in the area of the main cutter shaft bearing and motor of a slicer, temperatures of 25ºC+ can be found when material at below 4ºC is being processed in an area operating at 7–10ºC, providing an ideal growth temperature for many bacteria. During operation, such warm spaces in machines may fill with product debris, which is retained, warmed or incubated and then released back into or onto product. Ideally, such hot spots should have been eliminated during the design of the machine, but in practice in many existing machines, these risks can only be minimised. For safe, hygienic operation, the critical areas of machines, i.e. warm ones retaining product, must be identified, and then controlled by suitable cleaning or cooling or other preservation techniques to ensure that microbial growth and product contamination are minimised. In some cases, simple engineering modifications may be used to improve the hygiene of machines by reducing the quantities of product retained and its retention time.

11.7.3 Transport and transfers

From the time that a product is cooked, it can undergo many transfer stages from one production area to another before it is finally put into its primary packaging and contamination is excluded.

Containers

In the most simple process lines, products that have been cooked in open vessels will be unloaded into trays or other containers for cooling. Whatever sort of containers are used they should not contaminate the product, and their shape, size and loading should ensure that rapid cooling is possible. Generally, stainless steel or aluminium trays are more hygienic than plastic ones, because the latter type become more difficult to clean as their surfaces are progressively scratched in use. Plastic also has slower rates of heat conduction than either stainless steel or aluminium, and therefore metal trays are to be preferred.

Belts

In most complex production lines, transfer or conveyor belts may be used for transporting both unwrapped product or product in intermediate wrappings from one process stage to another. Belts may also be an integral part of tunnel or spiral equipment, such as cookers, ovens and coolers. Although many types of belting material are used, they can be broken down into two broad groups: fabric or solid, and mesh or link. For hygienic operation, fabric conveyors should be made of a material that does not absorb water and has a smooth surface finish, so that it is easy to clean and disinfect. Fabric belts are generally used only for transport under chill or ambient conditions, as they are not often heat-resistant. Solid belts used for heating or cooling (by conduction) are usually made of stainless steel, which is hygienic and can be heated or cooled indirectly.

Fabric and other solid belts can be cleaned in-place by in-line spray cleaners, which can provide both cleaning and rinsing and may incorporate a drying stage using an air knife or vacuum system. The hygienic problems of conveyor belts are usually associated with either unhygienic design of drive axles and the beds or frames supporting them or poor maintenance.

Mesh or plastic link belts are used for a wide variety of transport and other functions. Their main advantages are that they can carry heavy loads and form corners or curves. Metal mesh belts are widely used in ovens and chillers where circulation of hot or cold air is a part of the process, for example in spiral coolers or cookers. Belts that are regularly heated, or pasteurised, do not present a hygiene problem, as any material retained between the links will be freed of microbes by the heating. When such belts are used in coolers, or for the transport of unwrapped product at ambient temperatures, special attention must be paid to their potential to recontaminate product with micro-organisms growing in retained debris. Cleaning systems should be devised which remove the debris from between the links (for example, by high pressure spray cleaning on the return leg of the belt). After cleaning, the belt should be disinfected or preserved, for example by chilling, so that microbial growth does not occur during the processing period. With all forms of belt, hygienic performance becomes more difficult to achieve if routine engineering maintenance is not carried out correctly and the belt becomes damaged or frayed during use. Specialised belts are often used in product delivery, packaging and collation systems. In such equipment, incorrect setting, or use with fragile products, will increase the quantity of product waste generated, so that even a properly designed and operated cleaning system will not keep the system in a hygienic condition. Product characteristics, such as stickiness and crumbliness, should therefore be considered when transfer systems are designed, so that the generation of debris and, in turn, hygiene problems are minimised.

11.7.4 Dosing and pumping

Most chilled products are sold in weight-controlled packs, often with individual ingredients in a fixed ratio to one another, for example meat and sauce. Where the ingredients are liquid or include small (ca. 5 mm) particles suspended in a liquid, they may be dosed using filling heads or pump systems. If this type of equipment is used for dosing decontaminated materials, then its freedom from bacteria and its hygienic design and operation is critical to product safety and shelf-life.

Dosing and filling systems may be operated at cold (below 8–10ºC), intermediate (20–45ºC) or hot (above 60ºC) temperatures. The most hazardous operations are those run at intermediate temperatures, which allow the growth of food-poisoning bacteria, and unless the food producer is completely confident of the hygiene of his equipment and is prepared for frequent cleaning/disinfection breaks, such temperatures should not be used. Where hot filling is the preferred option, control of the minimum filling head and residual product temperatures is critical to safety. Often, ingredient target temperatures are set well above the growth maximum of food-borne pathogens (ca. 55–65ºC) to allow for cooling during dosing, and especially for breaks in production when the flow of product is halted. If product is supplied to the filling heads by pipework, it is necessary to ensure that an unacceptable temperature drop (leading to temperatures in the growth range) does not occur at the boundaries of the pipes or during breaks and stoppages. For this purpose a recirculatory loop returning to a heated tank or vessel may be used. Such dosing equipment is often cleaned by CIP (clean-in-place) systems and it is essential that the pumps, valves and couplings used are suitable for this form of cleaning as well as for their production function.

Weight control

Where in-pack pasteurisation systems are used, the reliability of dosing systems in delivering a consistent amount of product plays a major role in ensuring that packs with uniform heating characteristics and headspaces are controlled. Therefore, dosing accuracy should be carefully controlled.

11.7.5 Packaging

Primary packaging

Most chilled products are sold in a packaged form, the most common technical functions of packaging being to prevent contamination and retain the product. Packing materials may be chosen for a variety of technical reasons, such as machinability and heat resistance; but from a microbiological point of view, their most important attributes are the ability to exclude bacteria, physical strength and gas barrier properties. This last property is most important if packaging is part of the product's preservation system; for example, if it contains an inhibitory modified atmosphere or a vacuum. Hence packaging equipment can perform an important function in ensuring product safety, and it must be able to consistently produce strong, gas- and bacteria-tight (hermetic) packs.

Modified atmosphere packs

Many products designed for chilled storage rely on control of the composition of the gas surrounding the product as part of the preservation system. Exclusion of oxygen from the headspace by vacuum packing, or its replacement with either carbon dioxide or nitrogen or a blend of both gases during the packaging operation, gives a considerable extension of shelf-life at chilled temperatures, for example with chilled meats. The efficiency of flushing and replacement of oxygen with the gas mixture, control of vacuum level, and the frequency with which leaking packs are produced are key operational parameters for this type of packaging. Colour indicators have been advocated to indicate leaking packs (Ahvenainen *et al.* 1997).

There are many reasons for the production of leaking packs (leakers) which have a shortened shelf-life and also present an increased risk of contamination. Apart from the reasons (such as incorrect choice of packaging film, or equipment faults such as mis-setting of the sealing head temperature, its pressure, alignment or dwell time), soiling of the seal area of the pack with product during filling is the major cause of pack failure. If product or product residues remain in the seal area during the sealing operation, then the two layers of plastic of the lid and base cannot be welded together by the sealing head. Some sealing heads are profiled to move soiling out of the seal area during the sealing cycle. But with many combinations of sealing head profile and products (i.e. fills) this is not an effective solution to the problem, as food still remains in the seal area, preventing formation of a continuous weld or causing a bridge to be formed across it. Sealing problems are encountered especially when flat or unprofiled sealing heads are used, as these trap material under them during sealing. Problem fills are fat, water (which turns to steam on heating) and cellulosic fibres.

Where the pack is vacuumised, a faulty seal will be evident as the pack will not be gas tight; this will be seen, or can be felt, soon after manufacture. Where the pack has a headspace, a faulty one may feel soft, if squeezed. If packs with the incorrect headspace volume or with weakened seals are processed in systems such as retorts or ovens that generate a pressure differential between the pack and its surroundings, bursting or pack weakening may result.

Other types of pack

Short-shelf-life products may be an exception to the general preference for bacteria-tight packs. Some of these products are sold in packs where a crimped seal is formed between the lid and the base container (which may be made of aluminium, for example). In this type of pack there is a risk of product contamination, unless it is overwrapped in a sealed pouch. The function of such tray packs is to provide a container for oven cooking or reheating by the customer.

Additionally, it is important that the packaging films or trays coming into direct contact with the products do not contaminate them either chemically or microbiologically. Overwrapping of the primary or food contact packaging and its handling within the storage and production areas should be designed to minimise the chances of contamination. Hence handling and disinfection or overwrapping procedures used in the factory should take full account of the destination of the product and the hazards that must be controlled.

Secondary packaging
Once the product is in its primary packaging, it is either completely protected from recontamination (in a hermetically sealed pack) or well protected in a sealed, film wrap or pack with a crimped seal. Primary packs sometimes have additional, secondary packaging. The function of this packaging may merely be decorative, but in some cases it may be to protect the pack from damage or stress during handling or transport. For this latter function, control of the secondary pack characteristics should be a part of the factory QA system.

When chilled products are handled or marketed either in boxes or closely packed on pallets, there is only a limited opportunity for changing their temperature – as the surface area to volume ratio is unfavourable to rapid temperature changes, and rates of heat penetration through product and packaging are low. Therefore it is essential that product in its primary packaging is at the target temperature prior to secondary packaging and palletisation.

11.8 Control systems

11.8.1 Instrumentation and calibration

Wherever control limits are specified in the HACCP plan, it is essential that reliable instrumentation or measurement procedures are present and correctly located and calibrated. Their output can either be used for the control of process conditions (such as during pasteurisation, chilling or storage) or for monitoring compliance with specifications. Sensors and their associated instruments may be in-line (e.g. oven, heat exchanger or fridge thermometers), at the side of the line (e.g. drained weight apparatus, salt or pH meters) or in the laboratory (e.g. colour measurement or nitrogen determination). Wherever the instrument is situated, it needs to be maintained, with its sensor kept free of product debris – as this may produce erroneous signals, it needs to be calibrated and the operative needs to have a measurement or recording procedure.

11.8.2 Process monitoring, validation and verification

Manufacturing, storage and distribution operations within the supply chain should be controlled and monitored to ensure that the whole chain performs within the agreed limits. Wherever possible the data from control systems should be recorded and used to produce management and operative information and trend analyses. Thermocouples can be used to measure product temperatures. The sensors need to be prepared, installed and monitored, to prevent errors in temperature readings (Sharp, 1989). Often responsibilities for the safety and quality a range of products will be shared between several suppliers and producers. Concentration of businesses on their core activities means that vertical integration within the supply chain is uncommon, therefore product safety and quality systems rely on effectively managed and specified customer–supplier relationships. Even if safe process and product design principles originating from a HACCP study have been used and cover the realistic hazards, unsafe products can result if the conditions and procedures noted in the plan are not carried out correctly or are not working effectively.

Verification is an auditing activity that systematically analyses the working and implementation of the HACCP plan, by examining process and product-related data. These data may also be compared with specifications and other technical agreements which are not part of the HACCP but none the less form the customers' requirements. In-house QA departments, regulatory authorities, auditors and those who are inspecting suppliers on behalf of customers, are usually responsible for verification. They act as systems experts working to establish compliance with systems, procedures and the other outputs of the HACCP plan, or in some cases the ISO 9000 series documentation.

As a minimum, verification should focus on data showing the producer's performance at each CCP and use his existing procedures, systems and records, supplemented if necessary with sample analysis, record inspection and auditing to form an opinion on the consistency of product quality or how well the process is controlled. To establish what data should be included in verification, the series of steps in the HACCP plan, hazard analysis, identification of critical control points (CCPs), establishment of control criteria and critical limit values and monitoring of the CCPs, should be considered. Any process stages where the risks of microbial contamination, survival or growth are significant should be covered. The aim is to show how well the workforce and management is complying with the requirements of HACCP plan. Operational risks investigated by verification include poor training and management procedures, poor hygiene and segregation of raw and cooked material, inadequate management of heating, cooling or packaging and occurrence of defective products. It is done routinely after implementation and is part of the review of the scientific and technical content of a HACCP plan.

Reliable verification is based on validation. Validation examines the scientific basis of the HACCP plan and the range of hazards covered. It should be done prior to its implementation of the plan and regularly during production, when it becomes part of the HACCP review procedure. The function of validation is to determine whether realistic hazards have been identified, and suitable process control, hygiene and monitoring measures implemented, along with safe remedial actions for use when processes go outside their control limits. Because many chilled foods are sold as ready to eat and are unpreserved, except by chilled storage, the specification of correct heat processing, prevention of re-contamination and assurance of storage temperatures and times are essential activities and should be directed at specific hazards. Unsuitable process features which should be uncovered by validation include the specification or operation of unhygienic equipment, the recommendation of working practices, targets, controls or layouts, which may lead to product contamination or the design of unstable preservation or distribution systems.

11.8.3 Process and sample data

As chill foods have relatively short shelf-life and are often distributed immediately after manufacture, microbiological results cannot be used for assurance of safety, if they are to have the maximum time available for sale. This is for two reasons.

- Firstly, sufficient numbers of portions cannot be sampled to provide any confidence that processing and ingredient quality were under control for the duration of processing.
- Secondly, the time for a microbiological result will be longer than the pre-despatch time in the factory, even when rapid methods are used.

Microbiological results should be used only for supplier monitoring, trend analysis of process control and hygiene and for 'due diligence' purposes. Data showing the control of CCPs should be taken at a defined frequency and kept for a period equal to at least the shelf-life plus the period of use of the product to verify the performance of the supply chain. Process control records may be generated and retained according to the framework proposed in the ISO 9000 documents on Quality Management.

The provision of conformance samples, to track long-term changes in quality is a problem because of the short shelf-life of the products. Some manufacturers retain frozen samples and accept the quality change caused by freezing. Others take end-of-shelf-life samples and score their sensory attributes against fixed scales or parameters, or use physical measurements of colour or ingredient size.

11.8.4 Lot tracking

Packs of chilled product should be coded to allow production lots or batches of ingredients to be identified. This is a requirement of the EU General Hygiene Directive (93/43). Documentation and coding should allow any batch of finished product to be correlated with deliveries or batches of the raw materials used in its manufacture and with corresponding process data and laboratory records. In practice, the better defined the lot tracking system, the better the chances of identifying and minimising the impact of a manufacturing or ingredient fault. Consideration should be given to allocating each delivery or batch of ingredients a reference code to identify it in processing and storage. Deliveries of raw materials and packaging should be stored so that their identity does not become lost. If there is a fault, re-call or good coding and tracking procedures will facilitate responding to a complaint.

11.8.5 Training, operatives, supervisors and managers

The staff involved in the manufacture of chilled products should be adequately trained because they make an important contribution to the control of product quality and safety. The best way of achieving this is by a period of formal or standardised training at induction, which will enable them to make their contribution to assuring the safe manufacture of high quality products (Mortimore and Smith 1998, Engel 1998). After training they should at least understand the critical aspects of hygiene for food handlers, product composition, presentation and control and the prevention of re-contamination. Many manufacturing operations will be done under conditions of 'High Care' and will involve team work, therefore all staff must be trained in, and understand, the reasons for specified hygiene standards and procedures. Because the products are often ready to eat, or will only receive minimal heat processing by the customer, staff and supervisors should have defined responsibilities to ensure that quantities of low quality or unsafe product produced by the manufacturing process are minimised. The role of external HACCP consultants has been identified (White 1998) as being especially useful to small food companies in helping them put together effective HACCP programmes and training of staff. Health screening may be required to ensure that personnel have the standard of health and personal cleanliness required for the job.

11.8.6 Process auditing

Auditing is the collection of data or information about a process or factory by a visit to the premises involved. Inspection and auditing may be used to determine whether the HACCP plan is correctly established, effectively implemented, and suitable to achieve its objectives, this is especially important where safety relies on many aspects of the process (van Schothorst 1998, Sperber 1998). An integral part of an audit is to see the line operating; effective auditing does more than inspect records. Topics covered should include an assessment of the company structure typified by company policy on quality and safety, the resources and management of the Development department and in-take procedures, storage and handling of raw and packaging materials. Design, control and operation of the manufacturing should be assessed under the headings that include the safe design of products and processes, including matching the products to consumer use and the facilities available. Lastly, the operational aspects of manufacturing should be examined – production, hygiene and housekeeping. Because of the importance of chilled distribution, logistics should be given special attention. How the Quality Assurance Department and the Laboratory contribute to the management of the operation and training should be assessed.

Internal or external auditors may do the auditing, a checklist may be used and often a scoring system or noting of non-compliances may be applied if there is a customer–supplier relationship. A more recent development is supplier self-auditing. This involved the development of the supplier-customer relationship on a partnership basis. It starts with a clear statement of the trading objectives and the resources and materials involved. Next an evidence package covering specifications and records is agreed between the supplier and customer. Because of the variability of quality parameters and process control, a mechanism for challenge must also be agreed, to prevent false rejection of product and the realistic management of any risks or non-compliances. Successful operation of such a system relies on the identification and allocation of responsibilities and ownership of technical factors within both organisations, so that the person best placed to know can always be identified.

This type of audit system has a better cost/benefit than the traditional audit visit and produces information for decision-making on a continuing basis. In some cases it is carried out by externally accredited auditors; their ability to examine a process can be limited by their access to commercially sensitive data on processing. The basis of product safety in such systems is a continuing supply of data, and if properly handled this provides a better level of customer protection than end-product sampling and testing – as by the time microbiological results are available, the product is likely to have been consumed. Procedures for managing process breakdowns and other deviations must be developed and audited to ensure that risks to customers are minimised.

11.9 Conclusions

A major expansion in the sale of prepared chilled foods continues, particularly in Europe, to meet many of the constantly changing requirements of the consumer for convenience, a variety of flavours and tastes and less severe processing and preservation than hitherto. Such foods are processed using less heat, and they contain less preserving agents such as salt, sugar and acid. In addition they are less reliant on additives (e.g. less use of antimicrobial preservatives such as sorbate and benzoate). Most of these requirements are not immediately compatible with an improvement in microbial stability or safety. Indeed most of them lead to a lessening of the intrinsic preservation or stability of foods.

At the same time, assurance of the safety of such foods is an essential consumer requirement, and as such is paramount for the producer. It is here that modern chill food processing and distribution techniques backed up by HACCP procedures (Mayes 1992), properly applied, can more than compensate for the stability and safety that otherwise could be lost. Although complex in detail, as indicated above, the basic elements of effective and safe processing of chilled foods are few. They include:

- the reliable identification of hazardous micro-organisms for products, so that product designs will control them, plus the controlled and predictable inactivation of micro-organisms by processes, storage conditions and product formulations
- the avoidance of recontamination and cross-contamination after decontamination

- control of the survival or growth of any micro-organisms that remain in the food, for example by refrigeration
- the provision of clear consumer use instructions, that are compatible with customer expectations.

Safe effective processing, and distribution, of chilled foods depends on confident and robust control of these four elements, by the means detailed in the sections above.

Although these present means are effective if properly carried out, it is likely that effectiveness could be further improved in the future. This may come mainly from improvements in the control and management of process stages critical to the control of microbes in the product. To a smaller extent, improvements could come through the availability of new processing techniques, such as high hydrostatic pressure to 'pressure-pasteurise' foods. This is likely to be applicable only to products in which (pressure tolerant) bacterial spores are not a problem, e.g. low pH jams and fruit juices (Hoover *et al.* 1989, Smelt 1998). Irradiation (Haard 1992, Grant and Patterson 1992, McAteer *et al.* 1995) also has the potential to inactivate micro-organisms in chilled products, but the changes in sensory characteristics caused may limit the potential of irradiation to extend shelf-life and improve safety.

Finally, I would like to thank Grahame Gould, my co-author of this chapter in the previous edition, for his help during this revision.

11.10 References

ADAIR C and BRIGGS P A, (1993) The concept and application of expert systems in the field of microbiological safety, *Journal of Industrial Microbiology*, **12** 263–7.

ADAMS C E, (1991) Applying HACCP to *sous vide* products, *Food Technology*, **45** (4) 148–9, 151.

ADVISORY COMMITTEE ON DANGEROUS PATHOGENS, (1996) Microbiological Risk Assessment: an interim report, HMSO, London.

AHVENAINEN R, EILAMO M and HURME E, (1997) Detection of improper sealing and quality deterioration of modified atmosphere packed pizzas by a colour indicator, *Food Control*, **8** 177–84.

ANON, (1982) *Guidelines for the handling of chilled foods*, Inst. Food Sci. Technol., (UK), London, pp. 1–27.

ANON, (1984) *Guidelines on precooked chill foods*, HMSO, London.

ANON, (1986) *Health service catering hygiene*, Department of Health & Social Security, HMSO, London.

ANON, (1989) *Chilled and Frozen. Guidelines on cook-chill and cook-freeze catering systems*, Department of Health, HMSO, London.

BAIRD-PARKER A C, (1994) Use of HACCP by the chilled food industry, *Food-Control*, **5** (3) 167–70.

BEN-EMBAREK P K and HUSS H H, (1993) Heat resistance of *Listeria monocytogenes* in vacuum packaged pasteurised fish fillets, *International Journal of Food Microbiology*, **20** 85–95.

BETTS G D, (1996) Code of practice for the manufacture of vacuum and modified atmosphere packaged chilled foods with particular regard to the risks of botulism, Guideline No.11, Campden and Chorleywood Food Research Association.

BETTS G, (1997) Predicting microbial spoilage, *Food-Processing, UK*, **66** 23–4.

BETTS G D and GAZE J E, (1995) Growth and heat resistance of psychrotrophic *Clostridium botulinum* in relation to 'sous vide' products, *Food-Control*, **6** 57–63.

BOGH-SORENSON L and OLSSON P, (1990) The chill chain. In: Gormley, T. R. (ed.) *Chilled foods: the state of the art*, Elsevier Applied Science, London, pp. 245–67.

BRACKETT R E, (1992) Microbiological safety of chilled foods: current issues, *Trends in Food Science & Technology*, **3** 81–5.

BUCHANAN R L, (1995) The role of microbiological criteria and risk assessment in HACCP, *Food Microbiology*, **12** 421–4.

CFDRA, (1990) *Evaluation of the shelf-life for chilled foods*, Campden Food and Drink Research Association Technical Manual no. 28.

CFDRA, (1992) Food pasteurisation treatments: Part 1 Guidelines to the types of food products stabilised by pasteurisation treatments; Part 2 Recommendations for the design of pasteurisation processes, Campden Food and Drink Research Association Technical Manual no. 27.

CFDRA, (1997) Guidelines for Air Quality for the Food Industry, Guideline no. 12.

CHANDARY V, JHONS P, GUPTA R P and SINHA R P, (1990) Lethal effects of pulsed high electric fields on food-borne pathogens, *J. Dairy Sci.*, **73** (Suppl. 1) D92.

CHILLED FOOD ASSOCIATION (CFA), (1997) *Guidelines for Hygenic Practice in the Manufacture, Distribution and Retail Sale of Chilled Foods*, CFA, London.

CHURCH I J and PARSONS A L, (1993) Review: *sous vide* cook-chill technology, *International Journal of Food Science & Technology*, **28**: 563–74.

COOTE P J, HOLYOAK C D and COLE M B, (1991) Thermal inactivation of *Listeria monocytogenes* during simulated microwave heating, *J. Appl. Bact.*, **70** 489–94.

COWEY P, (1997) Food-Processing, UK, **66** 10–11.

DE ROEVER C, (1998) Microbiological safety evaluations and recommendations on fresh produce, *Food Control*, **9** 321–47.

DOYLE M P, (1988) Effect of environmental and processing conditions on *Listeria monocytogenes*, *Food Technol.*, **42** 169–71.

ELLIOT P H, (1996) Predictive microbiology and HACCP, *J. Fd Prot.*, Supplement 48–53.

ENGEL D, (1998) Teaching HACCP – theory and practice from the trainer's point of view, *Food Control*, 9 (2/3) 137–9.

EVANS J, RUSSELL S and JAMES S, (1996) Chilling of recipe dish meals to meet cook-chill guidelines, *International Journal of Refrigeration*, **19** 79–86.

FARQUHAR J and SYMONS H W, (1992) Chilled food handling and merchandising: a code of recommended practices endorsed by many bodies, *Dairy, Food and Environmental Sanitation*, **12** 210–13.

FELON D R, WILSON J and DONACHIE W, (1996) The incidence and level of *Listeria monocytogenes* contamination of food sources at primary production and initial processing, *J. appl. Bacteriol.*, **81** 641–50.

FRAILE P and BURG P, (1998a) Influence of convection heat transfer on the reheating of a chilled ready cooked dish in an experimental superheated steam cell, *J. Food Engineering*, **33** 263–80.

FRAILE P and BURG P, (1998b) Re-heating of a chilled dish of mashed potato in a superheated steam oven, *J. Food Engineering*, **33** 57–80.

GAZE J E and BETTS G D, (1992) Food pasteurisation treatments. CFDRA Microbiology Panel Pasteurisation Working Party Technical-Manual, Campden Food & Drink Research Association; No. 27. Campden Food & Drink Res. Association, Chipping Campden GL55 6LD, UK.

GAZE J E, BROWN G D, GASKELL D E and BANKS J G, (1989) Heat resistance of *Listeria monocytogenes* in homogenates of chicken, beef steak and carrot, *Food Microbiol.*, **6** 251–9.

GILBERT R J, MILLER K L and ROBERTS D, (1989) *Listeria monocytogenes* in chilled foods, *Lancet i*, 502–3.

GILL C O, RAHN K, SLOAN K and MCMULLEN L M, (1997) Assessment of the hygienic performances of hamburger patty production processes, *International Journal of Food Microbiology*, **36** 171–8.

GLEW G, (1990) Precooked chilled foods in catering. In: Zeuthen, P., Cheftel, J.C., Eriksson, C., Gormley, T.R., Linko, P. and Paulus, K. (eds) *Processing and quality of foods Vol. 3. Chilled foods: the revolution in freshness*, Elsevier Applied Science, London, pp. 3.31–3.41.

GOULD G W, (1989) Heat-induced injury and inactivation. In: Gould, G.W. (ed.) *Mechanisms of action of food preservation procedures*, Elsevier Applied Science, London, pp. 11–42.

GOULD G W, (1992) Ecosystem approaches to food preservation. In: *Ecosystems: Microbes: Food*. Soc. Appl. Bact. Symp. Series 20. Suppl. to, *J. Appl. Bact.*, **72** 585.

GRAHAM A F, MASON D R and PECK M W, (1996) Predictive model of the effect of temperature, pH and sodium chloride on growth from spores of non-proteolytic *Clostridium botulinum. International Journal of Food Microbiology*, **31** 69–85.

GRANT I R and PATTERSON M F, (1992) Sensitivity of foodborne pathogens to irradiation in the components of a chilled ready meal, *Food Microbiology*, **9** 95–103.

HAARD N F, (1992) Technological aspects of extending prime quality of seafood: a review, *Journal of Aquatic Food Product Technology*, **1** 9–27.

HANSEN T B and KNOCHEL S, (1996) Thermal inactivation of *Listeria monocytogenes* during rapid and slow heating in sous vide cooked beef. *Letters in Applied Microbiology*; 22: 425–8.

HEDDLESON R A, DOORES S, ANANTHESWARAN R C and KUHN G D, (1996) Viability loss of Salmonella species, *Staphylococcus aureus*, and *Listeria monocytogenes* in complex foods heated by microwave energy, *Journal-of-Food-Protection*, **59** 813–18.

HOLYOAK C D, TANSEY F S and COLE M B, (1993) An alginate bead technique for determining the safety of microwave cooking, *Letters in Applied Microbiology*, **16** 62–5.

HOOVER D G, METRICK C, PAPINEAU A M, FARKAS D and KNORR D, (1989) Biological effects of high hydrostatic pressure on food micro-organisms, *Food Technol.*, **3** 99–107.

ICMSF, (1988) Micro-organisms in foods. IV. HACCP in microbiological safety and quality, Blackwell Scientific Publications, London.

IMPE J F-VAN, NICOLAI B M, MARTENS T, BAERDEMAEKER J-DE and VANDEWALLE J, (1992) Dynamic mathematical model to predict microbial growth and inactivation during food processing, *Applied and Environmental Microbiology*, **58** 2901–9.

INSTITUTE OF FOOD SCIENCE AND TECHNOLOGY (UK), (1998) *Food and drink good manufacturing practice: a guide to its responsible management*, ISBN: 0-905367-15-4.

JAMES S J and BAILEY C, (1990) Chilling systems for foods. In: Gormley, T. R. (ed). *Chilled Foods: the state of the art*, Elsevier Applied Science, London, pp. 1–35.

JAMES S J, BURFOOT D and BAILEY C, (1987) The engineering aspects of ready meal production. In: *Engineering innovation in the food industry*, Inst. Chem. Eng., Bath, pp. 21–8.

JAYKUS L A, (1996) The application of quantitative risk assessment to microbial safety risks, *Critical Reviews in Microbiology* 22: 279–93.

JOUVE J L, STRINGER M F and BAIRD-PARKER A C, (1998) Food Safety Management Tools. Report ILSI Europe Risk Analysis in Microbiology Task Force, 83 Avenue E. Mounier, B-1200 Brussels, Belgium.

JUNEJA V K and MARMER B S, (1996) Growth of Clostridium perfringens from spore inocula in sous-vide turkey products, *International Journal of Food Microbiology*, **32** 115–23.

KENNEDY G, (1997) Application of HACCP to cook-chill operations, *Food Australia*, **49** 65–9.

KERR K G and LACEY R W, (1989) Listeria in cook-chill food, *Lancet ii*, 37–8.

LABUZA T P and BIN-FU, (1995) Use of time/temperature integrators, predictive microbiology, and related technologies for assessing the extent and impact of temperature abuse on meat and poultry products, *Journal of Food Safety*, **15** 201–27.

LEISTNER L, (1985) Hurdle technology applied to meat products of the shelf stable product and intermediate moisture types. In: Simatos, D and Multon, J L (eds) *Properties of water in foods*. Martinus Nijhoff Publishers,

Dordrecht, pp. 309–29.

LIVINGSTON G E, (1985) Extended shelf-life chilled prepared foods, *J. Foodserv. Syst.*, **3** 221–30.

LUND B M, KNOX M R and COLE M B, (1989) Destruction of *Listeria monocytogenes* during microwave cooking, *Lancet i*, 218.

MACKEY B M and BRATCHELL N, (1989) The heat resistance of *Listeria monocytogenes*, *Lett. Appl. Microbiol.*, **9** 89–94.

MACKEY B M and DERRICK C M, (1987) Changes in the heat resistance of *Salmonella typhimurium* during heating at rising temperatures, *Lett. Appl. Microbiol.*, **3** 1316.

MAYES T, (1992) Simple users' guide to the hazard analysis critical control point concept for the control of food microbiological safety, *Food Control*, **3** 14–19.

MCATEER N J, GRANT I R, PATTERSON M F, STEVENSON M H and WEATHERUP S T C, (1995) Effect of irradiation and chilled storage on the microbiological and sensory quality of a ready meal, *International Journal of Food Science & Technology*, **30** (6) 757–71.

MICHELS M J M, (1979) Determination of heat resistance of cold-tolerant spore formers by means of the 'Screw-cap tube' technique. In: *Cold tolerant microbes in spoilage and the environment*, SAB Technical series 13 Academic Press. London & NY, pp. 37–50.

MILLER A J, WHITING R C and SMITH J L, (1997) Use of risk assessment to reduce listeriosis incidence, *Food Technology*, **51** (4) 100–3.

MORTMORE S E and SMITH R, (1998) Standardized HACCP training: assurance for food authorities, *Food-Control*, **9** (2/3) 141–5.

MOSSELL D A A, VAN NETTEN P and PERALES I, (1987) Human listeriosis transmitted by food in a general medical-microbiological perspective, *J. Food Protect.*, **50** 894–5.

MOSSELL D A A and STRUIJK C B, (1991) Public health implication of refrigerated pasteurised ('*sous-vide*') foods, *Int. J. Food Microbiol.*, **13** 187–206.

NATIONAL FOOD PROCESSORS ASSOCIATION (NFPA), (1993) Guidelines for the Development, Production, Distribution and Handling of Refrigerated Foods, NFPA, New York.

NOTERMANS S, DUFRENNE J and LUND B M, (1990) Botulism risk of refrigerated, processed foods of extended durability, *J. Food Protect.*, **53** 1020–24.

NOTERMANS S, GALLHOFF G, ZWEITERING M H and MEAD G C, (1995) The HACCP concept: specification of criteria using risk assessment, *Food Microbiol*, **12** 81–90.

NOTERMANS S, MEAD G C and JOUVE J L, (1996) Food products and consumer protection: a conceptual approach and a glossary of terms, *Int. J. Food Microbiol.*, **30** 175–85.

PANISELLO P J and QUANTICK P C, (1998) Application of food MicroModel predictive software in the development of Hazard Analysis Critical Control Point (HACCP) systems, *Food Microbiology*, **15** (4) 425–39.

PECK M W, (1997) *Clostridium botulinum* and the safety of refrigerated processed foods of extended durability, *Trends in Food Science and Technology*, **8**

186–92.

PIN C and BARANYI J, (1998) Predictive models as means to quantify the interactions of spoilage organisms, *Int. J. Food Microbiol.*, **41** 59–72.

RUTHERFORD N, PHILLIPS B, GORSUCH T, MABEY M, LOOKER N and BOGGIANO R, (1995) How indicators can perform for hazard and risk management in risk assessment of food premises, *Food Science and Technology Today*, **9**(1) 19–30.

RYYNANEN S and OHLSSON T, (1996) Microwave heating uniformity of ready meals as affected by placement, composition, and geometry, *Journal of Food Science*, **61** 620–4.

SAGE J R and INGHAM S C, (1998) Survival of *Escherichia coli* O157:H7 after freezing and thawing in ground beef patties, *Journal of Food Protection*, **61** 1181–3.

SHARP A K, (1989) The use of thermocouples to monitor cargo temperatures in refrigerated freight containers and vehicles, *CSIRO Food Research Quarterly*, **49** 10–8,

SMELT J P P M, (1998) Recent advances in the microbiology of high pressure processing, *Trends in Food Science & Technology*, **9** 152–8.

SMITH J P, TOUPIN C, GAGNON B, VOYER R, FISET P P and SIMPSON M V, (1990) A hazard analysis critical control point approach (HACCP) to ensure the microbiological safety of *sous vide* processed meat/pasta product, *Food Microbiology*, **7** 177–98,

SNYDER O P Jr, (1992) HACCP – an industry food safety self-control program. XII. Food processes and controls. Dairy, *Food and Environmental Sanitation*, **12** 820–3.

SPERBER W H, (1998) Auditing and verification of food safety and HACCP, *Food Control*, **9** 157–62.

STOFOROS N G and TAOUKIS P S, (1998) A theoretical procedure for using multiple response time-temperature integrators for the design and evaluation of thermal processes, *Food Control*, **9** 279–87.

TASSINARI A D R and LANDGRAF M, (1997) Effect of microwave heating on survival of *Salmonella typhimurium* in artificially contaminated ready-to-eat foods, *Journal of Food Safety*, **17** 239–48.

TURNER B E, FOEGEDING P M, LARICK D K and MURPHY A H, (1996) Control of *Bacillus cereus* spores and spoilage microflora in sous vide chicken breast, *Journal of Food Science*, **61** 217–9, 234.

UNITED KINGDOM, ADVISORY COMMITTEE ON THE MICROBIOLOGICAL SAFETY OF FOOD, (1992) Report on vacuum packaging and associated processes ISBN 0-11-321558-4, HMSO, PO Box 276, London SW8 5DT, UK.

USDA, (1988) Time-temperature guidance – cooling heated produce, *Food Safety & Inspection Directive*, **71** 110.3.

VAN NETTEN P, MOSSEL D A A and VAN DE MOOSDIJK A, (1990) Psychrotrophic strains of *Bacillus cereus* producing enterotoxin, *J. Appl. Bact.*, **69** 73–9.

VAN REMMEN H H J, PONNE C T, NIJHUIS H H, BARTELS P V and KERKHIF P J A M, (1996) Microwave heating distributions in slabs, spheres and cylinders with relation to food processing, *Journal of Food Science*, **61** (6) 1105–13, 1117.

VAN SCHOTHORST M, (1998) Introduction to auditing, certification and inspection, *Food-Control*, **9** 127–8.

WALKER S J, (1990) Growth characteristics of food poisoning organisms at suboptimal temperatures. In: Zeuthen, P., Cheftel, J.C., Eriksson, C., Gormley, T.R., Linko, P. and Paulus, K. (eds) *Processing and quality of foods Vol. 3. Chilled foods: the revolution in freshness*, Elsevier Applied Science, London, pp. 3.159–162.

WALKER S J and STRINGER M F, (1987) Growth of *Listeria monocytogenes* and *Aeromonas hydrophila* at chill temperatures, Campden Food Preservation Research Association Tech. Memorandum No. 4652.

WALKER S J and STRINGER M F, (1990) Microbiology of chilled foods. In: Gormley, T.R. (ed.) *Chilled foods: the state of the art*, Elsevier Applied Science, London, pp. 269–304.

WHITE L, (1998) The role of the HACCP consultant, *International Food Hygiene*, **9** 29–30.

ZWIETERING M H and HASTING A P M, (1997) Modelling the hygienic processing of foods – a global process overview, *Food and Bioproducts Processing*, **75** (C3) 159–67.

ZWIETERING M H, KOOS J T-DE, HASENACK B E, WIT J C-DE and RIET K-VAN'T, (1991) Modeling of bacterial growth as a function of temperature, *Applied and Environmental Microbiology*, **57** 1094–101.

ZWIETERING M H, CUPPERS H G A M, WIT J C-DE and RIET, K-VAN't, (1994a) Evaluation of data transformations and validation of a model for the effect of temperature on bacterial growth, *Applied and Environmental Microbiology*, **60** 195–203.

ZWIETERING M H, CUPPERS H G A M, WIT J C-DE and RIET K-VAN'T, (1994B) Modeling of bacterial growth with shifts in temperature, *Applied and Environmental Microbiology*, **60** 204–13.

12

Quality and consumer acceptability

S. R. P. R. Durand, HP Foods Ltd

12.1 Introduction

Quality is an essential feature that will lead the consumer to select or not any food product. With numerous food scares that have hit the food market in the UK and Europe (E. coli, BSE, genetically modified organisms-based products, dioxins in animal feed in Belgium) consumers have become much more aware and therefore selective in their choice to food.

Quality takes many aspects from safety to nutrition, sensory characteristics to service qualities. Consumers integrate these concepts to decide which product to buy according to their own criteria. A safe product is an essential requirement but will never suffice to sell a product. Consumers want an attractive product in terms of organoleptic properties at a price they consider appropriate. It should satisfy their needs in terms of service provided (e.g. convenience, ease of opening) and more and more related to specific nutritional needs (vitamins, functional ingredients, low calories, low salt). In summary, quality is the combination of features in a product which ensure customer satisfaction.

From this definition of quality, it can be seen that customers will repeat the purchase of a chilled product if they are satisfied by its sensory quality, in one word if they are pleasurable to eat. Our senses are extremely sensitive and sophisticated to scrutinise our outside world. For example, our sense of smell, although not as sensitive as some other animal species, can detect hundreds of different odours. The topic of the first and second sections will give an overview of the way we measure quality of chilled products using our senses. It will also show how using a pool of people trained in the detection and description of sensory qualities will give the ability to product developers and quality managers to develop the best product, every time. However, understanding quality using trained people will only tell you one side of the story; how is a chilled product characterised?

In an increasingly competitive market, food manufacturers have realised the need to ask their existing and potential customers what they like and dislike through preference tests. This will form the second side of the story and the section dedicated to consumer acceptability will provide an exhaustive list of preference tests used to understand liking patterns. The final section will focus on combining these two sets of information, subjective and objective, to develop consumer-driven specifications that will fulfil their expectations.

12.2 What defines sensory quality?

The sensory quality of a product can be divided into its appearance, smell, texture, flavour and also aftertaste.

12.2.1 Appearance

The appearance of a product is the first assessment that a consumer will undertake to define the quality of a chilled product. Many aspects of the visual component of a food product can be used to assess its quality. The size, the shape, the distribution of pieces, the surface texture, the colour and the brightness are all determining factors to assess freshness and overall quality expectation. The dullness or sheen of the surface in red meat combined with the redness of the flesh are used by consumers to assess the freshness of meat products in supermarkets. In return, supermarkets use lighting and in some instance spraying a thin layer of water to improve the visual quality of meat behind counters.

The appearance can be divided into optical and visual structure components. The optical appearance is mainly related to colour, gloss and translucency. The visual structure is linked to the texture of the product, including particle size, smoothness and surface texture. Colour is probably the first and main characteristic that a customer will use to judge the quality of a chilled product, as the deterioration of food is often linked to a colour change (Piggott 1988). Colour is also involved with a psychological dimension; red is associated with power, orange and yellow with excitement and cheerfulness. The food industry is therefore spending a considerable amount of resources on eye-catching properties. In particular, farmed salmon are fed with carotenoids to give them a pink/orange colour, further enhanced when smoked, which is considered to be a sign of quality and freshness in the mind of the consumer.

12.2.2 Odour: type, intensity

The smell of chilled food gives a good indication of the freshness and quality. An odour is detected when volatiles are inhaled into the nasal cavity and make contact with the olfactory system. Our smelling system is more modest equipment than our vision system, detecting only 10,000 odours with 5 million receptors, versus millions of colours with more than 100 million receptors (Meilgaard *et al.* 1987). It is however highly efficient in the detection of spoilage, off-odours or taint and often more sensitive and accurate than many sophisticated instruments. When cutting a piece of meat or fish, the level of freshness can be assessed by smell. Jorgensen *et al.* (1988) has shown a strong relationship between the detection of malodorous volatiles and the spoilage of chilled fish during shelf-life as detected by a pool of trained people.

12.2.3 Flavour

Flavour is possibly the essential sensory component used to measure the quality of chilled products. If the flavour is undesirable or does not correspond to expectations, the product will be rejected. Flavour is defined as the sum of perceptions perceived in the mouth in the back of the throat and the nose via the retro-nasal route (Piggott 1988). Flavour includes the primary tastes (salty, sweet, acidic and bitter) caused by soluble substances, sensation factors such as astringency, heat or cooling effects and the aroma perception caused by volatile substances. Quality criteria associated with flavour relate to the expression of an expected flavour and the intensity of it.

12.2.4 Texture

Texture can be defined as the sensory perception on the physical structure of a food product. During the handling and preparation of food, texture properties can be measured by visual evaluation and touch to identify its overall quality. A hard cheese over matured will look dry with a sensation of mouthdrying and roughness in the mouth. Cheese makers use an agreed methodology to define the quality of cheese in relation to its texture characteristics.

Texture is a complex area of the evaluation of sensory quality, as described and classified in reviews by Bourne (1982) or Civille and Liska (1975). There are three different texture characteristics.

1. The mechanical dimension is related to the reaction of the food to stress, such as hardness, firmness, cohesiveness or chewiness, as measured by the muscles of the hand, fingers, lips, tongue or jaw.
2. The geometrical dimension is related to the arrangement of the physical components of a product such as size, shape, fibrousness, particles or lumps.
3. The surface dimension is related to the moisture and fat content of a product and how they are released during a chewing process.

A large part of enjoying a meat product involves its texture quality and consumers are well aware of words such as tenderness, chewiness or toughness. Slaughtering methods and storage conditions have an influence on the texture quality of fresh fish as described by Love (1988). For pastry-based products such as quiche or pizza, texture is the criterion that influences and therefore best predicts its quality. The pastry tends to adsorb moistness from the other components to become soggy prior to any development of off-flavour.

12.3 Sensory evaluation techniques

From the previous section, it is obvious that any test to measure quality of chilled food should involve the use of human subjects.

> Sensory evaluation is a scientific discipline involved with the measure, study and interpretation of responses to food properties as perceived by the senses of sight, smell, taste, touch and hearing (IFT 1975).

Sensory assessment is done either by a small pool of people (typically 8 to 30 people) who have received some training, known as 'objective testing' or by a larger pool of consumers who give their own opinion on a product or range of products without any prior training. This is known as 'subjective testing'. The objective and hypotheses defined in any project brief will determine the choice of the type of evaluation to be performed, and the type of panel, the test and the overall design of the experiment. Carpenter *et al.* (2000) defines in detail the criteria to consider.

Objective testing is carried out by qualified and trained people and can be used for discriminating and describing differences between chilled products. This pool of people, or sensory panel, can be employees of a company or a dedicated workforce. After initial recruitment to measure any sensory impairment such as anosmia (impairment in the sense of smell) or ageusia (impairment in the sense of taste), the panel members take part in an extensive and gradually more difficult training schedule in order to describe, discriminate and evaluate any subtle differences between the products under investigation. Many sensory professionals have published training schedules for general training sessions (Jellinek 1985) to more specific and precise sensory programmes (Civille and Szczesniak 1975). Tests used to characterise food properties by sensory experts can be either discriminative or descriptive.

12.3.1 Discriminative tests

These types of tests are used when it is required to identify if any difference exists between two or more products. It might be required to change an ingredient supplier for cost reduction or quality reasons and the brief is to confirm that the overall flavour, texture or appearance is not affected. It is more cost effective to ask a highly trained panel of experts in their field to assess any difference rather than asking consumers. The tests are very sensitive to any sensory variation as they involve a direct comparison. The triangle test is the most commonly used discriminative test. In this test, panellists are given three samples, in a pre-determined order, one is different from the other two. They are asked to identify the odd sample (BS 5929: part 5, 1988). For example, a ham manufacturer might want to assess the effectiveness of a controlled atmosphere packaging against a present system without changing the overall sensory properties over time. The test sensitivity can allow expensive capital expenditure requirements to be made with a high level of confidence.

When more than two products are compared, ranking tests or multiple comparison tests are used. Typically, a ranking test is used when no control is available or required and assessors are asked to rank products in order of intensity for a specific sensory characteristic. A multiple comparison test is used when a control sample is used as an anchor point and assessors are asked to evaluate the intensity of the difference if any.

Other tests might be required from time to time to evaluate the sensitivity of specific compounds and/or chemicals. These tests are important in taint evaluation or the detection of materials difficult to assess. Threshold tests are used to detect at which concentration level, a compound can be detected. Dilution technique determines the smallest amount of compound that can be detected in a product. Gillette *et al.* (1984) use this technique to assess the level of heat in red peppers.

12.3.2 Descriptive tests

These tests are used to identify sensory characteristics of a chilled product and to quantify them. Panellists are selected on their ability to describe and discriminate between samples. They are presented with variants of the product and asked to describe them. After a period dedicated to confirm and define agreed terms and scale, the panel is then presented with the samples, one at a time and asked to give a score. Stone *et al.* (1974) has described in detail one of these descriptive techniques called 'quantitative descriptive analysis'. Results are analysed by means of univariate and multivariate analyses. Analysis of variance (O'Mahony 1986) is used to measure any difference between samples for each attribute. Principal component analysis is a technique used to reduce the amount of dimension or sensory terms into a manageable format, usually two or three dimensions. Procustes analysis is mainly used for the assessment of individuals' performance and efficiency (Arnold and Williams 1986). Graphs are often used to summarise the results into condensed and meaningful information. Spider graphs (see Fig. 12.1), give a general overview of sensory differences by the overall shape of each product, providing an individual product fingerprint.

Other sensory techniques have emerged in recent years in the evaluation of food quality by sensory evaluation. In particular, the time-intensity technique takes into consideration the temporal dimension of tasting behaviour (Cliff *et al.*, 1993). This technique is particularly interesting for the assessment of spicy products or for primary taste evaluation, such as sweet sensation in artificially sweetened drinks and is used by Matysiak and Noble (1991).

Figure 12.1 shows an overview of how four samples of chilled smoked salmon differ in sensory terms from one another, using a spider graph. For each attribute, the samples are more intense away from the centre. In particular, salmon D had a more orangy colour and sample F had the least moist surface. Sample I was the least salty and smoky product.

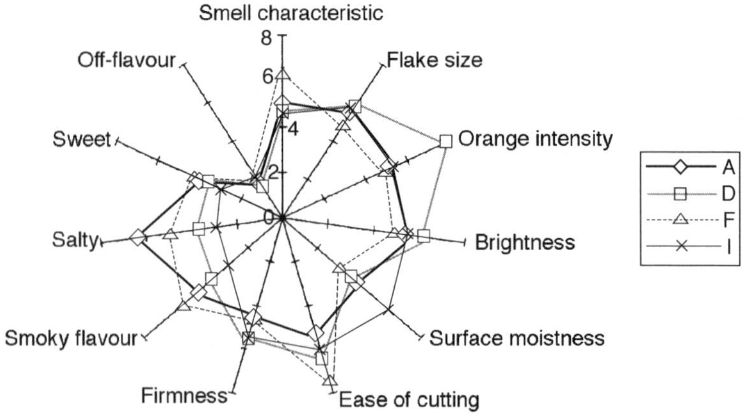

Fig. 12.1 Differences in sensory profile between four different chilled smoked salmon.

12.3.3 Use of a trained panel for measurement of sensory quality

Taint and off-flavours

In foods these are a major threat to manufacturers and retailers, as it can become extremely damaging and costly. A tainted product reaching the consumer can create problems for the food producer or retailer much more damaging than the complaint itself. Direct costs such as loss of production, cleaning of factory, damaged commercial relationships between suppliers, manufacturers and retailers (e.g. de-listing by retailers), litigation proceedings and even factory shutdown may occur. Other more substantial financial implications may arise, such as loss of customer goodwill and damaged brand image may manifest themselves through lower sales and loss of market share. Taint and off-flavours are caused by the presence of a chemical at very low concentration in a food product, usually a volatile organic compound, which imparts a flavour unacceptable to the consumer. Taint is an unpleasant odour or flavour imparted to food through external sources and off-flavour (or off-odour) is an unpleasant flavour (or odour) to food through internal changes, such as enzymatic or microbial activities. Unlimited sources of taint and off-flavours exist making it extremely difficult for manufacturers or retailers to control. Raw materials, packaging materials, factory environment (e.g. flooring, paint, cleaning agents), microbial spoilage are all known sources of taint and off-flavour. Water can also be a major source of contamination. It may contain chlorophenols or chloroanisol; this accounts for the largest number of known cases of taint. Sensory evaluation coupled with the use of sophisticated apparatus such as Gas Chromatography Mass Spectrometry (GCMS) is required to evaluate and identify the taint and its source. Such a combination was used by Farmer *et al.* (1995) to evaluate off-flavours in wild and farmed Atlantic salmon. The author found that the main difference was between wild salmon caught in rivers and those from the sea rather than between wild and farmed salmon. River-caught wild salmon showed enhanced earthy flavour and odour notes, with 2-methylisoborneol and geosmin being the major compounds involved.

Sensory monitoring of quality

The quality of chilled products should be consistent during production and food companies are using various techniques to ensure food quality conformity. These techniques have a common ground in that one or more parameters are measured against an agreed quality specification. For example, pH, salt or sugar levels might be measured to ensure no drift to the final composition of a product. These measurements, although important in their own right, will not allow the definition of flavour characteristic or texture quality. Sensory evaluation is therefore used to ensure integrity of a series of set quality parameters. The stages involved in setting up such routine sensory quality monitoring include the establishment of a product standard that represents the customer requirements. The product standard is translated into key sensory characteristics. The second stage involves the definition of acceptance ranges for each parameter, taking into account commercial risks against consumer loyalty. In the case of sensory quality assessment, a panel is selected and trained to recognise products that fall within and outside the acceptable range for each sensory parameter. It is essential that employees involved are committed to quality and that more than one assessor is used for each assessment to ensure precision and reliability. Standard procedures are set up and may include go/no-go quality rating or grading such as the CCFRA frozen and canned products specification (Rodway *et al.* 1999), difference test or more precise but lengthy descriptive analysis.

A specification sheet that includes acceptance ranges (Table 12.1) and a quality monitoring chart for a flavour parameter (Fig. 12.2) are provided as an example for the assessment of the quality of a chilled apple pie. The members of a panel are asked individually to assess each sample for each attribute using a 10-point scale, from not present to very strong. A consensus score or mean score is then entered into the collation sheet showing the non-acceptable range for each attribute (shaded areas). One or more scores outside the non-shaded area indicate a product that does not meet quality criteria and further investigation is required to confirm the findings. Figure 12.2 displays the trend of the change of strength of flavour characteristic for a chilled apple pie product over a production time span of six months. For each production batch evaluated, a score outside the two specification limits is considered as unacceptable. The strength of flavour varies considerably over this production period and an investigation should be organised to understand reasons for this variability and how to correct it.

Storage effect/definition of shelf-life

Shelf-life of chilled products can be defined as the period between manufacture and consumption during which the product is in an acceptable condition, both in

Table 12.1 Specification sheet for collation of data for flavour and texture of chilled apple pie

Flavour	0	1	2	3	4	5	6	7	8	9
Strength of apple										
Sweetness										
Acidic										
Oily/fatty pastry										
Off-flavours										
Texture										
Doughiness of pastry										
Crispness of apple										
Greasy mouthfeel										

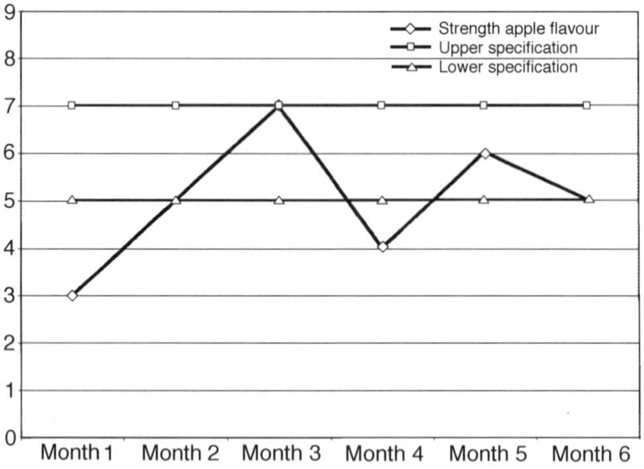

Fig. 12.2 Quality monitoring over a six month period for strength of flavour of a chilled apple pie.

terms of safety and quality. During storage, a chilled product will undergo changes linked to chemical, microbiological and physical reactions. It has been shown that for many products, sensory evaluation has proven to be the most sensitive technique to assess these changes. Similar techniques to those described for quality monitoring can be applied to sensory shelf-life evaluation. Claasen and Lawless (1992) found that sensory analysis by descriptive technique to measure the end of life of liquid milk was more sensitive than more traditional analytical techniques.

12.4 Determining consumer acceptability

There is no doubt that the taste of food is an essential criterion for the acceptability of a product by the consumer. The best marketing and advertising campaign for a new chilled product might convince a consumer to buy it for the first time. However, should this product not attain expected values, the consumer will not purchase the product again. Although taste undoubtedly plays an important part in food choice, it is not the only essential element. Attitudes, belief, nutritional awareness, brand and image, convenience, price and other socio-cultural aspects (e.g. religion, education) are all important criteria the market researcher will take into account. Although expert sensory panels are useful to understand the sensory properties of a chilled product, when consumer acceptability and behaviour is concerned, tests involving consumers must take place.

12.4.1 On-site trials for screening purposes

These tests are used by many food companies to assess the sensory quality of their products, including products under development versus competitors' ones. These tests are very useful for the screening of several samples prior to a full-scale assessment, in order to reduce costs and optimise tight deadlines. A minimum number of consumers are required in order to be confident of the result. It is advisable to involve at least 50 employees. The main negative aspect to this type of assessment is the knowledge employees have acquired working in contact with the product and inevitably biases may occur in their choice. The researcher should minimise these biases hiding recognisable clues when possible.

12.4.2 Home placement tests

These tests imply that consumers are asked to try and assess one or more products while at home. The advantage is that the assessment is carried out within a normal life situation and is therefore more likely to represent the true behaviour and liking for a product. It also has the advantage of getting an opinion from the whole family rather than just one individual. However, the drawback is that the control of the response is poor. In particular, different cooking procedures and processes may be used by different consumers, making a direct comparison more complicated. Also, when two or more products are assessed, the cooking procedure between products or the time interval between cooking and eating may differ.

In a home location test, samples may be delivered to homes or collected at a central location. After a certain period, the researcher will interview the consumer by phone or face-to-face by calling in. Questions such as liking for each product by each member of the household, type of cooking and serving, expectation, likelihood of purchase may be asked.

12.4.3 Central location or hall tests

When a consumer test must be controlled more closely, or the number of products exceeds two or three, a central location test in busy town centres is preferred. Interviewers stop potential participants and ask specific questions to ensure that they meet agreed criteria, i.e. that they fall within set quotas and to ensure their willingness to participate. This test tends to be preferred by food companies for being cheaper to run than home location tests. Hedonic scales are usually used to measure liking of products. Consumers are asked their liking for each product and also the preferred sample overall. The objective of the research will dictate the type of samples to be assessed and the questions asked.

12.4.4 Food choice and attitude

The above techniques all use hedonic scales that mainly reflect the preference for the product. However, when one is interested in the reasons for choosing certain foods, it is important to encompass other components to determine attitude towards foods. In particular, the cognitive and intention components. Many researchers have tried to incorporate more than one factor to explain what dictates choice. When consumers want to treat themselves to a high-quality piece of meat, they may be more inclined to go to a local butcher and select, often after discussion and advice, a quality sirloin steak.

The experience of buying and the context in which it is carried out becomes an integral part of the decision process and choice. Lange (1999) studied how name and packaging information could affect orange juice liking under economical constraint. The author observed differences between hedonic responses under blind conditions when various information and economical constraints were given. Other researchers have investigated the influence of health and nutrition information on liking (Kähkönen and Tuorila 1995). Contextual analysis has been extensively studied by Schutz (1994) to identify how, when, where and by whom specific food products are eaten.

12.4.5 Qualitative research

Qualitative research is carried out when the objective of the study is to understand behaviour and attitude rather than preference, although a combination of both is often considered. Focus groups and individual interviews are two types of qualitative techniques. In a focus group test, eight to ten consumers are recruited and asked to participate in a discussion around a specific topic. To manage such a group, a moderator is used who leads the discussion using specific elicitation techniques to gain information related to the topic. In particular, it is useful to use open-ended questions and probing questions to get the most useful information from the group. A carefully prepared questioning plan is used by the moderator to guide the discussion towards the main areas of interest.

In a recent study, a focus group technique was used to investigate whether a new concept of pre-cooked bacon would succeed in the market-place. Reactions to the idea, the product and various packaging formats formed the basis of the question guide. Convenience and healthiness were found to be the main aspects that would attract the consumer to this new concept. In particular, the issue was not whether bacon was healthy but rather the cooking method used, e.g. frying ...'anything's bad if you fry it'. Also, a microwavable bacon that would retain its sensory characteristics, i.e. crispness, was appealing to the housewives in particular.'I spend half my time in the kitchen and I think that's wonderful'.

Individual interviews are used when topics are more sensitive or when no group interaction is either wanted or needed. Various elicitation techniques are used, often similar to those used for focus groups. One technique that is used mainly in an interview environment is the Repertory Grid Method. Objects under study are arranged into groups of three, such that each object appears at least once and one object is carried over to the next trio. Two of these three objects are similar according to the criterion defined by the researcher and different from the last object. Each interviewee is asked to describe how one object differs from the other two. Scaling has sometimes been used to assess the level of difference or similarity for each consumer and each descriptor (Gains 1994).

12.5 Future trends and conclusion

Better understanding of consumer needs and expectations is an essential aspect to build consumer loyalty and generate new custom. This chapter has shown that both sensory and consumer assessments are required to fully understand consumer behaviour. It has recently emerged from researchers and also food companies that the use of an integrated approach to product quality and new product development lead to sustainable market success. Consumer-driven specification becomes very much a current trend in the food industry. After years of production-driven specification, food companies have realised that customer complaints increase while loyalty does not. The potential of consumer input into the development of food quality specifications has been described by McEwan (1999) to relate sensory characteristics (as defined by a trained panel) and the acceptability range as defined by the consumer. The research involved canned grapefruit, and it was shown that the main deterrent to quality acceptability was found to be any physical defect that was perceived to have a negative effect on the texture and flavour. On the other hand, ragged edges and broken segments were thought to have less adverse effects than previously thought by the industry. Bech *et al.* (1997) has put in place an overall consumer-driven quality tool, House of Quality, where consumer needs are translated into measurable sensory attributes using frozen peas as a vehicle for the demonstration. These principles can be applied to chilled foods.

Consumer integration at an early stage of product development is becoming widely spread amongst food manufacturers, reducing time and cost, often related to product failure when rejected at first glance by the consumer. In the UK, it is estimated that 90% of new products introduced to the market each year will be de-listed from the retailers after six months on the shelves.

New and exciting techniques have seen great potential in recent years. In particular, preference mapping used to relate sensory data to consumer acceptability has proved very useful to both marketing and product development teams alike (Risvik *et al.* 1997). This technique allows researchers to focus more closely to the optimum product from a sensory point of view as defined by the consumer.

Major food scares have had in recent years a huge and detrimental effect on food acceptability, and in particular beef and GMO-related products. Consumers have made their voice heard and this has resulted in major and essential policies developed by governments. Food manufacturers have realised the importance of understanding consumer behaviour better and identifying the mechanisms for food choice and acceptability. Traceability is becoming a very hot topic for food manufacturers and retailers. The integration of total quality systems and closer control is being put in place in ever growing international and complex markets (see Chapter 15).

12.6 References

ARNOLD G M and WILLIAMS A A, 'The use of Generalised Procustes Techniques in Sensory Analysis', in *Statistical Procedures in Food Research*, Piggott J R, 1986, Elsevier Applied science, London.

BECH A C, HANSEN M and WIENBERG L, (1997) 'Application of House of Quality in Translation of Consumer Needs into Sensory Attributes Measurable by Descriptive Sensory Analysis', *Food Quality and Preference*, **8** (5/6) 329–48.

BOURNE M, (1982) *Food Texture and Viscosity*, Academic Press, London.

BSI, 'Methods for Sensory Analysis of Foods: Part 5: Triangle Test', BS 5929, 1988.

CARPENTER R P, LYON D H and HASDELL T A, (2000) *Guidelines for Sensory Analysis in Food Product Development and Quality Control*, Aspen Publishers Inc., Maryland.

CIVILLE G and SZCZESNIAK A S, (1975) 'Guidelines to Training a Texture Profile Panel', *Journal of Texture Studies*, **4** 204–23.

CIVILLE G and LISKA I H, (1975) 'Modifications and Applications to Foods of the General Foods Sensory Texture Profile Technique', Journal of Texture Studies, **6** 19–31.

CLAASEN M and LAWLESS H T, (1992) 'Comparison of Descriptive Terminology Systems for Sensory Evaluation of Fluid Milk', Journal of Food Science, **57** 596–601.

CLIFF M and HEYMANN H, (1993) 'Development and Use of Time-Intensity Methodology for Sensory Evaluation: a Review', *Food Research International*, **26** 375–85.

FARMER L J, MCCONNELL J M, HAGAN T D J and HARPER D B, (1995) 'Flavour and

Off-Flavour in Wild and farmed Atlantic Salmon from Locations around Northern Ireland', *Water Science and Technology*, **31** (11) 259–64.

GAINS N, 'The Repertory Grid Approach', pp. 50–76, in *Measurement of Food Preferences*, ed. by MacFie H J H and Thomson D M H, 1994, Blackie Academic and Professional, Chapman and Hall, Glasgow.

GILLETTE M H, APPEL C E and LEGO M C, (1984) 'A new Method for Sensory Evaluation of Red Pepper Heat', *Journal of Food Science*, **49** (4) 1028–33.

INSTITUTE OF FOOD TECHNOLOGISTS, (1975) 'Shelf-life of Foods', *Food Technology in Australia*, **27** (8) 315–19.

JELLINEK G, (1985) *Sensory Evaluation of Food: Theory and practice*, Ellis Horwood, Chichester.

JORGENSEN B R, GIBSON D M and HUSS H H, (1988) 'Microbiological Quality and Shelf-life Prediction of Chilled Fish', *International Journal of Food Microbiology*, **6** 295–307.

KÄHKÖNEN P and TUORILA H, (1995) 'The Role of Expectations and Information in Sensory Perception of Low-Fat and Regular Fat Sausages', *Appetite*, **24** 298–9.

LANGE C, ROUSSEAU F and ISSANCHOU S, (1999) 'Expectations Liking and Purchase Behaviour under Economical Constraint', *Food Quality and Preference*, **10** 31–9.

LOVE R M, 1988, *The Food Fishes, Their Intrinsic Variation and Practical Implications*, Farrand Press, London.

MATYSIAK N L and NOBLE A C, (1991) 'Comparison of Temporal Perception of Fruitiness in Model Systems Sweetened with Aspartame, and Aspartame-Acesulfame K Blend or Sucrose', *Journal of Food science*, **56** (3) 823–6.

MCEWAN J A, (1999) Consumer Input into the Development of Food Quality Specifications, personal communication.

MEILGAARD M, CIVILLE G V and CARR B T, (1987) *Sensory Evaluation techniques*, Vols I and II, CRC Press, Florida.

PIGGOTT J R, (1988) *Sensory Analysis of Foods*, Elsevier Applied Science, London.

O'MAHONY M, (1986) *Sensory Evaluation of Food, Statistical Methods and Procedures*, Marcel Dekker Inc., New York.

RISVIK E, MCEWAN J A and RODBOTTEN M, (1997) 'Evaluation of Sensory Profiling and Projective Mapping Data', *Food Quality and Preference*, **8** (1) 63–71.

RODWAY E C and THE QUICK FROZEN SPECIFICATION WORKING PARTY, (1999) 'Quick Roasting Potatoes', Campden Food Specification: **L100**. Campden and Chorleywood Food Research Association, UK.

SCHUTZ H G, (1994) 'Appropriateness as Measure of the Cognitive Contextual Aspects of Food Acceptance, pp. 25–50, in *Measurement of Food Preferences*, ed. by MacFie H J H and Thomson D M H , Blackie Academic and Professional, Chapman and Hall, Glasgow.

STONE H, SIDEL J, OLIVER S, WOOLSEY A and SINGLETON R C, (1974) 'Sensory Evaluation by Quantitative Descriptive Analysis', *Food Technology*, **28** (11) 24, 26, 28–29, 32, 34.

13

The hygienic design of chilled foods plant

J. Holah and R. H. Thorpe, Campden and Chorleywood Food Research Association

13.1 Introduction

The primary concern of chilled food manufacturers is to produce a product that is both wholesome, i.e. it has all the fresh, quality attributes associated with a chilled food, and safe, i.e. free from pathogenic microorganisms and chemical and foreign body contamination. This is particularly important in this product sector as, due to the nature and method of production, many chilled foods are classified as high-risk products.

The schematic diagram shown in Fig. 13.1, which is typical for all food factories, shows that the production of safe, wholesome foods stems from a thorough risk analysis. Indeed this is now a legal requirement. The diagram also shows that given specified raw materials, there are four major 'building blocks' that govern the way the factory is operated to ensure that the safe, wholesome food goal is realised. Hygienic design dictates the design of the production facility and equipment whilst process development enables the design of safe, validated processes. Hygienic practices and process control subsequently ensure the respective integrity of these two dependables.

Risk analysis encompasses identifying the hazards that may affect the quality or safety of the food product and controlling them at all stages of the process such that product contamination is minimised. In the food industry this is commonly referred to as Hazard Analysis Critical Control Point (HACCP).

Such hazards are usually described as

- biological, e.g. bacteria, yeasts, moulds
- chemical, e.g. cleaning chemicals, lubricating fluids
- physical, e.g. glass, insects, pests, metal, dust.

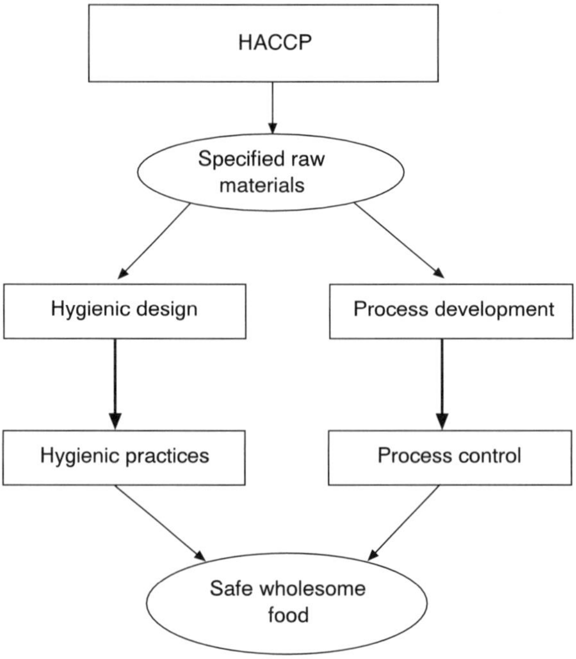

Fig. 13.1 Schematic stages required to ensure safe, wholesome chilled products.

A hazard analysis should be undertaken at the earliest opportunity in the process of food production and if possible, before the design and construction of the processing facility. This allows the design of the production facility to play a major role in hazard elimination or risk reduction.

Of the four building blocks illustrated in Figure 13.1, this chapter deals with hygienic design. For the food factory, hygienic design begins at the level of its siting and construction and is concerned with such factors as the design of the building structure, the selection of surface finishes, the segregation of work areas to control hazards, the flow of raw materials and product, the movement and control of people, the design and installation of the process equipment and the design and installation of services (air, water, steam, electrics, etc.).

With regard to legislation, there are some EEC Directives relating to the production of certain foodstuffs, such as meat, fish and egg products, in which requirements for the premises are specified. On the 14 June, 1993, however, a Council Directive *on the hygiene of foodstuffs* was adopted (Council Directive 93/43/EEC). This Directive applies to the production of all foodstuffs and it is more specific than any previous regulations. The first of ten chapters covers the general requirements for food premises and the second the specific requirements for room where foodstuffs are prepared, treated or processed; only dining areas and premises specified in Chapter 3, e.g. marquees, market stalls etc. are excluded. Within all of these documents, however, advice is at best, concise.

13.2 Segregation of work zones

Factories should be constructed as a series of barriers that aim to limit the entrance of contaminants. The number of barriers created will be dependent on the nature of the food product and will be established from the HACCP study. Figure 13.2 shows that there are up to three levels of segregation that are typical for food plants.

Level 1 represents the siting of the factory, the outer fence and the area up to the factory wall. This level provides barriers against environmental conditions e.g. prevailing wind and surface water run-off, unauthorised public access and avoidance of pest harbourage areas.

Level 2 represents the factory wall and other processes (e.g. UV flytraps) which should separate the factory from the external environment. Whilst it is obvious that the factory cannot be a sealed box, the floor of the factory should ideally be at a different level to the ground outside and openings should be designed to be pest proof when not in use.

Level 3 represents the internal barriers that are used to separate manufacturing processes of different risk e.g. pre and post-heat treatment. Such separation should seek to control the air, people and surfaces (e.g. the floor and drainage systems) and the passage of materials and utensils across the barrier.

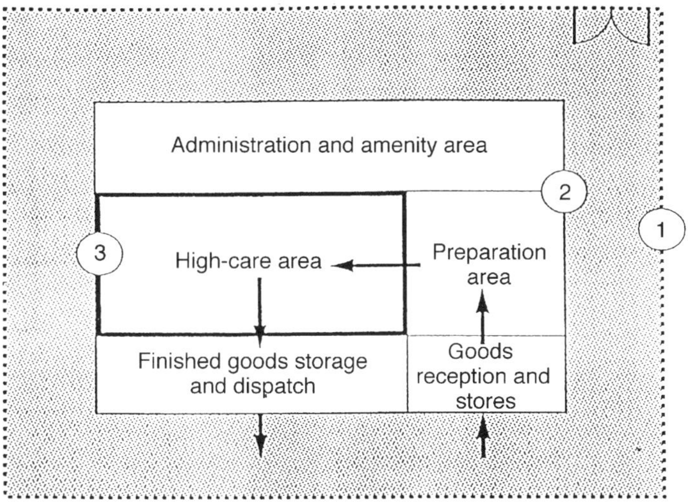

Fig. 13.2 Schematic layout of a factory site showing 'barriers' against contamination. (1) Perimeter fence; (2) Main factory buildings; (3) Walls of high-care area.

13.2.1 The factory site

Attention to the design, construction and maintenance of the site surrounding the factory provides an opportunity to set up the first (outer) of a series of barriers to protect production operations from contamination. It is a sound principle to take all reasonable precautions to reduce the 'pressures' that may build up on each of the barriers making up the overall protective envelope. A number of steps can be taken. For example, well-planned and properly maintained landscaping of the grounds can assist in the control of rodents, insects, and birds by reducing food supplies and breeding and harbourage sites.

The use of two lines of rodent baits located every 15–21m along the perimeter boundary fencing and at the foundation walls of the factory, together with a few mouse traps near building entrances is advocated by Imholte (1984). Both Katsuyama and Strachan (1980) and Troller (1983) suggest that the area immediately adjacent to buildings be kept grass free and covered with a deep layer of gravel or stones. This practice helps weed control and assists inspection of bait boxes and traps.

The control of birds is important, otherwise colonies can become established and cause serious problems. Shapton and Shapton (1991) state there should be a strategy of making the factory site unattractive by denying birds food and harbourage. They stress the importance of ensuring that waste material is not left in uncovered containers and that any spillages of raw materials are cleared up promptly.

Shapton and Shapton (1991) state that many insects are carried by the wind and therefore are inevitably present in a factory. They point out the importance of preventing the unauthorized opening of doors and windows and the siting of protective screens against flying insects. Imholte (1984) considers such screens present maintenance problems. These authors draw attention to lighting for warehouses and outdoor security systems attracting night-flying insects and recommend high pressure sodium lights in preference to mercury vapour lamps. Entrances that have to be lit at night should be lit from a distance with the light directed to the entrance, rather than lit from directly above. This prevents flying insects being attracted directly to the entrance. Some flying insects require water to support part of their life cycle e.g. mosquitoes, and experience has shown that where flying insects can occasionally be a problem, all areas where water could collect or stand for prolonged periods of time (old buckets, tops of drums, etc.) need to be removed or controlled,

Good landscaping of sites can reduce the amount of dust blown into the factory, as can the sensible siting of any preliminary cleaning operations for raw materials such as root vegetables, which are often undertaken outside the factory. Imholte (1984) advocates orientating buildings so that prevailing winds do not blow directly into manufacturing areas. The layout of vehicular routes around the factory site can affect the amount of soil blown into buildings. Shapton and Shapton (1991) suggest that for some sites it may be necessary to restrict the routes taken by heavily soiled vehicles to minimize dust contamination.

13.2.2 The factory building

The building structure is the second and a major barrier, providing protection for raw materials, processing facilities and manufactured products from contamination or deterioration. Protection is both from the environment, including rain, wind, surface runoff, delivery and dispatch vehicles, dust, odours, pests and uninvited people etc. and internally from microbiological hazards (e.g. raw material cross-contamination), chemical and physical hazards (e.g. from plantrooms and engineering workshops). Ideally, the factory buildings should be designed and constructed to suit the operations carried out in them and should not place constraints on the process or the equipment layout.

The type of building, either single- or multistorey, needs to be considered. Imholte (1984), comments that the subject has always been a controversial one and describes the advantages and disadvantages of both types of buildings. He also suggests a compromise may be achieved by having a single-storey building with varying headroom featuring mezzanine floors to allow gravity flow of materials, where this is necessary. Single-storey buildings are preferred for the majority of chilled food operations and generally allow the design criteria for high-risk areas to be more easily accommodated. However, it should be appreciated that where production is undertaken in renovated buildings, it may not be possible to capitalize on some of the advantages quoted by Imholte (1984). Of particular concern in multistorey buildings is leakage, of both air and fluids, from areas above and below food processing areas. The authors have undertaken investigative work in a number of factories in which contamination has entered high-risk areas via leakage from above, through both floor defects and badly maintained drains. In addition, on a number of occasions the drainage systems have been observed to act as air distribution channels, with air from low-risk areas (both above and below) being drawn into high risk. This can typically occur when the drains are little used and the water traps dry out.

The factory layout is paramount in ensuring both an economic and safe processing operation and should be such that processing operations are as direct as possible. Straight line flow minimises the possibility of contamination of processed or semi-processed product by unprocessed or raw materials and is more efficient in terms of handling. It is also easier to segregate clean and dirty process operations and restrict movement of personnel from dirty to clean areas. Whilst ideally the process line should be straight, this is rarely possible, but there must be no backtracking and, where there are changes in the direction of process flow, there must be adequate physical barriers.

The layout should also consider that provision is made for the space necessary to undertake the process and associated quality control functions, both immediately the factory is commissioned and in the foreseeable future. Space should also be allowed for the storage and movement of materials and personnel. Surrounding equipment, Imholte (1984) states 915 mm (3.0 feet) should be considered as the bare minimum for most units; however, he recommends 1830 mm (6.0 feet) as a more practical figure to allow production, cleaning and maintenance operations to be undertaken in an efficient manner.

In addition to process areas, provision may have to be made for a wide range of activities including raw material storage; packaging storage; water storage; wash-up facilities; plantroom; engineering workshop; cleaning stores; microbiology, chemistry and QC laboratories; test kitchens; pilot plant; changing facilities; restrooms; canteens; medical rooms; observation areas/viewing galleries and finished goods dispatch and warehousing.

Other good design principles given by Shapton and Shapton (1991) are:

- The flow of air and drainage should be away from 'clean' areas towards 'dirty' ones.
- The flow of discarded outer packaging materials should not cross, or run counter to, the flow of either unwrapped ingredients or finished products.

Detailed information on the hygienic design requirements for the construction of the external walls or envelope of the factory is not easily found. Much of the data available is understandably concerned with engineering specifications, which are not considered in this chapter. Shapton and Shapton (1991), Imholte (1984) and Timperley (1994) discuss the various methods of forming the external walls and give a large amount of advice on pest control measures, particularly for rodents. A typical example of a suitable outside wall structure is shown in Figure 13.3. The diagram shows a well sealed structure that resists pest ingress and is protected from external vehicular damage. The ground floor of the factory is also at a height above the external ground level. By preventing direct access into the factory at ground floor level, the entrance of contamination (mud, soil, foreign bodies etc.), particularly from vehicular traffic (forklift trucks, raw material delivery etc.) is restricted.

In addition, the above references provide considerable information on the hygienic requirements for the various openings in the envelope, particularly doors and windows. Points of particular interest are:

- Doors should be constructed of metal, glass reinforced plastic (GRP) or plastic, self-closing, designed to withstand the intended use and misuse and be suitably protected from vehicular damage where applicable.
- Exterior doors should not open directly into production areas and should remain closed when not in use. Plastic strip curtains may be used as inner doors.
- If possible, factories should be designed not to have windows in food processing areas. If this is not possible, e.g. to allow visitor or management observation, windows should be glazed with either polycarbonate or laminated. A glass register, detailing all types of glass used in the factory, and their location, should be composed.
- Metal or plastic frames with internal sills sloped (20°–40°) to prevent their use as 'temporary' storage places and with external sills sloped at 60° to prevent bird roosting, should be used.
- Opening windows must be screened in production areas and the screens be designed to withstand misuse or attempts to remove them.

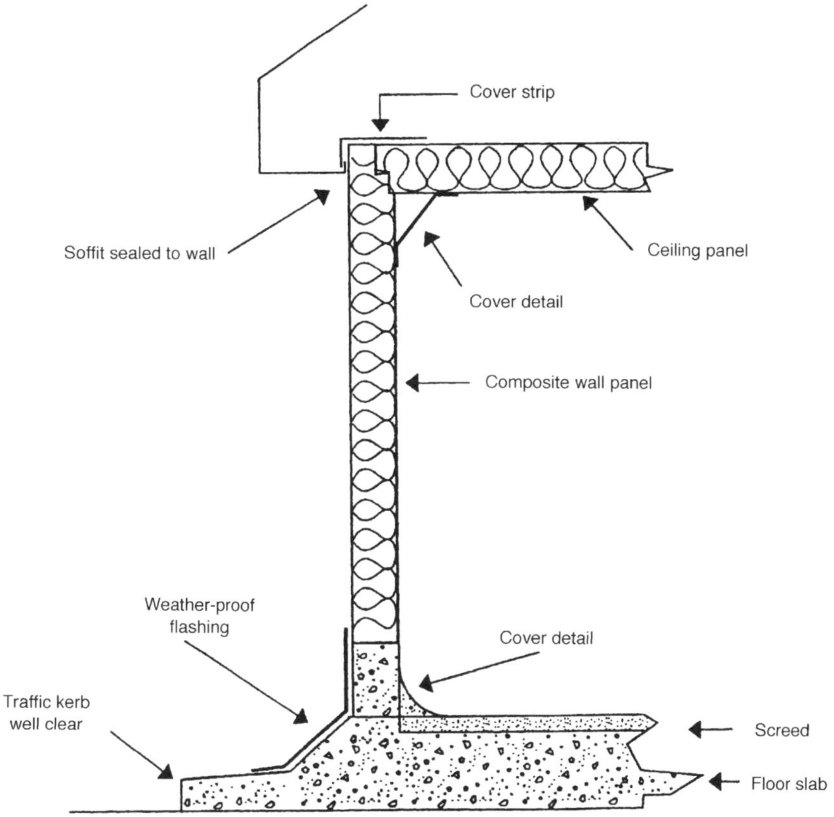

Fig. 13.3 Outside wall configuration showing a well sealed structure with elevated factory floor level.

13.2.3. High-risk production area

It is unfortunate that the term 'high risk', which is also used to describe other foods, for example low-acid canned foods, has become associated with the particular area of the factory where chilled foods are produced. The terms 'high-risk area' and 'low-risk area' are often used to describe parts of a chilled foods factory where different hygiene requirements apply.

It is considered that such terminology is misleading, and its use can imply to employees and other people that lower overall standards are acceptable in those areas where, for example, operations concerned with raw material reception, storage and initial preparation are undertaken. In practice, all operations concerned with food production should be carried out to the highest standard. Unsatisfactory practices in so-called low-risk areas may put greater pressures on the 'barrier system' separating the two areas. Whilst undesirable, however, it is probable that such terminology will remain for the near future. The advent of the use of more 'pharmaceutical' techniques in hygienic food manufacture may lead to the use of appropriate pharmaceutical terminology, e.g. 'clean' zones.

More recently, the Chilled Food Association in the UK (Anon. 1997a) established guidelines to describe the hygiene status of chilled foods and indicate the area status of where they should be processed after any heat treatment. Three levels were described, high-risk area (HRA), high-care area (HCA) and good manufacturing practice (GMP). Their definitions were:

HRA An area to process components, ALL of which have been heat treated to >90°C for 10 mins or >70°C for 2 mins, and in which there is a risk of contamination between heat treatment and pack sealing that may present a food safety hazard.

HCA An area to process components, SOME of which have been heat treated to >70°C for 2 mins, and in which there is a risk of contamination between heat treatment and pack sealing that may present a food safety hazard.

GMP An area to process components, NONE of which have been heat treated to >70°C for 2 mins, and in which there is a risk of contamination prior to pack sealing that may present a food safety hazard.

In practice, the definition of HCA has been extended to include an area to further process components that have undergone a decontamination treatment e.g. fruit and vegetables after washing in chlorinated water or fish after low temperature smoking and salting.

Most of the requirements for the design of HRA and HCA operations are the same, with the emphasis on preventing contamination in HRA and minimising contamination in HCA operations (Anon. 1997a). In considering whether a high risk or high care is required and therefore what specifications should be met, chilled food manufacturers need to carefully consider their existing and future product ranges, the hazards and risks associated with them and possible developments in the near future. If budgets allow, it is always cheaper to build to the highest standards from the onset of construction rather than try to retrofit or refurbish at a later stage. Guidance within this chapter is aimed at satisfying the requirements for high-risk operations.

Listeria philosophy

In terms of chilled food product safety, the major contamination risk is microbiological, particularly from the pathogen most commonly associated with the potential to grow in chilled foods, *Listeria monocytogenes*. For many chilled food products, *L. monocytogenes* could well be associated with the raw materials used and thus may well be found in the low-risk area. After the product has been heat processed or decontaminated (e.g. by washing), it is essential that all measures are taken to protect the product from cross-contamination from low risk, *L. monocytogenes* sources. Similarly, foreign body contamination that would jeopardize the wholesomeness of the finished product, could also be found in low risk. A three-fold philosophy has been developed by the authors to help reduce the incidence of *L. monocytogenes* in finished product and at the same time, control other contamination sources.

1. Provide as many barriers as possible to prevent the entry of *Listeria* into the high-risk area.
2. Prevent the growth and spread of any *Listeria* penetrating these barriers during production.
3. After production, employ a suitable sanitation system to ensure that all *Listeria* are removed from high risk prior to production recommencing.

13.3 High-risk barrier technology

The building structure, facilities and practices associated with the high-risk production and assembly areas provide the third and inner barrier protecting chilled food manufacturing operations from contamination. This final barrier is built up by the use of combinations of a number of separate components or sub-barriers, to control contamination that could enter high risk from the following routes:

- product entering high risk via a heat process
- product entering high risk via a decontamination process. Product entering high risk that has been heat processed/decontaminated off-site but whose outer packaging may need decontaminating on entry to high risk
- other product transfer
- packaging materials
- liquid and solid waste materials
- surfaces, usually associated with low/high-risk physical junctions and concerned with floors, walls, doors, and false or suspended ceilings
- food operatives entering high risk
- the air
- utensils, which may have to be passed between low and high risk

13.3.1. Heat treated product

Where a product heat treatment forms the barrier between low and high risk (e.g. an oven, fryer or microwave tunnel), two points are critical to facilitate its successful operation.

1. All product passing through the heat barrier must receive its desired cooking time/temperature combination. This means that the heating device should be performing correctly (e.g. temperature distribution and maintenance are established and controlled and product size has remained constant) and that it should be impossible, or very difficult, for product to pass through the heat treatment without a cook process being initiated.
2. The heating device must be designed such that as far as is possible, the device forms a solid, physical barrier between low and high risk. Where it is not physically possible to form a solid barrier, air spaces around the heating equipment should be minimised and the low/high-risk floor junction should be fully sealed to the highest possible height.

The fitting of heating devices that provide heat treatment within the structure of a building presents two main difficulties. Firstly, the devices have to be designed to load product on the low-risk side and unload in high risk. Secondly, the maintenance of good seals between the heating device surfaces, which cycle through expansion and contraction phases, and the barrier structure which has a different thermal expansion, is problematical. Of particular concern are ovens and the authors are aware of the following issues:

- Some ovens have been designed such that they drain into high risk. This is unacceptable for the following reason. It may be possible for pathogens present on the surface of product to be cooked (which is their most likely location if they have been derived from cross-contamination in low risk) to fall to the floor through the melting of the product surface layer (or exudate on overwrapped product) at a temperature that is not lethal to the pathogen. The pathogen could then remain on the floor or in the drain of the oven in such a way that it could survive the cook cycle. On draining, the pathogen would then subsequently drain into high risk. Pathogens have been found at the exit of ovens in a number of food factories.
- Problems have occurred with leakage from sumps under the ovens into high risk. There can also be problems in sump cleaning where the use of high pressure hoses can spread contamination into high risk.
- Where the floor of the oven is cleaned, cleaning should be undertaken in such a way that cleaning solutions do not flow from low to high risk. Ideally, cleaning should be from low risk with the high-risk door closed and sealed. If cleaning solutions have to be drained into high risk, or in the case of ovens that have a raining water cooling system, a drain should be installed immediately outside the door in high risk.

Other non-oven related issues to consider include the following:

- The design of small batch product blanchers or noodle cookers (i.e. small vessels with water as the cooking medium) does not often allow the equipment to be sealed into the low/high-risk barrier as room has to be created around the blancher to allow product loading and unloading. Condensation is likely to form because of the open nature of these cooking vessels and it is important to ventilate the area to prevent microbial build-up where water condenses. Any ventilation system should be designed so that the area is ventilated from low risk; ventilation from high risk can draw into high risk large quantities of low-risk air.
- Early installations of kettles as barriers between low and high risk, together with the associated bund walls to prevent water movement across the floor and barriers at waist height to prevent the movement of people, whilst innovative in their time, are now seen as hygiene hazards. It is virtually impossible to prevent the transfer of contamination, by people, the air and via cleaning, between low and high risk. It is now possible to install kettles within low risk and transfer product (by pumping, gravity, vacuum etc.)

through into high risk via a pipe in the dividing wall. The kettles need to be positioned in low risk at a height such that the transfer into high risk is well above ground level. Installations have been encountered where receiving vessels have had to be placed onto the floor to accept product transfer.

13.3.2 Product decontamination

Fresh produce to be processed in high care should enter high care via a decontamination operation, usually involving a washing process with the washwater incorporating a biocide. The use of chlorinated dips, mechanically stirred washing baths or 'jaquzzi' washers are the most common method, though alternative biocides are also used (e.g. bromine, chlorine dioxide, ozone, organic acids, peracetic acid, hydrogen peroxide).

In addition, it is now seen as increasingly important, following a suitable risk assessment, to decontaminate the outer packaging of various ingredients on entry into high risk (e.g. product cooked elsewhere and transported to be processed in the high-risk area, canned foods and some overwrapped processed ingredients). Where the outer packaging is likely to be contaminated with food materials, decontamination is best done using a washing process incorporating a disinfectant (usually a quaternary ammonium compound). If the packaging is clean, the use of UV light has the advantage that it is dry and thus limits potential environmental microbial growth.

Decontamination systems have to be designed and installed such that they satisfy three major criteria.

1. As with heat barriers, decontamination systems need to be installed within the low/high-risk barrier to minimise the free space around them. As a very minimum, the gap around the decontamination system should be smaller than the product to be decontaminated. This ensures that all ingredients in high risk must have passed through the decontamination system and thus must have been decontaminated (it is impossible to visually assess whether the outer surface of an ingredient has been disinfected, in contrast to whether an ingredient has been heat processed).
2. Prior to installation, the decontamination process should be established and verified. For a wet process, this will involve the determination of a suitable disinfectant that combines detergency and disinfectant properties and a suitable application temperature, concentration and contact time. Similarly, for UV light, a suitable wavelength, intensity and contact time should be determined. The same degree of decontamination should apply to all the product surfaces or, if this is not possible, the process should be established for the surface receiving the least treatment.
3. After installation, process controls should be established and may include calibrated, automatic disinfectant dosing, fix speed conveyors, UV light intensity meters etc. In process monitoring may include the periodic checking for critical parameters, for example blocked spray nozzles or UV

lamp intensity and, from the low-risk side, the loading of the transfer conveyor to ensure that product is physically separated such that all product surfaces are exposed.

13.3.3 Other product transfer

It is now poor practice to bring outer packaging materials into high risk. All ingredients and product packaging must, therefore, be de-boxed and transferred into high risk.

Some ingredients, such as bulk liquids that have been heat-treated or are inherently stable (e.g. oils or pasteurised dairy products), are best handled by being pumped across the low/high-risk barrier directly to the point of use. Dry, stable bulk ingredients (e.g. sugar) can also be transferred into high risk via sealed conveyors.

For non-bulk quantities, it is possible to open ingredients at the low/high-risk barrier and decant them through into high risk via a suitable transfer system (e.g. a simple funnel set into the wall), into a receiving container. Transfer systems should, preferably, be closeable when not in use and should be designed to be cleaned and disinfected, from the high-risk side, prior to use as appropriate.

13.3.4 Packaging

Packaging materials (film reels, cartons, containers, trays etc.) are best supplied to site 'double bagged'. This involves a cardboard outer followed by two plastic bag layers surrounding the packaging materials. The packaging is brought on site, de-boxed, and stored double bagged until use in a suitable packaging store. When called for in high risk, the packaging material is brought to the low/high-risk barrier, the outer plastic bag removed and the inner bag and packaging enters high risk through a suitable hatch. The second plastic bag keeps the packaging materials covered until they are loaded onto the line or the packaging machine.

The hatch, as with all openings in the low/high-risk barrier, should be as small as possible and should be closeable when not in use. This is to reduce airflow through the hatch and thus reduce the airflow requirements for the air handling systems to maintain high-risk positive pressure. For some packaging materials, especially heavy film reels, it may be required to use a conveyor system for moving materials through the hatch. An opening door or preferably, double door airlock, should only be used if the use of a hatch is not technically possible and suitable precautions must be taken to decontaminate the airlock after use.

13.3.5 Liquid and solid wastes

On no account should low-risk liquid or solid wastes be removed from the factory via high risk and attention is required to the procedures for removing high-risk wastes. The handling of liquid wastes from low and high risk is described later in this chapter in the section on drainage.

Solid wastes that have fallen on the floor or equipment, etc., through normal production spillages, should be bagged-up or placed in easily cleanable bins, on an on-going basis commensurate with good housekeeping practices. It may also be necessary to remove solid waste product from the line at break periods or to facilitate line product changes. Waste bags should leave high risk in such a way that they minimise any potential cross-contamination with processed product and should, preferably, not be routed in the reverse direction to the product. For small quantities of bagged waste, existing hatches should be used e.g. the wrapped product exit hatches or the packaging materials entrance hatch, as additional hatches increase the risk of external contamination and put extra demands on the air handling system. For waste collected in bins, it may be necessary to decant the waste through purpose built, easily cleanable from high risk, waste chutes that deposit directly into waste skips. Waste bins should be colour coded to differentiate them from other food containers and should only be used for waste.

13.3.6 Surfaces

In this context, surfaces are associated with sealed low/high-risk physical junctions and are concerned with floors, walls, doors and false or suspended ceilings. Detailed descriptions on the requirements for these areas are contained in the Construction section later in this chapter.

13.3.7 Personnel

Within the factory building, provision must be made for adequate and suitable staff facilities and amenities for changing, washing and eating. There should be lockers for storing outdoor clothing in areas that must be separate from those for storing work clothes. Toilets must be provided and must not open directly into food-processing areas, all entrances of which must be provided with handwashing facilities arranged in such a way that their ease of use is maximised. In addition, staff (including visitors and contractors etc.) have personal responsibilities which they should follow to ensure good hygienic practices. These are normally formulated as the factory hygiene policy and typically include the following:

1. Protective clothing, footwear and headgear issued by the company must be worn and must be changed regularly. When considered appropriate by management, a fine hairnet must be worn in addition to the protective headgear provided. Hair clips and grips should not be worn.
2. Protective clothing must not be worn off the site and must be kept in good condition.
3. Beards must be kept short and trimmed and a protective cover worn when considered appropriate by management.
4. Nail varnish, false nails and make up must not be worn in production areas.

5. False eyelashes, wrist watches and jewellery (except wedding rings, or the national equivalent, and sleeper earrings) must not be worn.
6. Hands must be washed regularly and kept clean at all times.
7. Personal items must not be taken into production areas unless carried in inside overall pockets (handbags, shopping bags, etc. must be left in the lockers provided).
8. Food and drink must not be taken into or consumed in areas other than the rest areas and the staff canteen/restaurant.
9. Sweets and chewing gum must not be consumed in production areas.
10. Smoking or taking snuff is forbidden in food production, warehouse and distribution areas where 'No Smoking' notices are displayed.
11. Spitting is forbidden in all areas on the site.
12. Superficial injuries (e.g. cuts, grazes, boils, sores and skin infections) must be reported to the medical department or the first aider on duty via the line supervisor and clearance obtained before the operative can enter production areas.
13. Dressings must be waterproof, suitably coloured to differentiate them from product and contain a metal strip as approved by the medical department.
14. Infectious diseases (including stomach disorders, diarrhoea, skin conditions and discharge from eyes, nose or ears) must be reported to the medical department or first aider on duty via the line supervisor. This also applies to staff returning from foreign travel where there has been a risk of infection.
15. All staff must report to the medical department when returning from both certified and uncertified sickness.

With regard to high-risk operatives, however, personnel facilities and requirements must be provided in a way that minimises any potential contamination of high-risk operations. The primary sources of potential contamination arise from the operatives themselves and from low-risk operations. This necessitates further attention to protective clothing and, in particular, special arrangements and facilities for changing into high-risk clothing and entering high risk. Best practice with respect to personnel hygiene is continually developing and has been recently reviewed by Guzewich and Ross (1999), Taylor and Holah (2000) and Taylor *et al.* (2000).

High-risk factory clothing does not necessarily vary from that used in low risk in terms of style or quality, though it may have received higher standards of laundry, especially related to a higher temperature process, sufficient to significantly reduce microbiological levels. Indeed some laundries now operate to the same low/high-risk principles as the food industry such that dirty laundry enters 'low risk', is loaded into a washing machine that bridges a physical divide, is cleaned and disinfected and exits into 'high risk' to be dried and packed.

Additional clothing may be worn in high risk, however, to further protect the food being processed from contamination arising from the operatives body (e.g. gloves, sleeves, masks, whole head coveralls, coats with hoods, boiler suits, etc.). All clothing and footwear used in the high-risk area is colour coded to distinguish it from that worn in other parts of the factory and to reduce the chance that a breach in the systems would escape early detection.

High-risk footwear should be captive to high risk; i.e. it should remain within high risk, operatives changing into and out of footwear at the low/high-risk boundary. This has arisen because research has shown that boot baths and boot washers are unable adequately to disinfect low-risk footwear such that they can be worn in both low and high risk and decontaminated between the two (Taylor *et al.* 2000). In addition, boot baths and boot washers can both spread contamination via aerosols and water droplets that, in turn, can provide moisture for microbial growth on high-risk floors. Bootwashers were, however, shown to be very good at removing organic material from boots and are thus a useful tool in low-risk areas both to clean boots and help prevent operative slip hazards.

The high-risk changing room provides the only entry and exit point for personnel working in or visiting the area and is designed and built to both house the necessary activities for personnel hygiene practices and minimise contamination from low risk. In practice, there are some variations in the layout of facilities of high-risk changing rooms. This is influenced by, for example, space availability, product throughput and type of products, which will affect the number of personnel to be accommodated and whether the changing room is a barrier between low- and high-risk operatives or between operatives arriving from outside the factory and high risk. Generally higher construction standards are required for low/high-risk barriers than outside/high-risk barriers because the level of potential contamination in low risk, both on the operatives hands and in the environment, is likely to be higher (Taylor and Holah 2000). In each case, the company must evaluate the effectiveness of the changing-room layout and procedure to ensure the high-risk area and products prepared in it are not being put at risk. This is best undertaken by a HACCP approach, so that data are obtained to support or refute any proposals regarding the layout or sequence.

Research at CCFRA has also proposed the following hand hygiene sequence to be used on entry to high risk (Taylor and Holah 2000). This sequence has been designed to maximise hand cleanliness, minimise hand transient microbiological levels, maximise hand dryness yet at the same time reduce excessive contact with water and chemicals that may both lead to dermatitis issues of the operatives and reduce the potential for water transfer into high risk.

1. Remove low-risk or outside clothing.
2. Remove low-risk/outside footwear and place in designated 'cage' type compartment.
3. Cross over the low- risk/high-risk dividing barrier.
4. WASH HANDS.
5. Put on in the following order:

   - high-risk captive footwear
   - hair net; put on over ears and covering *all* hair; (plus beard snood if needed) and hat (if appropriate)
   - overall (completely buttoned up to neck).

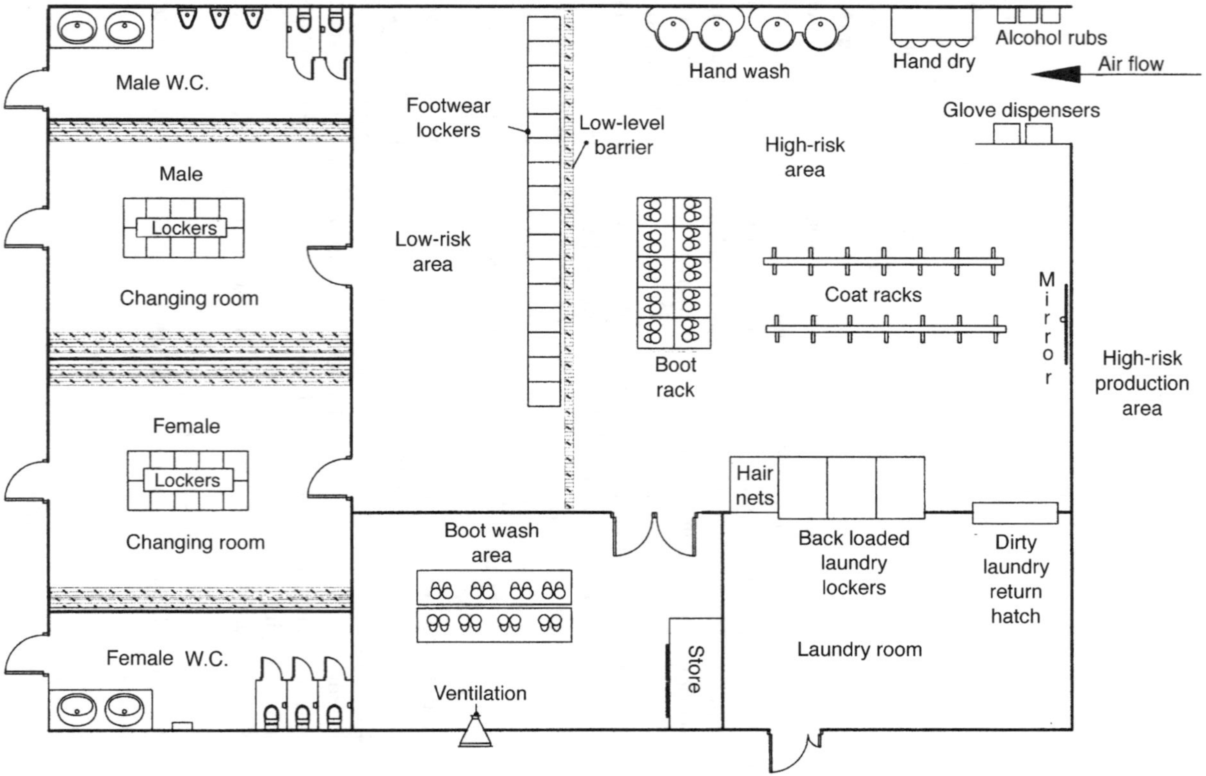

Fig. 13.4 Schematic layout for a high-risk changing room.

6. Check dress and appearance in the mirror provided.
7. Go into the high-risk production area and apply an alcohol-based sanitizer.
8. Draw and put on disposable gloves, sleeves and apron, if appropriate.

A basic layout for a changing room is shown in Fig. 13.4 and has been designed to accommodate the above hand hygiene procedure and the following requirements.

- An area at the entrance to store outside or low-risk clothing. Lockers should have sloping tops.
- A barrier to divide low- and high-risk floors. This is a physical barrier such as a small wall (approximately 60cm high), that allows floors to be cleaned on either side of the barrier without contamination by splashing, etc. between the two.
- Open lockers at the barrier to store low-risk footwear.
- A stand on which footwear is displayed/dried.
- An area designed with suitable drainage for bootwashing operations. Research has shown (Taylor *et al.* 2000) that manual cleaning (preferably during the cleaning shift) and industrial washing machines are satisfactory bootwashing methods.
- Hand wash basins to service a single, hand wash. Handwash basins must have automatic or knee/foot operated water supplies, water supplied at a suitable temperature (that encourages hand washing) and a waste extraction system piped directly to drain. It has been shown that hand wash basins positioned at the entrance to high risk, which was the original high-risk design concept to allow visual monitoring of hand wash compliance, gives rise to substantial aerosols of Staphylococcal strains that can potentially contaminate the product.
- Suitable hand-drying equipment e.g. paper towel dispensers or hot-air dryers and, for paper towels, suitable towel disposal containers.
- Access for clean factory clothing and storage of soiled clothing. For larger operations this may be via an adjoining laundry room with interconnecting hatches.
- Interlocked doors are possible such that doors only allow entrance to high risk if a key stage, e.g. hand washing has been undertaken.
- CCT cameras as a potential monitor of hand wash compliance.
- Alcoholic hand rub dispensers immediately inside the high-risk production area.

There may be the requirement to site additional hand washbasins inside the high-risk area if the production process is such that frequent hand washing is necessary. As an alternative to this, Taylor *et al.* (2000) demonstrated that cleaning hands with alcoholic wipes, which can be done locally at the operative's work station, is an effective means of hand hygiene.

13.3.8 Air

The air is an important, potential source of pathogens and the intake into the high-risk area has to be controlled. Air can enter high risk via a purpose-built air-handling system or can enter into the area from external uncontrolled sources (e.g. low-risk production, packing, outside). For high-risk areas, the goal of the air-handling system is to supply suitably filtered fresh air, at the correct temperature and humidity, at a slight overpressure to prevent the ingress of external air sources.

The cost of the air-handling systems is one of the major costs associated with the construction of a high-risk area and specialist advice should always be sought before embarking on an air-handling design and construction project. Following a suitable risk analysis, it may be concluded that the air-handling requirements for high-care areas may be less stringent, especially related to filtration levels and degree of overpressure. Once installed, any changes to the construction of the high-risk area (e.g. the rearrangement of walls, doors or openings) should be carefully considered as they will have a major impact on the air-handling system.

Air quality standards for the food industry were reviewed by a CCFRA Working Party and guidelines were produced (Brown 1996). The design of the air-handling system should consider the following issues:

- degree of filtration of external air
- overpressure
- air flow – concerned with operational considerations and operative comfort
- air movement
- temperature requirements
- local cooling and barrier control
- humidity requirements
- installation and maintenance.

The main air flows within a high-risk area are shown in Fig. 13.5 and a more detailed schematic of the air handling system is shown in Fig. 13.6.

A major risk of airborne contamination entering high risk is from low-risk processing operations, especially those handling raw produce that is likely to be contaminated with pathogens. The principal role of the air-handling system is thus to provide filtered air to high risk with a positive pressure with respect to low risk. This means that wherever there is a physical break in the low/high-risk barrier, e.g. a hatch, the air flow will be through the opening from high to low risk. Microbial airborne levels in low risk, depending on the product and processes being undertaken, may be quite high (Holah *et al.* 1995) and overpressure should prevent the movement of such airborne particles, some of which may contain viable pathogenic microorganisms, entering high risk.

To aid the performance of the air-handling system, it is also important to control potential sources of aerosols, generated from personnel, production and cleaning activities, in both low and high risk. Filtration of air is a complex matter and requires a thorough understanding of filter types and installations. The choice of filter will be dictated by the degree of microbial and particle removal required and filter types are described in detail in the CCFRA guideline document (Brown 1996). For high-risk applications, a series of filters is required

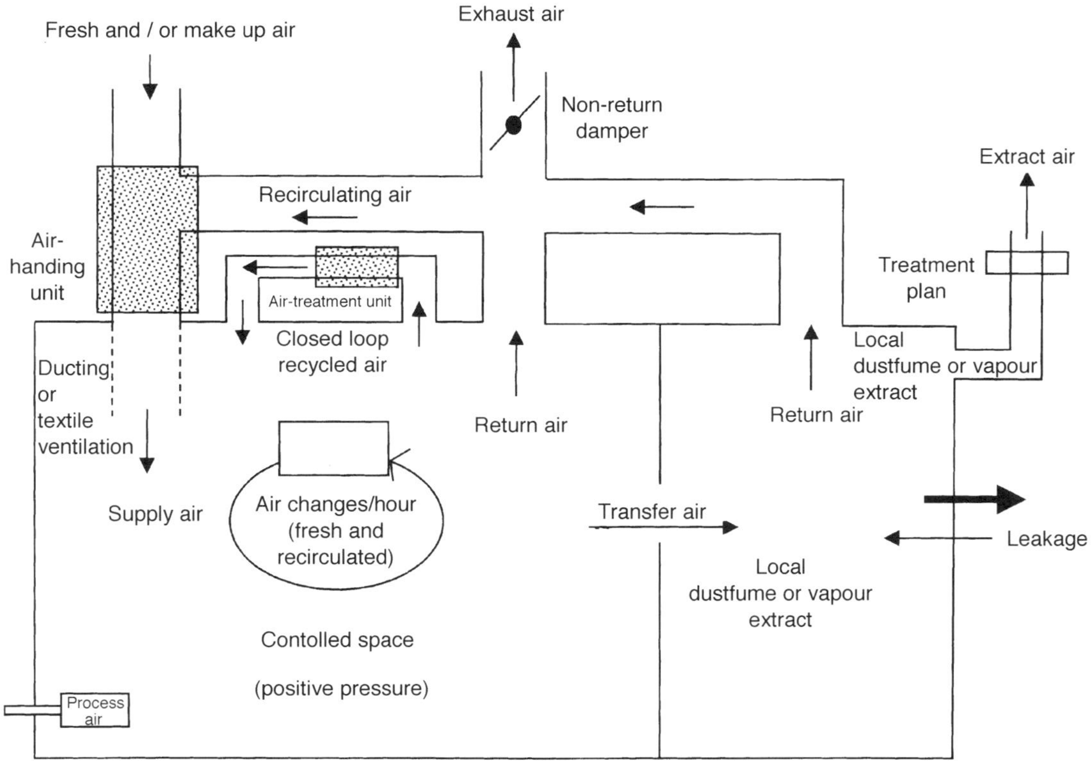

Fig. 13.5 Schematic diagram showing the airflows within a high-risk or high-care production area.

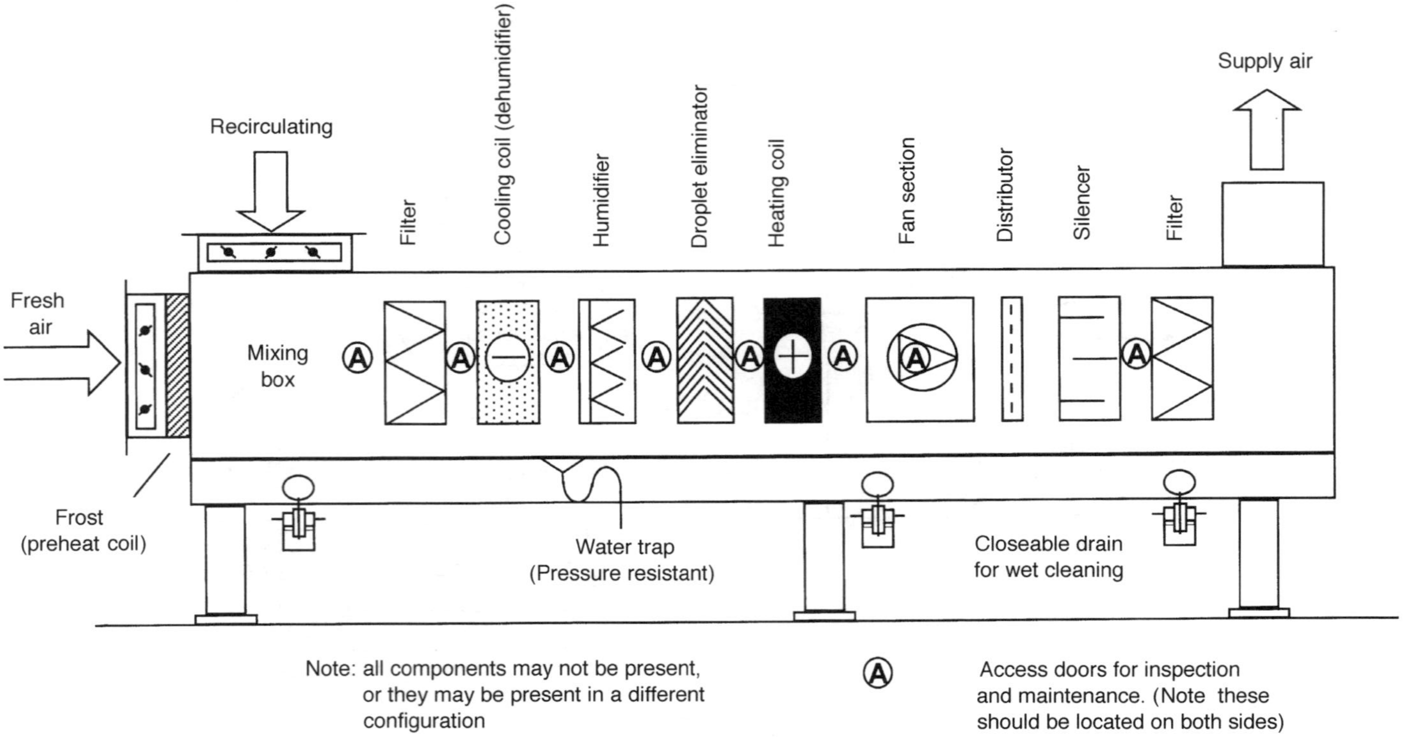

Fig 13.6 Schematic diagram of the components of a typical air-handling system.

(Fig. 13.6) to provide air to the desired standard and is usually made up of a G4/F5 panel or pocket filter followed by an F9 rigid cell filter. For some high-risk operations an H10 or H11 final filter may be desirable, whilst for high-care operations an F7 or F8 final filter may be acceptable.

To be effective, the pressure differential between low and high risk should be between 5–15 Pascals. The desired pressure differential will be determined by both the number and size of openings and also the temperature differentials between low and high risk. For example if the low-risk area is at ambient (20ºC) and the high-risk areas at 10ºC, hot air from low risk will tend to rise through the opening whilst cold air from high risk will tend to sink through the same opening, causing two-way flow. The velocity of air through the opening from high risk may need to be 1.5m/sec or greater to ensure one-way flow is maintained.

In addition to providing a positive over-pressure, the air-flow rate must be sufficient to remove the heat load imposed by the processing environment (processes and people) and provide operatives with fresh air. Generally 5–25 air changes per hour are adequate, though in a high-risk area with large hatches/doors that are frequently opened, up to 40 air changes per hour may be required.

Air is usually supplied to high risk by either ceiling grilles or textile ducts (socks), usually made from polyester or polypropylene to reduce shrinkage. Ceiling grilles have the advantage that they are cheap and require little maintenance but have limitations on velocity and flow rate without high noise levels or the potential to cause draughts. With respect to draughts, the maximum air speed close to workers to minimise discomfort through 'wind-chill' is 0.3m s^{-1}. Air socks have the ability to distribute air, at a low draught free velocity with minimal ductwork connections, though they require periodic laundering and spare sets are required. Ceiling mounted chillers that cool and recirculate the air are only really suitable for high-care operations unless additional air supplies are used to maintain positive pressures.

Joint work undertaken since 1995 by CCFRA and the Silsoe Research Institute, sponsored by the UK Ministry of Fisheries and Food (MAFF), has looked at the control of airborne microbial contamination in high-risk food production areas. The work has resulted in the production of a best practice guideline on air flows in high-risk areas, which will be published by MAFF in 2001. The work has centred on the measurement of both air flows and airborne microbiological levels in actual food factories and computational fluid dynamics (CFD) models have been developed by Silsoe to predict air and particle (including microorganism) movements. The work has led to innovations in two key areas.

Firstly, the influence on airflows of air intakes and air extracts, secondary ventilation systems in, e.g., washroom areas, the number of hatches and doors and their degree of openings and closings, can readily be visualised by CFD. This has led to the redesign of high-risk areas, from the computer screen, such that airflow balances and positive pressures have been achieved. Secondly, the CFD models allow the prediction of the movement of airborne microorganisms from known sources of microbial contamination, e.g., operatives. This has allowed the design of air-handling systems which provide directional air that moves particles away from the source of contamination, in a direction that does not compromise product safety. As an illustration, Fig. 13.7 (a) shows the predicted air flows in a real factory generated using a CFD software package developed by Silsoe following air flow measurements. The model was then used to predict the movement of 10μm particles (similar to shed skin squams) from line operatives (Fig. 13.7 (b)). The predicted tracks indicate that in some cases the airflow is good and moves shed particles away from the product whilst in other cases, particles move directly over or along the product conveyors, thus presenting a hygiene risk.

Chilled-foods manufacturers have traditionally chosen to operate their high-risk areas at low temperatures, typically around 10–12°C, both to restrict the general growth of microorganims in the environment and to prevent the growth of some (e.g. *Salmonella*) but not all (e.g. *Listeria*) food pathogens. Chilling the area to this temperature is also beneficial in reducing the heat uptake by the product and thus maintaining the chill chain. Moreover, chilled food manufacturers have to ensure that their products meet, in the UK, the requirements of the *Food Safety (Temperature Control) Regulations 1995* (Anon. 1995) as well as those imposed by their retail customers.

In the UK, *The Workplace (Health, Safety and Welfare) Regulations* (Anon. 1992a) require that the 'temperature in all workplaces inside buildings shall be reasonable', which, in the supporting Approved Code of Practice (1992b), is normally taken to be at least 16°C or at least 13°C where much of the work involves serious physical effort. To help solve this conflict of product and operative temperature, a Working Group comprising members of the Health and Safety Executive (HSE) and the chilled food industry was established at CCFRA in 1996. The Working Group produced a document *Guidance on achieving reasonable working temperatures and conditions during production of chilled foods* (Brown 2000) which extends the information provided in HSE Food Sheet No. 3 (Rev) *Workroom temperatures in places where food is handled* (Anon. 1999). The guidance document (Brown 2000) states that employers will first need to consider alternative ways of controlling product temperatures to satisfy the *Food Safety (Temperature Control) Regulations* (Anon. 1995) rather than simply adopting lower workroom temperatures. If the alternative measures are not practical then it may be justified for hygiene reasons for workrooms to be maintained at temperatures lower than 16°C (or 13°C). Where such lower temperatures are adopted, employers should be able to demonstrate that they have taken appropriate measures to ensure the thermal comfort of employees. Full guidance on these issues is given in the document.

Another joint CCFRA/Silsoe, MAFF sponsored project, has examined the use of localised cooling with the objectives of:

- Providing highly filtered (H11-12), chilled air directly over or surrounding product. This could reduce the requirement to chill the whole of the high-risk

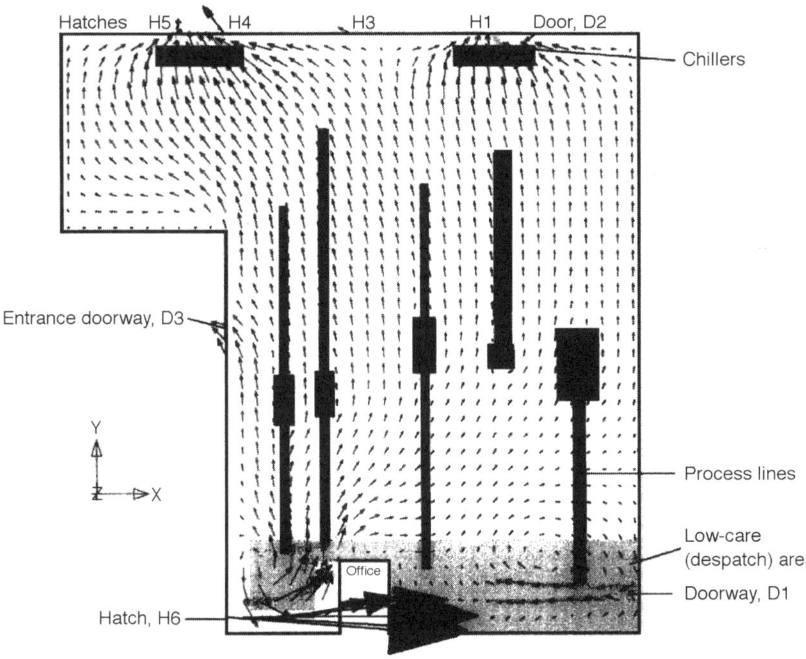

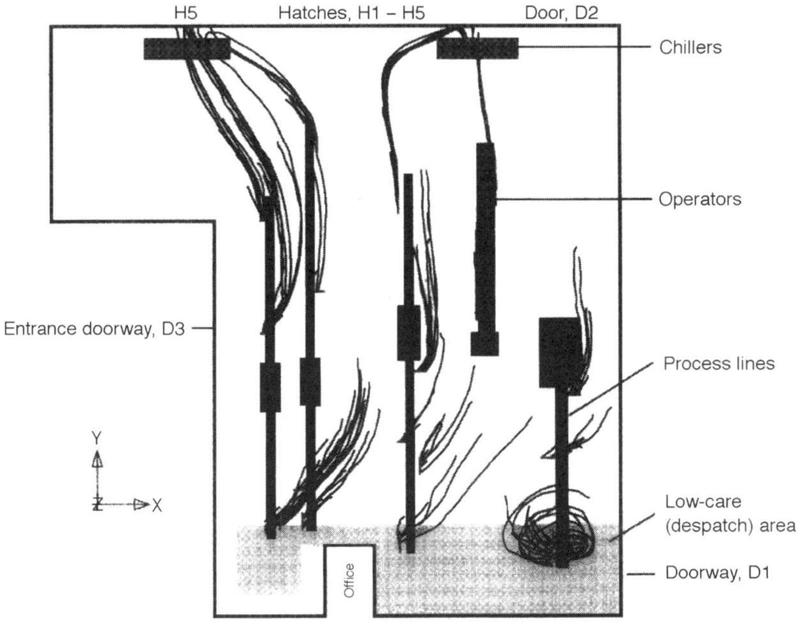

Fig 13.7 Schematic diagram of: (a) predicted airflows in an actual chilled food factory estalished from airflow measurements. The length and size of the arrow indicates air speed whilst the orientation shows flow direction; (b) predicted flow from line operatives. The flow of product down the five lines is in the direction Y to Z. (Courtesy of Silsoe Research Institute)

area to 10°C (13°C would be acceptable), and reduce the degree of filtration required (down to H8-9). The requirement for positive pressure in low risk is paramount, however, and the number of air changes per hour would remain unchanged.

- Using the flow of the air to produce a barrier that resists the penetration of aerosol particles, some of which would contain viable microorganisms

An example of such a technology is shown in Fig. 13.8 which shows a schematic diagram of a conveyor that has chilled, filtered air directed over it, sufficient to maintain the low temperature of the product. When a microbial aerosol was generated around the operational conveyor, microbiological air sampling demonstrated a 1–2 log reduction of microorganisms within the protected zone. This work has been reported in Burfoot *et al.* 2000.

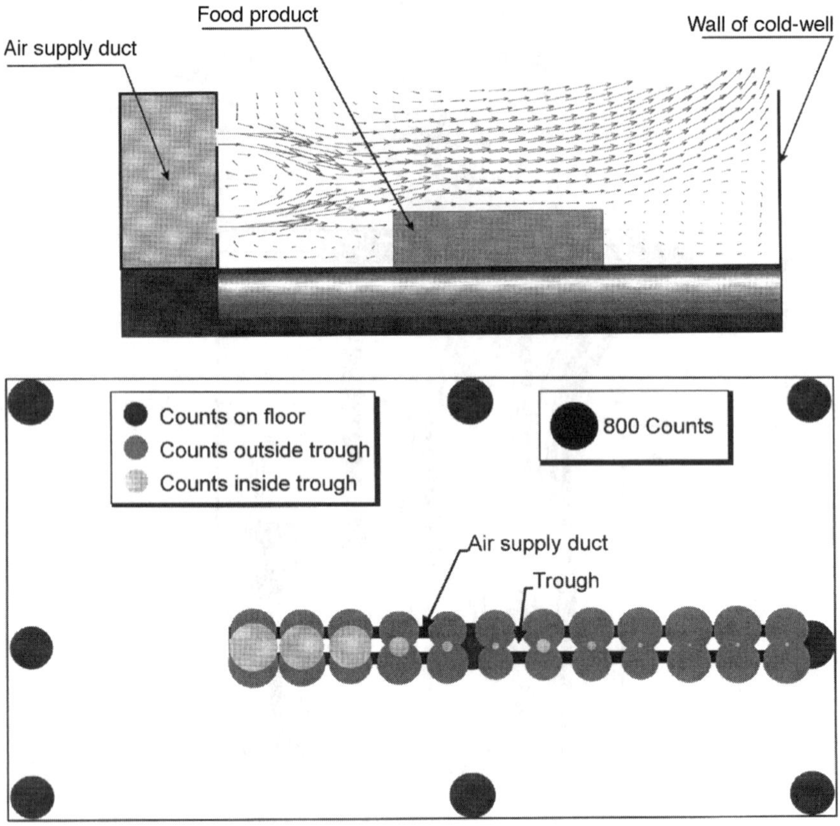

Fig. 13.8 Schematic diagram of (a) a conveyor belt with cooled, filtered air directed across the product and (b) the reduction in microbial counts (colony forming units, cfu) within the conveyor during operation. The diameter of the circles is directly proportional to the number of cfus recorded. (Courtesy of Silsoe Research Institute)

The choice of relative air humidity is a compromise between operative comfort, product quality and environmental drying. A relative humidity of 55–65% is very good for restricting microbial growth in the environment and increases the rate of equipment and environment drying after cleaning operations. Low humidities can, however, cause drying of the product with associated weight and quality loss, especially at higher air velocities. Higher humidities maintain product quality but may give rise to drying and condensation problems that increase the opportunity for microbial survival and growth. A compromise target humidity of 60–70% is often recommended, which is also optimal for operative comfort.

Finally, air-handling systems should be properly installed such that they can be easily serviced and cleaned and as part of the commissioning programme, their performance should be validated for normal use. The ability of the system to perform in other roles should also be established. These could include dumping air directly to waste during cleaning operations, to prevent air contaminated with potentially corrosive cleaning chemicals entering the air handling unit, and recirculating ambient or heated air after cleaning operations to increase environmental drying.

13.3.9 Utensils

Wherever possible, any equipment, utensils and tools, etc. used routinely within high risk, should remain in high risk. This may mean that requirements are made for the provision of storage areas or areas in which utensils can be maintained or cleaned. Typical examples include:

- The requirement for ingredient or product transfer containers (trays, bins, etc.) should be minimised but where these are unavoidable they should remain within high risk and be cleaned and disinfected in a separate wash room area.
- Similarly, any utensils (e.g. stirrers, spoons, ladles) or other non-fixed equipment (e.g. depositors or hoppers) used for the processing of the product should remain in high risk and be cleaned and disinfected in a separate washroom area.
- A separate washroom area should be created in which all within-production wet cleaning operations can be undertaken. The room should preferably be sited on an outside wall that facilitates air extraction and air make-up. An outside wall also allows external bulk storage of cleaning chemicals that can be directly dosed through the wall into the ring main system. The room should have its own drainage system that, in very wet operations, may include barrier drains at the entrance and exit to prevent water spread from the area. The wash area should consist of a holding area for equipment, etc. awaiting cleaning, a cleaning area for manual or automatic cleaning (e.g. traywash) as appropriate, and a holding/drying area where equipment can be stored prior to use. These areas should as segregated as possible.

- All cleaning equipment, including hand tools (brushes, squeegees, shovels, etc.) and larger equipment (pressure washers, floor scrubbers and automats, etc.) should remain in high risk and be colour coded to differentiate between high- and low-risk equipment if necessary. Special provision should be made for the storage of such equipment when not in use.
- Cleaning chemicals should preferably be piped into high risk via a ring main (which should be separate from the low-risk ring main). If this is not possible, cleaning chemicals should be stored in a purpose built area.
- The most commonly used equipment service items and spares, etc., together with the necessary hand tools to undertake the service, should be stored in high risk. For certain operations, e.g. blade sharpening for meat slicers, specific engineering rooms may need to be constructed.
- Provision should be made in high risk for the storage of utensils that are used on an irregular basis but that are too large to pass through the low/high-risk barrier, e.g. stepladders for changing the air distribution socks.
- Written procedures should be prepared detailing how and where items that cannot be stored in high risk but are occasionally used there, or new pieces of equipment entering high risk, will be decontaminated. If appropriate, these procedures may also need to detail the decontamination of the surrounding area in which the equipment decontamination took place.

13.4 Hygienic construction

13.4.1 Basic design concepts

The design of any food-processing area must allow for the accommodation of five basic requirements, i.e.

1. raw materials and ingredients
2. processing equipment
3. staff concerned with the operation of such equipment
4 packaging materials
5. finished products.

There is a philosophy which has considerable support, that states that all other requirements should be considered as secondary to these five basic requirements and, wherever possible, must be kept out of the processing area. These secondary requirements are:

- structural steel framework of the factory
- service pipework for water, steam and compressed air; electrical conduits and trunking; artificial lighting units; and ventilation ducts
- compressors, refrigeration units and pumps
- maintenance personnel associated with any of these services.

This philosophy is well suited to the requirements for high-risk production areas. Ashford (1986) describes it as the principle of building a 'box within a box' by

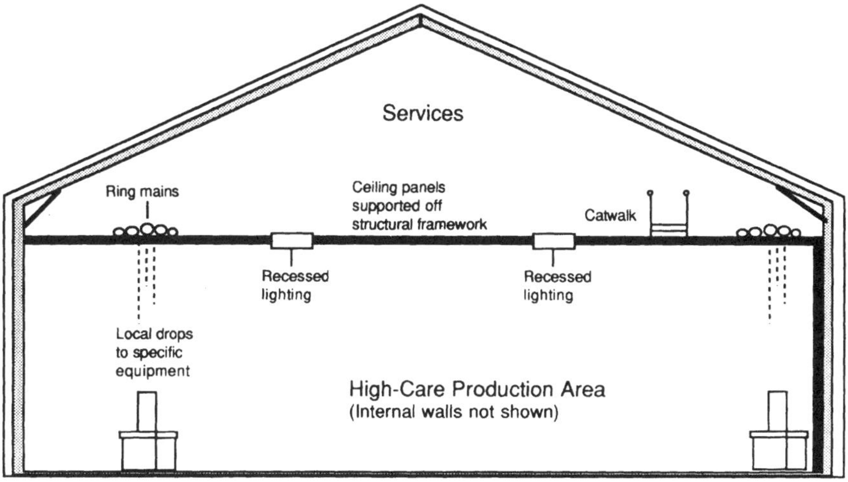

Fig. 13.9 Basic design concepts – the separation of production from services and maintenance operations.

creating insulated clean rooms within the structural box of the factory with the services and control equipment located in the roof void above the ceiling. Refrigeration equipment and ductwork is suspended from the structural frames and access to all services is provided by catwalks, as shown diagrammatically in Fig. 13.9. This arrangement, if properly undertaken, eliminates a major source of contamination from the process area.

13.4.2 Floors

The floor may be considered as one of the most important parts of a building because it forms the basis of the entire processing operation. It is thus worthy of special consideration and high initial capital investment. Guidelines for the design and construction of floors have been prepared by Timperley (1993).

Unsatisfactory floors increase the chances of accidents, cause difficulties in attaining the required hygiene standards and put up sanitation costs. The failure of a floor can result in lengthy disruptions of production and financial loss whilst repairs are completed. Many problems with floors have arisen because insufficient attention to detail was taken, at the design stage, with the overall specification. This should cover requirements for:

- the structural floor slab
- the waterproof membrane, which should extend up walls to a height above the normal spillage level
- movement joints in the subfloor and final flooring, around the perimeter of the floor, over supporting walls, around columns and machinery plinths
- drainage, taking into account the proposed layout of equipment

- screeds, either to give a flat enough surface to accept the flooring or to form the necessary falls when these are not incorporated in the concrete slab
- floor finish, either tiles or a synthetic resin
- processing considerations including trucking; impact loads from proposed operations and equipment and machinery to be installed; degree of product spillage and associated potential problems with corrosion, thermal shock, and drainage requirements; types of cleaning chemicals to be used and requirements for slip resistance

The choice of flooring surfaces can be broadly grouped into three categories – concretes, fully vitrified ceramic tiles and seamless resin screeds. Concrete flooring, including the high-strength granolitic concrete finishes, whilst being suitable and widely used in other parts of a factory, is not recommended for high-risk production areas. This is because of its ability to absorb water and nutrients and hence allow microbial growth below the surface, where it is extremely difficult to apply effective sanitation programmes.

Pressed or extruded ceramic tiles have been used by the food industry for many years and are still extensively used in processing areas. In recent years they have been partially replaced on grounds of cost by the various seamless resin floors now widely available. Provided tiles of a suitable specification (fully vitrified ceramics) are selected and properly laid – an important prerequisite for all types of flooring – they are perfectly suitable for high-risk production areas and will give a long-life floor available in a very wide choice of colours.

Tiles are laid on sand and cement mortar-bonded to the subfloor (thin bed), or on a semi-dry sand and cement mix (thick bed). A tile thickness of approximately 20 mm will provide adequate strength with either of the bedding methods. Thinner tiles (12 mm) are used for bedding into a resin bed by a vibratory method. Tile surfaces may be smooth, studded or incorporate silicon carbide granules to improve slip resistance. Studded tiles are not recommended because of the greater difficulty of cleaning such surfaces. Ideally, surfaces that offer the greatest ease in cleaning should be used. However, in practice, the requirements for anti-slip conditions cannot be ignored and as a result the final choice should reflect a balance of the relevant factors and the emphasis placed on them.

Joints should be grouted as soon as practical, otherwise the joint faces may become contaminated. Cementitious grouts are not considered suitable for hygienic applications and resin grouts are normally used. These should not be applied for at least three days after the tiles have been laid, so that water from the bed can evaporate. Epoxy resins are widely used for grouting but do have limited resistance to very high concentrations of sodium hypochlorite and soften at temperatures above 80°C. Polyester and Furan resins are more resistant to chemical attack. Shapton and Shapton (1991) cite data for the chemical resistance of different resins given by Beauchner and Reinert (1972). The grouting material should fill the joints completely to a depth of at least 12 mm and be finished flush with the tile surface. Thinner joints (1 mm) are achieved when tiles are vibrated into a resin bed. The procedure ensures a flat plane and reduces the possibility of damage to the tile edges in use. One advantage of tile floors that is not always fully appreciated is that sections or local areas of damaged surface can be replaced and colour-matched with relative ease, so that the overall standard and appearance of the floor can be maintained.

Resin-based seamless floors offer a good alternative means of attaining a hygienic surface provided they are laid on a sound concrete base. The choice of finish can be made either from various resin-based systems (primarily epoxy or polyurethane) or from polymer-modified cementitious systems. The resin-based systems can be broadly grouped under three headings:

- *Heavy duty:* heavily filled trowel-applied systems 5–12 mm thick. Such screeds are of high strength and are normally slip-resistant.
- *Self-levelling:* 'poured and floated' systems applied at 2–5 mm thickness. These systems are sometimes more correctly described as 'self-smoothing'. They generally give smooth glossy surfaces.
- *Coatings:* usually 0.1–0.5 mm thick. They are not recommended for high-risk or other production areas because of their poor durability. Failures of such floors have been associated with microbial contamination, including *Listeria monocytogenes*, becoming trapped under loosened areas where the coating has flaked.

A further aspect that needs to be considered is whether the proposed floor meets legislative requirements. Statements in UK and EC legislation are of a general nature but do call for floors to be 'waterproof' or 'impervious' and 'cleanable'. Work at CCFRA (Taylor and Holah 1996) has developed a simple technique to assess the water absorption of flooring materials and materials can be quickly accepted or rejected on any water uptake recorded. Water uptake is unacceptable because if fluids are able to penetrate into flooring materials, microorganisms can be transported to harbourage sites that are impossible to chemically clean and disinfect.

Cleanability is more difficult to interpret but both Taylor and Holah (1996) and Mettler and Carpentier (1998) have proposed suitable test methods in which the cleanability of attached microorganisms are assessed. Differences in cleanability between materials have been found but these do not necessarily correlate to surface roughness, traditionally measured as μmRa. Both sets of authors have shown that microbial cleanability is defined at a magnitude of surface imperfection 100–1000-fold below that as recognized as being important in terms of slip resistance. For example, flooring materials designed to be slip resistant may have an average peak to valley height measured in millimetres, whilst the size of imperfection that could harbour microorganisms would be measured in microns. It is possible, therefore, to obtain 'rough' surfaces that are good for slip resistance but, at a micron level, are also good for microbial cleanability and vice versa 'smooth' surfaces with limited slip resistance and also poor cleanability at the micron level. When considering the selection of flooring materials, therefore, evidence for imperviousness and cleanability should be sought. The floor should be coved where it meets walls or other vertical surfaces such as plinths or columns as this facilitates cleaning.

With the considerable choice of resin materials/systems available it is clearly important that the processor reviews the end-user requirements carefully and discusses them in detail with the flooring contractor. Imholte (1984), states that, in general, the higher-quality materials may be more expensive but last longer and have lower maintenance costs. The use of established flooring contractors and the viewing of an existing floor of the type under consideration would also seem to be sensible steps in the selection process.

13.4.3 Drainage

Ashford (1986) states that drainage is often neglected and badly constructed. Detailed consideration of the drainage requirements is an important aspect of floor design. Ideally, the layout and siting of production equipment should be finalised before the floor is designed to ensure that discharges can be fed directly into drains. In practice, this is not always possible, and in the food industry in particular, there is a greater chance that the layout of lines will be frequently changed. Equipment should not be located directly over drainage channels as this may restrict access for cleaning.

Discharges from equipment, however, should be fed directly into drains to avoid floor flooding. Alternatively, a low wall may be built around the equipment from which water and solids may be drained. Where the channels are close to a wall they should not be directly against it to avoid flooding of the wall to floor junction. An indirect advantage of channels near a wall is that the siting of equipment hard up to the wall is prevented, thus providing access for cleaning.

Satisfactory drainage can be achieved only if adequate falls to drainage points are provided. A number of factors should be taken into consideration when establishing the optimum or practical fall, for example:

- *Volume of water:* wet processes require a greater fall.
- *Floor finish:* trowelled resin surface finishes require a greater fall than self-levelling ones. Otherwise 'puddles' created by small depressions in the surface may remain.
- *Safety:* falls greater than 1 in 40 may introduce operator safety hazards and also cause problems with wheeled vehicles.

Timperley (1993) states that floors should have a fall to drain of between 1 in 50 and 1 in 100, depending upon the process operation and surface texture whilst Cattell (1988), suggests a compromise figure of 1 in 80 for general purposes and safety.

The type of drain used depends to a great extent upon the process operation involved. For operations involving a considerable amount of water and solids, channel drains are often the most suitable (Fig. 13.10). For operations generating volumes of water but with little solids, aperture channel drains are more

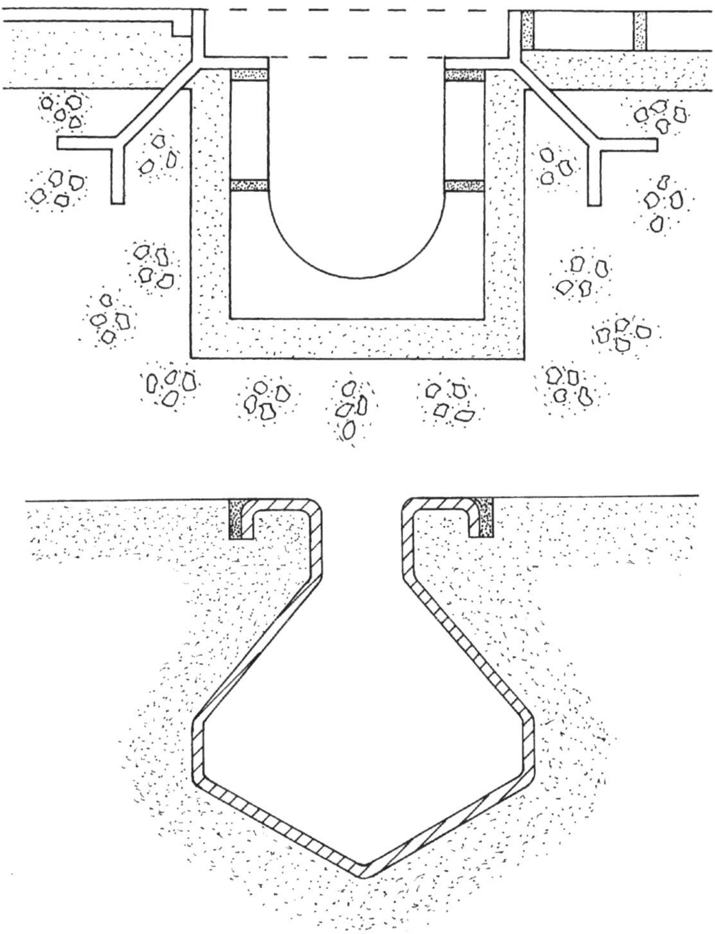

Fig. 13.10 Half-round drainage channel with reinforced rebate for grating and stainless steel aperture channel drain.

favourable (Fig. 13.10). Many chilled food operations, however, do not require extensive high-volume drainage systems and, in fact, fewer drains lead to less water use and thus increased control of environmental microbial contamination. In such cases, a small number of gulley type drains within the processing area is appropriate.

In most cases, channels should have a fall of at least 1 in 100, have round bottoms and be no deeper than 150 mm for ease of cleaning and must be provided with gratings for safety reasons. The channel gratings must be easily removable, with wide apertures (20 mm minimum) to allow solids to enter the drain. In recent years there has been a marked increase in the use of corrosion-resistant materials of construction, such as stainless steel for drain gratings. Stainless steel is also finding a wider use in other drain fittings, e.g. various designs of traps, and for the channels of shallower (low-volume) drainage systems. The profile of aperture channel drains is such that all internal surfaces can be easily cleaned.

The edge of the channel rebate must be properly designed and constructed to protect it, by an angle, from damage (Fig. 13.10). This is particularly so if wheeled vehicles are in use, to prevent damage to the channel/floor interface. This is critical in terms of the control of microbial pathogens and the authors are aware of a number of cases where the seal has been broken between the drain channel and floor structure so as to leave an uncleanable void between the two. When the channel is subsequently walked upon or traversed by wheeled vehicles, a small volume of foul liquid and microorganisms is 'pumped' to the floor surface.

The drainage system should flow in the reverse direction of production (i.e. from high to low risk) and whenever possible, backflow from low-risk to high-risk areas should be impossible. This is best achieved by having separate low- and high-risk drains running to a master collection drain with an air-break between each collector and master drain. The drainage system should also be designed such that rodding points are outside high risk areas. Solids must be separated from liquids as soon as possible, by screening, to avoid leaching and subsequent high effluent concentrations. Traps should be easily accessible, frequently emptied and preferably outside the processing area.

13.4.4 Walls

Guidelines for the design and construction of walls, ceilings and services have been prepared by Timperley (1994). A number of different types of materials may be used to construct walls forming the boundaries of a high-risk area and of the individual rooms within the area. When considering the alternative systems, a number of technical factors such as hygiene characteristics, insulation properties, and structural characteristics need to be taken into consideration.

Modular insulated panels are now used very widely for non-load-bearing walls. The panels are made of a core of insulating material between 50 and 200 mm thick, sandwiched between steel sheets, which are bonded to both sides of the core. Careful consideration must be given, not only to the fire retardation of the wall insulation or coating material, but also to the toxicity of the fumes emitted in the event of a fire as these could hamper a fire-fighting operation. The steel cladding is generally slightly ribbed to provide greater rigidity and can be finished with a variety of hygienic surface coatings, ready for use. The modules are designed to lock together and allow a silicone sealant to provide a hygienic seal between the units. The modules can be mounted either directly (in a U-shaped channel) onto the floor or on a concrete upstand or plinth (Fig. 13.11). The latter provides useful protection against the possibility of damage from vehicular traffic, particularly fork-lift trucks. However, it should be appreciated that this arrangement reduces the possibility of relatively easy and inexpensive changes to room layout to meet future production requirements. Sections fixed

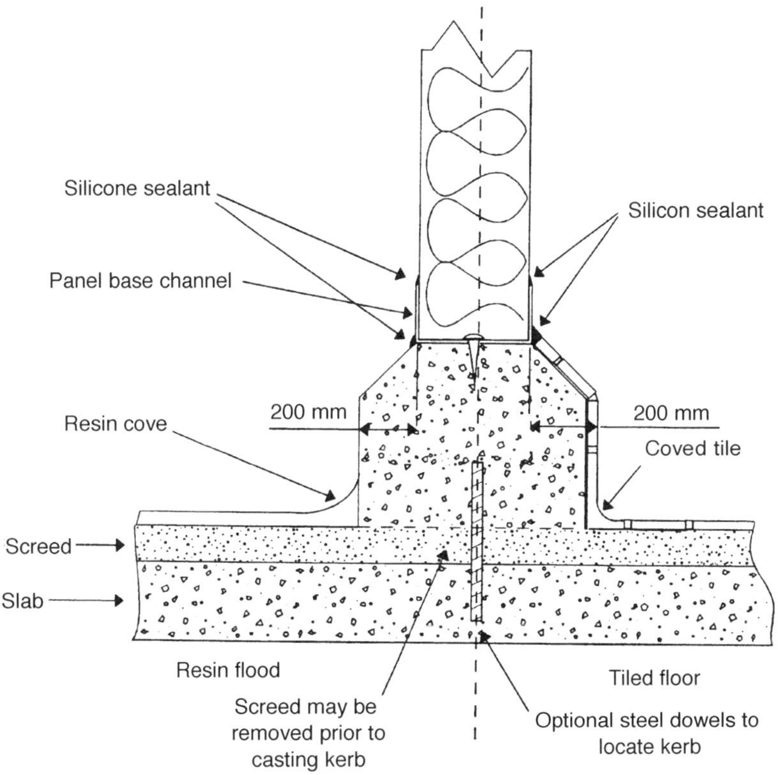

Fig. 13.11 Modular insulated panel located in U-channel and fixed to a concrete plinth.

directly onto the floor must be properly bedded in silicone sealant and coved to provide an easily cleanable and watertight junction. As with wall-to-floor joints, it is also good practice to cove wall-to-ceiling junctions to assist cleaning.

To ensure continuity in the appearance and surface characteristics of walling throughout a high-risk area, thin sections (50 mm) of insulated panel are sometimes used to cover external or load-bearing walls. When such a practice is adopted, there is a possibility of introducing harbourage sites for pests between the two walling materials. The chances of problems occurring are greatly increased if openings for services are made in the insulated panels without effective sealing.

In the UK, load-bearing and fire-break walls are often constructed from brick or blockwork. Walls made from such materials do not generally provide a smooth enough surface to allow the direct application of the various types of coatings. A common practice is to render the brickwork with a cement and sand screed to achieve the desired surface smoothness for the coating layer. The walls may be covered by other materials such as tiles or sheets of plastics. The former is preferred, provided each tile is fully bedded and an appropriate resin is used for grouting. In very wet or humid areas, where there is a strong possibility of mould growth, the application of a fungicidal coating may be considered; there is evidence that some such coatings remain effective for many years.

Hygiene standards for walls as defined in various EC Directives require that they must be constructed of impervious, non-absorbent, washable, non-toxic materials and have smooth crack-free surfaces up to a height appropriate for the operations. For high-risk areas the standard of construction and finish must apply right up to ceiling level. The same hygienic assessment techniques as described for flooring materials are also directly applicable to wall coverings and finishes.

Openings in the walls of the high-risk area need to be limited and controlled and openings for product, packaging and personnel have already been considered. In addition:

- Emergency exits; such doors must be fitted with 'out-only' operating bars. The doors must remain closed except in the case of an emergency.
- Larger 'engineering' doors required for the occasional access of equipment in and out of high risk; these doors must also remain closed and should be sealed when not in use.

13.4.5 Ceilings

When considering the basic design concepts for high-risk areas, the idea of using ceilings to separate production and service functions was discussed. In practice this is often achieved by either using suitable load-bearing insulation panels or suspending sections of insulated panels, as used for the internal walls, from the structural frame of the building. The use of such insulated panels meets legislative requirements by providing a surface that is easily cleanable and will not shed particles.

It is important to ensure that drops from services passing through the ceiling are sealed properly to prevent ingress of contamination. Cables may be run in trunking or conduit but this must be effectively sealed against the ingress of vermin and water. All switchgear and controls, other than emergency stop buttons, should, whenever possible, be sited in separate rooms away from processing areas, particularly if wet operations are taking place.

Lighting may be a combination of both natural and artificial. Artificial lighting has many advantages in that, if properly arranged, it provides even illumination over inspection belts and a minimum of 500–600 lux is recommended. Fluorescent tubes and lamps must be protected by shields, usually of polycarbonate, to protect the glass and contain it in the event of breakage. Suspended units should be smooth, easily cleanable and designed to the appropriate standards to prevent the ingress of water. It is suggested that lighting units are plugged in so that in the event of a failure the entire unit can be replaced and the faulty one removed from the processing areas to a designated workshop for maintenance. Ideally, recessed lighting flush with the ceiling is recommended from the hygienic aspect (Fig. 13.9) but this is not always possible and maintenance may be difficult.

13.5 Equipment

The manufacture of a large proportion of chilled foods generally involves some element of batch or assembly operations or both. The equipment used for such operations is predominantly of the open type, that is, it cannot be cleaned by recirculation (CIP) procedures, and must be of the highest hygienic design standards. Hygienic equipment design provides three major benefits to food manufacturers.

1. Quality – good hygienic design maintains product in the main product flow. This ensures that product is not 'held-up' within the equipment where it could deteriorate and affect product quality on rejoining the main product flow. Or, for example in flavourings manufacture, one batch could not taint a subsequent batch.
2. Safety – good hygienic design prevents the contamination of the product with substances that would adversely affect the health of the consumer. Such contamination could be microbiological (e.g. pathogens), chemical (e.g. lubricating fluids, cleaning chemicals) and physical (e.g. glass).
3. Efficiency – good hygienic design reduces the time required for an item of equipment to be cleaned. This reduction of cleaning time is significant over the lifetime of the equipment such that hygienically designed equipment which is initially more expensive (compared to similarly performing poorly designed equipment), will be more cost effective in the long term. In addition, savings in cleaning time may lead to increased production.

A relatively few academic texts have been published on hygienic design, though texts by Anon (1983), Timperley and Timperley (1993), the European Hygienic Design Group (EHEDG 1995), Timperley (1997) and Holah (1998) are appropriate to chilled foods. Within Europe (the EHEDG) and the USA (the 3-A Standards and the National Sanitation Foundation – NSF), a number of organisations exist to foster consensus in hygienic design and the use of these organisations' guidelines can have a quasi-legal status. It should be noted that in Europe, hygienic design guidelines tend to be more generic in nature than the more prescriptive requirements American readers may be familiar with.

In the EC, the Council Directive *on the approximation of the laws of Member States relating to machinery* (89/392/EEC) was published on 14 June 1989. The Directive includes a short section dealing with hygiene and design requirements which states that machinery intended for the preparation and processing of foods must be designed and constructed so as to avoid health risks and consists of seven hygiene rules that must be observed. These are concerned with materials in contact with food; surface smoothness; preference for welding or continuous bonding rather than fastenings; design for cleanability and disinfection; good surface drainage; prevention of dead spaces which cannot be cleaned and design to prevent product contamination by ancillary substances, e.g. lubricants. The Directive requires that all machinery sold within the EC shall meet these basic standards and be marked accordingly to show compliance (the 'CE' mark).

Subsequent to this Directive, a European Standard EN 1672-2 *Food processing machinery-Safety and hygiene requirements-Basic concepts-Part 2; Hygiene requirements* (Anon. 1997b) has recently been adopted to further clarify the hygiene rules established in 89/392/EEC. In addition to this, a number of specific standards on bakery, meat, catering, edible oils, vending and dispensing, pasta, bulk milk coolers, cereal processing and dairy equipment are in preparation. The basic hygienic design requirements as presented in EN 1672-2 can be summarised under eleven headings and are described below:

1. Construction materials. Materials used for product contact must have adequate strength over a wide temperature range, a reasonable life, be non-tainting, corrosion and abrasion resistant, easily cleaned and capable of being shaped. Stainless steel usually meets all these requirements and there are various grades of stainless steel which are selected for their particular properties to meet operational requirements, e.g. Type 316 which contains molybdenum is used where improved corrosion resistance is necessary.
2. Surface finish. Product contact surfaces must be finished to a degree of surface roughness that is smooth enough to enable them to be easily cleaned. Surfaces will deteriorate with age and wear (abrasion) such that cleaning will become more difficult.
3. Joints. Permanent joints, such as those which are welded, should be smooth and continuous. Dismountable joints, such as screwed pipe couplings must be crevice-free and provide a smooth continuous surface on the product side. Flanged joints must be located with each other and be sealed with a gasket because, although metal/metal joints can be made leak tight, they may still permit the ingress of microorganisms.
4. Fasteners. Exposed screw threads, nuts, bolts, screws and rivets (Fig. 13.12) must be avoided wherever possible in product contact areas. Alternative methods of fastening can be used (Fig. 13.13) where the washer used has a rubber compressible insert to form a bacteria-tight seal.
5. Drainage. All pipelines and equipment surfaces should be self draining because residual liquids can lead to microbial growth or, in the case of cleaning fluids, result in contamination of product.
6. Internal angles and corners. These should be well radiused, wherever possible, to facilitate cleaning.
7. Dead spaces. As well as ensuring that there are no dead spaces in the design of equipment, care must be taken that they are not introduced during installation.
8. Bearings and shaft seals. Bearings should, wherever possible, be mounted outside the product area to avoid possible contamination of product by lubricants, unless they are edible, or possible failure of the bearings due to the ingress of the product. Shaft seals must be of such design so as to be easily cleaned and if not product lubricated, then the lubricant must be edible. Where a bearing is within the product area, such as a foot bearing

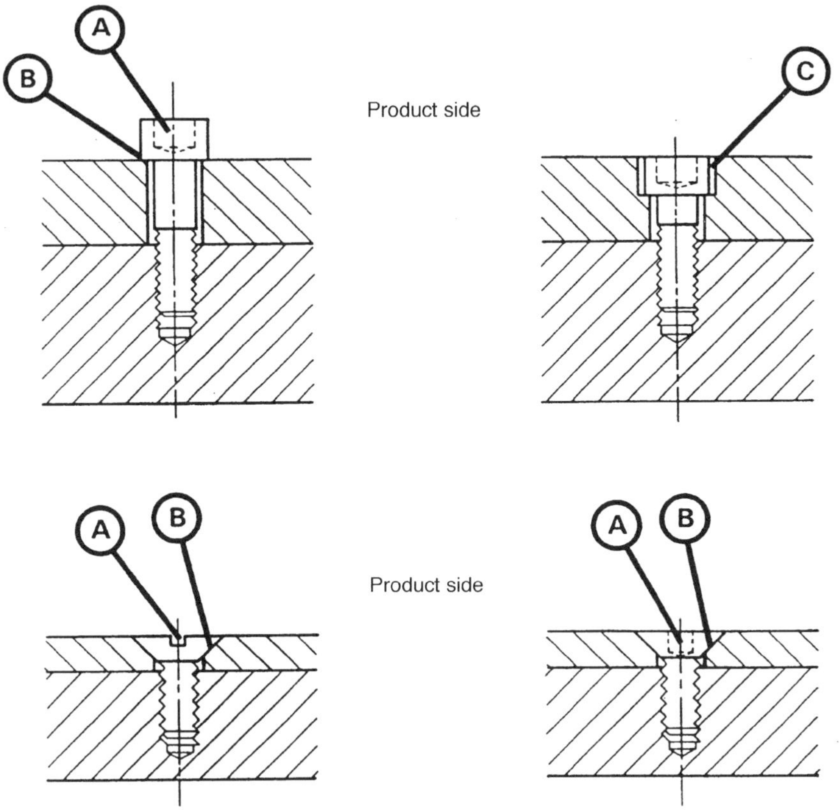

Fig. 13.12 Examples of unhygienic fasteners. A = Soil trap points, B = Metal to metal, C = Dead spaces.

for an agitator shaft in a vessel, it is important that there is a groove completely through the bore of the bush, from top to bottom to permit the passage of cleaning fluid.

9. Instrumentation. Instruments must be constructed from appropriate materials and if they contain a transmitting fluid, such as in a bourdon tube pressure gauge, then the fluid must be approved for food contact. Many instruments themselves are hygienic but often they are installed unhygienically.
10. Doors, covers and panels. Doors, covers and panels should be designed so that they prevent the entry of and/or prevent the accumulation of soil. Where appropriate they should be sloped to an outside edge and should be easily removed to facilitate cleaning.
11. Controls. These should be designed to prevent the ingress of contamination and should be easily cleanable (Fig. 13.14), particularly those that are repeatedly touched by food handlers to allow process operation.

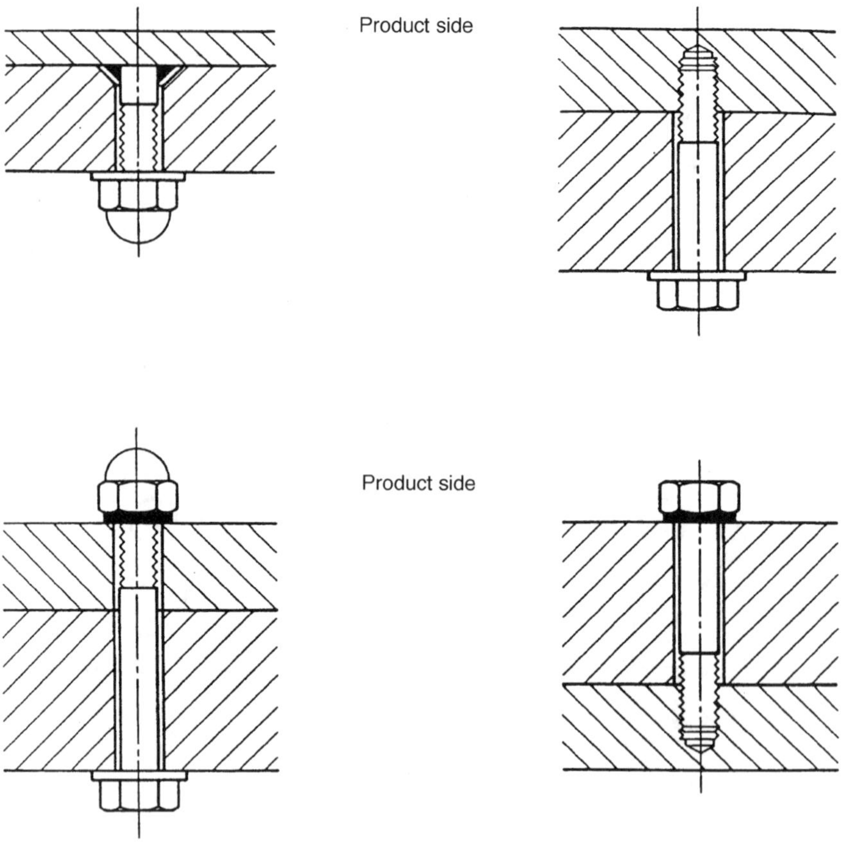

Fig. 13.13 Examples of hygienic fasteners.

The importance of good hygienic design in chilled foods operations can be illustrated with reference to a sliced-meat factory which had slicers whose action was initiated by pressing a control switch identical to that shown in Fig. 13.14. The factory concerned was having problems due to product contamination with *Listeria monocytogenes*, and was eventually forced to stop production for a few days with a subsequent financial loss in excess of £1 million. The problem was finally traced to a source of *L. monocytogenes* that was being harboured within the body of the slicer switches. At the beginning of production the slicing operative picked up a log of meat, placed it on the slicer and pressed the control switch to start slicing. From this point on, and every time he subsequently repeated this procedure, *L. monocytogenes* was transferred from his hand to the slicer and, by the middle of the shift, sufficient *L. monocytogenes* was present on the slicer to be detected in the product. The conclusion to the incident was the purchase of a number of rubber switch covers as shown in Fig. 13.14, for the cost of a few pounds.

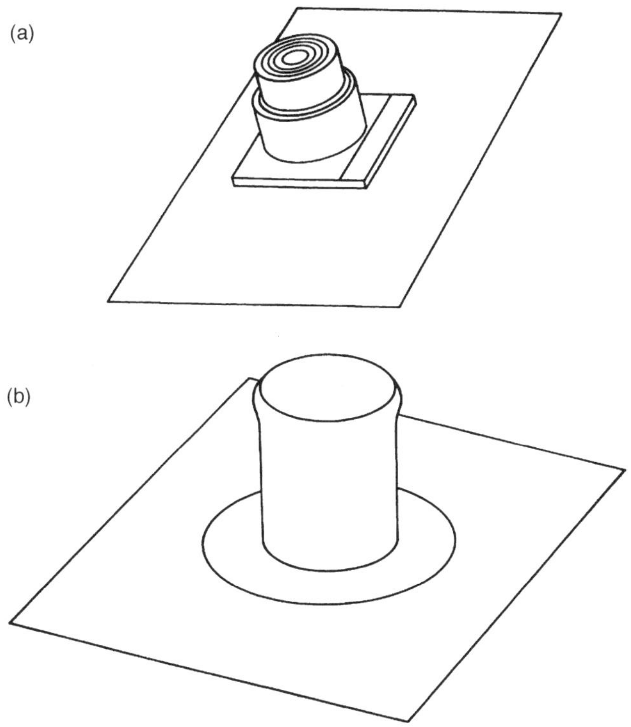

Fig. 13.14 Typical operating switch with inherent crevices (a). Hygienic rubber-capped alternative allowing easy cleaning (b).

13.5.1 Installation of equipment

The potential for well designed and constructed equipment to be operated in a hygienic manner may be easily vitiated by inadequate attention to its location and installation. Timperley (1997), when considering the accessibility of equipment, recommended that it is more effective to consider complete lines instead of individual items of equipment and recommended the following:

- There should be sufficient height to allow adequate access for inspection, cleaning and maintenance of the equipment and for the cleaning of floors.
- All parts of the equipment should be installed at a sufficient distance from walls, ceilings and adjacent equipment to allow easy access for inspection, cleaning and maintenance, especially if lifting is involved.
- Ancillary equipment, control systems and services connected to the process equipment should be located so as to allow access for maintenance and cleaning.
- Supporting framework, wall mountings and legs should be kept to a minimum. They should be constructed from tubular or box section material which should be sealed to prevent ingress of water or soil. Angle or channel section material should not be used.

- Base plates used to support and fix equipment should have smooth, continuous and sloping surfaces to aid drainage. They should be coved at the floor junction. Alternatively, ball feet should be fitted.
- Pipework and valves should be supported independently of other equipment to reduce the chance of strain and damage to equipment, pipework and joints.

13.6 Conclusion

As a food preservation method, chilling technology already provides the consumer with a range of products. Both the range and volume of products can be expected to grow considerably in the future together with the need for higher hygiene standards. However, it should be remembered that whilst we may have had over ten years experience in this product sector, chilled foods, like any other forms of preserved foods, have been developed to a high degree of commercial success before all the technical aspects of the system have been fully established or understood.

Within the terms of reference of this chapter there is clearly a need for more information on the routes of contamination into high-risk areas and for a greater understanding of the effectiveness of different procedures that are currently used to minimize the ingress of contamination. Similarly, any hygienic design aspects that impinge on the ability of pathogens to survive, grow or be transported around within the high-risk area should be explored. It is also just as important that new or alternative procedures that may be advocated are thoroughly evaluated. Such work would provide the opportunity to bring together the application of the principles of hygienic design with both current microbiological knowledge related to cleaning and disinfection procedures and industry sectors where microbial or particle control is critical (e.g. pharmaceuticals and microelectronics). This could be further advanced by the use of HACCP and mathematical modelling derived from work on predictive microbiology, and a more fundamental knowledge of the microbiological strains capable of growth and survival in chilled environments.

13.7 References

ANON, (1983) *Hygienic design of food processing equipment*. Campden and Chorleywood Food Research Association, Technical Manual No. 7.

ANON, (1992a), *The Workplace (Health, Safety and Welfare) Regulations*, HMSO, ISBN 0-11-886333-9.

ANON, (1992b) *Workplace Health, Safety and Welfare*. Approved Code of Practice and Guidance on Workplace (Health, Safety and Welfare) Regulations 1992, L24 HSE Books ISBN 0-7176-0413-6.

ANON, (1995) *Food Safety (Temperature Control) Regulations 1995*, HMSO, ISBN 0-11-053383-6.

ANON, (1997a) *Guidelines for good hygienic practices in the manufacture of chilled foods*. Chilled Food Association, 6, Catherine Street, London WC2B 5JJ.

ANON, (1997b) EN 1672-2. *Food processing machinery. Basic requirements. Part 2; Hygiene requirements*, ISBN 0 580 27957 X.

ANON, (1999), *Workroom temperatures in places where food is handled*, HSE Food Sheet No. 3 (Revised).

ASHFORD M J, (1986) *Factory design principles in the food processing industry*. In: *Preparation, Processing and Freezing in Frozen Food Production*, The Institution of Mechanical Engineers, London.

BEAUCHNER F R and REINERT D G, (1972) How to select sanitary flooring, *Food Engineering*, **44** (10) 120–2 and 126.

BROWN K, (1996) *Guidelines on air quality for the food industry*, Campden and Chorleywood Food Research Association, Guideline No. 12.

BROWN K L, (2000) *Guidance on achieving reasonable working temperatures and conditions during production of chilled foods*, Campden and Chorleywood Food Research Association, Guideline No. 26.

BURFOOT D, BROWN K, REAVELL S and XU Y, (2000) Improving food hygiene through localised air flows. *Proceedings International Congress on Engineering and Food*, April 2000, Puebla, Mexico, Technomic Publishing Co. Inc., Pensylvania, USA (in press).

CATELL D, (1988) *Specialist floor finsihes: Design and installation*, Blackie and Son Ltd., Glasgow and London.

EC, (1993) Hygiene of Foodstuffs, Off. J. European Communities, L175, 1–11.

EHEDG, Document No. 13. (1995). *Hygienic design of equipment for open processing of foods*. Campden and Chorleywood Food Research Association, Chipping Campden, Glos GL55 6LD and as an extended abstracts in *Trends in Food Science and Technology*, 6:305–10.

GUZEWICH J and ROSS P, (1999) *Evaluation of risks related to microbiological contamination of ready-to-eat food by food preparation workers and the effectiveness of interventions to minimise those risks*, Food and Drug Administration, White Paper, Section One, USA.

HOLAH J T, (1998) Hygienic design: International issues, *Dairy, Food and Environmental Sanitation*, **18** 212–20.

HOLAH J T, HALL K E, HOLDER J, ROGERS S J, TAYLOR J and BROWN K L, (1995) *Airborne microorganism levels in food processing environments*, Campden Food and Drink Research Association, R&D report No. 12.

IMHOLTE T J, (1984) *Engineering for Food Safety and Sanitation*, Technical Institute of Food Safety, Crystal, Minnesota.

KATSUYARNA A M and STRACHAN J P, (eds) (1980) *Principles of Food Processing Sanitation*, The Food Processors Institute, Washington DC.

METTLER E and CARPENTIER B, (1998) Variations over time of microbial load and physicochemical properties of floor materials after cleaning in food industry premises, *Journal of Food Protection*, **61** 57–65.

SHAPTON D A and SHAPTON N F, (eds) (1991) *Principles and Practices for the*

Safe Processing of Foods. Butterworth Heinemann.

TAYLOR J and HOLAH J, (1996) A comparative evaluation with respect to bacterial cleanability of a range of wall and floor surface materials used in the food industry, *Journal of Applied Bacteriology*, **81** 262–7.

TAYLOR J H and HOLAH J T, (2000) *Hand hygiene in the food industry: a review*. Review No. 18, Campden and Chorleywood Food Research Association.

TAYLOR J H, HOLAH J T, WALKER H and KAUR M, (2000) *Hand and footwear hygiene: An Investigation to define best practice*, Campden and Chorleywood Food Research Association, R&D Report No. 112.

TIMPERLEY A W, (1994) *Guidelines for the design and construction of walls, ceilings and services for food production areas*, Campden and Chorleywood Food Research Association, Technical Manual No. 44.

TIMPERLEY A W, (1997) *Hygienic design of liquid handling equipment for the food industry*, (second edn) Campden and Chorleywood Food Research Association, Technical Manual No. 17.

TIMPERLEY D A, (1993) *Guidelines for the design and construction of floors for food production areas*, Campden and Chorleywood Food Research Association, Technical Manual No. 40.

TIMPERLEY D A and TIMPERLEY A W, (1993) *Hygienic design of meat slicing machines*, Campden and Chorleywood Food Research Association, Technical Memorandum No. 679, Chipping Campden, UK.

TROLLER J A, (1983) *Sanitation in Food Processing*, Academic Press, New York.

14

Cleaning and disinfection

J. Holah, Campden and Chorleywood Food Research Association

14.1 Introduction

Chapter 13 has outlined the concept of 'hygienic design' and 'hygienic practices' in controlling the safety of chilled food products. This chapter deals with hygienic practices, specifically those related to cleaning and disinfection.

Contamination in food products may arise from four main sources: the constituent raw materials, surfaces, people (and other animals) and the air. Control of the raw materials is addressed elsewhere in this book and is the only non-environmental contamination route. Food may pick up contamination as it is moved across product contact surfaces or if it is touched or comes into contact with people (food handlers) or other animals (pests). The air acts as both a source of contamination, i.e. from outside the processing area, or as a transport medium, e.g. moving contamination from non-product to product contact surfaces.

Provided that the process environment and production equipment have been hygienically designed (Chapter 13), cleaning and disinfection (referred to together as 'sanitation') are the major day-to-day controls of the environmental routes of food product contamination. When undertaken correctly, sanitation programmes have been shown to be cost-effective and easy to manage, and, if diligently applied, can reduce the risk of microbial or foreign body contamination. Given the intrinsic demand for high standards of hygiene in the production of short shelf-life chilled foods, together with pressure from customers, consumers and legislation for ever-increasing hygiene standards, sanitation demands the same degree of attention as any other key process in the manufacture of safe and wholesome chilled foods.

This chapter is concerned with the sanitation of 'hard' surfaces only – equipment, floors, walls and utensils – as other surfaces, e.g. protective clothing or skin, have been dealt with under personal hygiene (Chapter 13). In this context, surface sanitation is undertaken to:

- remove microorganisms, or material conductive to microbial growth. This reduces the chance of contamination by pathogens and, by reducing spoilage organisms, may extend the shelf-life of some products.
- remove materials that could lead to foreign body contamination or could provide food or shelter for pests. This also improves the appearance and quality of product by removing food materials left on lines that may deteriorate and re-enter subsequent production runs.
- extend the life of, and prevent damage to equipment and services, provide a safe and clean working environment for employees and boost morale and productivity.
- present a favourable image to customers and the public. On audit, the initial perception of an 'untidy' or 'dirty' processing area, and hence a 'poorly managed operation' is subsequently difficult to overcome.

14.2 Sanitation principles

Sanitation is undertaken primarily to remove all undesirable material (food residues, microorganisms, foreign bodies and cleaning chemicals) from surfaces in an economical manner, to a level at which any residues remaining are of minimal risk to the quality or safety of the product. Such undesirable material, generally referred to as 'soil', can be derived from normal production, spillages, line-jams, equipment maintenance, packaging or general environmental contamination (dust and dirt). To undertake an adequate and economic sanitation programme, it is essential to characterise the nature of the soil to be removed.

The product residues are readily observed and may be characterised by their chemical composition, e.g. carbohydrate, fat, protein or starch. It is also important to be aware of processing and/or environmental factors, however, as the same product soil may lead to a variety of cleaning problems dependent primarily on moisture levels and temperature. Generally, the higher the product soil temperature (especially if the soil has been baked) and the greater the time period before the sanitation programme is initiated (i.e. the drier the soil becomes), the more difficult the soil is to remove.

Microorganisms can either be incorporated into the soil or can attach to surfaces and form layers or biofilms. There are a number of factors that have been shown to affect attachment and biofilm formation such as the level and type of microorganisms present, surface conditioning layer, substratum nature and roughness, temperature, pH, nutrient availability and time available. Several reviews of biofilm formation in the food industry have been published including Pontefract (1991), Holah and Kearney (1992), Mattila-Sandholm and Wirtanen (1992), Carpentier and Cerf (1993), Zottola and Sasahara (1994), Gibson *et al.*

(1995) and Kumar and Anand, (1998). In general, however, biofilm formation is usually found only on environmental surfaces, and progression of attached cells through microcolonies to extensive biofilm is limited by regular cleaning and disinfection.

Gibson *et al.*(1995) in studies of attached microorganisms in 17 different processing environments, recorded 79% of isolates as Gram negative rods, 8.6% Gram positive cocci, 6.5% Gram positive rods and 1.2% yeast strains. The most common organisms were *Pseudomonas*, *Staphylococcus* and *Enterobacter* spp. Pseudomonads are environmental psychrotrophic organisms that readily attach to surfaces and are common spoilage organsisms in chilled foods. Other common Gram negatives that have been associated with surfaces are coliform organisms that are widely distributed in the environment and may also be indicators of inadequate processing or post process contamination. Staphylococci are associated with human skin and therefore their presence on surfaces may be as a result of transfer from food handlers. In addition, Mettler and Carpentier (1998) studied the microflora associated with the surfaces in milk, meat and pastry sites and concluded that the micro-flora was specific to the processing environment.

Bacteria adhering to the food product contact surfaces may be an important source of potential contamination leading to serious hygienic problems and economic losses due to food spoilage. For example, pseudomonads and many other Gram negative organisms detected on surfaces are the spoilage organisms of concern in chilled foods. The survival of organisms in biofilms may be a source of post process contamination, resulting in reduced shelf life of the product. In addition, *Listeria monocytogenes* has been isolated from a range of food processing surfaces (Walker *et al.* 1991, Lawrence and Gilmore 1995 and Destro *et al.* 1996) and is usually looked for in high-risk processing areas via the company environmental sampling plan.

Following HACCP principles, if the food processor believes that biofilms are a risk to the safety of the food product, appropriate control steps must be taken. These would include providing an environment in which the formation of the biofilm would be limited, undertaking cleaning and disinfection programmes as required, monitoring and controlling these programmes to ensure their success during their operation and verifying their performance by a suitable (usually microbiological) assessment.

Within the sanitation programme, the cleaning phase can be divided up into three stages, following the pioneering work of Jennings (1965) and interpreted by Koopal (1985), with the addition of a fourth stage to cover disinfection. These are described below.

1. The wetting and penetration by the cleaning solution of both the soil and the equipment surface.
2. The reaction of the cleaning solution with both the soil and the surface to facilitate: peptisation of organic materials, dissolution of soluble organics and minerals, emulsification of fats and the dispersion and removal from the surface of solid soil components.

3. The prevention of redeposition of the dispersed soil back onto the cleansed surface.
4. The wetting by the disinfection solution of residual microorganisms to facilitate reaction with cell membranes and/or penetration of the microbial cell to produce a biocidal or biostatic action. Dependent on whether the disinfectant contains a surfactant and the disinfectant practice chosen (i.e. with or without rinsing), this may be followed by dispersion of the microorganisms from the surface.

To undertake these four stages, sanitation programmes employ a combination of four major factors as described below. The combinations of these four factors vary for different cleaning systems and, generally, if the use of one energy source is restricted, this short-fall may be compensated for by utilising greater inputs from the others.

1. mechanical or kinetic energy
2. chemical energy
3. temperature or thermal energy
4. time.

Mechanical or kinetic energy is used to remove soils physically and may include scraping, manual brushing and automated scrubbing (physical abrasion) and pressure jet washing (fluid abrasion). Of all four factors, physical abrasion is regarded as the most efficient in terms of energy transfer (Offiler 1990), and the efficiency of fluid abrasion and the effect of impact pressure has been described by Anon. (1973) and Holah (1991). Mechanical energy has also been demonstrated to be the most efficient for biofilm removal (Blenkinsopp and Costerton 1991, Wirtanen and Mattila Sandholm 1993, 1994, Mattila-Sandholm and Wirtanen 1992 and Gibson *et al.* 1999).

In cleaning, chemical energy is used to break down soils to render them easier to remove and to suspend them in solution to aid rinsability. At the time of writing, no cleaning chemical has been marketed with the benefit of aiding microorganism removal. In chemical disinfection, chemicals react with microorganisms remaining on surfaces after cleaning to reduce their viability. The chemical effects of cleaning and disinfection increase with temperature in a linear relationship and approximately double for every 10°C rise. For fatty and oily soils, temperatures above their melting point are used, to break down and emulsify these deposits and so aid removal. The influence of detergency in cleaning and disinfection has been described by Dunsmore (1981), Shupe *et al.* (1982), Mabesa *et al.* (1982), Anderson *et al.* (1985) and Middlemiss *et al.* (1985). For cleaning processes using mechanical, chemical and thermal energies, generally the longer the time period employed, the more efficient the process. When extended time periods can be employed in sanitation programmes, e.g. soak-tank operations, other energy inputs can be reduced (e.g. reduced detergent concentration, lower temperature or less mechanical brushing).

Soiling of surfaces is a natural process which reduces the free energy of the system. To implement a sanitation programme, therefore, energy must be added to the soil to reduce both soil particle-soil particle and soil particle-equipment surface interactions. The mechanics and kinetics of these interactions have been discussed by a number of authors (Jennings 1965, Schlussler 1975, Loncin 1977, Corrieu 1981, Koopal 1985, Bergman and Tragardh 1990), and readers are directed to these articles since they fall beyond the scope of this chapter. In practical terms, however, it is worth looking at the principles involved in basic soil removal, as they have an influence on the management of sanitation programmes.

Soil removal from surfaces decreases such that the log of the mass of soil per unit area remaining is linear with respect to cleaning time (Fig. 14.1(a)) and thus follows first-order reaction kinetics (Jennings 1965, Schlusser 1975). This approximation, however, is only valid in the central portion of the plot and, in practice, soil removal is initially faster and ultimately slower (dotted line in Fig. 14.1(a)) than that which a first-order reaction predicts. The reasons for this are unclear, though initially, unadhered, gross oil is usually easily removed (Loncin 1977) whilst ultimately, soils held within surface imperfections, or otherwise protected from cleaning effects, would be more difficult to remove (Holah and Thorpe 1990).

Routine cleaning operations are never, therefore, 100% efficient, and over a course of multiple soiling/cleaning cycles, soil deposits (potentially including microorganisms) will be retained. As soil accumulates, cleaning efficiency will decrease and, as shown in plot A, Fig. 14.1(b), soil deposits may for a period grow exponentially. The timescale for such soil accumulation will differ for all

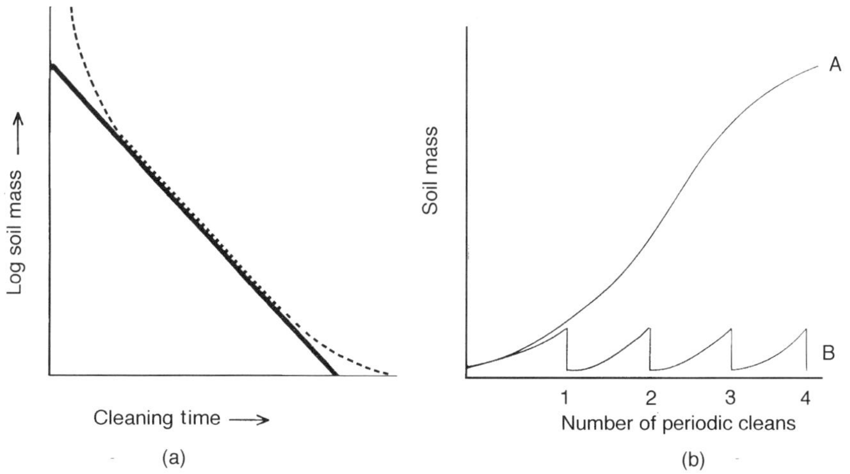

Fig. 14.1 Soil removal and accumulation. (a) Removal of soil with cleaning time. Solid line is theoretical removal, dotted line is cleaning in practice. (b) Build up of soil (and/or microorganisms); A, without periodic cleans and B, with periodic cleans. (After Dunsmore *et al.* 1981).

processing applications and can range from hours (e.g. heat exchangers) to typically several days or weeks, and in practice is controlled by the application of a 'periodic' clean (Dunsmore *et al.* 1981). Periodic cleans are employed to return the surface-bound soil accumulation to an acceptable base level (plot B, Fig. 14.1(b)) and are achieved by increasing cleaning time and/or energy input, e.g. higher temperatures, alternative chemicals or manual scrubbing. A typical example of a periodic clean is the 'week-end clean down' or 'bottoming'.

14.3 Sanitation chemicals

In many instances, management view the costs of cleaning and disinfection as the price of the chemicals purchased, primarily because this is the only 'invoice' that they see. In reality, however, sanitation chemicals are likely to represent approximately only 5% of the true costs, with labour and water costs being the most significant. The purchase of a good quality formulated cleaning product, whilst being initially more expensive, will more than cover its costs by increasing both the standard of clean and cleaning efficiency.

Within the sanitation programme it has traditionally been recognised that cleaning is responsible for the removal of not only the soil but also the majority of the microorganisms present. Mrozek (1982) showed a reduction in bacterial numbers on surfaces by up to 3 log orders whilst Schmidt and Cremling (1981) described reductions of 2–6 log orders. The results of work at CCFRA on the assessment of well constructed and competently undertaken sanitation programmes on food processing equipment in eight chilled food factories is shown in Table 14.1. The results suggest that both cleaning and disinfection are equally responsible for reducing the levels of adhered microorganisms. It is important, therefore, not only to purchase quality cleaning chemicals for their soil removal capabilities but also for their potential for microbial removal.

Unfortunately no single cleaning agent is able to perform all the functions necessary to facilitate a successful cleaning programme; so a cleaning solution, or detergent, is blended from a range of typical characteristic components:

- water
- surfactants
- inorganic alkalis

Table 14.1 Arithmetic and log mean bacterial counts on food processing equipment before and after cleaning and after disinfection

	Before cleaning	After cleaning	After disinfection
Arithmetic mean	1.32×10^6	8.67×10^4	2.5×10^3
log mean	3.26	2.35	1.14
No. of observations	498	1090	3147

- inorganic and organic acids
- sequestering agents.

For the majority of food processing operations it may be necessary, therefore, to employ a number of cleaning products, for specific operations. This requirement must be balanced by the desire to keep the range of cleaning chemicals on site to a minimum so as to reduce the risk of using the wrong product, to simplify the job of the safety officer and to allow chemical purchase to be based more on the economics of bulk quantities. The range of chemicals and their purposes is well documented (Anon. 1991, Elliot 1980, ICMSF 1980, 1988, Hayes 1985, Holah 1991, Koopal 1985, Russell *et al.* 1982) and only an overview of the principles is given here.

Water is the base ingredient of all 'wet' cleaning systems and must be of potable quality. Water provides the cheapest readily available transport medium for rinsing and dispersing soils, has dissolving powers to remove ionic-soluble compounds such as salts and sugars, will help emulsify fats at temperatures above their melting point, and, in high-pressure cleaning, can be used as an abrasive agent. On its own, however, water is a poor 'wetting' agent and cannot dissolve non-ionic compounds.

Organic surfactants (surface-active or wetting agents) are amphipolar and are composed of a long non-polar (hydrophobic or lyophilic) chain or tail and a polar (hydrophilic or lyophobic) head. Surfactants are classified as anionic (including the traditional soaps), cationic, or non-ionic, depending on their ionic charge in solution, with anionics and non-ionics being the most common. Amphipolar molecules aid cleaning by reducing the surface tension of water and by emulsification of fats. If a surfactant is added to a drop of water on a surface, the polar heads disrupt the water's hydrogen bonding and so reduce the surface tension of the water and allow the drop to collapse and 'wet' the surface. Increased wettability leads to enhanced penetration into soils and surface irregularities and hence aids cleaning action. Fats and oils are emulsified as the hydrophilic heads of the surfactant molecules dissolve in the water whilst the hydrophobic end dissolves in the fat. If the fat is surface-bound, the forces acting on the fat/water interface are such that the fat particle will form a sphere (to obtain the lowest surface area for its given volume) causing the fat deposit to 'roll-up' and detach itself from the surface.

Alkalis are useful cleaning agents as they are cheap, break down proteins through the action of hydroxyl ions, saponify fats and, at higher concentrations, may be bactericidal. Strong alkalis, usually sodium hydroxide (or caustic soda), exhibit a high degree of saponification and protein disruption, though they are corrosive and hazardous to operatives. Correspondingly, weak alkalis are less hazardous but also less effective. Alkaline detergents may be chlorinated to aid the removal of proteinaceous deposits, but chlorine at alkaline pH is not an effective biocide. The main disadvantages of alkalis are their potential to precipitate hard water ions, the formation of scums with soaps, and their poor rinsability.

Acids have little detergency properties, although they are very useful in making soluble carbonate and mineral scales, including hard water salts and proteinaceous deposits. As with alkalis, the stronger the acid the more effective it is; though, in addition, the more corrosive to plant and operatives. Acids are not used as frequently as alkalis in chilled food operations and tend to be used for periodic cleans.

Sequestering agents (sequestrants or chelating agents) are employed to prevent mineral ions precipitating by forming soluble complexes with them. Their primary use is in the control of water hardness ions and they are added to surfactants to aid their dispersion capacity and rinsability. Sequestrants are most commonly based on ethylene diamine tetracetic acid (EDTA), which is expensive. Although cheaper alternatives are available, these are usually polyphosphates which are environmentally unfriendly.

A general-purpose food detergent may, therefore, contain a strong alkali to saponify fats, weaker alkali 'builders' or 'bulking' agents, surfactants to improve wetting, dispersion and rinsability and sequestrants to control hard water ions. In addition, the detergent should ideally be safe, non-tainting, non-corrosive, stable, environmentally friendly and cheap. The choice of cleaning agent will depend on the soil to be removed and on its solubility characteristics, and these are summarised for a range of chilled products in Table 14.2 (modified from Elliot 1980).

Because of the wide range of food soils likely to be encountered and the influence of the food manufacturing site (temperature, humidity, type of equipment, time before cleaning, etc.), there are currently no recognised laboratory methods for assessing the efficacy of cleaning compounds. Food manufacturers have to be satisfied that cleaning chemicals are working appropriately, by conducting suitable field trials. Although the majority of the microbial contamination is removed by the cleaning phase of the sanitation

Table 14.2 Solubility characteristics and cleaning procedures recommended for a range of soil types

Soil type	Solubility characteristics	Cleaning procedure recommended
Sugars, organic acids, salt	Water-soluble	Mildly alkaline detergent
High protein foods (meat, poultry, fish)	Water-soluble Alkali-soluble Slightly acid-soluble	Chlorinated alkaline detergent
Starchy foods, tomatoes, fruits	Partly water-soluble Alkali-soluble	Mildly alkaline detergent
Fatty foods (fat, butter, margarine, oils)	Water-insoluble Alkaline-soluble	Mildly alkaline detergent; if ineffective, use strong alkali
Heat-precipitated water hardness, milk stone, protein scale	Water-insoluble Alkaline-insoluble Acid-soluble	Acid cleaner, used on a periodic basis

programme, there are likely to be sufficient viable microorganisms remaining on the surface to warrant the application of a disinfectant. The aim of disinfection is therefore to further reduce the surface population of viable microorganisms, via removal or destruction, and/or to prevent surface microbial growth during the inter-production period. Elevated temperature is the best disinfectant as it penetrates into surfaces, is non-corrosive, is non-selective to microbial types, is easily measured and leaves no residue (Jennings 1965). However, for open surfaces, the use of hot water or steam is uneconomic, hazardous or impossible, and reliance is, therefore, placed on chemical biocides.

Whilst there are many chemicals with biocidal properties, many common disinfectants are not used in food applications because of safety or taint problems, e.g. phenolics or metal-ion-based products. In addition, other disinfectants are used to a limited extent only in chilled food manufacture and/or for specific purposes, e.g. peracetic acid, biguanides, formaldehyde, glutaraldehyde, organic acids, ozone, chlorine dioxide, bromine and iodine compounds. Of the acceptable chemicals, the most commonly used products are:

- chlorine-releasing components
- quaternary ammonium compounds
- amphoterics
- quaternary ammonium/amphoteric mixtures.

Chlorine is the cheapest disinfectant and is available as hypochlorite (or occasionally as chlorine gas) or in slow releasing forms (e.g. chloramines, dichlorodimethylhydantoin). Quaternary ammonium compounds (Quats or QACs) are amphipolar, cationic detergents, derived from substituted ammonium salts with a chlorine or bromine anion and amphoterics are based on the amino acid glycine, often incorporating an imidazole group.

In a (CCFRA) survey undertaken of the UK food industry in 1987, of 145 applications of disinfectants 52% were chlorine based, 37% were quaternary ammonium compounds and 8% were amphoterics. Of these biocides there were, respectively, 44, 30 and 8 branded products used. In a (CCFRA) European survey of 1993, the most common disinfectants used in the UK and Scandinavian countries were QACs for open surfaces and peracetic acid and chlorine for closed, liquid handling surfaces. The survey also showed that open surfaces were usually cleaned with alkaline detergents which were foamed and then rinsed with medium pressure water (250psi) and closed systems were CIP cleaned with caustic followed by acidic detergents with a suitable rinse in-between. A survey of the approved disinfectant products in Germany (DVG listed) in 1994 indicated that 36% were QACs, 20% were mixtures of QACs with aldehydes or biguanides and 10% were amphoterics (Knauer-Kraetzl 1994). More recently the synergistic combinations of QACs and amphoterics have been explored in the UK and these compounds are now widely used in chilled food plants. The characteristics of the most commonly used are compared in Table 14.3. The properties of QAC/amphoteric mixes will be similar to their parent compounds with often enhanced microorganism control.

Table 14.3 Characteristics of some universal disinfectants

Property	Chlorine	QAC	Amphoteric	Peracetic acid
Microorganism control				
Gram-positive	++	++	++	++
Gram-negative	++	+	++	++
Spores	+	–	–	++
Yeast	++	++	++	++
Developed microbial resistance	–	+	+	–
Inactivation by organic matter	++	+	+	+
water hardness	–	+	–	–
Detergency properties	–	++	+	–
Surface activity	–	++	++	–
Foaming potential	–	++	++	–
Problems with taints	+/–	–	–	+/–
Stability	+/–	–	–	+/–
Corrosion	+	–	–	–
Safety	+	–	–	++
Other chemicals	–	+	–	–
Potential environmental impact	++	–/+	–/+	–
Cost	–	++	++	+

– no effect (or problem).
+ effect.
++ large effect.

Within the chilled food industry, particularly for mid-shift cleaning and disinfection in high-risk areas, alcohol based products are commonly used. This is primarily to restrict the use of water for cleaning during production as a control measure to prevent the growth and spread of any food pathogens that penetrate the high-risk area barrier controls. Ethyl alcohol (ethanol) and isopropyl alcohol (isopropanol) have bactericidal and virucidal (but not sporicidal) properties (Hugo and Russell 1999), though they are only active in the absence of organic matter i.e. the surfaces need to be wiped clean and then alcohol reapplied. Alcohols are most active in the 60–70% range, and can be formulated into wipe and spray based products. Alcohol products are used on a small, local scale because of their well recognised health and safety issues.

The efficacy of disinfectants is generally controlled by five factors: interfering substances (primarily organic matter), pH, temperature, concentration and contact time. To some extent, and particularly for the oxidative biocides, the efficiency of all disinfectants is reduced in the presence of organic matter. Organic material may react chemically with the disinfectant such that it loses its biocidal potency, or spatially such that microorganisms are protected from its effect. Other interfering substances, e.g. cleaning chemicals, may react with the disinfectant and destroy its antimicrobial properties, and it is therefore essential to remove all soil and chemical residues prior to disinfection.

Disinfectants should be used only within the pH range as specified by the manufacturer. Perhaps the classic example of this is chlorine, which dissociates in water to form HOCl and the OCl ion. From pH 3–7.5, chlorine is predominantly present as HOCl, which is a very powerful biocide, though the potential for corrosion increases with acidity. Above pH 7.5, however, the majority of the chlorine is present as the OCl ion which has about 100 times less biocidal action than HOCl.

In general, the higher the temperature the greater the disinfection. For most food manufacturing sites operating at ambient conditions (around 20ºC) or higher this is not a problem as most disinfectants are formulated (and tested) to ensure performance at this temperature. This is not, however, the case in the chilled food industry. Taylor *et al.* (1999) examined the efficacy of 18 disinfectants at both 10ºC and 20ºC and demonstrated that for some chemicals, particularly quaternary ammonium based products, disinfection was much reduced at 10ºC and recommended that in chilled production environments, only products specifically formulated for low-temperature activity should be used.

In practice, the relationship between microbial death and disinfectant concentration is not linear but follows a sigmoidal curve. Microbial populations are initially difficult to kill at low concentrations, but as the biocide concentration is increased, a point is reached where the majority of the population is reduced. Beyond this point the microorganisms become more difficult to kill (through resistance or physical protection) and a proportion may survive regardless of the increase in concentration. It is important, therefore, to use the disinfectant at the concentration as recommended by the manufacturer. Concentrations above this recommended level may thus not enhance biocidal effect and will be uneconomic whilst concentrations below this level may significantly reduce biocidal action.

Sufficient contact time between the disinfectant and the microorganisms is perhaps the most important factor controlling biocidal efficiency. To be effective, disinfectants must find, bind to and transverse microbial cell envelopes before they reach their target site and begin to undertake the reactions which will subsequently lead to the destruction of the microorganism (Klemperer 1982). Sufficient contact time is therefore critical to give good results, and most general-purpose disinfectants are formulated to require at least five minutes to reduce bacterial populations by five log orders in suspension. This has arisen for two reasons. Firstly five minutes is a reasonable approximation of the time taken for disinfectants to drain off vertical or near vertical food processing surfaces. Secondly, when undertaking disinfectant efficacy tests in the laboratory, a five-minute contact time is chosen to allow ease of test manipulation and hence timing accuracy. For particularly resistant organisms such as spores or moulds, surfaces should be repeatedly dosed to ensure extended contact times of 15–60 minutes.

Ideally, disinfectants should have the widest possible spectrum of activity against microorganisms, including bacteria, fungi, spores and viruses, and this should be demonstrable by means of standard disinfectant efficacy tests. The range of currently available disinfectant test methods was reviewed by Reybrouck (1998) and fall into two main classes, suspension tests and surface tests. Suspension tests are useful for indicating general disinfectant efficacy and for assessing environmental parameters such as temperature, contact time and interfering matter such as food residues. In reality however, microorganisms disinfected on food contact surfaces are those that remain after cleaning and are therefore likely to be adhered to the surface. A surface test is thus more appropriate.

A number of authors have shown that bacteria attached to various surfaces are generally more resistant to biocides than are organisms in suspension (Dhaliwal *et al.* 1992, Frank and Koffi 1990, Holah *et al.* 1990a, Hugo *et al.* 1985, Le Chevalier *et al.* 1988, Lee and Frank 1991, Ridgeway and Olsen 1982, Wright *et al.* 1991, Andrade *et al.* 1998, Das *et al.* 1998). In addition, cells growing as a biofilm have been shown to be more resistant (Frank and Koffi 1990, Lee and Frank 1991, Ronner and Wong 1993). The mechanism of resistance in attached and biofilm cells is unclear but may be due to physiological differences such as growth rate, membrane orientation changes due to attachment and the formation of extracellular material which surrounds the cell. Equally, physical properties may have an effect e.g. protection of the cells by food debris or the material surface structure or problems in biocide diffusion to the cell/material surface. To counteract such claims of enhanced surface adhered resistance, it can be argued that in reality, surface tests do not consider the environmental stresses the organisms may encounter in the processing environment prior to disinfection (action of detergents, variations in temperature and pH and mechanical stresses) which may affect susceptibility. Both suspension and surface tests have limitations, however, and research based methods are being developed to investigate the effect of disinfectants against adhered microorganisms and biofilms *in-situ* and in real time. Such methods have been reviewed by Holah *et al.* (1998).

In Europe, CEN TC 216 is currently working to harmonise disinfectant testing and has produced a number of standards. The current food industry disinfectant test methods of choice for bactericidal and fungicidal action in suspension are EN 1276 (Anon. 1997) and EN 1650 (Anon. 1998a) respectively and food manufacturers should ensure that the disinfectants they use conform to these standards as appropriate. A harmonised surface test is expected in 2000. Because of the limitations of disinfectant efficacy tests, however, food manufacturers should always confirm the efficacy of their cleaning and disinfection programmes by field tests either from evidence supplied by the chemical company or from in-house trials.

As well as having demonstrable biocidal properties, disinfectants must also be safe (non-toxic) and should not taint food products. Disinfectants can enter food products accidentally e.g. from aerial transfer or poor rinsing, or deliberately e.g. from 'no rinse status' disinfectants. The practice of rinsing or not rinsing has yet to be established. The main reason for leaving disinfectants on surfaces is to provide an alleged biocide challenge (this has not been proven) to any subsequent microbial contamination of the surface. It has been argued, however, that the low biocide concentrations remaining on the surface, especially if the biocide is a QAC, may lead to the formation of resistant surface populations.

In Europe, legislation is confusing surrounding whether or not disinfectants can be left on surfaces without rinsing. The Meat Products Directive (95/68/EC) allows disinfectants to remain on surfaces (no rinse status) 'when the directions for use of such substances render such rinsing unnecessary', whilst the Egg Products (89/437/EEC) and Milk Products (92/46/EEC) Directives require that disinfectants must be rinsed off by potable water. There is no specific guidance for other food product categories although the general Directive on the hygiene of foodstuffs (93/43/EEC) requires 'Food business operators shall identify any step in their activities which is critical to ensuring food safety and ensure that adequate safety procedures are identified, implemented, maintained and reviewed. . .'.

In terms of the demonstration of non-toxicity, legislation will vary in each country although in Europe, this will be clarified with the implementation of Directive 98/9/EC concerning the placing of biocidal products on the market, which contains requirements for toxicological and metabolic studies. A recognised acceptable industry guideline for disinfectants is a minimum acute oral toxicity (with rats) of 2,000 mg/Kg bodyweight.

Approximately 30% of food taint complaints are thought to be associated with cleaning and disinfectant chemicals and are described by sensory scientists as 'soapy', 'antiseptic' or 'disinfectant' (Holah 1995). CCFRA have developed two taint tests in which foodstuffs which have and have not been exposed to disinfectant residues are compared by a trained taste panel using the standard triangular taste test (Anon. 1983a). For assessment of aerial transfer, a modification of a packaging materials odour transfer test is used (Anon. 1964) in which food products, usually of four types (high moisture e.g. melon, low moisture e.g. biscuit, high fat e.g. cream, high protein e.g. chicken) are held above disinfectant solution of distilled water for 24 hours. To assess surface transfer, a modification or a food container transfer test is used (Anon. 1983b) in which food products are sandwiched between two sheets of stainless steel and left for 24 hours. Disinfectants can be sprayed onto the stainless steel sheets and drained off, to simulate no rinse status, or can be rinsed off prior to food contact. Control sheets are rinsed in distilled water only. The results of the triangular test involve both a statistical assessment of any flavour differences between the control and disinfectant treated sample and a description of any flavour changes.

14.4 Sanitation methodology

Cleaning and disinfection can be undertaken by hand using simple tools, e.g. brushes or cloths (manual cleaning), though as the area of open surface requiring cleaning and disinfection increases, specialist equipment becomes necessary to dispense chemicals and/or provide mechanical energy. Chemicals may be applied as low pressure mists, foams or gels whilst mechanical energy is provided by high and low pressure water jets or water or electrically powered scrubbing brushes. These techniques have been well documented (Anon. 1991, Marriott 1985, Holah 1991) and this section considers their use in practice.

The use of cleaning techniques can perhaps be described schematically following the information detailed in Fig. 14.2. The figure details the different energy source inputs for a number of cleaning techniques and shows their ability to cope with both low and high (dotted line) levels of soiling. For the manual cleaning of small items a high degree of mechanical energy can be applied directly where it is needed and with the use of soak tanks (or clean-out-of-place techniques) contact times can be extended and/or chemical and temperature inputs increased such that all soil types can be tackled.

Alternatively, dismantled equipment and production utensils may undergo manual gross soil removal and then be cleaned and disinfected automatically in tray or tunnel washers. As with soak tank operations, high levels of chemical and thermal energy can be used to cope with the majority of soils. The siting of tray washes in high-risk chilled production areas should be carefully considered, however, as they are prone to microbial aerosol production which may lead to aerial product contamination (see Chapter 13).

In manual cleaning of larger areas, for reasons of operator safety, only low levels of temperature and chemical energy can be applied, and as the surface area requiring cleaning increases, the technique becomes uneconomic with respect to time and labour. Labour costs amount to 75% of the total sanitation programme and for most food companies, the cost of extra staff is prohibitive. Only light levels of soiling can be economically undertaken by this method.

The main difference between the mist, foam and gel techniques is in their ability to maintain a detergent/soil/surface contact time. For all three techniques, mechanical energy can be varied by the use of high or low-pressure water rinses, though for open surface cleaning, temperature effects are minimal. Mist spraying is undertaken using small hand-pumped containers, 'knapsack' sprayers or pressure washing systems at low pressure. Misting will only 'wet' vertical smooth surfaces; therefore only small quantities can be applied and these will quickly run off to give a contact time of five minutes or less. Because of the nature of the technique to form aerosols that could be an inhalation hazard, only weak chemicals can be applied, and so misting is useful only for light soiling. On cleaned surfaces, however, misting is the most commonly used method for applying disinfectants.

Foams can be generated and applied by the entrapment of air in high-pressure equipment or by the addition of compressed air in low-pressure systems. Foams work on the basis of forming a layer of bubbles above the surface to be cleaned which then collapses and bathes the surface with fresh detergent contained in the bubble film. The critical element in foam generation is for the bubbles to collapse at the correct rate: too fast and the contact time will be minimal; too

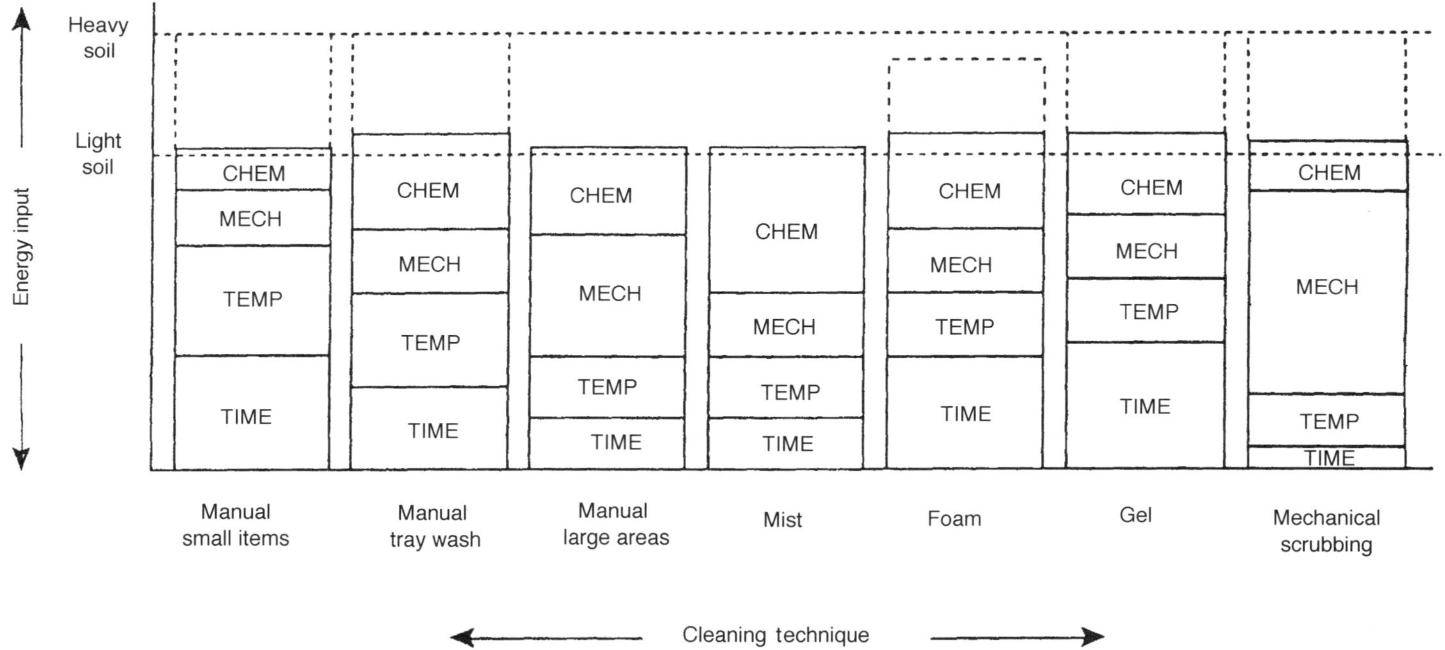

Fig. 14.2 Relative energy source inputs for a range of cleaning techniques. (Modified from Offiler 1990).

slow and the surface will not be wetted with fresh detergent. Gels are thixotropic chemicals which are fluid at high and low concentrations but become thick and gelatinous at concentrations of approximately 5–10%. Gels are easily applied through high- and low-pressure systems or from specific portable electric pumped units and physically adhere to the surface.

Foams and gels are more viscous than mists, are not as prone to aerosol formation and thus allow the use of more concentrated detergents, and can remain on vertical surfaces for much longer periods (foams 10–15 minutes, gels 15 minutes to an hour or more). Foams and gels are able to cope with higher levels of soils than misting, although in some cases rinsing of surfaces may require large volumes of water, especially with foams. Foams and gels are well liked by operatives and management, because of the nature of the foam, a more consistent application of chemicals is possible and it is easier to identify areas that have been 'missed'.

Fogging systems have been traditionally used in the chilled food industry to create and disperse a disinfectant aerosol to reduce airborne microorganisms and to apply disinfectant to difficult to reach overhead surfaces. The efficacy of fogging was recently examined in the UK and has been reported (Anon. 1998b). Providing a suitable disinfectant is used, fogging is effective at reducing airborne microbial populations by 2–3 log orders in 30–60 minutes. Fogging is most effective using compressed air driven fogging nozzles producing particles in the 10–20 micron range. For surface disinfection, fogging is effective only if sufficient chemical can be deposited onto the surface. This is illustrated in Figure 14.3 which shows the log reductions achieved on horizontal, vertical and upturned (underneath) surfaces arranged at five different heights from just below the ceiling (276 cm) to just above the floor (10 cm) within a test room. It can be seen that disinfection is greatest on surfaces closest to the floor and that disinfection is minimal on upturned surfaces close to the ceiling. To reduce inhalation risks, sufficient time (45–60 minutes) is required after fogging to

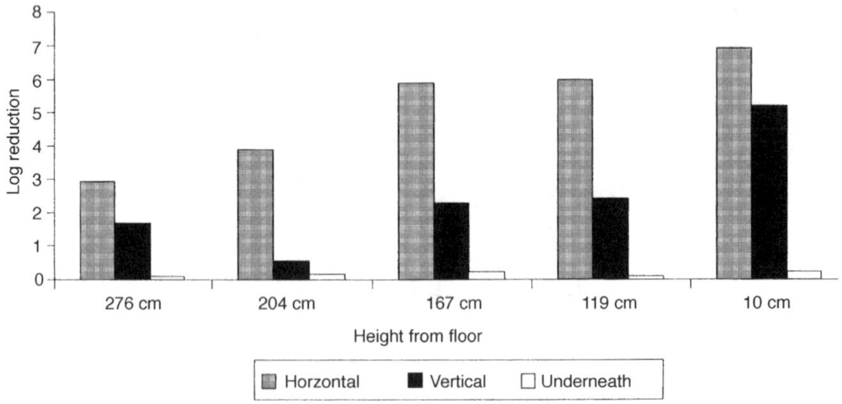

Fig. 14.3 Comparative log reductions of microorganisms adhered to surfaces and positioned at various heights and orientations.

allow the settling of disinfectant aerosols before operatives can re-enter the production area.

Cleaning chemicals are removed from surfaces by low-pressure/high-volume hoses operating at mains water pressure or by high-pressure/low volume pressure washing systems. Pressure washing systems typically operate at between 25–100 bar through a 15° nozzle and may be mobile units, wall mounted units or centralised ring-mains. Water jets confer high mechanical energy, can be used on a wide range of equipment and environmental surfaces, will penetrate into surface irregularities and are able to mix and apply chemicals.

Mechanical scrubbers include traditional floor scrubbers, scrubber/driers (automats) for floors, and water-driven attachments to high pressure systems and electrically operated small-diameter brushes that can be used on floors, walls and other surfaces. Contact time is usually limited with these techniques (though can be increased), but the combination of detergency with high mechanical input allows them to tackle most soil types. The main limitation is that food-processing areas have not traditionally been designed for their use, though this can be amended in new or refurbished areas.

The hygienic implications of the design and use of cleaning equipment should be carefully considered. Sanitation equipment should be constructed out of smooth, non-porous, easily cleanable materials such as stainless steel or plastic. Mild steel or other materials subject to corrosion may be used but must be suitably painted or coated, whilst the use of wood is unacceptable. Frameworks should be constructed of tubular or box section material, closed at either end and properly jointed, e.g. welds should be ground and polished and there should be no metal-to-metal joints. Crevices and ledges where soil could collect should be avoided and exposed threads should be covered or dome nuts used. Tanks for holding cleaning chemicals or recovered liquids should be self-draining, have rounded corners and should be easily cleaned. Shrouds around brush heads or hoods and rotary scrubbing heads should be easily detachable to facilitate cleaning. Brushes should have bristles of coloured, impervious material, e.g. nylon, embedded into the head with resin so no soil trap points are apparent. Alternatively, brushes with the head and bristles moulded as one unit may be used.

Cleaning equipment is prone to contamination with *Listeria* spp. and other pathogenic microoganisms and, by the nature of its use, provides an excellent way in which contamination can be transferred from area to area. Cleaning equipment should be specific to high risk and after use, equipment should be thoroughly cleaned and, if appropriate, disinfected and dried. The potential for cleaning equipment to disperse microbial contamination by the formation of aerosols has been reported (Holah *et al.* 1990b) and it was shown that all cleaning systems tested produced viable bacterial aerosols from test surfaces contaminated with attached biofilms.

The degree of contamination impinging on a surface was graded from total coverage to the minimum level thought likely to give concern if a proportion of the droplets contained viable microorganisms and the maximum height and

Table 14.4 Maximum height and distance of aerosol impingement for a number of cleaning techniques

Cleaning technique	Height (cm)	Distance (cm)
High-pressure/low-volume spray lance	309	700
Low-pressure/high-volume hose	210	350
Floor scrubber/drier	47	80
Manual brushing	24	75
Manual wiping	23	45

distance travelled by this contamination level is shown in Table 14.4. Assuming an average food contact surface height of 1 m, the results suggest that both the high pressure low-volume (HPLV) and low pressure high-volume (LPHV) techniques disperse a significant level of aerosol to this height and should not, therefore, be used during production periods. The other techniques, however, are acceptable for use in clean-as-you-go operations as the chance of contamination to product is low, though care is needed when using floor scrubber/driers (these are useful in that the cleaning fluid is removed from the floor) if product is stored in racks close to the floor. After production, HPLV and LPHV techniques may be safely used (and are likely to be the appropriate choice), but it is required that disinfection of food contact surfaces is the last operation to be performed within the sanitation programme. Subsequent work has shown that reducing water pressure or changing impact angle made little difference to the degree of aerosol spread for HPLV and LPHV systems, dispersal to heights > 1 m still being achieved.

14.5 Sanitation procedures

Sanitation procedures are concerned with both the stage at which the sanitation programme is implemented and the sequence in which equipment and environmental surfaces are cleaned and disinfected within the processing area. Sanitation programmes are so constructed as to be efficient with water and chemicals, to allow selected chemicals to be used under their optimum conditions, to be safe in operation, to be easily managed and to reduce manual labour. In this way an adequate level of sanitation will be achieved, economically and with due regard to environmental friendliness. The principal stages involved in a typical sanitation programme are described below.

Production periods. Production staff should be encouraged to consider the implications of production practices on the success of subsequent sanitation programmes. Product should be removed from lines during break periods and this may be followed by manual cleaning, usually undertaken by wiping with alcohol (to avoid the use of water during production periods). Production staff should also be encouraged to operate good housekeeping practices (this is also an aid to ensuring acceptable product quality) and to leave their work stations in a reasonable condition. Soil left in hoppers and on process lines, etc. is wasted product! Sound sanitation practices should be used to clean up large product spillages during production.

Preparation. As soon as possible after production, equipment should be dismantled as far as is practicable or necessary to make all surfaces that microorganisms could have adhered to during production accessible to the cleaning fluids. All unwanted utensils/packaging/equipment should be covered or removed from the area. Dismantled equipment should be stored on racks or tables, not on the floor! Machinery should be switched off, at the machine and at the power source, and electrical and other sensitive systems protected from water/chemical ingress. Preferably, production should not occur in the area being cleaned, but in exceptional circumstances if this is not possible, other lines or areas should be screened off to prevent transfer of debris by the sanitation process.

Gross soil removal. Where appropriate, all loosely adhered or gross soil should be removed by brushing, scraping, shovelling or vacuum, etc. Wherever possible, soil on floors and walls should be picked up and placed in suitable waste containers rather than washed to drains using hoses.

Pre-rinse. Surfaces should be rinsed with low pressure cold water to remove loosely adhered small debris. Hot water can be used for fatty soils, but too high a temperature may coagulate proteins.

Cleaning. A selection of cleaning chemicals, temperature and mechanical energy is applied to remove adhered soils.

Inter-rinse. Both soil detached by cleaning operations and cleaning chemical residues should be removed from surfaces by rinsing with low pressure cold water.

Disinfection. Chemical disinfectants (or occasionally heat) are applied to remove and/or reduce the viability of remaining microorganisms to a level deemed to be of no significant risk. In exceptional circumstances and only when light soiling is to be removed, it may be appropriate to combine stages 5–7 by using a chemical with both cleaning and antimicrobial properties (detergent-sanitiser).

Post-rinse. Disinfectant residues should be removed by rinsing away with low pressure cold water of known potable quality. Some disinfectants, however, are intended to be left on surfaces until the start of subsequent production periods and are thus so formulated to be both surface-active and of low risk, in terms of taint or toxicity, to foodstuffs.

Inter-production cycle conditions. A number of procedures may be undertaken, including the removal of excess water and/or equipment drying, to prevent the growth of microorganisms on production contact surfaces in the period up until the next production process. Alternatively, the processing area may be evacuated and fogged with a suitable disinfectant.

Periodic practices. Periodic practices increase the degree of cleaning for specific equipment or areas to return them to acceptable cleanliness levels. They include weekly acidic cleans, weekend dismantling of equipment, cleaning and disinfection of chillers and sanitation of surfaces, fixtures and fittings above two metres.

A sanitation sequence should be established in a processing area to ensure that the applied sanitation programme is capable of meeting its objectives and that cleaning programmes, both periodic and for areas not cleaned daily, are implemented on a routine basis. In particular, a sanitation sequence determines the order in which the product contact surfaces of equipment and environmental surfaces (walls, floors, drains etc.) are sanitised, such that once product contact surfaces are disinfected, they should not be re-contaminated.

Based on industrial case studies, the following sanitation sequence for chilled food production areas, has been demonstrated to be useful in controlling the proliferation of undesirable microorganisms. The sequence must be performed at a 'room' level such that all environmental surfaces and equipment in the area are cleaned at the same time. It is not acceptable to clean and disinfect one line and then move onto the next and start the sequence again as this merely spreads contamination around the room.

1. Remove gross soil from production equipment.
2. Remove gross soil from environmental surfaces.
3. Rinse down environmental surfaces (usually to a minimum of 2 m in height for walls).
4. Rinse down equipment and flush to drain.
5. Clean environment surfaces, usually in the order of drains, walls then floors.
6. Rinse environmental surfaces.
7. Clean equipment.
8. Rinse equipment.
9. Disinfect equipment and rinse if required.
10. Fog (if required)

14.6 Evaluation of effectiveness of sanitation systems

Assessment of the effectiveness of the sanitation programme's performance is part of day-to-day hygiene testing and, as such, is linked to the factory environmental sampling plan. The control of the environmental routes of contamination is addressed via the development of a thorough risk analysis and management strategy, typically undertaken as part of the factory HACCP study, resulting in the development of the factory environmental sampling plan. The development of environmental sampling plans has recently been established by a CCFRA industrial working party and is reported in Holah (1998).

Environmental sampling is directly linked with both process development and product manufacture and as such, has three distinct phases;

1. process development to determine whether a contamination route is a risk and assessing whether procedures put in place to control the risk identified are working
2. routine hygiene assessment
3. troubleshooting to identify why products (or occasionally environmental samples) may have a microbiological count that is out of specification or may contain pathogens.

Related to chilled food manufacture, routine hygiene testing is concerned with assessing the performance of the high-risk barrier systems in preventing pathogen access during production and, after production has finished, the performance of the sanitation programme.

Routine hygiene testing is an important aspect of due diligence and is used for two purposes, monitoring to check sanitation process control, and verification to assess sanitation programme success. Monitoring is a planned sequence of observations or measurements to ensure that the control measures within the sanitation programme are operating within specification and are undertaken in a time frame that allows sanitation programme control. Verification is the application of methods in a longer time frame to determine compliance with the sanitation programme's specification.

Monitoring the sanitation programme is via physical, sensory and rapid chemical hygiene testing methods. Microbiological testing procedures are never fast enough to be used for process monitoring. Physical tests are centred on the critical control measures of the performance of sanitation programmes and include, for example, measurement of detergent/disinfectant contact time; rinse water, detergent and disinfectant temperatures; chemical concentrations; surface coverage of applied chemicals; degree of mechanical or kinetic input; cleaning equipment maintenance and chemical stock rotation.

Sensory evaluation is usually undertaken after each of the sanitation programme stages and involves visual inspection of surfaces under good lighting, smelling for product or offensive odours, and feeling for greasy or encrusted surfaces. For some product soils, residues can be more clearly observed by wiping the surface with paper tissues. Rapid hygiene methods are defined as monitoring methods whose results are generated in a time frame (usually regarded as within approximately 10 minutes) sufficiently quickly to allow process control. Current methodology allows the quantification of microorganisms (ATP), food soils (ATP, protein) or both (ATP). No technique is presently available which will allow the detection of specific microbial types within this time frame.

The most popular and established rapid hygiene monitoring technique is that based on the detection of adenosine triphosphate (ATP) by bioluminescence and is usually referred to as ATP testing. ATP is present in all living organisms, including microorganisms (microbial ATP), in a variety of foodstuffs and may also be present as free ATP (usually referred to together as non-microbial ATP). The bioluminescent detection system is based on the chemistry of the light reaction emitted from the abdomen of the North American firefly *Photinus pyralis,* in which light is produced by the reaction of luciferin and luciferase in the presence of ATP. For each molecule of ATP present, one photon of light is emitted which are then detected by a luminometer and recorded as relative light units (RLU). The reaction is very rapid and results are available within seconds of placing the sample to be quantified in the luminometer. The result, the amount of light produced, is also directly related to the level of microbial and non-microbial ATP present in the sample and is often referred to as the 'hygienic' status of the sample.

ATP has been successfully used to monitor the hygiene of surfaces for approximately 15 years and many references are available in the literature citing its proficiency and discussing its future potential e.g. Bautista *et al.* (1992), Poulis *et al.* (1993), Bell *et al.*(1994), Griffiths *et al.* (1994), Hawronskyj and Holah (1997). It is possible to differentiate between the measurement of microbial and non-microbial ATP but for the vast majority of cases, the measurement of total ATP (microbial and non-microbial) is preferred. As there is more inherent ATP in foodstuffs than in microorganisms, the measurement of total ATP is a more sensitive technique to determine remaining residues. Large quantities of ATP present on a surface after cleaning and disinfection, regardless of their source, is an indication of poor cleaning and thus contamination risk (from microorganisms or materials that may support their growth).

Many food processors typically use the rapidity of ATP to allow monitoring of the cleaning operation such that if a surface is not cleaned to a predetermined level it can be recleaned prior to production. Similarly, pieces of kit can be certified as being cleaned prior to use in processing environments where kit is quickly recycled or when the manufacturing process has long production runs. Some processors prefer to assess the hygiene level after the completion of both the cleaning and disinfection phases whilst others monitor after the cleaning phase and only go onto the disinfection phase if the surfaces have been adequately cleaned.

Techniques have also been developed which use protein concentrations as markers of surface contamination remaining after cleaning operations. As these are dependent on chemical reactions, they are also rapid but their applicability is perhaps less widespread as they can only be used if protein is a major part of the food product processed. As with the ATP technique, a direct correlation between the degree of protein remaining after a sanitation programme has been completed and the number of microorganisms remaining as assessed by traditional microbiological techniques, is not likely to be useful. They are cheaper in use than ATP based systems as the end point of the tests is a visible colour change rather than a signal which is interpreted by an instrument e.g. light output measured by a luminometer.

Protein hygiene tests were further developed and recently reintroduced by Konica as the 'Swab' n' Check' hygiene monitoring kit. This assay detects the presence of protein on a surface by an enhanced Biuret reaction, the end point of which is a colour change from green through to purple. The surface to be assessed is swabbed and the swab placed into a tube of resuspension fluid containing the reagents necessary to activate the Biuret reaction. After ten minutes any colour change is compared to a supplied colour card and the degree of colour change used as an indication of the hygienic nature of the surface. However, there is currently little published data on both the efficacy of this system (Griffith *et al.* 1997) and the food-processing environments to which it is best suited. Other manufacturers have also recently launched competing products.

Verification of the performance of the sanitation programmes is usually undertaken by microbiological methods in the chilled food industry, though ATP levels are also used (especially in low risk). Microbiological sampling is typically for the total number of viable microorganisms remaining after cleaning and disinfection, i.e. total viable count (TVC), both as a measurement of the ability of the sanitation programme to control all microorganisms and to maximise microbial detection. Sampling targeted at specific pathogens or spoilage organisms, which are thought to play a major role in the safety or quality of the product, is undertaken to verify the performance of the sanitation programme designed for their control. Microbiological assessments have also been used to ensure compliance with external microbial standards, as a basis for cleaning operatives' bonus payments, in hygiene inspection and troubleshooting exercises, and to optimise sanitation procedures.

Traditional microbiological techniques appropriate for food factory use involving the removal or sampling of microorganisms from surfaces, and their culture using standard agar plating methods have been reviewed by Holah (1998). Microorganisms may be sampled via sterile cotton or alginate swabs and sponges, after which the microorganisms are resuspended by vortex mixing or dissolution into suitable recovery or transport media, or via water rinses for larger enclosed areas (e.g. fillers). Representative dilutions are then incubated in a range of microbial growth media, depending on which microorganisms are being selected for, and incubated for 24–48 hours. Alternatively, microorganisms may be sampled directly onto self-prepared or commercial ('dip slides') agar contact plates.

The choice of sampling site will relate to risk assessment. Where there is the potential for microorganisms remaining after (poor) cleaning and disinfection to, via e.g. direct product contact, infect large quantities of product, these sources would require sampling much more frequently than other sites which, whilst they may be more likely to be contaminated, pose less of a direct risk to the product. For example, it is more sensible, and gives more confidence, to sample the points of the equipment that directly contact the product and that are difficult to clean than to sample non-direct contact surfaces, e.g. underneath of the equipment framework.

In relation to microorganism numbers, it is difficult to suggest what is an 'acceptable' number of microorganisms remaining on a surface after cleaning and disinfection as this is clearly dependent on the food product, process, 'risk area' and degree of sanitation undertaken. A number of figures have been quoted in the past (as total viable count per square decimeter) including 100 (Favero *et al.* 1984), 540 (Thorpe and Barker 1987) and 1000 (Timperley and Lawson 1980) for dairies, canneries and general manufacturing respectively. The results in Table 14.1 show that in chilled food production, sanitation programmes should achieve levels of around 1000 microorganisms per swab, which on flat surfaces approximately equates to a square decimeter. Expressing counts arithmetically is always a problem, however, as single counts taken in areas where cleaning has been inadequate (which may be in excess of 10^8 per swab) produce an artificially high mean count, even over thousands of samples. It is better, therefore, to express counts as log to the base 10, a technique that places less emphasis on a relatively few high counts, and Table 14.1 shows that log counts of approximately 1 should be obtained.

Because of the difficulty in setting external standards, it is best to set internal standards as a measurement of what can be achieved by a given sanitation programme. A typical approach would be to assess the level of microorganisms or ATP present on a surface after a series of ten or so sanitation programmes in which the sanitation programme is carefully controlled (i.e. detergent and disinfectant concentrations are correct, contact times are adhered to, water temperatures are checked, pressure hoses are set to specified pressures, sanitation schedules are followed, etc.). The mean result will provide an achievable standard (or standards if specific areas differ significantly in their cleanability) which can be immediately used and can be reviewed as subsequent data points are obtained in the future. A review of the standard would be required if either the food product or process or the sanitation programme were changed.

As part of the assessment of sanitation programmes, it is worthwhile looking how the programme is performing over a defined time period (weekly, monthly, quarterly etc.) as individual sample results are only an estimate of what is happening at one specific time period. This may be to ensure that the programme remains within control, to reduce the variation within the programme or, as should be encouraged, to try and improve the programme's performance. An assessment of the performance of the programme with time, or trend analysis, can be undertaken simply, by producing a graphical representation of the results on a time basis, or can be undertaken from a statistical perspective using Statistical Process Control (SPC) techniques as described by Harris and Richardson (1996). Generally, graphical representation is the most widely used approach, though SPC techniques should be encouraged for more rigorous assessment of improvement in the programme's performance.

14.7 Management responsibilities

Senior management must take full responsibility for the successful operation of the sanitation programme; ultimately, failures in the programme generally reflect poor management. For the majority of chilled food processing operations, the following is a guide to the responsibilities of senior management.

- Always seek to improve hygiene standards in line with the high-risk philosophies adopted in Chapter 13. Hygiene has traditionally not had the same research support as other areas of importance in food manufacture and is thus a new and developing science. It is only relatively recently that new concepts have been developed, based on scientific assessments, and management must be flexible enough to try out and to encourage such concepts when they emerge.
- Lead by example by being both always properly attired in food production areas and (occasionally) present in production areas when sanitation is being undertaken (usually in the early hours of the morning!)
- Provide the required equipment (including maintenance), the staffing levels and the time to undertake the sanitation programme effectively. Cleaning operatives must be a dedicated labour pool whose priority is to sanitation (i.e. not production). Similarly, operatives should not join in the cleaning team as an 'introduction to production'.
- Management should be capable of giving praise when sanitation is undertaken correctly, as well as discipline when it is not. In companies where bonus systems have been employed based on microbiological assessments of equipment after cleaning, results have indicated that hygiene has generally been improved and bonuses are rarely missed.
- Appoint or nominate a manager to be responsible for the day-to-day implementation of the sanitation programme.

The manager who assumes responsibility for the sanitation programme must have technical hygiene expertise and has a range of job functions including the following:

- the selection of a suitable chemical supplier
- the selection of sanitation chemicals, equipment and methodology
- the training of cleaning operatives
- the development of cleaning schedules
- the implementation of sanitation programme monitoring systems
- representation of hygiene issues to senior management.

Good chemical suppliers are able to do much more than simply supply detergents and disinfectants. They should be chosen on their abilities to undertake site hygiene audits, supply suitable chemical dosing and application equipment, undertake operative training and help with the development of cleaning schedules and sanitation monitoring and verification systems. Good chemical companies respond quickly to their customers' needs, periodically review their customers' requirements and visit during sanitation periods to ensure that their products are being used properly and are working satisfactorily. The cleaning manager may also need to visit the chemical supplier's site to audit their quality systems.

Whilst in theory systems and/or chemicals could seem appropriate for the required task, every factory, with its water supply, food products, equipment, materials of construction and layout, etc., is unique. All sanitation chemicals, equipment and methodology must, therefore, be proven in the processing environment. New products and equipment are always being produced and a good working relationship with hygiene suppliers is beneficial. Only disinfectants that have been approved to the relevant European Standards should be used.

The cleaning operatives' job is both technical and potentially hazardous, and all steps should be undertaken to ensure that sufficient training is given. By the nature of the job, training is likely to be comprehensive and should include:

- a knowledge of basic food hygiene
- the importance of maintaining low/high-risk barriers during cleaning
- the implications to product safety/spoilage of poor sanitation practices
- an understanding of the basic function and use of sanitation chemicals and equipment and of their sequence of operation
- a thorough knowledge of the safe handling of chemicals and their application and the safe use of sanitation equipment.

For each piece of equipment or for each processing area, a cleaning or sanitation schedule should be developed, preferably in a loose-leaf format so that it can readily be updated and which should always be available for inspection by cleaning operatives or auditors. The schedule must show clearly each stage of the cleaning and disinfection process (diagrammatically if this would help), all pertinent information on safety, and the key inspection points and how these should be assessed. It is difficult to produce a list of requirements that should be found in a cleaning schedule, but the following is a typical, non-exhaustive list:

- a description, hazard code, in-use concentration, method of make up, storage conditions, location and amount to be drawn of all chemicals used
- type, use of, set parameters (pressure, nozzle type, etc.), maintenance and location of sanitation equipment
- description of the equipment to be cleaned, need for fitters, need to disconnect from services, dismantling and reassembly procedures
- full description of the cleaning process, its frequency and requirement for periodic measures
- staff requirements and their responsibilities
- key points for assessment of the sanitation procedure and description of evaluation procedures for programme monitoring and verification.

When new equipment is purchased or processing areas designed or refurbished, insufficient attention is usually placed on sanitation requirements.

Equipment or areas of poor hygienic design will be more expensive to clean (and maintain) and may not be capable of being cleaned to an acceptable standard in the time available. If improperly cleaned, adequate disinfection is impossible and thus contamination will not be controlled. Hygiene management must be strongly represented, thus ensuring that hygiene requirements are considered alongside those of engineering, production and accounts, etc.

Three types of sanitation programme can be implemented by management and each has its advantages and disadvantages: at the end of production, production operatives clean their workstations and then (a), they form a cleaning crew and undertake the sanitation programme; (b), a separate, dedicated cleaning gang complete the sanitation programme; or (c) cleaning and disinfection is undertaken by contract cleaners. Whilst each option will place different demands on the food manufacturer, the principles as mentioned above should always be incorporated and the sanitation programme effectively managed.

14.8 References

ANON, (1964) Assessment of odour from packaging material used for foodstuffs. British Standard 3744. British Standards Institute, London.

ANON, (1973) The effectiveness of water blast cleaning in the food industry. *Food Technology in New Zealand* 8, **15** and **21**.

ANON, (1983a) Sensory analysis of food. Part 5: Triangular test. ISO 4120. International Standards Organisation.

ANON, (1983b) Testing of container materials and containers for food products. DIN 10955. Deutsches Institut für Normung e.V., Berlin.

ANON, (1991) Sanitation. In: Shapton, D.A. and N.F. (eds) *Principles and Practices for the Safe Processing of Foods*. Butterworth-Heinemann Ltd., Oxford, pp. 117–99.

ANON, (1997) EN 1276:1997 Chemical disinfectants and antiseptics – Qunatitative suspension test for the evaluation of bactericidal activity of chemical disinfectants and antiseptics used in food, industrial, domestic, and institutional areas – Test method and requirements (phase 2, step 1).

ANON, (1998a) EN 1650:1998 Chemical disinfectants and antiseptics – Quantitative suspension test for the evaluation of fungicidal activity of chemical disinfectants and antiseptics used in food, industrial, domestic, and institutional areas – Test method and requirements (phase 2, step 1).

ANON, (1998b) A practical guide to the disinfection of food processing factories and equipment using fogging. Ministry of Agriculture Fisheries and Food, 650, St. Catherine Street, London, SE1 0UD.

ANDERSON M E, HUFF H E and MARSHALL R T, (1985) Removal of animal fat from food grade belting as affected by pressure and temperature of sprayed water, *Journal of Food Protection*, **44** 246–8.

ANDRADE N J, BRIDGEMAN T A and ZOTTOLA E A, (1998) Bacteriocidal activity of sanitizers against *Enterococcus faecium* attached to stainless steel as determined by plate count and impedance methods. *Journal of Food Protection*, **661**, 833–8.

BAUTISTA D A, MCINTYRE L, LALEYE L and GRIFITHS M W, (1992) The application of ATP bioluminescence for the assessment of milk quality and factory hygiene, *Journal Rapid Methods Autom Microbiology*, **1** 179–93.

BELL C, STALLARD P A, BROWN S E and STANDLEY J T E, (1994) ATP-Bioluminescence techniques for assessing the hygienic condition of milk transport tankers, *International Dairy Journal*, **4** 629–40.

BERGMAN B-O and TRAGARDH C, (1990) An approach to study and model the hydrodynamic cleaning effect, *Journal of Food Process Engineering*, **13** 135–54.

BLENKINSOPP S A and COSTERTON J W, (1991) Understanding bacterial biofilms, *Trends in Biotechnology*, **9** 138–43.

CARPENTIER B and CERF O, (1993) Biofilms and their consequences, with particular reference to the food industry, *Journal of Applied Bacteriology*, **75** 499–511.

CORRIEU G, (1981) State-of-the-art of cleaning surfaces. In *Proceedings of Fundamentals and Application of Surface Phenomena Associated with Fouling and Cleaning in Food Processing*, April 6–9, Lund University, Sweden, pp. 90–114.

DAS J R, BHAKOO M, JONES M V and GILBERT P, (1998) Changes in biocide susceptibility of *Staphylococcus epidermidis* and *Escherichia coli* cells associated with rapid attachment to plastic surfaces, *Journal of Applied Microbiology*, **84** 852–8.

DESTRO M T, LEITAO M F F and FARBER J M, (1996) Use of molecular typing methods to trace the dissemination of *Listeria monocytogenes* in a shrimp processing plant, *Applied and Environmental Microbiology*, **62** 705–11.

DHALIWAL D S, CORDIER J L and COX L J, (1992) Impedimetric evaluation of the efficacy of disinfectants against biofilms, *Letters in Applied Microbiology*, **15** 217–21.

DUNSMORE D G, (1981) Bacteriological control of food equipment surfaces by cleaning systems. 1. Detergent effects, *Journal of Food Protection*, **44** 15–20.

DUNSMORE D G, TWOMEY A, WHITTLESTONE W G and MORGAN H W, (1981) Design and performance of systems for cleaning product contact surfaces of food equipment: A review, *Journal of Food Production*, **44** 220–40.

ELLIOT R P, (1980) Cleaning and sanitation. In Katsuayama, A.M. (ed.) *Principles of food processing sanitation*, The Food Processors Institute, USA, pp. 91–129.

FAVERO M S, GABIS D A and VESELEY D, (1984) Environmental monitoring procedures. In: Speck, M. L. (ed.) *Compendium of Methods for the Microbiological Examination of Foods*, American Public Health Association, Washington, DC.

FRANK J F and KOFFI R A, (1990) Surface-adherent growth of *Listeria monocytogenes* is associated with increased resistance to surfactant sanitizers and heat, *Journal of Food Protection*, **53** 550–4.

GIBSON H, TAYLOR J H, HALL K E and HOLAH J T, (1995) *Biofilms and their detection in the food industry*, CCFRA R&D Report No. 1, CCFRA, Chipping Campden, Glos., GL55 6LD, UK.

GIBSON H, TAYLOR J H, HALL K E and HOLAH J T, (1999) Effectiveness of cleaning techniques used in the food industry in terms of the removal of bacterial biofilms. *Journal of Applied Microbiology*, **87** 41–8.

GRIFFITH C J, DAVIDSON C A, PETERS A C and FIELDING L M, (1997) Towards a strategic cleaning assessment programme: hygiene monitoring and ATP luminometry, an options appraisal, *Food Science and Technology Today*, **11** 15–24.

GRIFFITHS J, BLUCHER A, FLERI J and FIELDING L, (1994) An evaluation of luminometry as a technique in food microbiology and a comparison of six commercially available luminometers, *Food Science and Technology Today*, **8** 209–16.

HARRIS C S M and RICHARDSON P S, (1996) *Review of Principles, Techniques and Benefits of Statistical Process Control*. Review No. 4, Campden and Chorleywood Food Research Association, Chipping Campden, UK.

HAWRONSKYI J-M and HOLAH J T, (1997) ATP: A universal hygiene monitor, *Trends in Food Science and Technology*, **8** 79–84

HAYES P R, (1985) Cleaning and Disinfection: Methods. In: *Food Microbiology and Hygiene*, Elsevier Applied Science Publishers, Barking, pp. 268–305.

HOLAH J T, (1991) Food Surface Sanitation. In: Hui, Y H (ed.) *Encyclopedia of Food Science and Technology*, John Wiley and Sons, New York.

HOLAH J T, (1995) Special needs for disinfectants in food-handling establishments, *Revue Scientific et Technique Office International des Épizooties*, **14** 95–104.

HOLAH J T, (1998) *Guidelines for Effective Environmental Microbiological Sampling*, Guideline No. 20, Campden and Chorleywood Food Research Association, Chipping Campden, UK.

HOLAH J T and KEARNEY L R, (1992) Introduction to biofilms in the food industry. In *Biofilms – Science and Technology*, edited by L.F. Melo, T.R. Bott, M. Fletcher and B. Capdeville, pp. 35–45. Dordrecht: Kluwer.

HOLAH J T and THORPE R H, (1990) Cleanability in relation to bacterial retention on unused and abraded domestic sink materials, *Journal of Applied Bacteriology*, **69** 599–608.

HOLAH J T, HIGGS C, ROBINSON S, WORTHINGTON D and SPENCELEY H, (1990a) A Malthus based surface disinfection test for food hygiene, *Letters in Applied Microbiology*, **11** 255–9.

HOLAH, J T, LAVAUD A, PETERS W and DYE K A, (1998) Future techniques for disinfectant testing, *International Biodeterioration and Biodegradation*, **41** 273–9.

HOLAH J T, TIMPERLEY A W and HOLDER J S, (1990b) *The spread of Listeria by cleaning systems. Technical Memorandum No. 590*, Campden Food and

Drink Research Association, Chipping Campden, UK.

HUGO W B, PALLENT L J, GRANT D J W, DENYER S P and DAVIES A, (1985) Factors contributing to the survival of *Pseudomonas cepacia* in chlorhexidine, *Letters in Applied Microbiology*, **2** 37–42.

HUGO W B and RUSSELL A D, (1999) Types of antimicrobial agent. In *Principles and Practices of Disinfection, Preservation and Sterilization*. Eds A.D. Russell, W.B. Hugo and G.A.J. Ayliffe, Blackwell Sciences, Oxford, pp. 5–95.

ICMSF, (1980) Cleaning, disinfection and hygiene. In *Microbial Ecology of Foods, Volume 1, Factors Affecting Life and Death of Microorganisms*. The International Commission on Microbiological Specifications for Foods, Academic Press, London, pp. 232–58.

ICMSF, (1988) Cleaning and disinfecting. In *Microorganisms in foods, Volume 4, Application of the Hazard Analysis Critical Control Point (HACCP) System to Ensure Microbiological Safety and Quality*. The International Commission on Microbiological Specifications for Foods, Blackwell Scientific, Oxford, pp. 93–116.

JENNINGS W G, (1965) Theory and practice of hard-surface cleaning, *Advances in Food Research*, **14** 325–459.

KLEMPERER R, (1982) Tests for disinfectants: principles and problems. In: *Disinfectants: their assessment and industrial use*, Scientific Symposia Ltd, London.

KNAUER-KRAETZL B, (1994) Reinigung und Desinfektion. 12. Fleischwarenforum: Fleischhygiene, BEHR's Seminar, Darmstadt, 14–15 April 1994.

KOOPAL L K, (1985) Physico-chemical aspects of hard-surface cleaning, *Netherlands Milk and Dairy Journal*, **39** 127–54.

KUMAR C G and ANAND S K, (1998) Significance of microbial biofilms in the food industry: a review, *International Journal of Food Microbiology*, **42** 9–27.

LAWRENCE L M and GILMORE A, (1995) Characterisation of *Listeria monocytogenes* isolated from poultry products and poultry-processing environment by random amplification of polymorphic DNA and multilocus enzyme electrophoresis, *Applied and Environmental Microbiology*, **61** 2139–44.

LE CHEVALIER M W, CAWTHORN C D and LEE R G, (1988) Inactivation of biofilm bacteria, *Applied and Environmental Microbiology*, **44** 972–87.

LEE S-L and FRANK J F, (1991) Effect of growth temperature and media on inactivation of *Listeria monocytogenes* by chlorine, *Journal of Food Safety*, **11** 65–71.

LONCIN M, (1977) Modelling in cleaning, disinfection and rinsing. In: *Proceedings of Mathematical modelling in food processing*, 7–9 September, Lund. Lund Institute of Technology.

MABESA R C, CASTILLO M M, CONTRERAS E A, BANNAAD L and BANDIAN V, (1982) Destruction and removal of microorganisms from food equipment and utensil surfaces by detergents, *The Philippine Journal of Science*, **3** 17–22.

MARRIOT N G, (1985) Sanitation Equipment and Systems. In *Principles of Food Sanitation*, AVI Publishing Co., Westport, CT, pp. 117–51.

MATTILA-SANDHOLM T and WIRTANEN G, (1992) Biofilm formation in the food industry: a review, *Food Reviews International*, **8**, 573–603.

METTLER E and CARPENTIER B, (1998) Variations over time of microbial load and physicochemical properties of floor materials after cleaning in food industry premises, *Journal of Food Protection*, **61** 57–65.

MIDDLEMISS N E, NUNES C A, SORENSEN J E and PAQUETTE G, (1985) Effect of a water rinse and a detergent wash on milkfat and milk protein soils, *Journal of Food Protection*, **48** 257–60.

MROZEK H, (1982) Development trends with disinfection in the food industry, *Deutsche-Molkerei-Zeitung*, **12** 348–52.

OFFILER M T, (1990) Open plant cleaning: equipment and methods. In *Proceedings of 'Hygiene for the 90s'*, November 7–8, Campden Food and Drink Research Association, Chipping Campden, Glos., UK, pp. 55–63.

PONTEFRACT R D, (1991) Bacterial adherence: its consequesnces in food processing, *Canadian Institute of Food Science and Technology Journal*, **24** 113–17.

POULIS J A, DE PIJPER M, MOSSEL D A A and DEKKERS P P H A, (1993). Assessment of cleaning and disinfection in the food industry with the rapid ATP-bioluminescence technique combined with tissue fluid contamination test and a conventional microbiological method, *International Journal of Food Microbiology*, **20** 109–16.

REYBROUCK G, (1998) The testing of disinfectants, *International Biodeterioration and Biodegradation*, **41** 269–72.

RIDGEWAY H F and OLSEN B H, (1982) Chlorine resistance patterns of bacteria from two drinking water distribution systems, *Applied and Environmental Microbiolog*, **44** 972–87.

RONNER A B and WONG A C L, (1993) Biofilm development and sanitizer inactivation of *Listeria monocytogenes* and *Salmonella typhimurium* on stainless steel and Buna-N rubber, *Journal of Food Protection*, **56** 750–8.

RUSSELL A D, HUGO W B and AYLIFFE G A J, (1982) *Principles and Practice of Disinfection, Preservation and Sterilization*, Blackwell Scientific Publications, London.

SCHLUSSLER H J, (1975) Zur Kinetic von Reinigungsvorgangen an festen Oberflachen. *Symposium über Reinigen und Desinfizieren lebensmittelverarbeitender Anlagen*, Karlsruhe, 1975.

SCHMIDT U and CREMLING K, (1981) Cleaning and disinfection processes. IV. Effects of cleaning and other measures on surface bacterial flora, *Fleischwirtschaft*, **61** 1202–7.

SHUPE W L, BAILEY J S, WHITEHEAD W K and THOMPSON J E, (1982) Cleaning poultry fat from stainless steel flat plates, *Transactions of the American Society of Agricultural Engineers*, **25** 1446–9.

TAYLOR J H, ROGERS S J and HOLAH J T, (1999) A comparison of the bactericidal efficacy of 18 disinfectants used in the food industry against *Escherichia coli* 0157:H7 and *Pseudomonas aeruginosa* at 10°C and 20°C, *Journal of Applied Microbiology*, **87** 718–25.

THORPE R H and BARKER P M, (1987) *Hygienic Design of Liquid Handling Equipment for the Food Industry. Technical Manual No. 17*, Campden Food and Drink Research Association, Chipping Campden, UK.

TIMPERLEY D A and LAWSON G B, (1980) Test rigs for evaluation of hygiene in plant design. In Jowitt, R. (ed.), *Hygienic Design and Operation of Food Plant*. Ellis Horwood, Chichester.

WALKER R L, JENSEN L H, KINDE H, ALEXANDER A V and OWENS L S, (1991) Environmental survey for *Listeria* species in frozen milk product plants in California, *Journal of Food Protection*, **54**, 178–82.

WIRTANEN G and MATTILA-SANDHOLM T, (1993) Epifluorescence image analysis and cultivation of foodborne bacteria grown on stainless steel surfaces, *Journal of Food Protection*, **56** 678–83.

WIRTANEN G and MATTILA-SANDHOLM T, (1994) Measurement of biofilm of *Pediococcus pentosacceus* and *Pseudomonas fragi* on stainless steel surfaces, *Colloids and Surfaces B: Biointerfaces*, **2** 33–9.

WRIGHT J B, RUSESKA I and COSTERTON J W, (1991) Decreased biocide susceptibility of adherent *Legionella pneumophila*, *Journal of Applied Microbiology*, **71** 531–8.

ZOTTOLA E A and SASAHARA K C, (1994) Microbial biofilms in the food processing environment – should they be a concern? *International Journal of Food Microbiology*, **23** 125–48.

15

Total quality management

D. J. Rose, Campden and Chorleywood Food Research Association

15.1 Introduction

Organisations looking to develop their business operations through the current volatile economic climates need to establish clear objectives as to how the various elements of the business need to perform to ensure continuing growth and viability. In order to achieve these objectives it is further imperative to have mechanisms in place to monitor performance and also to provide a process by which change can be implemented in those areas of activity which need strengthening. Total Quality Management (TQM) is a management tool which provides that opportunity.

In its broadest sense TQM provides a business system by which the whole organisation can be harnessed to meet the needs of customer requirements. It is important to emphasise that TQM is not merely a technical standard but encompasses both the technical and business operations. The fundamental requirement for a successful TQM system is to have good management practices, TQM alone cannot provide this and any systems implemented will only ever be as successful as the staff involved.

The purpose of this chapter is to describe the key elements that need to be considered when setting up a Total Quality Management system. It provides information on the typical range of quality systems that may already be in place within an organisation and looks at how these can be used to bring together all of the requirements necessary to achieve a TQM system. It further explains the key steps necessary to begin development of the system and the implementation process required. Finally the key monitoring processes needed to confirm successful implementation and for continued improvement and development of the system are explained.

15.1.1 Defining quality

Despite the preponderance of quality assurance texts, quality standards, and definitions of quality, many people are still confused by the term quality. In the early incarnations of quality management, quality assurance and quality control were often used synonymously. However the importance of differentiating between reactive quality management (quality control) and pro-active quality management (quality assurance) was quickly realised. More importantly the benefits to be derived from the wide ranging implications of quality assurance were soon realised and capitalised upon by practitioners. The concept of TQM takes the now more familiar quality assurance requirements, as exemplified by BS5750/ISO9000, one step further and seeks to view ALL operations and processes that a company utilises as being inherently important to their overall business performance and quality of service parameters.

According to BS 7850, Total Quality Management may be defined as follows – 'Management philosophy and company practices that aim to harness the human and material resources of an organisation in the most effective way to achieve the objectives of the organisation.' On a slightly different tack, Margaret Thatcher once paraphrased quality very succinctly, 'The combinations of features in a product which ensures that customers come back for a product which does not.'

However, it is important to realise that the objectives of the organisation can be multifaceted and reflect other primary business needs as well as the more obvious product quality issues. TQM systems should therefore be capable of incorporating objectives as diverse as customer satisfaction, business growth, profit maximisation, market leadership, environmental concerns, health and safety issues and reflect the company's position and role within the local community. One over-riding principle must be for the TQM system to ensure compatibility with the needs of current legislation in all its guises – food safety, business practices, environmental and waste, employment rights and health and safety.

The need to meet the ever-increasing demands of customers for improved reliability and quality of product have fuelled the need to consider TQM systems. Supplying 'just-in-time' manufactured products with short shelf-lives to the retail outlet in a reliable and dependable manner, pressure on margins to provide cheap yet wholesome foods, and the continuing need to provide evidence of safe food production have all added to the requirement to consider the totality of the chilled food business operation.

Unfortunately for staff tasked with considering TQM systems there has been much confusing literature produced on the subject. Various titles have been used to describe TQM systems, e.g. Continuous Quality Improvement, Total Quality, Total Business Management, Company Wide Business Management, Cost Effective Quality Management, Integrated Management Systems. Suffice it to say that the objectives of the various schemes have all been synonymous and I refer the reader back to the definition of TQM given earlier from BS 7850. The challenge to practitioners of TQM is usually not with the title given to the system, but rather to understanding their business well enough to identify all of the key elements required to be set up and managed within the umbrella of TQM.

This analysis of the key business processes may be achieved by a variety of different means. Most critical to the analysis is the ability to collect suitable and useable data which reflects the process. The use of data collection forms, performance data, market research, productivity information or financial data may all be appropriate. Analysis of the data to extract useful and usable outputs may be performed by a variety of different techniques. BS 7850 recommends affinity diagrams, brainstorming sessions, cause and effect diagrams, flow charts and tree diagrams to analyse non-numerical data. Control charts, histograms, Pareto diagrams and scatter diagrams may be useful for numerical data. By understanding all of its business processes companies are able to define the process, implement controls, monitor performance and measure improvements. This is the fundamental basis of Total Quality Management.

15.1.2 Quality assurance systems

The foundation for any quality system is to be found in the fundamental principles of Good Manufacturing Practice (GMP). The technical requirements for GMP for chilled food operations are addressed elsewhere in this book. There are also many guidelines available for the manufacture, handling and preparation of chilled foods (Department of Health, 1989; Institute of Food Science and Technology, 1990; Chilled Food Association, 1995, and 1997; National Cold Store Federation, 1989). All focus on the key technical requirements for safe, hygienic, good manufacturing practices, allied to good storage, handling and distribution practices. In this context, these can be considered the fundamental technical objectives or standards to be achieved. Currently a large number of targeted quality assurance systems have found favour throughout the Chilled Food Industry.

The most prevalent of the formal quality systems is still the BS EN ISO 9000 (BS5750) suite of standards incorporating specifically BS EN ISO 9002 for production facilities and BS EN ISO 9001 for production operations incorporating new product development activities. ISO 9000 systems provide the advantage of laying down formal management controls for production activities, but also can easily be extended to other critical business activities such as purchasing, sales and distribution. Many operations have already extended their ISO 9000 systems into a TQM system by encompassing their other critical business processes.

Manufacturing production sites have now also been forced to consider the requirements necessary to meet the requirements of Hazard Analysis Critical Control Point (HACCP) systems based on *Codex Alimentarius* definitions. As well as providing the key control measures necessary to understand the mechanics of producing safe food, HACCP systems also provide the basis around which to build production control systems and to ensure product quality in the operation.

Documentation of HACCP plans to meet the seven fundamental principles of HACCP as laid out by *Codex Alimentarius* is also required. These documented plans, together with associated operational procedures, records of operation and evidence of maintenance of the critical control points, often form enough of a basis for production activities to be controlled and managed by using the HACCP plans as a quality system – see principle 7: 'Establish documentation concerning all procedures and records appropriate to these principles and their application.'

More recently businesses have needed to consider the impact of their operations on the environment. Moves to standardise environmental control and management have been formalised within ISO 14001. This international standard 'Environmental Management Systems – Specification for Guidance and Use' has strong links to ISO 9001 and covers issues such as policy statements, process control, system structure, training, awareness and competence, system documentation, checking and corrective action, preventive action, record keeping, system auditing and management review. The stated aim of ISO 14001 is to 'provide organisations with the elements of an effective environmental management system which can be integrated with other management requirements'. This approach is an obvious lead in to the concept of incorporating environmental objectives within a TQM system.

Yet more recently, safety systems have been targeted for incorporation within the suite of quality system functions and BS 8800 ('Guide to Occupational Health and Safety Management Systems') provides a framework within which to manage safety systems and safety training activities. Given the increasing importance of staff occupational safety and the need to minimise exposure to potential litigation, manufacturers are well advised to treat this area of activity seriously. Companies may also have an interest in other systems related to staff training – i.e. the Investors in People standard within the UK, organised through local Training and Enterprise Councils, which requires proper evidence of structured training programs for staff, records of all training activities and clear benefits being derived from both staff and employers from their training programme.

In a critical key development the British Retail Consortium (BRC) has now issued its core Technical Standard for Companies Supplying Retailer Branded Food Products. This standard is being used by a large number of UK retailers as the definitive standard for suppliers and terms of business are being agreed which include the requirement for companies to meet this standard. The BRC standard itself focuses on a large number of essential and recommended good manufacturing practices and is underpinned by the need to establish supporting management systems to back up these manufacturing practices. In essence 6 key areas are involved, HACCP systems, quality management systems, factory environmental standards, product control, process control and personnel. Implementation of the standard is being handled through third-party inspection bodies whose remit is to ensure compliance of the operating site with the standard. In some cases, as for the European Food Safety Inspection Service (EFSIS), the inspection bodies have incorporated the BRC standard within their own inspection standard to provide and even more rigorous examination of the operating site.

All of the quality systems mentioned above have essential core elements and similarities. Most importantly the critical elements of control can easily be related to the core business functions of the site. These, taken hand in hand with the key legal, safety and environmental control measures all sites are obliged to adopt, offer a comprehensive and complete set of frameworks within which to develop a total quality management system.

15.2 The scope of a quality system

This section summarises the essential business processes to be considered when addressing Total Quality Management systems concerned with the production of chilled foods. The next section deals with the necessary administrative detail of developing the quality system.

15.2.1 Raw materials, purchasing and control

- Raw and packaging materials should be purchased to agreed specifications, and from suppliers capable of achieving those specifications. Suppliers should be audited for quality and safety.
- Raw materials should be stored under hygienic conditions that prevent contamination by microorganisms, insects and other pests.
- Stock control systems should be used for minimising storage times. Coding systems should be used to ensure traceability.
- Inspection of raw and packaging materials should complement the suppliers' quality systems.
- Control and release should be under the responsibility of a competent technical person.
- Non-conforming raw materials should be recorded and investigated to identify and rectify problems.

15.2.2 Process control

- The HACCP approach should be used to identify critical control points as part of developing process specifications and to determine monitoring programmes.
- The HACCP plan must be suitably verified and the control points demonstrated to be sufficient to control the product.
- The arrangement of plant should minimise the likelihood of cross-contamination.
- Plant should be hygienically designed.

- Plant cleaning schedules should be developed and implemented.
- Critical measures such as time, temperature and quantity must be recorded throughout the production process.
- Sampling regimes must be set up to measure product quality and safety throughout the production process.
- Contingency plans need to be in place to cover any possible major safety issues that may arise.

15.2.3 Premises

- Premises should be constructed to minimise the risk of contamination.
- Premises should be maintained to a hygienic standard commensurate with the degree of risk.
- Where used, high care areas must be suitably constructed and all necessary control measures for their operation implemented.
- Suitable waste disposal facilities need to be in place.

15.2.4 Quality Control

- Clearly defined product specifications and quality standards should be used to supplement HACCP analysis in identifying non-safety quality issues.
- Product quality (in terms of sensory characteristics) should be defined to meet the specifications given above, and agreed with clients.
- Product quality should be verified to ensure acceptability before release and on-going monitoring checks should be in place to prevent major defects arising.

15.2.5 Personnel

- Personnel should be trained in hygienic practices and other quality requirements of the job.
- High standards of personal hygiene are essential.
- Clothing appropriate to the task is required.
- Appropriate sanitary facilities are required.
- Medical screening is required.

15.2.6 Final product

- Inspection must take place to determine conformance with the product specification and freedom from any foreign body contamination.
- A system for isolating non-conforming product is required.
- The type and level of inspection should be determined from HACCP.
- Critical testing and inspection should be done by competent laboratories.
- Where technically important, or for legal reasons, checks on packaging should be done.
- Records of inspection must be kept.

- Shelf-life validation is required.
- A system for monitoring complaint trends is required.
- Product release should be by positive approval.

15.2.7 Distribution

- Arrangements must be made to maintain product integrity in the distribution chain.
- The level of batch traceability must be commensurate with the risk of recall.
- A recall system should be developed and tested.

The above list is by no means comprehensive, but indicates the breadth of considerations to be addressed in chilled food operations. The task is complex and requires a high degree of skilled management. It should be developed and implemented as part of cohesive quality system.

15.3 Developing a quality system

Developing a quality management system to meet the requirements of your business is a complex task. Not only do the elements described above need to be considered, but also such factors as management responsibility, documentation and auditing. The standard model for quality systems for some years now has been the ISO 9000 series of standards, the international standards for quality systems. The general applicability of ISO 9000 to the food industry has been demonstrated by its successful application in many production facilities. However ISO 9000 has suffered from criticism over the years due to its unfriendly nature and the perception that it does not lead to quality improvement, only control and standardisation of processes. Consequently quality management, as exemplified by ISO 9000, has often been seen as being on a parallel track to business management, and not as an all embracing TQM system.

Forthcoming changes to the ISO 9000 standard, due to be published in the year 2000 are believed to address this failing by focusing the system back towards quality improvements, process development and customer satisfaction. The above comments notwithstanding, the fundamentals of ISO 9000 still provide the basis around which to start the development of the quality system, a TQM system being built by encompassing all of the other business process requirements onto this model.

15.3.1 Management responsibility

The importance of senior management commitment to the quality system cannot be over-emphasised. If quality is established as a board room priority, all other parts of the organisation will follow and become intimately involved in the process. Similarly, by defining key responsibilities for all levels of activity, those staff whose actions can influence the quality of the food or the process under which it is manufactured can be identified and made aware of their responsibilities. This is so that errors do not occur through it not being clear who is responsible for various actions, for example, who monitors a chill room temperature, or who should carry out a particular quality control test.

Senior management must review the continuing effectiveness of the quality system at periodic intervals. Key information sources would include data from internal audits, non-conforming product records, quality control records on conformance to specifications, and customer complaints trends.

The second key role of the management review process is to establish mechanism for improvements and new initiatives. The evaluation of key data such as quality markers, which can be used to measure client satisfaction, and performance measures, which can be used to measure the efficiency of the delivery process, should be encouraged. Based on the analysis of these data, together with the data mentioned above, senior management can identify business processes which would benefit from improvement or re-design.

15.3.2 Documentation of the system

Effective documentation of the procedures and actions required to achieve the required quality is an essential part of the quality system. Such documentation can be used for reference and for training purposes. It reduces the risk of misunderstandings arising from oral communication. All documents should also be controlled so that personnel do not work from obsolete documents. There must be a means of circulating new procedures and withdrawing superseded ones, and a controlled means of making changes to procedures. Staff discipline with documentation also needs to be instilled so that only the current versions of documents are used.

15.3.3. Customer requirements

A clear understanding of customer requirements is essential for any business operation. Mechanisms to ensure that sufficient information is captured from clients prior to manufacturing, need to be set up. These will include fundamentals such as manufacturing details, supplier arrangements, product specifications, delivery times, quantities and packaging. However consideration must also be given to other matters such as legal requirements, environmental, employee and safety considerations.

15.3.4 Raw material control and supplier quality assurance

The quality of raw materials and the suitability of packaging materials has a considerable effect on the final quality of chilled foods. However, relationships with suppliers go well beyond these criteria and extend to the quality of service, prices and financial stability of the supplier. These factors must be combined together to achieve a smooth and profitable relationship between vendor and purchaser. The objective must be to build a confident partnership between the two so that the purchaser can rely on the vendor as much as on 'in-house' departments.

There are a number of stages to go through in achieving this objective. It should be noted that all purchased materials which can affect product or service quality should be included in this programme. Often capital and services items (e.g. motors, pumps etc.) are omitted, and faults are only detected after installation. Clearly this does have an impact on the overall efficiency and quality of the operation and should be included.

Supplier quality policy

There should be a stated and preferably written policy. This usually takes the form of a summary of the principles involved:

- mutual co-operation; the partnership
- prior contractual understanding; agreeing specifications
- methods of evaluation
    - audit
    - inspection at source
    - inspection on receipt (the policy is to accept only material meeting the requirements)
- plans for settling disputes
- feedback on performance
- vendor responsible for delivery to standard.

Contractual understanding

There is little point in trying to develop a partnership with suppliers unless there is a clear understanding about the objectives to be achieved. This usually takes the form of a contract covering for example, material specifications, delivery parameters, responsibilities for quality including those for verification, access to supplier, procedures for settling disputes. It is important that all these parameters are agreed and verified prior to signing the contract and entering a supplier onto an approved list.

List of approved suppliers

The purchaser should maintain a list of approved suppliers. Lying behind this should be a set of procedures which describe the stages of approval. There are two main criteria to be considered here:

1. financial capability and stability
2. ability to meet specification.

The latter can be assessed in a number of ways:

- auditing supplier's quality system
- vendor's previous performance

- vendor's reputation
- tests on representative samples.

Auditing suppliers
The objective here is to establish the supplier's ability to meet agreed requirements. Auditors should be trained to conduct this activity promptly and efficiently. The auditors observe the manufacturing facilities, buildings environment, plant, quality procedures and implementation of such procedures. Other evidence to collect includes management attitudes, workforce attitudes, quality control records and so on. Often auditors will also look at financial and technological aspects.

Feedback on performance
It is absolutely essential in building the partnership that the vendor receives prompt and accurate feedback on performance.

Performance data can be collected from a number of sources.

- *Raw material conformance:* sources of information here include the vendor's own inspection records, incoming inspection records. Most non-conformances in this area are clear and are well 'flagged' because usually they result in a delay in deliveries or production.
- *Process conformance:* non-conformances here are less easily detected but at least should be reviewed during audit. It may be written into contractual requirements that process non-conformances are communicated to the purchaser.
- *Procedural non-conformance:* similar comments appertain here as for process non-conformance.
- *Raw material unfit for use:* this is the worst scenario where a non-conformance is not detected until it fails either on the production line or in distribution or in use (complaints). The impact is usually severe, affecting ability to sell the final product. Despite the severity of the problem, it is often difficult to gather sufficient evidence to inform the vendor of the fault.

Feedback should be given on a regular basis so that each non-conformance is not seen by the vendor as a 'complaint'. The main message here is to transmit good as well as bad news. Where possible, evidence should be incontrovertible. The best evidence is records and samples. Regular meetings with suppliers will ensure that the positive feedback is given. This helps to support the partnership when exceptional communication of non-conformances is necessary.

15.3.5 Process control
All aspects of the production of chilled foods having a direct bearing on the quality of the final product must be specified, documented and recorded to ensure that failures due to inadequate control are eliminated. Critical control point monitoring as identified by HACCP forms part of this requirement. Action

when results are outside specifications must be clearly identified with responsibility allocated. The HACCP principle should be used throughout the production process and include raw materials and final storage and distribution. It can be used for all potential hazards including inadequate quality as well as safety.

HACCP includes the assessment of potential hazards, prescribes for the elimination of available hazards and sets tolerances for the hazards that cannot be eliminated in the processing of a food. It defines the appropriate control measures, the frequency of their application, the sampling programme, the specific tests to be applied and the criteria for product acceptance. Since HACCP is an ongoing dynamic process, analyses will need to be reviewed in the light of new hazards and changes in the process parameters. HACCP has the potential to identify areas of control where failure has not yet been experienced, making it particularly useful for new operations.

The following definitions are used in HACCP:

- Hazard analysis is the identification of potentially hazardous ingredients, storage conditions, packaging, critical process points and relevant human factors which may affect product safety or quality.
- Critical control points (CCP) are the processing factors of which loss of control would result in an unacceptable food safety or quality risk.

Carrying out a HACCP analysis

To carry out a HACCP analysis, a formalised and structured approach is needed. A broad base of information is required and will therefore require specialist knowledge from many disciplines, since safety and quality assurance cannot be categorised by a single discipline. The first stage of an analysis is to obtain a detailed flow diagram for the process under consideration, including methods and schedules of production, preparation and transport of raw materials. Many of the considerations will be influenced by issues specific to the factory.

The second stage of an analysis identifies the essential characteristics of the product and its use, enabling definitive conclusions to be drawn about the hazards or potential risks which will threaten either the consumer or the product. Consideration is given to food storage conditions, formulation of the product, the packaging used, the expected customer handling practices and the target consumer group.

The third part of an analysis is consideration of all the stages in the process, taking into account realistic process deviations. Critical stages in the process are identified which must be controlled adequately to assure safety – the critical control points (CCP). A judgement of risk must be made using one of three basic methods: probabilistic, comparative or pragmatic. The choice of method depends upon circumstances and the basis for any judgement should be recorded. Such judgements require a high degree of expertise and experience and should only be made by suitably qualified people. Ideally, the opinion of more than one 'expert' should be sought. If process details are incomplete, the most unfavourable assumptions must be made unless, for example, there is a long, proven history of the raw materials presenting no hazard to the process or the product. The final stage of an analysis is to devise standards for and effective procedures to monitor critical control points and appropriate corrective action as mentioned earlier.

Monitoring of critical control points

Monitoring of CCP may be best accomplished through the use of physical, microbiological and chemical tests, visual observations and sensory evaluations. Monitoring procedures, including those which take the form of a visual inspection only and do not involve measurements, should be recorded on suitable checklists. These checklists should show details of the location of the CCP, the monitoring procedures, the frequency of monitoring and satisfactory compliance criteria. For chilled foods, the cleanliness of equipment is a CCP. Therefore a hygiene maintenance schedule must be devised that specifies what should be cleaned, how it should be cleaned, when it should be cleaned and who should clean it.

When monitoring of CCP takes the form of inspection, particular attention should be given to temperatures of food, hygienic practices and techniques of handling foods by workers, whether employees are ill or have infections which can be transmitted to the food and opportunities for cross-contamination from raw to cooked foods. Control options also include arrangement of plant to minimise cross contamination, building maintenance and cleaning, and staff training.

15.3.6 Inspection and testing

From HACCP, a schedule of testing for raw materials, intermediate and final products is developed. Methods of tests must be defined, responsibility for testing and the acceptance criteria drawn from appropriate specifications. At each stage, product should not be released until inspection is complete. If release takes place earlier, a traceability system must exist for recall purposes. The time required to complete microbiological tests on chilled product is problematical here. However, most microbiological tests are used to monitor the success of process control rather than for testing product characteristics. Untested, tested, approved or rejected materials need to be clearly marked to avoid any possibility of confusion.

All test equipment used to demonstrate compliance with a defined specification or to control a critical process should be of known accuracy. Required measurements should be identified, the measuring equipment calibrated at defined intervals, against acceptable physical or nationally recognised standard references. Calibration methods should be described and adhered to, and the calibrated equipment must be identified as such. Records of calibration should be kept, and if a calibrated instrument is found to be inaccurate, then a designated person must review the situation and decide what action should be taken in respect of materials previously measured with that instrument. Where necessary, critical tests should be performed by a suitably accredited laboratory, either in-house or external.

Any product which is found to be outside specification should be segregated to prevent inadvertent use. The product should then be destroyed, re-worked, or re-graded. In exceptional circumstances, customers may be prepared to accept the product, but not if safety is in question. Re-work must be controlled strictly. Causes of non-conforming product should be identified and action taken to prevent recurrence. Complaint trends should also be monitored and corrective action taken as appropriate.

15.3.7 Handling, storage, packaging, delivery

This is extremely important for chilled foods. Precautions must be taken to protect product quality throughout production and the chill chain. Hygiene precautions, including vehicles and chill storage, pest control and restrictions on access would be included here. The legislative requirements for 'food handlers' and legal constraints on labelling, date coding and food contact materials should be addressed here. The means of temperature control, monitoring and recording are critical. Determination and control of shelf-life through stock rotation must be included.

Decisions on the extent of, and method of, traceability must be reached with respect to the risks of recall. A fully documented and workable recall system must be implemented. The system should extend to distribution centres, the trade and in extreme cases, consumers. The recall plan must be tested to ensure its effectiveness.

15.3.8 Records

An effective record system is essential. The control of records, including their identification, safe storage, retrieval and disposal, should be defined. It should be clear when records can be disposed of, and who is responsible. The most important records are those which demonstrate that what has been specified has, in fact, been achieved. These include process control and inspection records. However, in order to satisfy the legislative requirement for 'due diligence', other evidence will be required such as records of internal audits, management review, supplier audits, HACCP records, temperatures in distribution, corrective action, cleaning and training.

15.3.9 Quality audits

A scheduled system of internal quality audits is essential to ensure that all procedures are implemented and working effectively and that instructions are written down and followed. These audits are an effective management tool for monitoring the success of the quality system and ensuring that everyone is working to the system.

System audits should be undertaken by trained personnel independent of the area being audited. Audits are carried out by a process of observation, interview and examination of records. Any non-compliances should be recorded and referred to the responsible manager for timely corrective action. A follow-up to ensure that the action has taken place should be carried out, and records of the action kept. The results of audits should be reviewed by senior management. The audit schedule should cover all aspects of the quality system and include compliance with legislative requirements and voluntary Codes of Practice.

15.3.10 Training

All staff must be trained to fulfil their responsibilities with regard to tasks undertaken which affect quality. Training needs should be reviewed, the needs identified should be fulfilled and records kept. Staff education and training is often a most useful option for control of hazards such as microbial contamination. In addition to hygiene training, there is another special training requirement for the food industry to be considered here: training for sensory analysis. An attempt to ensure that judgement of product quality in this respect is objective must be made.

15.4 Implementation

Implementation of any quality initiative is difficult. Change, often perceived to be change for changes sake, is not always acceptable to staff. It is therefore imperative that the correct empowerment is given to the implementation and that it is introduced and explained to staff in the right manner.

15.4.1 Chief executive commitment

The ramifications of a TQM strategy are too large for them to be considered at anything other than the highest levels within a company. Ideally the idea to implement the system should come from the chief executives themselves. Alternatively it may come from other sources. Whatever the source it is not worth starting the exercise until the right level of approval and commitment is achieved. Once the senior management are on board with the idea they must throw their whole weight behind the initiative, any perceived weaknesses will be exploited by opponents of the scheme.

15.4.2 Steering group

As a first step, set up a steering group to manage the implementation programme. This group should consist of staff drawn from each of the principal areas of operation in the company: this should include sales and marketing, purchasing, production, distribution, technical and finance. The group should be headed by a member of staff with sufficient managerial experience and should be accountable to the board or directly to the chief executive. The steering group should also appoint the person to be designated as the 'management representative', whoever named, who will be responsible for the maintenance and control of the system in the future. This person is concerned with ensuring that documentation proceeds smoothly and that documentation is controlled. He/she is also responsible for the internal audit system. This so-called 'Quality Manager' may have other duties within the quality system, but this should not be to the detriment of 'ownership' of the system by all the constituent parts.

15.4.3 Initial status

The steering group should arrange for the two key activities to be carried out: (a) a definition of current business processes carried out in all parts of the business and (b) based on this information define the scope within which you wish to implement the TQM system. This is a key decision and must be based on a sound understanding not only of the essential processes which support delivery of your products to your customers, but must also understand all of the support functions which help to maintain that delivery mechanism (e.g. finance, maintenance).

Once the scope of activity is defined it is imperative to carry out an exercise to establish the level of benefit that may be obtained from the introduction of the TQM system. Typically this can best be achieved by carrying out a quality costing exercise. Quality costing will determine the operational costs of not doing things right, such as wastage in manufacturing, loss due to non-conforming product, down-time on equipment. Based on the findings of this exercise it is possible to estimate the possible company wide benefit of introducing the quality system. These potential savings can then be reviewed by senior management, and a firm commitment to establish the system made. If necessary the scope of the TQM can be reviewed at this point to ensure that the areas covered will lead to maximum return.

15.4.4 Planning

The steering group should draw up the implementation sequence and agree timescales with all appropriate parties so that a plan can be made. The plan must cover all elements of the implementation including process analysis, documentation, implementation, training and PR elements. The group should monitor the plan of implementation. If there are problems to overcome in achieving the plan, the steering group must be sufficiently senior to overcome blockages. If the plan cannot be achieved for unavoidable reasons, the steering group must give an account of this to the chief executive.

15.4.5 Quality policy

It is important that the chief executive writes a quality policy for the company. This can be anything from a relatively simple statement to something more complex. At its simplest it states that the company is dedicated to meeting customer requirements. If it is the intention to work to a recognised quality system, then a statement to the effect that the company wishes to comply with the requirement of, for example, ISO 9002 should also be included. However, other statements about business culture and objectives can also be included. For example, employee welfare, environmental policies, position in market place and so on, can all be considered. It is best not to make the policy too lengthy or complex.

15.4.6 Briefing

It is the steering group's responsibility to initiate and co-ordinate briefings throughout the project. During the initial phases, this would be an announcement (from the chief executive) about TQM or ISO 9000 explaining what it is and the reasons for pursuing the course of action. Such briefing should be to all employees, but would be more detailed for some depending on the level of involvement envisaged. Also the substance of the briefing will depend on the seniority of the audience.

It is best to keep the briefings short and to the point initially; more detailed training can follow later. A big 'launch' package with trumpets blaring is not the best course. Small and informed focusing on the facts, the importance of the initiative and not underestimating the amount of work involved will get the message home. The seniority of the person carrying out the briefing speaks volumes about the importance of the mission.

15.4.7 Structure of the quality system

It is essential that the structure of the quality system is agreed at an early stage. This is best embodied in the documentation. There should be three levels to the documentation, although for a small company this may be kept under one cover. The three levels are:

1. policies
2. procedures
3. work instructions.

Policies should be used to state the company's intent with respect to key elements of the system, e.g. policies with respect to purchasing arrangements or staff training.

Procedures will form the bulk of the system and will provide the detailed instructions as to how principle operations are carried out. These form the bible from which the company will be expected to operate and will be judged (audited) against.

Work Instructions provide the 'shop-floor' level of instruction needed by staff. These should be formulated so that anyone coming to a job for the first time can, with a small amount of training, carry out the job effectively. Examples of work instructions could be how to make out an invoice, a purchase order or a customer order. Other examples include those in production for the basic operation of a machine or on how to carry out weight checks.

All levels of the manual may need to be supported by record forms and these should either be incorporated with procedures/work instructions, or clearly separated off and identified. Setting the structure of the system and document numbering and cross-referencing will save a lot of trouble and retracing later. It is clear that documenting the quality system is a major task and it needs to be thought through clearly. Experience also teaches that the best systems are those written by the staff actually involved in the task being described. This simple device also ensures a wide level of personal involvement with the development of the quality system and helps to provide ownership throughout.

15.4.8 Quality manual(s)

The essential parts of this need to be written in draft form at an early stage because it sets in writing the structure referred to above. It usually contains the policy and headline procedures covering each key area of the quality system. It is usually used for overall guidance and should be available to customers. Therefore it should not contain anything of an overly sensitive or confidential nature. Each operational function within the company must agree to the contents of the quality manual as it applies to them. Better still, to gain ownership, if they actually write those parts which apply to them. The organisation and management authority must be clearly defined at this stage.

15.4.9 Quality improvement

Once the plan has proceeded far enough to ensure a reasonable understanding of the current business processes, the key task of planning for quality improvements can begin. The mechanisms for achieving this are varied and will need to be tailored to suit individual circumstances. Most critically it is essential to be able to measure the process, either in terms of inputs, outputs or throughputs. BS 7850 (ISO 9004-4) deals with a variety of techniques used for quality improvement and these will need to be considered. It should be remembered that improvements may need long-term solutions and that the implications to other elements of the business must be considered. However, failure to maintain momentum in this area will impact significantly on the usefulness of your TQM system.

15.4.10 Staff Training

The system designed will be of no use whatsoever unless sufficient time and resource is allocated to training and educating staff in the requirements of the new system. Therefore sufficient time must be allocated and planned into staff training. To meet both new skills training and new working practices, but also any implications that the new TQM ethos may bring to the business, i.e. the need to participate in quality circles, ability to generate improvement suggestions and the need for all staff to be committed to the ideals of the system.

15.4.11 Launch

The system can be launched when it is felt that all key elements are in place. These do not have to include all of the proposed quality improvements. Remember the system is designed to be a continually changing system and evolution to new procedures and practices should be a natural progression. Staff should not be concerned if parts of the system are not perfect, again improvements will be identified as the system evolves. It is much more important that the system accurately reflects the current business processes. Often the benefit of TQM comes with time when the quality system is used to collate data and information about the performance of the business and these are used to target and develop improvements. Once launched internal mechanisms to monitor and control changes to the system should be made operational as well (e.g. internal auditing systems, document control and document change systems).

15.5 Performance measuring and auditing

As indicated above, once the system is launched it is imperative to measure performance and to seek quality improvements. In fact, the most powerful part of any quality system is its ability to measure performance and gain improvement through corrective action. There are a number of tools available to do this within the quality system.

15.5.1 The internal audit system

Regularly measuring compliance with the stated system is a powerful means of determining its effectiveness. The corrective action that ensues from an audit ensures that the system is kept fresh and up to date, reflecting the latest requirements of the company.

15.5.2 The external audit system

Once formal certification and approval of the system is sought, external auditors for the inspection body will visit regularly to ensure on-going compliance with the relevant standards. Again useful information and corrective actions can be obtained from these visits. It is also possible that key customers may wish to audit the systems to give themselves confidence in your ability to provide safe wholesome food. These visits combine the benefits of an external inspection of the system with the specific requirements that the customer may have, enabling these requirements to be built into the overall operation.

15.5.3 Non-conforming product

Any such incidents must be investigated thoroughly. The reasons for non-conformance must be investigated and corrections made to the quality system and operating processes to prevent recurrence.

15.5.4 Conformance to specification

It is necessary not only to ensure that all product conforms to its final specification, but also to determine proximity to target of all measured parameters-both product and process. Clearly this serves two functions: to detect deteriorating trends early, and to detect persistent divergence from target while still within specification. Both may lead to corrective action.

15.5.5 Customer complaints

These should be treated like non-conforming product. They should be investigated thoroughly. Any deleterious trends must lead to corrective action.

15.5.6 Quality action initiatives

One of the key ingredients of any quality system, often referred to unkindly as the weakest link, is an organisation's own staff. However, staff also provide a company's greatest innovative resource. Involvement of all staff with the development of new ideas, process discussion groups and quality circles will enhance process efficiency, detect poor performance and lead to corrective action.

15.5.7 Performance measures

By setting performance measures for each key business area, or delivery process mechanism, the efficiency and performance of each key area can be monitored. Improvements, as well as declining performance, can be tracked and all elements of the business put into a measurable framework. Typical performance measures can extend beyond simple productivity related issues and may incorporate elements as diverse as; energy monitoring, waste management, sales lead successes, new product development time and customer satisfaction surveys. All these sources of information on performance should be subjected to senior management review. It is essential that senior management have the opportunity to review and take appropriate action at regular intervals.

In addition, feedback of performance to staff is an essential means of motivating staff to further improvement. It is quite easy for staff to be involved in performance measurement but not see a clear picture, because they see individual results rather than trends. Such feedback can be in the form of summaries of audits (based for example on a points system), trend graphs for conformance to specification or information on customer complaints.

15.6 Benefits

The achievement of total quality management or a good quality system is a never-ending road to improvement. Those who have embarked on this journey have found a number of benefits.

15.6.1 Economic

Generally, the operation is more cost-effective. This is achieved by getting it 'right first time'. There is a reduction in the amount of wasted material; productivity is increased as a result of the orderliness created. There is a reduction in the number of customer complaints. Machine efficiency improves and manufacturing capacity increases.

15.6.2 Marketing

By meeting customer needs consistently, there is an opportunity to secure the customer base, and to build sales success. Customers are more confident in the consistency of product and they see a commitment to quality.

15.6.3 Internal

A number of benefits are achieved within the operation. Staff morale improves because staff know what is expected of them. There is increased awareness of quality and a commitment to quality. Communication improves and staff are better trained. There is much improved management control with greater confidence in the operation, a reduced amount of 'fire-fighting', a uniformity of approach to procedures and a mechanism for continuous improvement.

15.6.4 Fulfilling legislative requirements

A good-quality system is of great benefit in demonstrating that attention has been paid to complying with legislative requirements, particularly those of due diligence. The quality system provides documented evidence of its functioning through written procedures, of its success through the records, and of its ability to improve through audits and review.

15.7 Future trends

A key change is nearly upon us at this point in time, the BS EN ISO 9000 series of standards are being revised and the new ISO 9000 (year 2000) version will be published in 2000. The new standard will mark a significant change to many areas of the old ISO 9000 standard and will address many of the key criticisms of the old standard. In particular there will be a change to using the process model approach so that individual businesses can suit the application of the standard to their own operations, rather than have the 20 key system elements imposed on them. Secondly the emphasis of the new standard will be firmly routed in the principles of continual improvement and meeting customer requirements. To this end specific requirements will be needed to measure and ascertain performance of the business with respect to quality and the ability to meet its customers' demands and requirements, howsoever defined. Finally the new standard will include the requirements to effectively communicate with customers and to manage all relevant streams of information passing through an organisation. In summary the essence of the new standard should help to ensure that you keep your existing customers by focusing on their needs, rather than the internal needs of the business.

It can also be predicted for the future that the involvement of customers, and particularly retailers, in the production supply chain will continue to grow. Requirements on production facilities to ensure that their products meet the needs of retailers is therefore imperative and the moves towards synergistic business relationships between suppliers and retailers should be encouraged. The continuing use of third party schemes to audit and assess production premises will obviously continue and the current standards being applied will develop with time. The challenge to all involved in this process is to ensure that the inspection standards are demanding but technically achievable to ensure safe and reliable food production.

Given the higher publicity now given to any food safety problem by the media, it is an inevitable consequence that governments will react to the media attention by raising standards through increased legislative input. In the UK the launch of the new Food Standards Agency (FSA) will enhance this process. It will be interesting to note whether the FSA will develop a highly prescriptive approach to safety matters or whether it will continue to place the emphasis of control onto manufactures themselves.

Finally, the environmental pressures being placed on the whole of society will impact on chilled food businesses like all others. TQM systems which seek to address environmental issues as well as production management issues should be applauded and encouraged. The European Union is keen to progress the ideas of an Integrated Product Policy (IPP) within all manufacturing areas. This approach takes a more holistic total life-cycle approach to products and looks into the total environmental effect on all elements of production, packaging, delivery. This ensures that the environmental impacts of the individual elements are acceptable and that any changes made do not create savings in one area by passing on the impact to another area. Backing IPP up by focused environmental audits and certified product labelling will ensure that consumer marketing can also be focused on this key area.

15.8 References

BRITISH RETAIL CONSORTIUM, (2000) *Technical Standard and Protocol for Companies Supplying Retailer Branded Food Products* BRC, London.

CHILLED FOOD ASSOCIATION, (1995) *Class A (high risk) area best hygienic practice guidelines*, CFA, London.

CHILLED FOOD ASSOCIATION, (1997) *Guidelines for good hygienic practice in the manufacture of chilled foods*, 3rd edn. CFA, London.

DEPARTMENT OF HEALTH, (1989) *Chilled and Frozen. Guidelines on cook-chill and cook-freeze catering systems*, HMSO, London.

FAO CODEX ALIMENTARIUS COMMISSION, (1997) *Food Hygiene Basic Texts. HACCP Principles* FAO/WHO, Rome.

INSTITUTE OF FOOD SCIENCE AND TECHNOLOGY, (UK) (1990) *Guidelines for the handling of chilled foods*, 2nd edn. IFST, London.

INTERNATIONAL COMMISSION ON MICROBIOLOGICAL SPECIFICATIONS FOR FOODS (ICMSF), (1988) *Micro-organisms in Foods. 4: Application of the hazard analysis critical control Point (HACCP) system to ensure microbiological safety and quality*, Blackwell Scientific, Oxford.

NATIONAL COLD STORAGE FEDERATION, (1989) *Guidelines for handling and distribution of chilled foods*, NCSF, London.

16

Legislation

K. Goodburn, Chilled Food Association

16.1 Introduction

This chapter focuses on the key elements of international and national regulatory controls and associated guidance on the manufacture, storage and distribution of chilled foods. European countries in membership of EU are covered only insofar as where national legislation or guidance exists in addition to general EU rules. Since chilled foods are a relatively recent development and the sector is highly dynamic and innovative, comprising an ever-increasing and vast heterogeneous range of products, there is a wide range of legislation that impacts on the sector but little is directly focused on it. It is very important, at any given time, to check with the appropriate official body in the particular country of interest what the precise regulatory position is.

16.2 Food law is reactive

All food law aims fundamentally to protect consumers' interests (health and fraud protection) and, to a degree, facilitate fair trade. Food laws are not static, developing to maintain an adequate level of consumer protection as new knowledge reveals new hazards. For example, analytical developments in the late 19th century revealed the significant extent to which food at that time was being adulterated, resulting in the foundation of modern food law. Later, when the link between food poisoning and bacterial contamination was established, regulated hygiene requirements were introduced.

To reflect new knowledge, technical innovation and changes in the pattern of the distribution, legislation must be kept continually under review. In many cases, changing consumer attitudes and social behaviour provide the innovatory and marketing driving forces. Current consumer preference for convenience and 'fresher', more 'natural' and less processed foods, and the use of fewer or even no additives, is based on a number of developments and is satisfied to a great extent by the chilled prepared food sector. However, the general absence of chemical preservatives and use of minimal preservation techniques designed to preserve safety without sacrificing quality, has brought new challenges in distribution systems (demanding handling procedures, strict temperature control, and shelf-life limitation) and consumer advice and behaviour. In particular, new knowledge of new food poisoning risks, e.g. listeriosis, or the re-introduction of old risks, e.g. botulism, through a new route, has invited legislative attention, primarily promoting HACCP-based systems, and quasi-legislative activity. The latter is particularly evident through industry codes of practice and guidelines, which are usually voluntary, and can be often industry-enforced standards. Such standards are generally the preferred route in the chilled food sector owing to its diversity and rate of innovation. Trading structures such as retailer own label can facilitate uptake of such standards through their endorsement by retailer customers.

16.3 Food laws and international trade

National food laws, although designed to facilitate fair trading within countries, may create barriers to international trade. Some degree of international agreement on food standards is desirable to provide a reasonably uniform level of protection in terms of public health and food standards, and also to minimise, if not remove completely, technical barriers to trade across frontiers.

The FAO/WHO Codex Alimentarius Commission was established in 1962 charged with pursuing these objectives. Codex is jointly funded by the Food and Agriculture Organization (FAO) and the World Health Organisation (WHO). The Commission is an intergovernmental body with 166 governments, as at June 1999, taking part in its work. At European level, the European Economic Community was set up in 1957 under the Treaty of Rome, having as one of its prime objectives the free movement of goods. Now called the European Union (EU) since the Maastricht Treaty of 1992, it consists of 15 European Member States, with other central and eastern European countries such as Estonia, Hungary, Poland, the Czech Republic, Slovenia and Cyprus seeking membership. Consumer protection and the movement of food between Member States are two of the EU's key priorities regarding food.

International trade in chilled foods is limited by the relatively short shelf-life of these products, and by differing national recipe and presentation preferences. However, in Continental Europe there is trade in chilled foods, particularly between neighbouring Member States where there can be the greatest cultural commonality. There is a small amount of trade between the UK and the Continent in short shelf-life own label chilled foods destined for UK retailers' stores in other Member States. However, there is greatest intercommunity trade in the EU in 'international' products such as fresh pasta (i.e. required to be kept chilled to maintain shelf-life) which have a relatively long shelf-life, compared with other chilled foods.

16.4 Chilled foods are...

Before looking at the regulatory framework applicable to chilled foods the definition of these products must be addressed. The definition adopted is that used by the UK Chilled Food Association, which focuses on retail foods (CFA 1997): 'Chilled foods are prepared foods, that for reasons of safety and/or quality are designed to be stored at refrigeration temperatures (at or below 8ºC, but not frozen) throughout their entire life.' This definition excludes non-prepared foodstuffs such as raw meat, poultry and fish portions sold alone and which require cooking prior to consumption. Similarly, commodity dairy products such as milk, butter and cheese are excluded from the definition, as they are not considered to be 'prepared'.

Chilled prepared foods are manufactured using a wide variety of raw materials, including vegetables, fruits and ingredients of animal origin. These materials are either used in their raw state or they are subjected to various treatments, e.g. blanching, freezing, and cooking (i.e. equivalent to a time-temperature combination of 70ºC for two minutes). Cross-contamination during manufacturing is avoided by the use of Good Hygienic Practice, as set out in the CFA Guidelines (1997) and European Chilled Food Federation Guidelines (ECFF 1996).

Chilled prepared foods can be manufactured from a variety of raw materials in terms of level of processing (Table 16.1) and can be designed to be ready to eat, to be reheated (minimal heat application before serving, for organoleptic purposes) or to be cooked (thorough and prolonged heating before serving).

Even though chilling extends shelf-lives without prejudicing safety or quality, it must be recognised that it is a relative extension in shelf-life. Cooked chilled foods are often erroneously referred to as 'cook-chill'. These foods have been defined as a catering system based on the full cooking of food followed by fast chilling and storage in controlled temperature conditions (0ºC–3ºC) and subsequent thorough reheating before consumption. A maximum shelf-life of

Table 16.1

Ingredients	Further Processing
Raw	None or reheated
Raw + cooked	None or reheated
Raw and/or cooked	Cooked, then packed
Raw and/or cooked	Cooked in package

five days, inclusive of the day of cooking, is recommended since these products are not packed prior to distribution.

16.5 Approaches to legislation

Many aspects of chilled foods are common to all foods, which is reflected in the range of legal instruments applying to them: food composition, additive usages, residues, contaminants, labelling, packaging, and so on.

Where chilled foods differ from foods generally is in their greater vulnerability to microbiological contamination. There is relatively little legislation that is specifically directed at 'chilled foods' owing to the great range of product types encompassed by this term. However, legislation in relation to 'vertical' segments (such as meat-, poultry- and fish-based products) that are part of the chilled food and other sectors is, at the time of writing, being consolidated into a European Regulation on the hygiene of foodstuffs based on HACCP. This proposed Regulation will replace vertical rules and draw in elements from the General Food Hygiene Directive 93/43/EEC (EC 1993), which relates to all food production including those foods not containing protein ingredients falling under vertical legislation.

However, there remains little in the way of clear legislated international or European standards for the manufacture of certain categories of chilled products such as those based on produce, or those using a range of raw materials, e.g. pizzas. It is for this reason that industry hygiene standards were first established in Europe in 1989 when the industry associations in the UK and France (CFA and SYNAFAP (Syndicat National des Fabricants de Plats Préparés Frais, the French ready meal manufacturers association), respectively established the first editions of national guidelines and in 1996 as part of the ECFF produced European industry guidelines.

In the UK, the context of food safety legislation was changed when the Food Safety Act (FSA) was brought into effect in 1990 (HMSO 1990) to enable a wide range of legislation. The FSA introduced the concept of the 'due diligence' defence which enabled operators, if taken to court, to offer in their defence measures they had taken which were designed to avoid an issue arising. Coupled with the implementation of the General Food Hygiene Directive through the Food Safety (General Food Hygiene) Regulations 1995 and greater prominence of HACCP and risk assessment, food safety legislation is placing greater onus on operators' own knowledge of their systems, of potential food safety hazards and on the introduction of internal controls. This approach is now being adopted in Codex and potentially the EU and brings with it a greater than ever need for education in food science, food microbiology and food technology.

Temperature control requirements are set out in national legislation, but these vary greatly across the EU, with virtually no commonality. Attempts have been made to harmonise these national rules at EU level, but political considerations and the differing performance of the chill chain in the various EU Member

States has, to date, prevented this from taking place. It is expected that with the production of the consolidated hygiene Directives this topic will again come under review.

16.6 Codex

The key role of Codex in the development of international trade standards was recognised when the World Trade Organisation (WTO) was established in January 1995. The WTO updated and replaced the General Agreement on Tariffs and Trade. The 'General Agreement' setting up the WTO was supplemented by several more detailed agreements including the Agreement on Sanitary and Phytosanitary Measures (the 'SPS' Agreement) and the Agreement on Technical Barriers to Trade (the 'TBT' Agreement). Codex standards are recognised as the basic standard upon which national measures will be judged. It is accepted that 'higher standards' may be deemed appropriate but there are restrictions based on them and they must be developed using risk assessment techniques. At its 22nd session in June 1997, the Codex Alimentarius Commission adopted a 'Statement of Principle Relating to the Role of Food Safety Risk Assessment'. This includes the statements that: 'Health and safety aspects of Codex decisions and recommendations should be based on a risk assessment, as appropriate to the circumstances.' and 'Food safety risk assessment should be soundly based on science, should incorporate the four steps of the risk assessment process, and should be documented in a transparent manner.'

Members of the WTO (i.e. most countries of Codex) are obliged to consider Codex standards as the basis for their national controls. The approaches of the WTO and EU to free trade are similar in that they both allow imports of products which may not comply with the strict legal requirements of the importing country but which meet the requirements of the Codex standard (in the case of WTO's 'free distribution') or another EU Member State (in the case of 'mutual recognition').

The Codex Alimentarius Commission produces food standards and codes of good manufacturing and hygienic practice. Responsibility for the development of codes of hygienic practice is mostly within the Codex Committee on Food Hygiene (CCFH), which works in conjunction with the other Codex committees that specifically develop codes and standards for particular food commodities. The hygiene codes are mostly directed at food commodities and deal with aspects that must be addressed during, for example, the production, processing, storage and distribution stages of foodstuffs. The *Recommended International Code of Practice General Principles of Food Hygiene* (Codex 1997a) sets out the approach to be followed in the production of all foods (from production on-farm to final preparation), other CODEX Codes supplementing details. Unlike previous versions of the *General Principles*, the 1997 code is not prescriptive in laying down design elements for factories or transport and storage facilities. Instead, it recommends a HACCP-based approach to enhance food safety as described in *Principles of HACCP* (Codex 1997b) and *Hazard Analysis and Critical Control Point (HACCP) System and Guidelines for its Application* (Codex 1997c). The *General Principles* concentrates on what is needed at each step to prevent or reduce risks of contamination and leaves a significant degree of flexibility to manufacturers or operators and regulatory bodies on how to achieve these objectives.

The HACCP approach in the above Codex texts sets out seven principles that must be followed to prepare an effective HACCP plan. A comprehensive review of a HACCP plan must include consideration of these principles. They are:

1. conduct a hazard analysis
2. determine the Critical Control Points (CCPs)
3. establish critical limit(s)
4. establish a system to monitor control of the CCP
5. establish the corrective action to be taken when monitoring indicates that a particular CCP is not under control
6. establish procedures for verification to confirm that the HACCP system is working effectively
7. establish documentation concerning all procedures and records appropriate to these principles and their application.

The implementation of HACCP in small and medium sized businesses has been the focus of much attention in both Codex and many individual countries. In the EU, however, the approach to date has utilised 'HACCP-type systems' that seek to avoid documentation burdens, but this approach has been criticised for presenting a weakened interpretation of HACCP.

Together with increased prominence of HACCP in Codex documents risk assessment has become seen as a tool of great potential. The Codex *Principles for the Establishment and Application of Microbiological Criteria for Foods* (Codex 1997b) are to be supplemented by *Principles and Guidelines for the Conduct of Microbiological Risk Assessment* (Codex 1998a), which was adopted by the 32nd Session of the Codex Committee on Food Hygiene at the end of 1999.

Especially relevant to chilled foods is the Codex *Code of Hygienic Practice for Refrigerated Packaged Foods with Extended Shelf Life* (Codex 1997c). This Code is at the time of writing due for final approval and publication. It was developed following agreement at CCFH that a separate code of practice should be developed covering sous vide products. As work progressed, the scope was extended to all refrigerated pre-prepared, extended shelf-life foods with a shelf life of more than five days at the suggested temperature of 4°C (or greater, depending on hazard analysis). The Code was based primarily on existing French legislation, explaining the reference to 4°C.

Shorter-shelf-life products fall within the scope of the Codex *Code of Hygienic Practice for Precooked and Cooked Meals in Mass Catering* (Codex 1989). However, this does not refer to chilled foods for retail sale. This code deals with the hygienic requirements for cooking raw foods and handling cooked and pre-cooked foods intended for feeding large groups of people and is not intended to be applied to the industrial production of complete meals. Chilled foods are defined in this code as 'product maintained at temperatures not exceeding 4°C in any part of the product and stored for no longer than five days'.

Other proposed Codes of relevance to chilled food production include a Code of Hygienic Practice for Primary Production, Harvesting and Packaging of Fresh Produce/Fruits and Vegetables (including an Annex for seed sprouts) (Codex 1999), and a Code of Hygienic Practice for Pre-Cut Raw Fruits and Vegetables (Codex 1998c). Work on the area of produce has come about since the international community has most recently recognised the potential for pathogens to be present on produce. Indeed, much attention is now being paid to the extension of food hygiene measures from 'farm to fork', particularly where raw agricultural products are used in foods that are to be eaten raw.

16.7 ATP

An Agreement on the International Carriage of Perishable Foodstuffs and on the Special Equipment to be used for such Transport was drawn up by the Inland Transport Committee of the UN Economic Committee for Europe in 1970–71. It is known as the ATP agreement, after the initials of its French title, and its purpose is to facilitate international traffic in certain perishable foodstuffs by setting common and centrally recognized standards (ATP 1987).

The foodstuffs to be carried in accordance with the Agreement are quick (deep) frozen and frozen foodstuffs, and also certain other perishable foodstuffs that fall into neither of these two categories but which need to be carried at chill temperatures. The foods in this latter category are red offal, butter, game, milk, dairy products (yoghurt, kefir, cream and fresh cheese), fresh fish, meat products, meat, poultry and rabbits. Maximum temperatures in the range 2–7°C are specified.

The Agreement lays down common standards for the temperature-controlled equipment (road vehicles, railway wagons and containers) in which these foodstuffs are carried. Over 20 countries have acceded to the Agreement. It was designed to apply primarily to all means of surface transport within Europe and is not applicable to air transport or to sea journeys exceeding 150 km. Despite its specified storage temperature requirements not having been updated for some time and their not being directly related to international legislation, the ATP is still often referred to by distributors.

Further advice on the carriage of chilled foods is to be found in *The Transport of Perishable Foodstuffs*, a handbook compiled by the Shipowners Refrigerated Cargo Research Association (SRCRA) at the request of the UK's Ministry of Agriculture, Fisheries and Food (SRCRA 1991). As well as directing attention to the requirements of the ATP Agreement, the handbook deals in a lucid and practical way with, for example, the conditions, which affect perishable foods, stowage, packaging, atmospheres, vehicles, refrigeration systems and commodities.

16.8 Canada

There are no Canadian regulations specifically dealing with chilled foods. Manufacturers of relevant products are self-regulated through the *Canadian Code of Recommended Practices for Pasteurised/Modified Atmosphere Packaged/Refrigerated Food* (Canada 1990a). The code is intended to assist and encourage compliance with the applicable federal, provincial or municipal legislation that deals with the safety of food. It was developed by the Agri-Food Safety Division of Agriculture Canada in consultation with relevant Government, industry and academic bodies. The code relates to pasteurised modified atmosphere (including vacuum) packed products that require refrigeration (−1°C to 4°C) throughout their shelf-life.

Like the more recently developed, broadly comparable guidelines available in other countries, the code is strongly HACCP oriented. Recommended microbiological criteria for the final product are given. The code addresses all of the components of the manufacturing practices concerned with chilled food technology as well as providing advice on food service and retail handling practices. It also includes a helpful example as to the steps to be followed in the development of a sous vide meat product. The Code includes the relevant parts of the *Canadian Code of Recommended Handling Practices for Chilled Foods* which was developed by the Food Institute of Canada (Canada 1990b).

The HACCP Implementation Manual, issued by Agriculture and Agri-Food Canada as part of the Canadian Food Safety Enhancement Program, includes guidelines and principles for the development of generic HACCP models. The aim is to encourage the establishment and maintenance of HACCP-based systems in federally registered agri-food processing. Thirty-eight generic HACCP models have been developed at the time of writing including one on 'Assembled meat', which includes multi-commodity food products with or without meat, e.g. pizzas and sandwiches.

16.9 European Union (EU)

The European Community (now EU) sought for many of its early years to remove technical barriers to intra-Community trade in food by harmonising the food laws of the individual Member States. However, owing to the significant differences in their legal approaches and the requirements of the different national food laws, this approach met with limited success. The Community therefore adopted a revised strategy in the mid-1980s under the 1985 White Paper Programme for processed foodstuffs. This combined the adoption of five Framework Directives, covering food labelling (79/112/EEC), additives (89/107/EEC), materials in contact with food (89/109/EEC), official controls (89/307/EEC), and foods for particular nutritional uses (89/398/EEC), with the principle of mutual recognition of national regulations and standards not needing to be legislated for by the Community.

EC vertical (product/sector-specific) legislation on food hygiene and controlling food temperature has been directed mainly at protein-based commodities such as meat, milk and poultry. However, it is only in the last ten years or so that the more extensive application of hygiene requirements and measures for monitoring and controlling pathogenic organisms in the food chain have come under consideration.

The 1993 General Food Hygiene Directive sets out the basic hygiene regulations for food production and introduced the application of HACCP principles in food businesses, the concept of Guides to good hygienic practice related to the Directive and intended to provide more detailed sector-specific information, encouragement of the application of ISO 9000 standards on quality assurance in order to implement general hygiene rules, making Member States responsible for enforcement in accordance with established EU rules and official controls.

Work on the consolidation of food hygiene legislation began in earnest in April 1996 with the publication of the *Guide to certain rules governing production, marketing and importation of products of animal origin intended for human consumption*. This Guide was essentially a consolidation of 14 Directives laying down animal and public health rules for the production and the placing on the market of products of animal origin. The EU Green Paper on Food law, published by the European Commission in April 1997 (EC 1997), launched a public debate on the future of European food legislation. The Green Paper stated that besides protection of public health, free movement of goods and European competitiveness, Community food law must ensure that legislation is primary based on scientific evidence and risk assessment. In addition, legislation was required to be coherent, rational and user friendly. The Paper raised questions of whether there was a need for hygiene legislation to be extended to primary agriculture, whether it should be applicable to retailing and whether it would be appropriate to introduce a general obligation on food businesses to ensure that food is safe, wholesome and fit for human consumption. A defence of due diligence (as allowed by the UK Food Safety Act 1990) was proposed to accompany this obligation. In addition, the question of whether unprocessed agricultural foodstuffs should be brought within the scope of the product liability Directive 85/374/EEC was raised but has since been addressed by the publication of Directive 1999/34/EC, which amends the 1985 Directive to include agricultural products within its scope.

At the time of writing – summer 2000 – new regulatory hygiene proposals (EC 2000) are under discussion that are intended to replace current various vertical hygiene-focused Directives for trade in a wide range of protein-based foods. In addition, the General Food Hygiene Directive 93/43/EEC is incorporated into the draft consolidated text. These proposals are directed at the wholesomeness of raw starting materials (excluding farm rules) and the conditions and practices to which they are exposed during subsequent preparation and processing operations, up to the point at which they are ready for placing on the market, but excluding retailer sale. The aim is the uniform

Table 16.2

Country	Chilled storage
Belgium	max. 7°C
Denmark	5°C
Finland	meat-based products: 6°C; other chilled products: 8°C
France	depends on stage of production, e.g. in retail, storage at $\leq$4°C
Italy	meat products: -1 to 7°C; fish products: 0 to 4°C
Spain	0–3°C
Sweden	$<$8°C
The Netherlands	max. 7°C
UK	max. 8°C

imposition of good manufacturing and good hygienic practices on food operators. Temperature control elements have not yet been harmonised, the draft consolidated text simply giving the requirements from regulations being consolidated. For example, various maximum temperatures are given for milk being held for collection and during its transport, and on certain aspects of the processing of meat and poultry meat. Examples of temperatures stipulated in national legislation are given in Table 16.2.

The EC food labelling directive (79/112/EEC, as amended) already impinges on chilled foods inasmuch as foodstuffs, which are microbiologically highly perishable, must be labelled with a 'use by' date. The date must be followed by a description of the storage conditions that must be observed (EC 1989a). In addition, foods whose shelf life has been extended by being packaged in any so authorised packaging gas are required by Directive 94/54 (EC 1994b) to be labelled with the phrase 'packaged in a protective atmosphere'. EU labelling rules are being consolidated with the aim of simplifying and clarifying Community laws to make it clearer and more accessible to consumers (EC 1999b).

The future of EU food safety activity is currently a matter of discussion, with support being evident for the creation of a European food standards agency. Currently, food matters are divided between DGIII (Internal Market), DGVI (Agriculture) and DGXXIV (Consumer Protection). In terms of European-level industry standards, there is little support in the chilled food sector for the production of Guides under 93/43/EEC since these are viewed as being over-simplified and bureaucratic to produce, potentially resulting in them not reflecting the highly technical detail required by the sector. ECFF is the representative body for national organisations or manufacturers of chilled prepared foods in European countries. ECFF in 1996 published Guidelines for Good Hygienic Practice for a wide spectrum of chilled foods which are based on the 1993 edition of the Chilled Food Association's (CFA) guidelines that were revised and updated in 1997. The approach in both texts is similar, i.e. application of full HACCP (in accordance with Codex 1997b), separation of the manufacturing area into segregated areas of which there are three categories (GMP, High Care and High Risk Areas) and specification of thermal processes. High Care Areas are designed for the handling of raw (non-protein) and cooked materials to be combined into a single product and are managed to minimise contamination by microorganisms. High Risk Areas are designed for the handling of cooked materials only (i.e. having been subjected to a time temperature combination equivalent to at least 70ºC for two minutes). Heat treatments for long-shelf-life chilled foods have been addressed by the Report of the ECFF *Botulinum* Working Party (ECFF 1998) which concentrated on the risk of psychrotrophic (non-proteolytic) *C. botulinum* for vacuum packaging and associated processes. The scope of the code and that of ECFF covers retail chilled prepared foods that are capable of supporting the growth of pathogenic organisms.

16.10 Australia/New Zealand

The Australia New Zealand food standards development system is a cooperative arrangement between Australia, New Zealand and the Australian States and mainland Territories to develop and implement uniform food standards. The system for the development of joint food standards was established under a December 1995 treaty between the two countries. Within Australia, the system is based upon a 1991 Commonwealth, State and Territory Agreement in relation to the adoption of uniform food standard.

The system is implemented by food legislation in each State and Territory and in New Zealand, and by the Australia New Zealand Food Authority Act 1991 of the Commonwealth of Australia. The ANZFA Act establishes the mechanisms for the development of joint food standards and creates the ANZFA as the agency responsible for the development and maintenance of a joint Australia New Zealand Food Standards Code.

Although food standards are developed by the ANZFA, responsibility for enforcing and policing food standards rests with the States and Territories in Australia and the New Zealand government in New Zealand. Each government has one or more agencies responsible for food surveillance within their health administration charged with the task of ensuring the requirements of the Food Standards Code are met.

All food sold in the two countries must comply with the Code which is adopted without amendment into the countries' food laws. Owing to this legal approach the Code must be used in conjunction with the relevant local food legislation. The Code is a collection of individual food standards, collated into a number of Parts. Part A deals with standards applicable to all foods. Parts B-Q deal with standards affecting particular classes of foods. Part R deals with special purpose foods. Parts S and T deal respectively with miscellaneous and transitional issues.

The draft Standards Code is currently under consultation and any changes will be made before October 2000 when the draft is to be recommended to the Australia New Zealand Food Standards Council (ANZFSC) for adoption in November 2000. Initially (November 2000 to May 2002) the Australia New

Zealand Food Standards Code will be in force in parallel with the Food Standards Code (Australia) the New Zealand Food Regulations 1984 and New Zealand Food Standards. During this period, a food will be able to be sold if it complies either with the current standards or with the new Joint Code. In May 2002, Food Standards Code (Australia), relevant parts of the New Zealand Food Regulations 1984 and New Zealand Food Standards will be revoked, leaving the Australia New Zealand Food Standards Code as the sole Code.

In broad terms, the standards require food businesses to introduce adequate food safety systems to control hazards, rather than specifying how this should be achieved.

In New Zealand, standards apply as part of a system of dual standards, where the Australian Food Standards Code is an alternative to the New Zealand Food Regulations.

The Code was designed to ensure that safety measures are applied at all stages of the food supply chain, from primary producers through to retailers. The standards required are risk based. Like the current Australian State and Territory food laws and regulations, the Code will apply to anyone who handles or sells food in any sector of the food industry. Primary production is covered insofar as if a farm sells directly to the public or processes food on site, it will have to meet the requirements of the Code.

There will be a staged approach to the introduction of the Code's requirements:

- Year 1: food businesses should meet the requirements of essential food safety practices.
- Years 2 and 3: high-risk businesses will have to demonstrate that they are meeting these requirements through a food safety programme.
- Year 4: medium-risk businesses attracting a lower risk classification will have to have developed a food safety programme.
- Year 6: low-risk operations will have to adopt food safety programmes.

Food industry associations have the option of choosing to develop guides or advice for businesses (referred to as guidelines) within a food industry sector to assist them to comply with the non-prescriptive requirements of the standards and the State and Territory Food Acts.

The Governmental organisation, the Australian Quarantine and Inspection Service (AQIS 1992) published in 1992 a *Code of practice for heat-treated refrigerated foods packaged for extended shelf life* (i.e. those with a shelf-life of more than five days) to provide guidance for the Australian industry.

16.11 France

Food legislation in France is extensive and highly specified. With regard to chilled foods, EC Regulations are reflected in French food law and France also supports the ATP Agreement with its associated temperature controls.

Regulations prescribe microbiological standards for food raw materials and ingredients, e.g. meat, meat products, poultry meats, seafood products, egg products, pastries, milk products, vegetables and spices. Amongst the various hygiene requirements, particular provisions apply to pre-cooked foods in respect of the manufacturing and preparation conditions, hygiene facilities and refrigeration, for example. For chilled foods, the refrigeration installation should be capable of holding temperatures between 0°C and 4°C.

The scope of French legislation covers only meals made from products from animal origin other than fish and meat products (e.g. milk, eggs). For ready-to-eat meals made from meat and fish products, the shelf life is determined by the manufacturer. A protocol for the validation of shelf life was developed by SYNAFAP (SYNAFAP 1995), stipulating shelf-life testing at 4°C for one-third of the total envisaged shelf life, and at 8°C for the remaining two-thirds of the total shelf-life. If the chill chain is 'fairly well' maintained, the conditions for shelf-life testing are at 4°C for two-thirds of the total envisaged shelf-life and at 8°C for one-third of the total shelf-life. Maximum acceptable numbers of pathogens and indicator organisms are specified at the end of shelf-life for refrigerated prepared meals.

The SYNAFAP *Guidelines of Good Hygiene Practice for Prepared Refrigerated Meals* (Syndicat National des Fabricants de Plats Préparés (SYNAFAP 1997) are based on the HACCP approach. Products falling outside their scope are those covered by other professional codes of good hygienic practices, meats subject to EEC Directive 88/657/EEC, and raw products or raw product mixes not ready to be eaten as such, i.e. cooking required before consumption. These guidelines are in line with those of ECFF (ECFF 1996).

16.12 The Netherlands

Like all EU Member States, The Netherlands observes EU rules (see earlier), as well as applying its own extensive controls on food safety, composition, hygiene, manufacture, transport, sale, etc. National Regulations govern cooked foods which are kept and transported chilled at no more than 7°C. However, during transport to the retailer, the maximum temperature limit is marginally relaxed to 10°C, but only in the least cold unit of the food, and only for a short – but unspecified – period.

If it is evident that the food product does not need to be heated or re-heated in order to prepare it for consumption, it must not be kept in such a way that its temperature falls between 7°C and 55°C. Microbiological criteria are prescribed for various categories of foods that are cooked and chilled, so becoming subject to the Cooked Foodstuffs Decree of 1 October 1979, as amended in 1985. The standards vary according to the type of foodstuff or food product, whether or not it is heated before consumption, or ready for eating without further heating, or, in some instances, whether or not it has undergone a preserving process other than heating.

In 1994, a Dutch/Belgian industry group produced a code of practice for the production, distribution and sale of chilled long shelf life pasteurised meals (TNO 1994). This Code sets out heat treatments and manufacturing hygiene standards which are in line with those adopted by the ECFF (ECFF 1996) and applies to hermetically sealed chilled products with shelf life between 11–42 days. Microbiological criteria are given for cook-chill products immediately after processing and at the end of shelf life.

16.13 United Kingdom

As the United Kingdom is a member of the EU, UK food law gives effect to EU Directives, Regulations and amendments of them. The UK implementing legislation relating to the General Food Hygiene Directive is the Food Safety (General Food Hygiene) Regulations 1995 (DH-UK 1995a). These Regulations do not apply to primary production nor to those activities regulated by other UK hygiene Regulations. The 1995 Regulations require the proprietor of a food business to ensure that food handlers are supervised and instructed and/or trained in food hygiene matters commensurate with their work activities unless other Regulations specify alternative requirements; and use of a HACCP-type system to analyse potential food hazards, identify points at which food hazards may occur, determine which of the points are critical to ensuring food safety, control and monitor those points and review the analysis of food hazards, critical points and the control and monitoring procedures periodically, and whenever the food business's operations change.

Regarding UK temperature control legislation, much-criticised Regulations were introduced at the beginning of 1990 (DH-UK 1990) specifying two maximum chill temperatures (5°C and 8°C, plus tolerances) for certain foods. The lower temperature was restricted mainly to those perishable foods considered to present higher risks from the growth of pathogenic organisms, e.g. cooked products containing meat, fish or eggs, meat and fish pâtés, quiches, sandwich fillings. Foods to be kept at or below 8°C were mould-ripened soft cheeses, prepared vegetable salads – such as coleslaws – and uncooked or partly cooked pastry products containing meat or fish. Official guidelines allowing tolerances of up to 4°C for thermometer accuracy and display cabinet defrost cycles were published (DH-UK 1991) and have not been withdrawn despite the Regulations being replaced in 1995 by the adoption of a single control temperature of 8°C maximum in the Food Safety (Temperature Control) Regulations 1995 (DH-UK 1995b). These later Regulations set specific chill and hot holding temperatures (63°C minimum) for certain foods and contain a general overall temperature requirement that 'no person shall keep any raw materials, ingredients, intermediate products and finished products likely to support the growth of harmful bacteria or the formation of toxins at temperatures which would result in a risk to health'.

The UK Government has been concerned for some time about the extent to which some food processes may rely excessively on impracticably low chill temperatures for long periods of time in order to prevent botulism. Thus the 1995 Regulations note that there will be situations where it is appropriate to keep foods at chill temperatures lower than 8ºC for safety reasons. MAFF-funded work on modified atmosphere (including vacuum) packaging resulted in guidelines on Vacuum Packed and Modified Atmosphere Packed Foods produced by Campden and Chorleywood Food RA (CCFRA 1996) which took into account the work of CFA (CFA 1997) the ACMSF (1992) and ECFF Botulinum Working Party (ECFF 1998).

In 1989 the UK Chilled Food Association (CFA) was formed and has since published three editions of its authoritative guidelines and supplementary technical information. The latest guidelines edition was published in 1997, being entitled *Guidelines for Good Hygienic Practice in the Manufacture of Chilled Foods* (CFA 1997). The guidelines were drawn up through wide consultation involving technologists from leading UK businesses concerned with chilled foods, as well drawing on the expertise of Government food scientists and microbiologists, food law enforcement officials and major multiples. The third edition was endorsed by major UK retailers and compliance with the *Guidelines* is a prerequisite of CFA membership. Compliance can be determined by specific bodies approved by CFA or, when the system is established, by any organisation accredited by UKAS, the UK Accreditation Service.

The 1997 CFA Guidelines were based on the 1996 ECFF Guidelines and were revised and restructured to improve accessibility by all levels of reader. The Guidelines uses contains a decision tree to determine the minimum hygiene status (GMP, High Care or High Risk) required for products taking into account the degree of heat treatment applied and potential for recontamination. Guidelines for the handling of chilled foods were first issued in 1982 by the UK Institute of Food Science and Technology, and the document was updated in 1990 (IFST 1990), referring the reader to the CFA Guidelines for specific technical information on the manufacture of chilled foods. Guidelines specifically directed at cook-chill catering systems were separately produced by the Department of Health (DH-UK 1989).

Although there is no legal compulsion in the UK for producers and handlers of chilled foods to follow these advisory guidelines, it is considered prudent for them to do so in terms of providing the basis of a due diligence defence under the 1990 Food Safety Act, and may, in the case of the CFA Guidelines and certain retailers, be a condition of supply.

16.14 United States

A number of federal and state bodies apply formal controls over the manufacture, distribution and retailing of chilled foods in the US. The US Department of Agriculture's (USDA) Food Safety and Inspection System Service (FSIS) in 1994 consolidated USDA's food safety-related functions and has a 'farm to table' mandate. The Food and Drug Administration (FDA) is authorised by Congress to enforce the Federal Food, Drug, and Cosmetic Act and several other public health laws. The FDA monitors the manufacture, import, transport, storage, and sale of foods which are required to be in compliance with the relevant Food Code - the current for 'retail' having been published in 1997 and that for catering in 1999 (FDA 1999). USDA oversees meat and poultry safety, while the FDA regulates fish. Setting safety standards for milk and eggs is the FDA's provenance, but once a cow is slaughtered for meat, or eggs are processed to make other foods, regulatory authority shifts to the USDA.

In July 1996, President Clinton announced the so-called 'Mega-Reg', the Pathogen Reduction Hazard Analysis Critical Control Points (HACCP) System final rule (Federal Register 1996) which applies to those food processors that are inspected by USDA or similar state agencies, i.e. mostly meat and poultry product processors. These regulations require such establishments to take preventive and corrective measures at each stage of the food production process where food safety hazards occur, using a variant of HACCP as defined by Codex. Each plant has the responsibility and flexibility to base its food safety controls on a bespoke HACCP plan, which must identify the CCPs, detailed in the Regulations and use the controls prescribed. Sanitation Standard Operating Procedures (SOPs) are required that describe daily procedures that are sufficient to prevent direct contamination or adulteration of products. Additional requirements are mandatory *E. coli* O157 testing by slaughter operations, and compliance with performance standards for *Salmonella*. The final rule is the starting point for the US HACCP approach as officials are also considering extending similar rules to the farm and through distribution.

HACCP implementation has begun initially in large meat and poultry operations, which had 18 months to comply. Small plants have 30 months to comply and very small plants have 42 months. However, following FSIS surveillance of *Listeria monocytogenes* in products and from outbreak studies, FSIS in May 1999 announced a requirement for reassessed HACCP plans for ready-to-eat livestock and poultry products to be submitted including *Listeria monocytogenes* as a specific hazard. Facilities failing to implement 'proper HACCP programs' will face enforcement action that could mean withdrawal of USDA's inspectors and plant shut down.

In January 1997, President Clinton announced a Food Safety Initiative (FSI) with the aim to reduce the incidence of foodborne illness to the greatest extent possible by improving the system for detecting outbreaks of foodborne illness, promoting research on emerging pathogens such as *E. coli* O157:H7 and *Cyclospora*, and to better educate consumers and the industry on safe food-handling practices. Related to this, in autumn 1997, the President introduced the Produce Safety Initiative (PSI) under the FSI. The PSI addresses the entire produce food chain from grower to table. Voluntary guidance was developed in the form of the *Guide to Minimize Microbial Food Safety Hazards For Fresh Fruits and Vegetables* (FDA 1998) that takes a HACCP-type approach on agricultural practices.

The US Department of Health and Human Services in August 1997 put forward the Food Safety Enforcement Enhancement Act of 1997 (DHHS 1997) which gave the FDA the authority to require the recall of food that presents a threat to public health and allowed it to levy civil monetary penalties for food-related violations of the Federal Food, Drug and Cosmetic Act. In addition, the development and adoption of model prevention programs is another technique used in the US with the aim of advancing food safety, an example of which is the 1997 Food Code which is applicable to 'retail' products, i.e. those which are sold through catering outlets. In the Code, as well as including general provisions, there are two items particularly impacting on chilled foods in reduced oxygen packages (ROP), which include vacuum-packed and modified-atmosphere-packed products. These foods are required to be held at or below 5°C (41°F). Also, the minimum time–temperature combination of 74°C (165°F) for 15 seconds is required, 74°C to be reached within two hours. The Code requires that 'potentially hazardous' food shall be cooled within four hours to 5°C (41°F) or less, or to 7°C (45°F) if prepared from ingredients at ambient temperature, such as reconstituted foods and canned tuna. Labelling of ROP foods relying on chilled storage as a hurdle must be labelled with the phrase 'Important – Must be kept refrigerated at 5°C (41°F)', be marked with a use-by date and consumed by the date required by the Code for that particular product.

The Food Code specifies that ROP foods that are intended for refrigerated storage beyond 14 days must be maintained at or below 3°C (38°F). ROP foods which have lower refrigeration requirements as a condition of safe shelf-life must be monitored for temperature history and must not be offered for retail sale if the temperature and time specified are exceeded.

Ready-to-eat 'potentially hazardous' commercially prepared food handled in a food establishment is required to be clearly marked with the time the original container was opened and to indicate the date by which the food should be consumed, which includes the day the original container was opened. Limits are set as follows for the shelf-life of such foods:

- Seven calendar days or less after the original container was opened, if the food is maintained at 5°C (41°F) or less; or
- Five calendar days or less from the day the original container was opened, if the food is maintained at 7°C (45°F) or less as specified under the Code.

Processed fish and smoked fish may not be packed by ROP unless establishments are approved for the activity and inspected by the regulatory authority.

The Code's ROP foods requirements are based on those of the Guidelines on Refrigerated Foods in Reduced Oxygen Packages which were developed by the Association of Food and Drug Officials (AFDO 1990a).

With regard to industry initiatives, guidelines relating to refrigerated foods, including chilled foods, have been published by the National Food Processors' Association (NFPA 1989) and are due to be updated in the next year or so. The now-defunct US Chilled Food Association issued a *Technical Handbook for the Chilled Foods Industry* (US CFA 1990). Both publications cover the essential aspects that need to be addressed by companies actively concerned with chilled foods. Emphasis is given to the importance of the HACCP approach in achieving the desired end-results for product safety and quality.

16.15 Summary

Chilled foods have been made subject to detailed regulatory controls at national level, particularly with respect to temperature requirements, and at international level where HACCP-based approaches to hygiene have been established. Such general, HACCP-based approaches to food hygiene legislation are the world-wide trend and are being developed at all levels, led by Codex. HACCP-based codes of practice and guidelines have been developed by several organisations, providing the necessary technical detail (e.g. food microbiology and practical hygiene management) which in practical terms indicate how those legislated standards are to be met. Given the highly dynamic and innovative nature of the chilled-food sector, there is a significant role for easily updated best practice guidelines focusing on the practical implementation of HACCP. All that is required at national and international level is a commitment to ensure that operators and enforcers have a clear understanding of food microbiology and technology in order to make HACCP a reality in practice, irrespective of the operation in question.

16.16 References and further reading

General

EC, (1997) *Harmonisation of Safety Criteria for Minimally processed Foods: Inventory Report*, FAIR Concerted Action CT97-1020.

Australia/New Zealand

AQIS, (1992) *Code of practice for heat-treated refrigerated foods packaged for extended shelf life*, Australian Government Publishing Service, GPO Box 84, Canberra ACT 2601, Australia.

Codex Alimentarius

CODEX, (1989) *Code of Hygienic Practice for Precooked and Cooked Meals in Mass Catering*, 1989, Codex Alimentarius Commission, FAO/WHO, Rome.

CODEX, (1997a) *Recommended international code of practice: General principles of food hygiene*, CAC/RCP 1-1969, Rev 3, 1997, Codex Alimentarius Commission, FAO/WHO, Rome.

CODEX (1997b) *Hazard Analysis and Critical Control (HACCP) System and Guidelines for its Application*, CAC/RCP 1-1969, Rev. 3, 1997, Codex Alimentarius Commission, FAO/WHO, Rome.

CODEX, (1997c) *Code of Hygienic Practice for Refrigerated Packaged Foods with Extended Shelf Life*, Alinorm 97/13, 1997, Codex Alimentarius Commission, FAO/WHO, Rome.

CODEX, (1998a) *Draft Principles and Guidelines for the Conduct of Microbiological Risk Assessment*, Alinorm 98/3, 1998, Codex Alimentarius Commission, FAO/WHO, Rome.

CODEX, (1998b) *Report of the 31st session of the Codex Committee on Food Hygiene*, Alinorm 98/12, 1998, Codex Alimentarius Commission, FAO/WHO, Rome.

CODEX, (1998c) *Proposed Draft of Hygienic Practice for Pre-Cut Raw Fruits and Vegetables*, Alinorm 98/8, 1998, Codex Alimentarius Commission, FAO/WHO, Rome.

CODEX, (1999) *Proposed Draft Code of Hygienic Practice for Primary Production, Harvesting and Packaging of Fresh Produce/Fruits and Vegetables*, ALINORM 99/13A page 7, 1999, Codex Alimentarius Commission, FAO/WHO, Rome.

Belgium/Netherlands

TNO, (1994) Code Voor de Produktie, Distributie en Verkoop van Gekoelde, Lang Houdbare Gepasteuriseerde Maaltijde, TNO Voeding, Zeist.

Canada

CANADA, (1990a) *Canadian code of recommended manufacturing practices for pasteurised, modified atmosphere packaged, refrigerated food*, Agri-Food Safety Division, Agriculture Canada, Ottawa K1A 0YA, Canada.

CANADA, (1990b) *The Canadian code of recommended handling practices for chilled food*, The Food Institute of Canada, 130 Albert St., Suite 1409, Ottawa, Ontario K1P 5G4.

European Union

EC, (1964) Council Directive 64/433/EEC on health conditions for the production and marketing of fresh meat, *Off J European Communities*, (121), 21001/64 (last amended by Directive 95/23/EC *ibid*, (L243, 1995, p. 7)).

EC, (1971) Council Directive 71/118/EEC laying down health rules for the production and placing on the market of fresh poultry meat, *ibid.*, (L55, 1971, p23) (last amended by Directive 96/23/EC, *ibid.*, L125, 1996, p. 10)).

EC, (1977a) Council Directive 77/96/EEC on the examination of trichinae (*Trichinella spiralis*) upon importation from third countries of fresh meat derived from domestic swine, *ibid.*, (L26, 1977, p. 8) (last amended by Directive 94/59/EC, *ibid.*, (L315, 1994, p. 18)).

EC, (1977b) Council Directive 77/99/EEC on health problems affecting the production and marketing of meat products and certain other products of animal origin, *ibid.*, (L26, 1977, p. 85) (last amended by Directive 95/68/EC, *ibid.*, (L332, 1995, p. 10)).

EC, (1979) Framework Directive 79/112/EEC on the labelling, presentation and advertising of foodstuffs, *ibid.*, (L33, 1979).

EC, (1989a) Council Directive 89/395/EEC amending Directive 79/112/EEC on the labelling of foodstuffs, 89/395/EEC, *ibid.*, 32 (L 186), 1720.

EC, (1989b) Council Directive 89/437/EEC on hygiene and health problems affecting the production and the placing on the market of egg products, *ibid.*, (L212, 1989, p. 87) (last amended by Directive 96/23/EC, *ibid.*, L125, 1996, p. 10).

EC, (1989c) Council Directive 89/397/EEC on the official control of foodstuffs, *ibid.*, (L186, 1989, p23).

EC, (1990) Directive 90/496/EEC laying down nutrition labelling rules for foodstuffs for sale to the consumer. *Ibid.*, (L276, 1990).

EC, (1992a) Council Directive 92/48/EEC laying down health conditions for the production and the placing on the market of fishery products, *ibid.*, (L268, 1991, p. 15) (last amended by Directive 96/23/EC, *ibid.*, L125, 1996, p. 10)).

EC, (1992b) Council Directive 92/117/EEC concerning measures for protection against specified zoonoses and specified zoonotic agents in animals and products of animal origin in order to prevent outbreaks of foodborne infections and intoxications, *ibid.*, (L62, 1992).

EC, (1992c) Council Directive 92/46/EEC laying down the health rules for the production and placing on the market of raw milk, heat-treated milk and milk-based products, *ibid.*, (L268, 1992, p. 1) (last amended by Directive 96/34/EC, *ibid*, (L187, 1992, p. 41))

EC, (1992d) Council Directive 92/118/EEC laying down animal health and public health requirements governing trade in and imports to the Community of products not subject to the said requirements laid down in specific Community rules referred to in Annex A (I) to Directive 89/662/EEC (*ibid.*, (L395, 1989) and, as regards pathogens, to Directive 90/425/EEC (*ibid.*, (L62, 1993, p49)) (last amended by Directive 96/90/EC (*ibid.*, L13, 1997, p. 24)).

EC, (1993) Council Directive 93/43/EEC on the general hygiene of foodstuffs, *ibid.*, (L175, 1993, p. 1).

EC, (1994a) Council Directive 94/65/EC laying down the requirements for the production and placing on the market of minced meat and meat preparations, *ibid.*, (L368, 1994, p. 10).

EC, (1994b) Council Directive 94/54/EC concerning the compulsory indication on the labelling of certain foodstuffs of particulars, *ibid.*, (L300, 1994, p. 14).

EC, (1996) Proposal for an European Parliament and Council Directive amending Directive 85/374/EEC on liability of defective products (COM(97) 478 final), *ibid*, (C337, 1996).

EC, (1997) Green Paper. The general principles of food law in the European Union. April 1997.

EC, (1999a) *Draft Proposal for a European Parliament and Council Directive on the Hygiene of Foodstuffs* (III/5227/98 rev 3., VI/1881/98 rev 1 (1999).

EC, (1999b) *Proposal for a European Parliament and Council Directive on the approximation of food laws of the Member States relating to the labelling, presentation and advertising standards* (COM (1999) 113 final 99/0090 (COD)).

EC, (2000) Commission Proposes New Food Safety Hygiene Rules (IP/00/791) Press Release, 17 July 2000.

ECFF, (1996) Guidelines for the Hygenic Manufacture of Chilled Foods, European Chilled Food Federation, c/o PO Box 14811, London NW10 9ZR, UK.

ECFF, (1998) *Sous Vide Foods: conclusions of an ECFF Botulinum Working Party*, Food Control, 10 (1999), 47–51.

France

ICACQ, (1987) *La régime juridique de la cuisson sous vide, L'Institut de Chimie Analytique et du Contrôle de la Qualité, Option Qualité*, No. 37 (February), 12–20.

SYNAFAP, (1995) *Aide à la maîtrise de l'hygiène alimentaire pour les plats préparés frais et réfrigérés, Syndicat National des Fabricants de Plats Préparés*, 44 rue d'Alesia, 75682 Paris Cedex 14.

SYNAFAP, (1997) *Guidelines of good hygiene practices for prepared refrigerated meals, Syndicat National des Fabricants de Plats Préparés*, 44 rue d'Alesia, 75682 Paris Cedex 14.

International Transport

ATP, (1987) Consolidated text of the Agreement on the international carriage of perishable foodstuffs and on special equipment to be used for such carriage, *Command No. 250*, November 1987, HMSO, London.

DT-UK, (1988) *A guide to the international transport of perishable foods*, Department of Transport, Publications Sales Unit, Building 1, Victoria Road, South Ruislip, Middlesex HA4 0NE.

SRCRA, (1991) *The transport of perishable foods*, Shipowners Refrigerated Cargo Research Association, 140 Newmarket Road, Cambridge, CB5 8HE, UK. ISBN 0 95102441 8.

USA

AFDO, (1990a) *Retail guidelines: refrigerated foods in reduced oxygen packages*, Quarterly Bulletin of the Assn. of Food and Drug Officials, 54 (5), 80–4.

AFDO, (1990b) *Guidelines for the transportation of food, ibid.*, 54 (5), 85–90.

DHHS, (1997) *Food Safety Enforcement Enhancement Act*, US Dept of Health and Human Services, 200 Independence Ave., SW, Washington DC 20201, USA.

FDA, (1998) *Guide to Minimize Microbiological Food Safety Hazards for Fresh Fruit and Vegetables*, US Dept of Health and Human Services, US FDA and Centre for Food Safety and Nutrition.

FDA, (1999) *Food Code*, US Dept of Health and Human Services, Public Health Service. Food and Drug Administration, Washington DC 20204, USA.

FEDERAL REGISTER, (1996) *Pathogen Reduction Hazard Analysis Critical Control Points (HACCP) System final rule*, 61 FR38806.

NFPA, (1989) *Guidelines for the development, production, distribution and handling of refrigerated foods*, National Food Processors' Association, 1401 New York Ave., NW, Washington DC 20005, USA.

US CFA, (1990) *Technical handbook for the chilled foods industry*, Chilled Foods Association, 5775 Peach Tree-Dunwoody Road, Suite 500G, Atlanta, Ga. 30342.

USDA, (1988) *Interim guidelines for the preparation of a partial quality control programme*, Food Safety and Inspection Service, US Dept. of Agriculture, Washington, DC 20250.

Sweden

SWEDEN, (1984) *Swedish Food Regulations: General implementing regulations*, The National Food Administration, Box 622, S751 26 Uppsala, Sweden.

United Kingdom

ACMSF, (1992) *Report on vacuum packaging and associated processes*, ISBN 0 11 321558 4, HMSO, London.

CCFRA, (1996) *A Guide to the Safe production of Vacuum and Gas Packed Foods, Guideline No. 11*, Campden and Chorleywood Food RA, Chipping Campden, GL55 6LD UK.

CFA, (1997) *Guidelines for good hygienic practice in the manufacture of chilled foods*, ISBN. 1 901798 00 3, Chilled Food Association, P O Box 14811, London NW10 9ZR.

DH-UK, (1989) *Chilled and Frozen – Guidelines on cook-chill and cook-freeze catering systems*, Dept. of Health, HMSO, 49 High Holborn, London WC1V 6HB.

DH-UK, (1990) *Food Hygiene (Amendment) Regulations* 1990, SI 1431, Dept of Health, HMSO.

DH-UK, (1991) *Guidelines on the Food Hygiene (Amendment) Regulations* 1990, SI 1431, ISBN 0 11 321369 7. Dept. of Health, Eileen House, London SE1 6EF.

DH-UK, (1995a) *Food Safety (General Food Hygiene) Regulations* 1995, SI 1763, ISBN 0-11-053227-9, HMSO, London.

DH-UK, (1995b) *Food Safety (Temperature Control) Regulations* 1995, SI 2200, HMSO, London.

DH-UK, (1995a) *Guidelines on the Food Safety (General Food Hygiene) Regulations* 1995, Dept. of Health, Eileen House, London SE1 6EF.

DH-UK, (1995b) *Guidelines on the Food Safety (Temperature Control) Regulations* 1995, Dept. of Health, Eileen House, London SE1 6EF.

DH-UK, (1995c) *Code of Practice on Food Hygiene Inspections (Code of Practice No. 9)*, ISBN 0-11-321931-8, Stationery Office, 49 High Holborn, London WC1V 6HB.

HMSO, (1990) *Food Safety Act*, ISBN 0-10-541690-8, HMSO, London.
IFST, (1990) *Guidelines for the handling of chilled foods*, Institute of Food Science and Technology (UK), 5 Cambridge Court, 210 Shepherds Bush Road, London W6 7NL.
UK, (1995a) *Food Safety (General Food Hygiene) Regulations 1995* (SI 1995 No 1763), Stationery Office, 49 High Holborn, London WC1V 6HB.
UK, (1995b) *Food Safety (Temperature Control) Regulations 1995* (SI 1995 No 2200), Stationery Office, 49 High Holborn, London WC1V 6HB.

Index